T0189840

LONDON MATHEMATICAL SOCIETY LECTURE NOTE SERIES

The titles below are available from booksellers, or from Cambridge University Press at www.cambridge.org/mathematics

325 Lectures on the Ricci flow, P. TOPPING
326 Modular representations of finite groups of Lie type, J.E. HUMPHREYS
327 Surveys in combinatorics 2005, B.S. WEBB (ed)
328 Fundamentals of hyperbolic manifolds, R. CANARY, D. EPSTEIN & A. MARDEN (eds)
329 Spaces of Kleinian groups, Y. MINSKY, M. SAKUMA & C. SERIES (eds)
330 Noncommutative localization in algebra and topology, A. RANICKI (ed)
331 Foundations of computational mathematics, Santander 2005, L.M PARDO, A. PINKUS, E. SÜLI & M.J. TODD (eds)
332 Handbook of tilting theory, L. ANGELERI HÜGEL, D. HAPPEL & H. KRAUSE (eds)
333 Synthetic differential geometry (2nd Edition), A. KOCK
334 The Navier–Stokes equations, N. RILEY & P. DRAZIN
335 Lectures on the combinatorics of free probability, A. NICA & R. SPEICHER
336 Integral closure of ideals, rings, and modules, I. SWANSON & C. HUNEKE
337 Methods in Banach space theory, J.M.F. CASTILLO & W.B. JOHNSON (eds)
338 Surveys in geometry and number theory, N. YOUNG (ed)
339 Groups St Andrews 2005 I, C.M. CAMPBELL, M.R. QUICK, E.F. ROBERTSON & G.C. SMITH (eds)
340 Groups St Andrews 2005 II, C.M. CAMPBELL, M.R. QUICK, E.F. ROBERTSON & G.C. SMITH (eds)
341 Ranks of elliptic curves and random matrix theory, J.B. CONREY, D.W. FARMER, F. MEZZADRI & N.C. SNAITH (eds)
342 Elliptic cohomology, H.R. MILLER & D.C. RAVENEL (eds)
343 Algebraic cycles and motives I, J. NAGEL & C. PETERS (eds)
344 Algebraic cycles and motives II, J. NAGEL & C. PETERS (eds)
345 Algebraic and analytic geometry, A. NEEMAN
346 Surveys in combinatorics 2007, A. HILTON & J. TALBOT (eds)
347 Surveys in contemporary mathematics, N. YOUNG & Y. CHOI (eds)
348 Transcendental dynamics and complex analysis, P.J. RIPPON & G.M. STALLARD (eds)
349 Model theory with applications to algebra and analysis I, Z. CHATZIDAKIS, D. MACPHERSON, A. PILLAY & A. WILKIE (eds)
350 Model theory with applications to algebra and analysis II, Z. CHATZIDAKIS, D. MACPHERSON, A. PILLAY & A. WILKIE (eds)
351 Finite von Neumann algebras and masas, A.M. SINCLAIR & R.R. SMITH
352 Number theory and polynomials, J. MCKEE & C. SMYTH (eds)
353 Trends in stochastic analysis, J. BLATH, P. MÖRTERS & M. SCHEUTZOW (eds)
354 Groups and analysis, K. TENT (ed)
355 Non-equilibrium statistical mechanics and turbulence, J. CARDY, G. FALKOVICH & K. GAWEDZKI
356 Elliptic curves and big Galois representations, D. DELBOURGO
357 Algebraic theory of differential equations, M.A.H. MACCALLUM & A.V. MIKHAILOV (eds)
358 Geometric and cohomological methods in group theory, M.R. BRIDSON, P.H. KROPHOLLER & I.J. LEARY (eds)
359 Moduli spaces and vector bundles, L. BRAMBILA-PAZ, S.B. BRADLOW, O. GARCÍA-PRADA & S. RAMANAN (eds)
360 Zariski geometries, B. ZILBER
361 Words: Notes on verbal width in groups, D. SEGAL
362 Differential tensor algebras and their module categories, R. BAUTISTA, L. SALMERÓN & R. ZUAZUA
363 Foundations of computational mathematics, Hong Kong 2008, F. CUCKER, A. PINKUS & M.J. TODD (eds)
364 Partial differential equations and fluid mechanics, J.C. ROBINSON & J.L. RODRIGO (eds)
365 Surveys in combinatorics 2009, S. HUCZYNSKA, J.D. MITCHELL & C.M. RONEY-DOUGAL (eds)
366 Highly oscillatory problems, B. ENGQUIST, A. FOKAS, E. HAIRER & A. ISERLES (eds)
367 Random matrices: High dimensional phenomena, G. BLOWER
368 Geometry of Riemann surfaces, F.P. GARDINER, G. GONZÁLEZ-DIEZ & C. KOUROUNIOTIS (eds)
369 Epidemics and rumours in complex networks, M. DRAIEF & L. MASSOULIÉ
370 Theory of p-adic distributions, S. ALBEVERIO, A.YU. KHRENNIKOV & V.M. SHELKOVICH
371 Conformal fractals, F. PRZYTYCKI & M. URBAŃSKI
372 Moonshine: The first quarter century and beyond, J. LEPOWSKY, J. MCKAY & M.P. TUITE (eds)
373 Smoothness, regularity and complete intersection, J. MAJADAS & A. G. RODICIO
374 Geometric analysis of hyperbolic differential equations: An introduction, S. ALINHAC
375 Triangulated categories, T. HOLM, P. JØRGENSEN & R. ROUQUIER (eds)
376 Permutation patterns, S. LINTON, N. RUŠKUC & V. VATTER (eds)
377 An introduction to Galois cohomology and its applications, G. BERHUY
378 Probability and mathematical genetics, N. H. BINGHAM & C. M. GOLDIE (eds)
379 Finite and algorithmic model theory, J. ESPARZA, C. MICHAUX & C. STEINHORN (eds)
380 Real and complex singularities, M. MANOEL, M.C. ROMERO FUSTER & C.T.C WALL (eds)
381 Symmetries and integrability of difference equations, D. LEVI, P. OLVER, Z. THOMOVA & P. WINTERNITZ (eds)

382 Forcing with random variables and proof complexity, J. KRAJÍČEK
383 Motivic integration and its interactions with model theory and non-Archimedean geometry I, R. CLUCKERS, J. NICAISE & J. SEBAG (eds)
384 Motivic integration and its interactions with model theory and non-Archimedean geometry II, R. CLUCKERS, J. NICAISE & J. SEBAG (eds)
385 Entropy of hidden Markov processes and connections to dynamical systems, B. MARCUS, K. PETERSEN & T. WEISSMAN (eds)
386 Independence-friendly logic, A.L. MANN, G. SANDU & M. SEVENSTER
387 Groups St Andrews 2009 in Bath I, C.M. CAMPBELL et al. (eds)
388 Groups St Andrews 2009 in Bath II, C.M. CAMPBELL et al. (eds)
389 Random fields on the sphere, D. MARINUCCI & G. PECCATI
390 Localization in periodic potentials, D.E. PELINOVSKY
391 Fusion systems in algebra and topology, M. ASCHBACHER, R. KESSAR & B. OLIVER
392 Surveys in combinatorics 2011, R. CHAPMAN (ed)
393 Non-abelian fundamental groups and Iwasawa theory, J. COATES et al. (eds)
394 Variational problems in differential geometry, R. BIELAWSKI, K. HOUSTON & M. SPEIGHT (eds)
395 How groups grow, A. MANN
396 Arithmetic differential operators over the p-adic integers, C.C. RALPH & S.R. SIMANCA
397 Hyperbolic geometry and applications in quantum chaos and cosmology, J. BOLTE & F. STEINER (eds)
398 Mathematical models in contact mechanics, M. SOFONEA & A. MATEI
399 Circuit double cover of graphs, C.-Q. ZHANG
400 Dense sphere packings: a blueprint for formal proofs, T. HALES
401 A double Hall algebra approach to affine quantum Schur–Weyl theory, B. DENG, J. DU & Q. FU
402 Mathematical aspects of fluid mechanics, J.C. ROBINSON, J.L. RODRIGO & W. SADOWSKI (eds)
403 Foundations of computational mathematics, Budapest 2011, F. CUCKER, T. KRICK, A. PINKUS & A. SZANTO (eds)
404 Operator methods for boundary value problems, S. HASSI, H.S.V. DE SNOO & F.H. SZAFRANIEC (eds)
405 Torsors, étale homotopy and applications to rational points, A.N. SKOROBOGATOV (ed)
406 Appalachian set theory, J. CUMMINGS & E. SCHIMMERLING (eds)
407 The maximal subgroups of the low-dimensional finite classical groups, J.N. BRAY, D.F. HOLT & C.M. RONEY-DOUGAL
408 Complexity science: the Warwick master's course, R. BALL, V. KOLOKOLTSOV & R.S. MACKAY (eds)
409 Surveys in combinatorics 2013, S.R. BLACKBURN, S. GERKE & M. WILDON (eds)
410 Representation theory and harmonic analysis of wreath products of finite groups, T. CECCHERINI-SILBERSTEIN, F. SCARABOTTI & F. TOLLI
411 Moduli spaces, L. BRAMBILA-PAZ, O. GARCÍA-PRADA, P. NEWSTEAD & R.P. THOMAS (eds)
412 Automorphisms and equivalence relations in topological dynamics, D.B. ELLIS & R. ELLIS
413 Optimal transportation, Y. OLLIVIER, H. PAJOT & C. VILLANI (eds)
414 Automorphic forms and Galois representations I, F. DIAMOND, P.L. KASSAEI & M. KIM (eds)
415 Automorphic forms and Galois representations II, F. DIAMOND, P.L. KASSAEI & M. KIM (eds)
416 Reversibility in dynamics and group theory, A.G. O'FARRELL & I. SHORT
417 Recent advances in algebraic geometry, C.D. HACON, M. MUSTAŢĂ & M. POPA (eds)
418 The Bloch–Kato conjecture for the Riemann zeta function, J. COATES, A. RAGHURAM, A. SAIKIA & R. SUJATHA (eds)
419 The Cauchy problem for non-Lipschitz semi-linear parabolic partial differential equations, J.C. MEYER & D.J. NEEDHAM
420 Arithmetic and geometry, L. DIEULEFAIT et al. (eds)
421 O-minimality and Diophantine geometry, G.O. JONES & A.J. WILKIE (eds)
422 Groups St Andrews 2013, C.M. CAMPBELL et al. (eds)
423 Inequalities for graph eigenvalues, Z. STANIĆ
424 Surveys in combinatorics 2015, A. CZUMAJ et al. (eds)
425 Geometry, topology and dynamics in negative curvature, C.S. ARAVINDA, F.T. FARRELL & J.-F. LAFONT (eds)
426 Lectures on the theory of water waves, T. BRIDGES, M. GROVES & D. NICHOLLS (eds)
427 Recent advances in Hodge theory, M. KERR & G. PEARLSTEIN (eds)
428 Geometry in a Fréchet context, C. T. J. DODSON, G. GALANIS & E. VASSILIOU
429 Sheaves and functions modulo p, L. TAELMAN
430 Recent progress in the theory of the Euler and Navier–Stokes equations, J.C. ROBINSON, J.L. RODRIGO, W. SADOWSKI & A. VIDAL-LÓPEZ (eds)
431 Harmonic and subharmonic function theory on the real hyperbolic ball, M. STOLL
432 Topics in graph automorphisms and reconstruction (2nd Edition), J. LAURI & R. SCAPELLATO
433 Regular and irregular holonomic D-modules, M. KASHIWARA & P. SCHAPIRA
434 Analytic semigroups and semilinear initial boundary value problems (2nd Edition), K. TAIRA
435 Graded rings and graded Grothendieck groups, R. HAZRAT
436 Groups, graphs and random walks, T. CECCHERINI-SILBERSTEIN, M. SALVATORI & E. SAVA-HUSS (eds)
437 Dynamics and analytic number theory, D. BADZIAHIN, A. GORODNIK & N. PEYERIMHOFF (eds)
438 Random walks and heat kernels on graphs, M.T. BARLOW
439 Evolution equations, K. AMMARI & S. GERBI (eds)
440 Surveys in combinatorics 2017, A. CLAESSON et al. (eds)
441 Polynomials and the mod 2 Steenrod algebra I, G. WALKER & R.M.W. WOOD
442 Polynomials and the mod 2 Steenrod algebra II, G. WALKER & R.M.W. WOOD
443 Asymptotic analysis in general relativity, T. DAUDÉ, D. HÄFNER & J.-P. NICOLAS (eds)
444 Geometric and cohomological group theory, P.H. KROPHOLLER, I.J. LEARY, C. MARTÍNEZ-PÉREZ & B.E.A. NUCINKIS (eds)

London Mathematical Society Lecture Note Series: 443

Asymptotic Analysis in General Relativity

Edited by

THIERRY DAUDÉ
Université de Cergy-Pontoise, France

DIETRICH HÄFNER
Université Grenoble Alpes, France

JEAN-PHILIPPE NICOLAS
Université de Bretagne Occidentale, France

CAMBRIDGE
UNIVERSITY PRESS

University Printing House, Cambridge CB2 8BS, United Kingdom

One Liberty Plaza, 20th Floor, New York, NY 10006, USA

477 Williamstown Road, Port Melbourne, VIC 3207, Australia

314-321, 3rd Floor, Plot 3, Splendor Forum, Jasola District Centre, New Delhi - 110025, India

79 Anson Road, #06-04/06, Singapore 079906

Cambridge University Press is part of the University of Cambridge.

It furthers the University's mission by disseminating knowledge in the pursuit of education, learning and research at the highest international levels of excellence.

www.cambridge.org
Information on this title: www.cambridge.org/9781316649404
DOI: 10.1017/9781108186612

First published 2018

A catalogue record for this publication is available from the British Library

Library of Congress Cataloging in Publication data
Names: Daudé, Thierry, 1977– editor. | Häfner, Dietrich, editor. |
Nicolas, J.-P. (Jean-Philippe), editor.
Title: Asymptotic analysis in general relativity / edited by Thierry Daudé
(Université de Cergy-Pontoise), Dietrich Häfner (Université Grenoble Alpes),
Jean-Philippe Nicolas (Université de Bretagne Occidentale).
Other titles: London Mathematical Society lecture note series ; 443.
Description: Cambridge, United Kingdom ; New York, NY :
Cambridge University Press, 2017. |
Series: London Mathematical Society lecture note series ; 443 |
Includes bibliographical references.
Identifiers: LCCN 2017023160 | ISBN 9781316649404 (pbk.) | ISBN 1316649407 (pbk.)
Subjects: LCSH: General relativity (Physics)–Mathematics.
Classification: LCC QC173.6 .A83 2017 | DDC 530.11–dc23 LC record available at
https://lccn.loc.gov/2017023160

ISBN 978-1-316-64940-4 Paperback

Contents

1	**Introduction to Modern Methods for Classical and Quantum Fields in General Relativity**	*page* 1
	Thierry Daudé, Dietrich Häfner and Jean-Philippe Nicolas	
1.1	Geometry of Black Hole Spacetimes	2
1.2	Quantum Field Theory on Curved Spacetimes	3
1.3	Conformal Geometry and Conformal Tractor Calculus	4
1.4	A Minicourse in Microlocal Analysis and Wave Propagation	5
References		6
2	**Geometry of Black Hole Spacetimes**	9
	Lars Andersson, Thomas Bäckdahl and Pieter Blue	
2.1	Introduction	9
2.2	Background	13
2.3	Black Holes	28
2.4	Spin Geometry	43
2.5	The Kerr Spacetime	53
2.6	Monotonicity and Dispersion	55
2.7	Symmetry Operators	65
2.8	Conservation Laws for the Teukolsky System	70
2.9	A Morawetz Estimate for the Maxwell Field on Schwarzschild	75
References		79
3	**An Introduction to Conformal Geometry and Tractor Calculus, with a view to Applications in General Relativity**	86
	Sean N. Curry and A. Rod Gover	
3.1	Introduction	86
3.2	Lecture 1: Riemannian Invariants and Invariant Operators	91

3.3 Lecture 2: Conformal Transformations and Conformal Covariance 94
3.4 Lecture 3: Prolongation and the Tractor Connection 104
3.5 Lecture 4: The Tractor Curvature, Conformal Invariants and Invariant Operators 120
3.6 Lecture 5: Conformal Compactification of Pseudo-Riemannian Manifolds 128
3.7 Lecture 6: Conformal Hypersurfaces 145
3.8 Lecture 7: Geometry of Conformal Infinity 151
3.9 Lecture 8: Boundary Calculus and Asymptotic Analysis 154
Appendix: Conformal Killing Vector Fields and Adjoint Tractors 160
References 168

4 An Introduction to Quantum Field Theory on Curved Spacetimes 171
Christian Gérard
4.1 Introduction 171
4.2 A Quick Introduction to Quantum Mechanics 175
4.3 Notation 178
4.4 CCR and CAR Algebras 179
4.5 States on CCR/CAR Algebras 183
4.6 Lorentzian Manifolds 190
4.7 Klein–Gordon Fields on Lorentzian Manifolds 192
4.8 Free Dirac Fields on Lorentzian Manifolds 197
4.9 Microlocal Analysis of Klein–Gordon Quasi-Free States 201
4.10 Construction of Hadamard States 209
References 217

5 A Minicourse on Microlocal Analysis for Wave Propagation 219
András Vasy
5.1 Introduction 219
5.2 The Overview 221
5.3 The Basics of Microlocal Analysis 230
5.4 Propagation Phenomena 285
5.5 Conformally Compact Spaces 315
5.6 Microlocal Analysis in the b-Setting 349
References 371

1

Introduction to Modern Methods for Classical and Quantum Fields in General Relativity

Thierry Daudé, Dietrich Häfner and Jean-Philippe Nicolas

The last few decades have seen major developments in asymptotic analysis in the framework of general relativity, with the emergence of methods that, until recently, were not applied to curved Lorentzian geometries. This has led notably to the proof of the stability of the Kerr–de Sitter spacetime by P. Hintz and A. Vasy [17]. An essential feature of many recent works in the field is the use of dispersive estimates; they are at the core of most stability results and are also crucial for the construction of quantum states in quantum field theory, domains that have a priori little in common. Such estimates are in general obtained through geometric energy estimates (also referred to as vector field methods) or via microlocal/spectral analysis. In our minds, the two approaches should be regarded as complementary, and this is a message we hope this volume will convey succesfully. More generally than dispersive estimates, asymptotic analysis is concerned with establishing scattering-type results. Another fundamental example of such results is asymptotic completeness, which, in many cases, can be translated in terms of conformal geometry as the well-posedness of a characteristic Cauchy problem (Goursat problem) at null infinity. This has been used to develop alternative approaches to scattering theory via conformal compactifications (see for instance F. G. Friedlander [11] and L. Mason and J.-P. Nicolas [22]). The presence of symmetries in the geometrical background can be a tremendous help in proving scattering results, dispersive estimates in particular. What we mean by symmetry is generally the existence of an isometry associated with the flow of a Killing vector field, though there exists a more subtle type of symmetry, described sometimes as hidden, corresponding to the presence of Killing spinors for instance. Recently, the vector field method has been adapted to take such generalized symmetries into account by L. Andersson and P. Blue in [2].

This volume compiles notes from the eight-hour mini-courses given at the summer school on asymptotic analysis in general relativity, held at the Institut

Fourier in Grenoble, France, from 16 June to 4 July 2014. The purpose of the summer school was to draw an up-to-date panorama of the new techniques that have influenced the asymptotic analysis of classical and quantum fields in general relativity in recent years. It consisted of five mini-courses:

- "Geometry of black hole spacetimes" by Lars Andersson, Albert Einstein Institut, Golm, Germany;
- "An introduction to quantum field theory on curved spacetimes" by Christian Gérard, Paris 11 University, Orsay, France;
- "An introduction to conformal geometry and tractor calculus, with a view to applications in general relativity" by Rod Gover, Auckland University, New Zealand;
- "The bounded L^2 conjecture" by Jérémie Szeftel, Paris 6 University, France;
- "A minicourse on microlocal analysis for wave propagation" by András Vasy, Stanford University, United States of America.

Among these, only four are featured in this book. The proof of the bounded L^2 conjecture having already appeared in two different forms [20, 21], Jérémie Szeftel preferred not to add yet another version of this result; his lecture notes are therefore not included in the present volume.

1.1. Geometry of Black Hole Spacetimes

The notion of a black hole dates back to the 18th century with the works of Simpson and Laplace, but it found its modern description within the framework of general relativity. In fact the year after the publication of the general theory of relativity by Einstein, Karl Schwarzschild [30] found an explicit non-trivial solution of the Einstein equations that was later understood to describe a universe containing nothing but an eternal spherical black hole. The Kerr solution appeared in 1963 [19] and, with the singularity theorems of Hawking and Penrose [15], black holes were eventually understood as inevitable dynamical features of the evolution of the universe rather than mere mathematical oddities. The way exact black hole solutions of the Einstein equations were discovered was by imposing symmetries. First Schwarzschild looked for spherically symmetric and static solutions in four spacetime dimensions, which reduces the Einstein equations to a non-linear ordinary differential equation (ODE). The Kerr solution appears when one relaxes one of the symmetries and looks for stationary and axially symmetric solutions. Roy Kerr obtained his solution by imposing on the metric the so-called "Kerr–Schild" ansatz that corresponds to assuming a special algebraic property for the Weyl tensor, namely that it has Petrov-type D, which is similar to the condition for a polynomial to have two double roots. This algebraic speciality of the Weyl tensor can be understood as another type of symmetry assumption about spacetime. This is a generalized symmetry that does not correspond to an isometry generated by the flow of a vector field, but is related to the existence of a Killing spinor. The Kerr family, which contains Schwarzschild's spacetime as the zero angular momentum case, is expected to be the unique family of asymptotically flat and stationary (perhaps pseudo-stationary, or locally stationary, would be more appropriate) black hole solutions of the Einstein vacuum equations (there is a vast literature on this topic, see for example the original paper by D. Robinson [27], his review article [28] and the recent analytic approach by S. Alexakis, A. D. Ionescu, and S. Klainerman [1]). Moreover it is believed to be stable (there is also an important literature on this question, the stability of Kerr–de Sitter black holes was established recently in [17], though the stability of the Kerr metric is still an open problem). These two conjectures play a crucial role in physics where it is commonly assumed that the long term dynamics of a black hole stabilizes to a Kerr solution. The extended lecture notes by Lars Andersson, Thomas Bäckdahl, and Pieter Blue take us through the many topics that are relevant to the questions of stability and uniqueness of the Kerr metric, including the geometry of stationary and dynamical black holes with a particular emphasis on the special features of the Kerr metric, spin geometry, dispersive estimates for hyperbolic equations and generalized symmetry operators. The type D structure is an essential focus of the course, with the intimate links between the principal null directions, the Killing spinor, Killing vectors and tensors, Killing–Yano tensors and symmetry operators. All these notions are used in the final sections where some conservation laws are derived for the Teukolsky system governing the evolution of spin $n/2$ zero rest-mass fields, and a new proof of a Morawetz estimate for Maxwell fields on the Schwarzschild metric is given.

1.2. Quantum Field Theory on Curved Spacetimes

In the 1980s, Dimock and Kay started a research program concerning scattering theory for classical and quantum fields on the Schwarzschild spacetime; see [9]. Their work was then pushed further by Bachelot, Häfner, and others, leading in particular to a mathematically rigorous description of the Hawking effect on Schwarzschild and Kerr spacetimes, see e.g. [4], [14]. In the Schwarzschild case there exists a global timelike Killing vector field in the exterior of the black hole that can be used to define vacuum and thermal states.

However, it is not clear how to extend these states to the whole spacetime. From a more conceptual point of view this is also quite unsatisfactory because the construction of vacuum states on the Minkowski spacetime uses the full Poincaré group. In addition general spacetimes will not even be locally stationary. On a curved spacetime, vacuum states are therefore replaced by so-called Hadamard states. These Hadamard states were first characterized by properties of their two-point functions, which had to have a specific asymptotic expansion near the diagonal. In 1995 Radzikowski reformulated the old Hadamard condition in terms of the wave front set of the two-point function; see [26]. Since then, microlocal analysis has played an important role in quantum field theory in curved spacetime, see e.g. the construction of Hadamard states using pseudodifferential calculus by Gérard and Wrochna [13]. The lectures given by Christian Gérard give an introduction to quantum field theory on curved spacetimes and in particular to the construction of Hadamard states.

1.3. Conformal Geometry and Conformal Tractor Calculus

Conformal compactifications were initially used in general relativity by André Lichnerowicz for the study of the constraints. It is Roger Penrose who started applying this technique to Lorentzian manifolds, more specifically to asymptotically flat spacetimes, in the early 1960s (see Penrose [25]). The purpose was to replace complicated asymptotic analysis by simple and natural geometrical constructions. To be precise, a conformal compactification allows one to describe infinity for a spacetime $(\mathcal{M}, g)$ as a finite boundary for the manifold $\mathcal{M}$ equipped with a well-chosen metric $\hat{g}$ that is conformally related to g. Provided a field equation has a suitably simple transformation law under conformal rescalings, ideally conformal invariance or at least some conformal covariance, the asymptotic behavior of the field on $(\mathcal{M}, g)$ can be inferred from the local properties at the boundary of the conformally rescaled field on $(\mathcal{M}, \hat{g})$. Penrose's immediate goal was to give a simple reformulation of the Sachs peeling property as the continuity at the conformal boundary of the rescaled field. But he had a longer term motivation which was to construct a conformal scattering theory for general relativity, allowing the setting of data for the spacetime at its past null conformal boundary and to propagate the associated solution of the Einstein equations right up to its future null conformal boundary. Since its introduction, the conformal technique has been used to prove global existence for the Einstein equations, or other non-linear hyperbolic equations, for sufficiently small data (see for example Y. Choquet-Bruhat and J. W. York [8]), to construct scattering theories for linear and non-linear test fields, initially on static backgrounds and, in recent years, in time dependent situations and on black hole spacetimes (see L. Mason and J.-P. Nicolas [22] and Nicolas [24] and references therein). It has also been applied to spacetimes with a non-zero cosmological constant. There is an important literature from the schools of R. Mazzeo and R. Melrose and more recently numerous studies using the tractor calculus approach by A. R. Gover and his collaborators. Tractor calculus in its conformal version started from the notion of a local twistor bundle on four-dimensional spin-manifolds as an associated bundle to the Cartan conformal connection, though it in fact dates back to T. Y. Thomas's work [31]. The theory in its modern form first appeared in the founding paper by T. Bailey, M. Eastwood, and Gover [6] where its origins are also thoroughly detailed. The extended lecture notes by Sean Curry and Rod Gover give an up-to-date presentation of the conformal tractor calculus: the first four lectures are mainly focused on the search for invariants; the second half of the course uses tractor calculus to study conformally compact manifolds with application to general relativity as its main motivation.

1.4. A Minicourse in Microlocal Analysis and Wave Propagation

One of the central questions in mathematical relativity is the stability of the Kerr or the Kerr–de Sitter spacetime. As mentioned above, stability has been established by Hintz and Vasy for the Kerr–de Sitter metric, and the question remains open for the Kerr metric. The advantage of the Kerr–de Sitter case is that the inverse of the Fourier transformed d'Alembert operator has a meromorphic extension across the real axis in appropriate weighted spaces. The poles of this extension are then called resonances. Resonances in general relativity were first studied from a mathematical point of view by Bachelot and Motet-Bachelot in [5]. Bony and Häfner gave a resonance expansion of the local propagator for the wave equation on the Schwarzschild–de Sitter metric [7] using the localization of resonances by Sá Barreto-Zworski [29]. Then Dyatlov, Hintz, Vasy, Wunsch, and Zworski made new progress leading eventually to a resonance expansion for the wave equation on spacetimes which are perturbations of the Kerr–de Sitter metric; see the work of Vasy [32]. The whole program culminated in the proof of the non-linear stability of the Kerr–de Sitter metric by Hintz and Vasy [17]. Many aspects come into this study. The first is trapping. Trapping situations were studied in the 1980s for the wave equation outside two obstacles by Ikawa who obtained local energy decay with loss of derivatives in this situation; see [18]. The trapping that appears on the Kerr (or the Kerr–de Sitter) metric is r-normally hyperbolic at least for small angular momentum. Suitable resolvent estimates for this kind of situation have been shown by Wunsch–Zworski [33] and Dyatlov [10]. Another important aspect is the presence of supperradiance due to the fact that there is no globally timelike Killing field outside a Kerr–de Sitter black hole. Whereas the cut-off resolvent can nevertheless be extended meromorphically across the real axis using the work of Mazzeo–Melrose [23] and several different Killing fields (see [12]), a more powerful tool to obtain suitable estimates is the Fredholm theory for non-elliptic settings developed by Vasy [32]. Microlocal analysis was first developed for linear problems. Nevertheless, as the work of Hintz–Vasy shows strikingly enough, it is also well adapted to quasilinear problems. In this context one needs to generalize some of the important theorems (such as the propagation of singularities) to very rough metrics. This program has been achieved by Hintz; see [16]. The last important aspect in the proof of the non-linear stability of the Kerr–de Sitter metric is the issue of the gauge freedom in the Einstein equations. Roughly speaking, a linearization of the Einstein equations can create resonances whose imaginary parts have the "bad sign," leading to exponentially growing modes. These resonances turn out to be "pure gauge" and can therefore be eliminated by an adequate choice of gauge; see [17]. The lectures notes by András Vasy introduce the essential tools used in the proof of the non-linear stability of the Kerr–de Sitter metric.

References

[1] S. Alexakis, A. D. Ionescu, S. Klainerman, *Rigidity of stationary black holes with small angular momentum on the horizon*, Duke Math. J. **163** (2014), 14, 2603–2615.

[2] L. Andersson, P. Blue, *Hidden symmetries and decay for the wave equation on the Kerr spacetime*, Ann. of Math. (2) **182** (2015), 3, 787–853.

[3] L. Andersson, P. Blue, *Uniform energy bound and asymptotics for the Maxwell field on a slowly rotating Kerr black hole exterior*, J. Hyperbolic Differ. Equ., **12** (2015), 4, 689–743.

[4] A. Bachelot, *The Hawking effect*, Ann. Inst. H. Poincaré Phys. Théor. **70** (1999), 1, 41–99.

[5] A. Bachelot, A. Motet-Bachelot, *Les résonances d'un trou noir de Schwarzschild*, Ann. Inst. H. Poincaré Phys. Théor. **59** (1993), 1, 3–68.

[6] T. N. Bailey, M. G. Eastwood, A. R. Gover, *Thomas's structure bundle for conformal, projective and related structures*, Rocky Mountains J. Math. **24** (1994), 4, 1191–1217.

[7] J.-F. Bony, D. Häfner, *Decay and non-decay of the local energy for the wave equation on the de Sitter–Schwarzschild metric*, Comm. Math. Phys. **282** (2008), 3, 697–719.

[8] Y. Choquet-Bruhat, J. W. York, *The Cauchy problem.* In A. Held, editor, General relativity and gravitation, Vol. 1, 99–172, Plenum, New York and London, 1980.

[9] J. Dimock, B. S. Kay, *Classical and quantum scattering theory for linear scalar fields on the Schwarzschild metric*, Ann. Physics **175** (1987), 2, 366–426.

[10] S. Dyatlov, *Spectral gaps for normally hyperbolic trapping*, Ann. Inst. Fourier (Grenoble) **66** (2016), 1, 55–82.

[11] F.G. Friedlander, *Radiation fields and hyperbolic scattering theory*, Math. Proc. Camb. Phil. Soc. **88** (1980), 483–515.

[12] V. Georgescu, C. Gérard, D. Häfner, *Asymptotic completeness for superradiant Klein–Gordon equations and applications to the De Sitter Kerr metric*, J. Eur. Math. Soc. **19** (2017), 2371–2444.

[13] C. Gérard, M. Wrochna, *Construction of Hadamard states by pseudo-differential calculus*, Comm. Math. Phys. **325** (2014), 2, 713–755.

[14] D. Häfner, *Creation of fermions by rotating charged black holes*, Mém. Soc. Math. Fr. (N.S.) **117** (2009), 158 pp.

[15] S. Hawking, R. Penrose, *The singularities of gravitational collapse and cosmology*, Proc. Roy. Soc. London Series A, Mathematical and Physical Sciences, **314** (1970), 1519, 529–548.

[16] P. Hintz, *Global analysis of quasilinear wave equations on asymptotically de Sitter spaces*, Ann. Inst. Fourier (Grenoble) **66** (2016), 4, 1285–1408.

[17] P. Hintz, A. Vasy, *The global non-linear stability of the Kerr–de Sitter family of black holes*, arXiv:1606.04014.

[18] M. Ikawa, *Decay of solutions of the wave equation in the exterior of two convex obstacles*, Osaka J. Math. **19** (1982), 3, 459–509.

[19] R. P. Kerr, *Gravitational field of a spinning mass as an example of algebraically special metrics*, Phys. Rev. Letters **11** (1963), 5, 237–238.

[20] S. Klainerman, I. Rodnianski, J. Szeftel, *The bounded L^2 curvature conjecture*, Invent. Math. **202** (2015), 1, 91–216.

[21] S. Klainerman, I. Rodnianski, J. Szeftel, *Overview of the proof of the bounded L^2 curvature conjecture*, arXiv:1204.1772v2.

[22] L. J. Mason, J.-P. Nicolas, *Conformal scattering and the Goursat problem*, J. Hyperbolic Differ. Equ., **1** (2) (2004), 197–233.

[23] R. Mazzeo, R. Melrose, *Meromorphic extension of the resolvent on complete spaces with asymptotically constant negative curvature*, J. Funct. Anal. **75** (1987), 2, 260–310.

[24] J.-P. Nicolas, *Conformal scattering on the Schwarzschild metric*, Ann. Inst. Fourier (Grenoble) **66** (2016), 3, 1175–1216.

[25] R. Penrose, *Zero rest-mass fields including gravitation: asymptotic behaviour*, Proc. Roy. Soc. London **A284** (1965), 159–203.

[26] M. Radzikowski, *Micro-local approach to the Hadamard condition in quantum field theory on curved space–time*, Comm. Math. Phys. **179** (1996), 3, 529–553.

[27] D. C. Robinson, *Uniqueness of the Kerr black hole*, Phys. Rev. Lett. **34** (1975), 905.

[28] D. C. Robinson, *Four decades of black hole uniqueness theorems.* In D.L. Wiltshire, M. Visser and S.M. Scott, editors, The Kerr space–time, 115–143, Cambridge University Press, 2009.

[29] A. Sá Barreto, M. Zworski, *Distribution of resonances for spherical black holes*, Math. Res. Lett. **4** (1997), 1, 103–121.

[30] K. Schwarzschild, *Über der Gravitationsfeld eines Massenpunktes nach der Einsteinschen Theorie*, K. Preus. Akad. Wiss. Sitz. **424** (1916).

[31] T. Y. Thomas, *On conformal geometry*, Proc. Nat. Acad. Sci. **12** (1926), 352–359.

[32] A. Vasy, *Microlocal analysis of asymptotically hyperbolic and Kerr–de Sitter spaces (with an appendix by Semyon Dyatlov)*, Invent. Math. **194** (2013), 2, 381–513.

[33] J. Wunsch, M. Zworski, *Resolvent estimates for normally hyperbolic trapped sets*, Ann. Henri Poincaré **12** (2011), 7, 1349–1385.

Laboratoire AGM, Département de Mathématiques, Université de Cergy-Pontoise, 95302 Cergy-Pontoise cedex
E-mail address: `thierry.daude@u-cergy.fr`

Université Grenoble Alpes, Institut Fourier, UMR 5582 du CNRS, 100, rue des maths, 38610 Gières, France
E-mail address: `Dietrich.Hafner@univ-grenoble-alpes.fr`

LMBA, Université de Brest, 6 avenue Victor Le Gorgeu, 29238 Brest Cedex 3, France
E-mail address: `jnicolas@univ-brest.fr`

2

Geometry of Black Hole Spacetimes

Lars Andersson, Thomas Bäckdahl and Pieter Blue

Abstract. These notes, based on lectures given at the summer school on Asymptotic Analysis in General Relativity, collect material on the Einstein equations, the geometry of black hole spacetimes, and the analysis of fields on black hole backgrounds. The Kerr model of a rotating black hole in a vacuum is expected to be unique and stable. The problem of proving these fundamental facts provides the background for the material presented in these notes.

Among the many topics which are relevant to the uniqueness and stability problems are the theory of fields on black hole spacetimes, in particular for gravitational perturbations of the Kerr black hole and, more generally, the study of nonlinear field equations in the presence of trapping. The study of these questions requires tools from several different fields, including Lorentzian geometry, hyperbolic differential equations, and spin geometry, which are all relevant to the black hole stability problem.

2.1. Introduction

A short time after Einstein published his field equations for general relativity in 1915, Karl Schwarzschild discovered an exact and explicit solution of the Einstein vacuum equations describing the gravitational field of a spherical body at rest. In analyzing Schwarzschild's solution, one finds that if the central body is sufficiently concentrated, light emitted from its surface cannot reach an observer at infinity. It was not until the 1950s that the global structure of the Schwarzschild spacetime was understood. By this time causality theory and the Cauchy problem for the Einstein equations were firmly established, although many important problems remained open. Observations of highly energetic phenomena occurring within small spacetime regions, eg. quasars, made it plausible that black holes played a significant role in astrophysics, and by the late 1960s these objects were part of mainstream astronomy. The term "black hole" for this type of object came into use in the 1960s. According to our current understanding, black holes are ubiquitous in the universe, in particular most galaxies have a supermassive black hole at their center, and these play an important role in the life of the galaxy. Our galaxy also has at its center a very compact object, Sagittarius A*, with a diameter of less than one astronomical unit, and a mass estimated to be $10^6\ M_\odot$. Evidence for this includes observations of the orbits of stars in its vicinity.

Recall that a solution to the Einstein vacuum equations is a Lorentzian spacetime $(\mathcal{M}, g_{ab})$, satisfying $R_{ab} = 0$, where R_{ab} is the Ricci tensor of g_{ab}. The Einstein equation is the Euler–Lagrange equation of the diffeomorphism invariant Einstein–Hilbert action functional, given by the integral of the scalar curvature of $(\mathcal{M}, g_{ab})$,

$$\int_{\mathcal{M}} R d\mu_g.$$

The diffeomorphism invariance, or general covariance, of the action has the consequence that Cauchy data for the Einstein equation must satisfy a set of constraint equations, and that the principal symbol of the Euler–Lagrange equation is degenerate.[1] After introducing suitable gauge conditions, the Einstein equations can be reduced to a hyperbolic system of evolution equations. It is known that, for any set of sufficiently regular Cauchy data satisfying the constraints, the Cauchy problem for the Einstein equation has a unique solution which is maximal among all regular, vacuum Cauchy developments. This general result, however, does not give any detailed information about the properties of the maximal development.

There are two main conjectures about the maximal development. The strong cosmic censorship conjecture (SCC) states that a generic maximal development is inextendible, as a regular vacuum spacetime. There are examples where the maximal development is extendible, and has non-unique extensions, which furthermore may contain closed timelike curves. In these cases, predictability fails for the Einstein equations, but if SCC holds, they are non-generic. At present, SCC is only known to hold in the context of families of spacetimes with symmetry restrictions; see [98, 7] and references therein. Further, some

[1] From the perspective of hyperbolic partial differential equations, the Einstein equations are both over and under-determined. Contracting the Einstein equation against the normal to a smooth spacelike hypersurface gives elliptic equations that must be satisfied on the hypersurface; these are called the constraint equations. After introducing suitable gauge conditions, the combination of the gauge conditions and the remaining Einstein equations form a hyperbolic system of evolution equations. Furthermore, if the initial data satisfies the constraint equations, then the solution to this hyperbolic system, when restricted to any spacelike hypersurface, also satisfies the constraint equations. If the initial hypersurface is null, the situation becomes more complicated to summarize but simpler to treat in full detail.

non-linear stability results without symmetry assumptions, including the stability of Minkowski space and the stability of quotients of the Milne model (also known as Löbell spacetimes, see [53, 18] and references therein), can be viewed as giving support to SCC. The weak cosmic censorship conjecture states that for a generic isolated system (i.e. an asymptotically flat solution of the Einstein equations), any singularity is hidden from observers at infinity. In this case, the spacetime contains a black hole region, i.e. the complement of the part of the spacetime visible to observers at infinity. The black hole region is bounded by the event horizon, the boundary of the region of spacetime which can be seen by observers at future infinity. Both of these conjectures remain wide open, although there has been limited progress on some problems related to them. The weak cosmic censorship conjecture is most relevant for the purpose of these notes; see [110].

The Schwarzschild solution is static, spherically symmetric, asymptotically flat, and has a single free parameter M which represents the mass of the black hole. By Birkhoff's theorem it is the unique solution of the vacuum Einstein equations with these properties. In 1963 Roy Kerr [68] discovered a new, explicit family of asymptotically flat solutions of the vacuum Einstein equations which are stationary, axisymmetric, and rotating. Shortly after this, a charged, rotating black hole solution to the Einstein–Maxwell equations, known as the Kerr–Newman solution, was found, cf. [87, 88]. Recall that a vector field ν^a is Killing if $\nabla_{(a}\nu_{b)} = 0$. A Kerr spacetime admits two Killing fields, the stationary Killing field $(\partial_t)^a$ which is timelike at infinity, and the axial Killing field $(\partial_\phi)^a$. The Kerr family of solutions is parametrized by the mass M and the azimuthal angular momentum per unit mass a. In the limit $a = 0$, the Kerr solution reduces to the spherically symmetric Schwarzschild solution.

If $|a| \leq M$, the Kerr spacetime contains a black hole, while if $|a| > M$, there is a ringlike singularity which is naked, in the sense that it fails to be hidden from observers at infinity. This situation would violate the weak cosmic censorship conjecture, and one therefore expects that an overextreme Kerr spacetime is unstable and, in particular, that it cannot arise through a dynamical process from regular Cauchy data.

For a geodesic $\gamma^a(\lambda)$ with velocity $\dot{\gamma}^a = d\gamma^a/d\lambda$, in a stationary axisymmetric spacetime,[2] there are three conserved quantities, the mass $\boldsymbol{\mu}^2 = \dot{\gamma}^a\dot{\gamma}_b$, energy $\boldsymbol{e} = \dot{\gamma}^a(\partial_t)_a$, and angular momentum $\boldsymbol{\ell}_z = \dot{\gamma}^a(\partial_\phi)_a$. In a general axisymmetric spacetime, geodesic motion is chaotic. However, as was discovered by Brandon Carter in 1968, there is a fourth conserved quantity

[2] We use the signature $+ - - -$; in particular timelike vectors have a positive norm.

for geodesics in the Kerr spacetime, the Carter constant $\boldsymbol{k}$; see Section 2.5 for details. By Liouville's theorem, this allows one to integrate the geodesic equations by quadratures, and thus geodesics in the Kerr spacetime do not exhibit a chaotic behavior.

The Carter constant is a manifestation of the fact that the Kerr spacetime is algebraically special, of Petrov type $\{2,2\}$, also known as type D. In particular, there are two repeated principal null directions for the Weyl tensor. As shown by Walker and Penrose [112] a vacuum spacetime of Petrov type $\{2,2\}$ admits an object satisfying a generalization of Killing's equation, namely a Killing spinor κ_{AB}, satisfying $\nabla_{A'(A}\kappa_{BC)} = 0$. As shown in the just cited paper, this leads to the presence of four conserved quantities for null geodesics.

Assuming some technical conditions, any asymptotically flat, stationary black hole spacetime is expected to belong to the Kerr family, a fact which is known to hold in the real-analytic case. Further, the Kerr black hole is expected to be stable in the sense that a small perturbation of the Kerr spacetime settles down asymptotically to a member of the Kerr family.

There is much observational evidence pointing to the fact that black holes exist in large numbers in the universe, and that they play a role in many astrophysically significant processes. For example, most galaxies, including our own galaxy, are believed to contain a supermassive black hole at their center. Further, dynamical processes involving black holes, such as mergers, are expected to be important sources of gravitational wave radiation, which could be observed by existing and planned gravitational wave observatories.[3] Thus, black holes play a central role in astrophysics.

Due to its conjectured uniqueness and stability properties, these black holes are expected to be modelled by the Kerr or Kerr–Newman solutions. However, in order to establish the astrophysical relevance of the Kerr solution, it is vital to find rigorous proofs of both of these conjectures, which can be referred to as the black hole uniqueness and stability problems, respectively. A great deal of work has been devoted to these and related problems, and although progress has been made, both remain open at present. The stability problem for the analog of the Kerr solution in the presence of a positive cosmological constant, the Kerr–de Sitter solution, has recently been solved for the case of small angular momenta [117].

Overview

Section 2.2 introduces a range of background material on general relativity, including a discussion of the Cauchy problem for the Einstein equations.

[3] At the time of writing, the first such observation has just been announced [1].

The discussion of black hole spacetimes is started in Section 2.3 with a detailed discussion of the global geometry of the extended Schwarzschild spacetime, followed by some background on marginally outer trapped surfaces and dynamical black holes. Section 2.4 introduces some concepts from spin geometry and the related Geroch–Held–Penrose (GHP) formalism. The Petrov classification is introduced and some properties of its consequential algebraically special spacetimes are presented. In Section 2.5 the geometry of the Kerr black hole spacetime is introduced.

Section 2.6 contains a discussion of null geodesics in the Kerr spacetime. A construction of monotone quantities for null geodesics based on vector fields with coefficients depending on conserved quantities, is introduced. In Section 2.7, symmetry operators for fields on the Kerr spacetime are discussed. Dispersive estimates for fields are the analog of monotone quantities for null geodesics, and in constructing these, symmetry operators play a role analogous to the conserved quantities for the case of geodesics.

2.2. Background

2.2.1. Minkowski Space

Minkowski space $\mathbb{M}$ is $\mathbb{R}^4$ with metric which in a Cartesian coordinate system $(x^a) = (t, x^i)$ takes the form[4]

$$d\tau_{\mathbb{M}}^2 = dt^2 - (dx^1)^2 - (dx^2)^2 - (dx^3)^2.$$

Introducing the spherical coordinates r, θ, ϕ we can write the metric in the form $-dt^2 + dr^2 + r^2 d\Omega_{S^2}^2$, where $d\Omega_{S^2}^2$ is the line element on the standard S^2,

$$d\Omega_{S^2}^2 = (g_{S^2})_{ab} dx^a dx^b = d\theta^2 + \sin^2\theta d\phi^2. \tag{2.1}$$

A tangent vector ν^a is timelike, null, or spacelike when $g_{ab}\nu^a\nu^b > 0, = 0$, or < 0, respectively. Vectors with $g_{ab}\nu^a\nu^b \geq 0$ are called causal. Let $p, q \in \mathbb{M}$. We say that p is in the causal (timelike) future of q if $p - q$ is causal (timelike). The causal and timelike futures $J^+(p)$ and $I^+(p)$ of $p \in \mathbb{M}$ are the sets of points which are in the causal and timelike futures of p, respectively. The corresponding past notions are defined analogously.

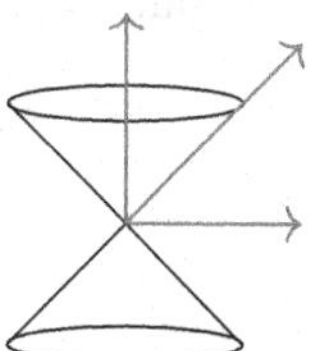

[4] Here and below we shall use line elements, eg. $d\tau_{\mathbb{M}}^2 = (g_{\mathbb{M}})_{ab} dx^a dx^b$, and metrics, eg. $(g_{\mathbb{M}})_{ab}$, interchangeably.

Let u, v be given by

$$u = t - r, \quad v = t + r.$$

In terms of these coordinates the line element takes the form

$$d\tau_{\mathbb{M}}^2 = dudv - r^2 d\Omega_{S^2}^2. \tag{2.2}$$

We see that there are no terms du^2, dv^2, which correspond to the fact that both u, v are null coordinates. In particular, the vectors $(\partial_u)^a$, $(\partial_v)^a$ are null. A complex null tetrad is given by

$$l^a = \sqrt{2}(\partial_u)^a = \frac{1}{\sqrt{2}}\left((\partial_t)^a + (\partial_r)^a\right), \tag{2.3a}$$

$$n^a = \sqrt{2}(\partial_v)^a = \frac{1}{\sqrt{2}}\left(((\partial_t)^a - (\partial_r)^a\right), \tag{2.3b}$$

$$m^a = \frac{1}{\sqrt{2}r}\left((\partial_\theta)^a + \frac{i}{\sin\theta}(\partial_\phi)^a\right) \tag{2.3c}$$

normalized so that $n^a l_a = 1 = -m^a \bar{m}_a$, with all other inner products of tetrad legs zero. Complex null tetrads with this normalization play a central role in the Newman–Penrose (NP) and GHP formalisms; see Section 2.4. In these notes we will use such tetrads unless otherwise stated.

In terms of a null tetrad, we have

$$g_{ab} = 2(l_{(a}n_{b)} - m_{(a}\bar{m}_{b)}). \tag{2.4}$$

Introduce compactified null coordinates $\mathcal{U}, \mathcal{V}$, given by

$$\mathcal{U} = \arctan u, \quad \mathcal{V} = \arctan v.$$

These take values in $\{(-\pi/2, \pi/2) \times (-\pi/2, \pi/2)\} \cap \{\mathcal{V} \geq \mathcal{U}\}$, and we can thus present Minkowski space in a *causal diagram*; see Figure 2.1. Here each point represents an S^2 and we have drawn null vectors at 45° angles. A compactification of Minkowski space is now given by adding the null boundaries[5] $\mathcal{I}^\pm$, spatial infinity i_0, and timelike infinity $i^\pm$ as indicated in the figure. Explicitly,

$$\begin{aligned}
\mathcal{I}^+ &= \{\mathcal{V} = \pi/2\} \\
\mathcal{I}^- &= \{\mathcal{U} = -\pi/2\} \\
i_0 &= \{\mathcal{V} = \pi/2, \mathcal{U} = -\pi/2\} \\
i_\pm &= \{(\mathcal{V}, \mathcal{U}) = \pm(\pi/2, \pi/2)\}.
\end{aligned}$$

[5] Here $\mathcal{I}$ is pronounced "Scri" for "script I."

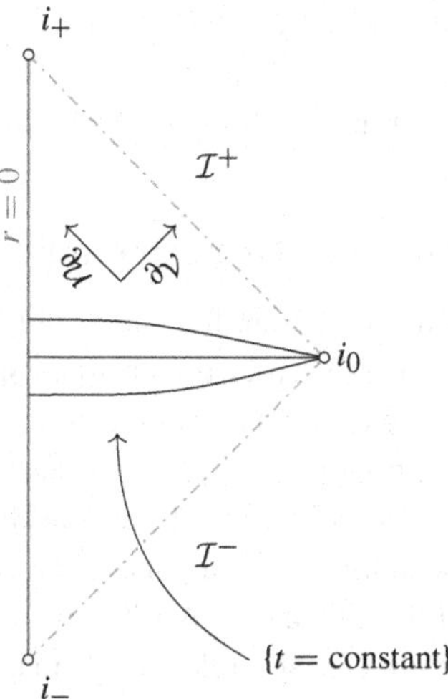

Figure 2.1. Causal diagram of Minkowski space

In Figure 2.1, we have also indicated schematically the t-level sets which approach spatial infinity i_0. Causal diagrams are a useful tool which, if applied with proper care, can be used to understand the structure of quite general spacetimes. Such diagrams are often referred to as Penrose or Carter–Penrose diagrams.

In particular, as can be seen from Figure 2.1, we have $\mathbb{M} = I^-(\mathcal{I}^+) \cap I^+(\mathcal{I}^-)$, i.e. any point in $\mathbb{M}$ is in the past of $\mathcal{I}^+$ and in the future of $\mathcal{I}^-$. This is related to the fact that $\mathbb{M}$ is *asymptotically simple*, in the sense that it admits a conformal compactification with a regular null boundary, and has the property that any inextendible null geodesic hits the null boundary. For massless fields on Minkowski space, this means that it makes sense to formulate a scattering map which takes data on $\mathcal{I}^-$ to data on $\mathcal{I}^+$; see [93].

Let

$$\mathcal{T} = \mathcal{V} + \mathcal{U}, \quad \mathcal{R} = \mathcal{V} - \mathcal{U}. \tag{2.5}$$

Then, with $\Phi^2 = 2\cos\mathcal{U}\cos\mathcal{V}$, the conformally transformed metric $\tilde{g}_{ab} = \Phi^2 g_{ab}$ takes the form

$$\begin{aligned}\tilde{g}^{\mathbb{M}}_{ab} &= d\mathcal{T}^2 - d\mathcal{R}^2 - \sin^2\mathcal{R} d\Omega^2_{S^2} \\ &= d\mathcal{T}^2 - d\Omega^2_{S^3}\end{aligned}$$

which we recognize as the metric on the cylinder $\mathbb{R} \times S^3$. This spacetime is known as the Einstein cylinder, and can be viewed as a static solution of the Einstein equations with dust matter and a positive cosmological constant [50].

2.2.2. Lorentzian Geometry and Causality

We now consider a smooth Lorentzian four-manifold $(\mathcal{M}, g_{ab})$ with signature $+---$. Each tangent space in a four-dimensional spacetime is isometric to Minkowski space $\mathbb{M}$, and we can carry intuitive notions of causality over from $\mathbb{M}$ to $\mathcal{M}$. We say that a smooth curve $\gamma^a(\lambda)$ is causal if the velocity vector $\dot{\gamma}^a = d\gamma^a/d\lambda$ is causal. Two points in $\mathcal{M}$ are causally related if they can be connected by a piecewise smooth causal curve. The concept of causal curves is most naturally defined for C^0 curves. A C^0 curve γ^a is said to be causal if each pair of points on γ^a are causally related. We may define a timelike curve and timelike related points in an analogous manner.

We now assume that $\mathcal{M}$ is time oriented, i.e. that there is a globally defined timelike vector field on $\mathcal{M}$. This allows us to distinguish between future and past directed causal curves, and to introduce a notion of the causal and timelike future of a spacetime point. The corresponding past notions are defined analogously. If q is in the causal future of p, we write $p \preccurlyeq q$. This introduces a partial order on $\mathcal{M}$. The causal future $J^+(p)$ of p is defined as $J^+(p) = \{q : p \preccurlyeq q\}$ while the timelike future $I^+(p)$ is defined in an analogous manner, with timelike replacing causal. A subset $\Sigma \subset \mathcal{M}$ is achronal if there is no pair $p, q \in \mathcal{M}$ such that $q \in I^+(p)$, i.e. Σ does not intersect its timelike future or past. The domain of dependence $D(S)$ of $S \subset \mathcal{M}$ is the set of points p such that any inextendible causal curve starting at p must intersect S.

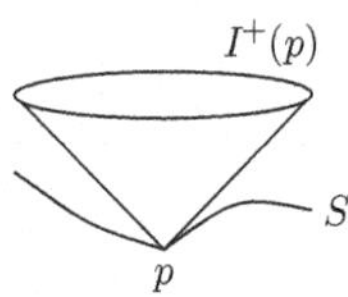

Definition 2.1 A spacetime $\mathcal{M}$ is globally hyperbolic if there is a closed, achronal $\Sigma \subset \mathcal{M}$ such that $\mathcal{M} = D(\Sigma)$. In this case, Σ is called a Cauchy surface.

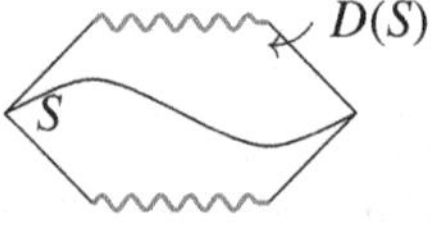

Due to the results of Bernal and Sanchez [28], global hyperbolicity is characterized by the existence of a smooth, Cauchy time function $\tau : \mathcal{M} \to \mathbb{R}$. A function τ on $\mathcal{M}$ is a time function if $\nabla^a\tau$ is timelike everywhere, and it is Cauchy if the level sets $\Sigma_t = \tau^{-1}(t)$ are Cauchy surfaces. If τ is smooth, its level sets are then smooth and spacelike. It follows that a globally hyperbolic spacetime $\mathcal{M}$ is globally foliated by Cauchy surfaces, and in particular is diffeomorphic to a product $\Sigma \times \mathbb{R}$. In the following, unless otherwise stated, we shall consider only globally hyperbolic spacetimes.

If a globally hyperbolic spacetime $\mathcal{M}$ is a subset of a spacetime $\mathcal{M}'$, then the boundary $\partial\mathcal{M}$ of $\mathcal{M}$ in $\mathcal{M}'$ is called the Cauchy horizon.

Example 2.1 Let O be the origin in Minkowski space, and let $\mathcal{M} = I^+(O) = \{t > r\}$ be its timelike future. Then $\mathcal{M}$ is globally hyperbolic with Cauchy time function $\tau = \sqrt{t^2 - r^2}$. Further, $\mathcal{M}$ is a subset of Minkowski space $\mathbb{M}$, which is a globally hyperbolic space with Cauchy time function t. Minkowski space is geodesically complete and hence inextendible. The boundary $\{t = r\}$ is the Cauchy horizon $\partial\mathcal{M}$ of $\mathcal{M}$. Past inextendible causal geodesics (i.e. past causal rays) in $\mathcal{M}$ end on $\partial\mathcal{M}$. In particular, $\mathcal{M}$ is incomplete. However, $\mathcal{M}$ is extendible, as a smooth flat spacetime, with many inequivalent extensions.

We remark that for a globally hyperbolic spacetime, which is extendible, the extension is in general non-unique. In the particular case considered in example 2.1, $\mathbb{M}$ is an extension of $\mathcal{M}$, which also happens to be maximal and globally hyperbolic. In the vacuum case, there is a unique maximal globally hyperbolic extension, cf. Section 2.2.5 below. However, a maximal extension is in general non-unique, and may fail to be globally hyperbolic.

2.2.3. Conventions and Notation

We will mostly use abstract indices, cf. [94], but will sometimes work with coordinate indices, and unless confusion arises we will not be too specific about this. We raise and lower indices with g_{ab}, e.g. $\xi^a = g^{ab}\xi_b$, with $g^{ab}g_{bc} = \delta^a{}_c$, where $\delta^a{}_c$ is the Kronecker delta, i.e. the tensor with the property that $\delta^a{}_c\xi^c = \xi^a$ for any ξ^a.

Let $\epsilon_{a\cdots d}$ be the Levi-Civita symbol, i.e. the skew symmetric expression which in any coordinate system has the property that $\epsilon_{1\cdots n} = 1$. The volume form of g_{ab} is $(\mu_g)_{abcd} = \sqrt{|g|}\epsilon_{abcd}$. Given $(\mathcal{M}, g_{ab})$ we have the canonically defined Levi-Civita covariant derivative ∇_a. For a vector ν^a, this is of the form

$$\nabla_a \nu^b = \partial_a \nu^b + \Gamma^b_{ac}\nu^c$$

where $\Gamma^b_{ac} = \frac{1}{2}g^{bd}(\partial_a g_{dc} + \partial_c g_{db} - \partial_d g_{ac})$ is the Christoffel symbol. In order to fix the conventions used here, we recall that the Riemann curvature tensor is defined by

$$(\nabla_a\nabla_b - \nabla_b\nabla_a)\xi_c = R_{abc}{}^d\xi_d.$$

The Riemann tensor R_{abcd} is skew symmetric in the pairs of indices ab, cd, $R_{abcd} = R_{[ab]cd} = R_{ab[cd]}$, is pairwise symmetric $R_{abcd} = R_{cdab}$, and satisfies the first Bianchi identity $R_{[abc]d} = 0$. Here square brackets $[\cdots]$ denote antisymmetrization. We shall similarly use round brackets $(\cdots)$ to denote symmetrization. Further, we have $\nabla_{[a}R_{bc]de} = 0$, the second Bianchi identity. A contraction gives $\nabla^a R_{abcd} = 0$. The Ricci tensor is $R_{ab} = R^c{}_{acb}$ and the scalar curvature $R = R^a{}_a$. We further let $S_{ab} = R_{ab} - \frac{1}{4}Rg_{ab}$ denote the tracefree part of the Ricci tensor. The Riemann tensor can be decomposed as follows,

$$\begin{aligned} R_{abcd} = & -\tfrac{1}{12}g_{ad}g_{bc}R + \tfrac{1}{12}g_{ac}g_{bd}R + \tfrac{1}{2}g_{bd}S_{ac} - \tfrac{1}{2}g_{bc}S_{ad} \\ & - \tfrac{1}{2}g_{ad}S_{bc} + \tfrac{1}{2}g_{ac}S_{bd} + C_{abcd}. \end{aligned} \tag{2.6}$$

This defines the Weyl tensor C_{abcd} which is a tensor with the symmetries of the Riemann tensor, and vanishing traces, $C^c{}_{acb} = 0$. Recall that $(\mathcal{M}, g_{ab})$ is locally conformally flat if and only if $C_{abcd} = 0$. It follows from the contracted second Bianchi identity that the Einstein tensor $G_{ab} = R_{ab} - \frac{1}{2}Rg_{ab}$ is conserved, $\nabla^a G_{ab} = 0$.

2.2.4. Einstein Equation

The Einstein equation in geometrized units with $G = c = 1$, where G, c denote Newton's constant and the speed of light, respectively, cf. [109, Appendix F], is the system

$$G_{ab} = 8\pi T_{ab}. \tag{2.7}$$

This equation relates geometry, expressed in the Einstein tensor G_{ab} on the left-hand side, to matter, expressed via the energy momentum tensor T_{ab} on the right-hand side. For example, for a self-gravitating Maxwell field F_{ab}, $F_{ab} = F_{[ab]}$, we have

$$T_{ab} = \frac{1}{4\pi}\left(F_{ac}F_{bc} - \frac{1}{4}F_{cd}F^{cd}g_{ab}\right).$$

The source-free Maxwell field equations

$$\nabla^a F_{ab} = 0, \quad \nabla_{[a}F_{bc]} = 0$$

imply that T_{ab} is conserved, $\nabla^a T_{ab} = 0$. The contracted second Bianchi identity implies that $\nabla^a G_{ab} = 0$, and hence the conservation property of T_{ab} is implied by the coupling of the Maxwell field to gravity. These facts can be seen to follow from the variational formulation of Einstein gravity, given by the action

$$I = \int_{\mathcal{M}} \frac{R}{16\pi} d\mu_g - \int_{\mathcal{M}} L_{\text{matter}} d\mu_g$$

where L_{matter} is the Lagrangian describing the matter content in spacetime. In the case of Maxwell theory, this is given by

$$L_{\text{Maxwell}} = \frac{1}{4\pi} F_{cd} F^{cd}.$$

Recall that in order to derive the Maxwell field equation, as an Euler–Lagrange equation, from this action, it is necessary to introduce a vector potential for F_{ab}, by setting $F_{ab} = 2\nabla_{[a} A_{b]}$, and to carry out the variation with respect to A_a. It is a general fact that for generally covariant (i.e. diffeomorphism invariant) Lagrangian field theories which depend on the spacetime location only via the metric and its derivatives, the symmetric energy momentum tensor

$$T_{ab} = \frac{1}{\sqrt{g}} \frac{\partial L_{\text{matter}}}{\partial g^{ab}}$$

is conserved when evaluated on solutions of the Euler–Lagrange equations.

As a further example of a matter field, we consider the scalar field, with action

$$L_{\text{scalar}} = \tfrac{1}{2} \nabla^c \psi \nabla_c \psi$$

where ψ is a function on $\mathcal{M}$. The corresponding energy-momentum tensor is

$$T_{ab} = \nabla_a \psi \nabla_b \psi - \tfrac{1}{2} \nabla^c \psi \nabla_c \psi g_{ab}$$

and the Euler–Lagrange equation is the free scalar wave equation

$$\nabla^a \nabla_a \psi = 0. \tag{2.8}$$

As (2.8) is another example of a field equation derived from a covariant action which depends on the spacetime location only via the metric g_{ab} or its derivatives, the symmetric energy-momentum tensor is conserved for solutions of the field equation.

In both of the just mentioned cases, the energy momentum tensor satisfies the dominant energy condition $T_{ab} \nu^a \zeta^b \geq 0$ for future directed causal vectors ν^a, ζ^a. This implies the null energy condition

$$R_{ab} \nu^a \nu^b \geq 0 \quad \text{if } \nu_a \nu^a = 0. \tag{2.9}$$

These energy conditions hold for most classical matter models.

There are many interesting matter systems which are worthy of consideration, such as fluids, elasticity, kinetic matter models including Vlasov, as well as fundamental fields such as Yang–Mills, to name just a few. We consider only spacetimes which satisfy the null energy condition, and for the most part we shall in these notes be concerned with the vacuum Einstein equations,

$$R_{ab} = 0. \tag{2.10}$$

2.2.5. The Cauchy Problem

Given a spacelike hypersurface[6] Σ in $\mathcal{M}$ with timelike normal T^a, induced metric h_{ab}, and second fundamental form k_{ab}, defined by $k_{ab}X^aY^b = \nabla_a T_b X^a Y^b$ for X^a, Y^b tangent to Σ, the Gauss and Gauss–Codazzi equations imply the constraint equations

$$R[h] + (k_{ab}h^{ab})^2 - k_{ab}k^{ab} = 16\pi T_{ab}T^aT^b \tag{2.11a}$$

$$\nabla[h]_a(k_{bc}h^{bc}) - \nabla[h]^b k_{ab} = T_{ab}T^b. \tag{2.11b}$$

A three-manifold Σ together with tensor fields h_{ab}, k_{ab} on Σ solving the constraint equations is called a Cauchy data set. The constraint equations for general relativity are analogs of the constraint equations in Maxwell and Yang–Mills theory, in that they lead to Hamiltonians which generate gauge transformations.

Consider a 3+1 split of $\mathcal{M}$, i.e. a one-parameter family of Cauchy surfaces Σ_t, with a coordinate system $(x^a) = (t, x^i)$, and let

$$(\partial_t)^a = NT^a + X^a$$

be the split of $(\partial_t)^a$ into a normal and tangential piece. The fields (N, X^a) are called lapse and shift. The definition of the second fundamental form implies the equation

$$\mathcal{L}_{\partial_t} h_{ab} = -2Nk_{ab} + \mathcal{L}_X h_{ab}.$$

In the vacuum case, the Hamiltonian for gravity can be written in the form

$$\int N\mathcal{H} + X^a \mathcal{J}_a + \text{ boundary terms}$$

where $\mathcal{H}$ and $\mathcal{J}$ are the densitized left-hand sides of (2.11). If we consider only compactly supported perturbations in deriving the Hamiltonian evolution equation, the boundary terms mentioned above can be ignored. However, for (N, X^a) not tending to zero at infinity, and considering perturbations compatible with asymptotic flatness, the boundary term becomes significant, cf. Section 2.2.6.4.

The resulting Hamiltonian evolution equations, written in terms of h_{ab} and its canonical conjugate $\pi^{ab} = \sqrt{h}(k^{ab} - (h^{cd}k_{cd}h^{ab}))$, are usually called the ADM (for Arnowitt–Deser–Misner) evolution equations.

Let $\Sigma \subset \mathcal{M}$ be a Cauchy surface. Given functions ϕ_0, ϕ_1 on Σ and F on $\mathcal{M}$, the Cauchy problem for the wave equation is that of finding solutions to

[6] Where there is no likelihood of confusion, we shall denote abstract indices for objects on Σ by $a, b, c, \ldots$.

$$\nabla^a \nabla_a \psi = F, \quad \psi\big|_\Sigma = \phi_0, \quad \mathcal{L}_{\partial_t} \psi\big|_\Sigma = \phi_1.$$

Assuming suitable regularity conditions, the solution is unique and stable with respect to the initial data. This fact extends to a wide class of non-linear hyperbolic PDEs including quasilinear wave equations, i.e. equations of the form

$$A^{ab}[\psi]\partial_a \partial_b \psi + B[\psi, \partial\psi] = 0$$

with A^{ab} a Lorentzian metric depending on the field ψ.

Given a vacuum Cauchy data set, (Σ, h_{ab}, k_{ab}), a solution of the Cauchy problem for the Einstein vacuum equations is a spacetime metric g_{ab} with $R_{ab} = 0$, such that (h_{ab}, k_{ab}) coincides with the metric and second fundamental form induced on Σ from g_{ab}. Such a solution is called a vacuum extension of (Σ, h_{ab}, k_{ab}).

Due to the fact that R_{ab} is covariant, the symbol of R_{ab} is degenerate. In order to get a well-posed Cauchy problem, it is necessary either to impose gauge conditions or to introduce new variables. A standard choice of gauge condition is the harmonic coordinate condition. Let $\widehat{g}_{ab}$ be a given metric on $\mathcal{M}$. The identity map $\mathbf{i} : \mathcal{M} \to \mathcal{M}$ is harmonic if and only if the vector field

$$V^a = g^{bc}(\Gamma^a_{bc} - \widehat{\Gamma}^a_{bc})$$

vanishes. Here Γ^a_{bc}, $\widehat{\Gamma}^a_{bc}$ are the Christoffel symbols of the metrics $g_{ab}, \widehat{g}_{ab}$. Then V^a is the tension field of the identity map $\mathbf{i} : (\mathcal{M}, g_{ab}) \to (\mathcal{M}, \widehat{g}_{ab})$. This is harmonic if and only if

$$V^a = 0. \tag{2.12}$$

Since harmonic maps with a Lorentzian domain are often called wave maps, the gauge condition (2.12) is sometimes called a wave map gauge condition. A particular case of this construction, which can be carried out if $\mathcal{M}$ admits a global coordinate system (x^a), is given by letting $\widehat{g}_{ab}$ be the Minkowski metric defined with respect to (x^a). Then $\widehat{\Gamma}^a_{bc} = 0$ and (2.12) is simply

$$\nabla^b \nabla_b x^a = 0, \tag{2.13}$$

which is usually called the wave coordinate gauge condition.

Going back to the general case, let $\widehat{\nabla}$ be the Levi-Civita covariant derivative defined with respect to $\widehat{g}_{ab}$. We have the identity

$$R_{ab} = -\tfrac{1}{2}\frac{1}{\sqrt{g}}\widehat{\nabla}_a \sqrt{g} g^{ab} \widehat{\nabla}_b g_{ab} + S_{ab}[g, \widehat{\nabla} g] + \nabla_{(a} V_{b)} \tag{2.14}$$

where S_{ab} is an expression which is quadratic in first derivatives $\widehat{\nabla}_a g_{cd}$. Setting $V^a = 0$ in (2.14) yields R^{harm}_{ab}, and (2.10) becomes a quasilinear wave equation

$$R^{\text{harm}}_{ab} = 0. \tag{2.15}$$

By standard results, the equation (2.15) has a locally well-posed Cauchy problem in Sobolev spaces H^s for $s > 5/2$. Using more sophisticated techniques, well-posedness can shown to hold for any $s > 2$ [71]. Recently a local existence has been proved under the assumption of curvature bounded in L^2 [73]. Given a Cauchy data set (Σ, h_{ab}, k_{ab}), together with initial values for lapse and shift N, X^a on Σ, it is possible to find $\mathcal{L}_t N, \mathcal{L}_t X^a$ on Σ such that the V^a are zero on Σ. A calculation now shows that, due to the constraint equations, $\mathcal{L}_{\partial_t} V^a$ is zero on Σ. Given a solution to the reduced Einstein vacuum equation (2.15), one finds that V^a solves a wave equation. This follows from $\nabla^a G_{ab} = 0$, due to the Bianchi identity. Hence, due to the fact that the Cauchy data for V^a is trivial, it holds that $V^a = 0$ on the domain of the solution. Thus, the solution to (2.15) is a solution to the full vacuum Einstein equation (2.10). This proves local well-posedness for the Cauchy problem for the Einstein vacuum equation. This fact was first proved by Yvonne Choquet-Bruhat [54]; see [99] for background and history.

Global uniqueness for the Einstein vacuum equations was proved by Choquet-Bruhat and Geroch [35]. The proof relies on the local existence theorem sketched above, patching together local solutions. A partial order is defined on the collection of vacuum extensions, making use of the notion of a common domain. The common domain U of two extensions $\mathcal{M}$, $\mathcal{M}'$ is the maximal subset in $\mathcal{M}$ which is isometric to a subset in $\mathcal{M}'$. We can then define a partial order by saying that $\mathcal{M} \leq \mathcal{M}'$ if the maximal common domain is $\mathcal{M}$. Given a partially ordered set, a maximal element exists by Zorn's lemma. This is proven to be unique by an application of the local well-posedness theorem for the Cauchy problem sketched above. For a contradiction, let $\mathcal{M}$, $\mathcal{M}'$ be two inequivalent extensions, and let U be the maximal common domain. Due to the Haussdorff property of spacetimes, this leads to a contradiction. By finding a partial Cauchy surface which touches the boundary of U (see Figure 2.2 and making use of local uniqueness) one finds a contradiction to the maximality of U. It should be noted that here uniqueness holds up to isometry, in keeping with the general covariance of the Einstein vacuum equations. These facts extend to the Einstein equations coupled to hyperbolic matter equations. See [101]

Figure 2.2. A partial Cauchy surface which touches the boundary of ∂U

for a construction of the maximal globally hyperbolic extension which does not rely on Zorn's lemma; see also [114]. The global uniqueness result can be generalized to Einstein-matter systems, provided the matter field equation is hyperbolic and that its solutions do not break down. General results on this topic are lacking, see however [92] and references therein. The minimal regularity needed for global uniqueness is a subtle issue, which has not been fully addressed. In particular, results on local well-posedness are known; see e.g. [72] and references therein, which require less regularity than the best results on global uniqueness.

2.2.6. Remarks

We shall now make several remarks relating to the above discussion.

2.2.6.1. Bianchi Identities as a Hyperbolic System

The vacuum Einstein equation $R_{ab} = 0$ implies that the Weyl tensor C_{abcd} satisfies the Bianchi identity $\nabla^a C_{abcd} = 0$. This is the massless spin-2 equation. In particular, this is a first order hyperbolic system for the Weyl tensor.

The spin-2 equation (i.e. the equation $\nabla^a W_{abcd}$ for a Weyl test field W_{abcd}, a tensor field with the symmetries and trace properties of the Weyl tensor) implies algebraic conditions relating the field and the curvature. In particular, in a sufficiently general background a Weyl test field must be proportional to the Weyl tensor C_{abcd} of the spacetime. This holds in particular for spacetimes of Petrov type D (cf. Section 2.4.6 below for the definition of Petrov types; see [10, §2.3] and references therein).

One may view the Bianchi identity for the Weyl tensor as the main gravitational field equation, and the vacuum Einstein equation as a type of "constraint" equation, which allows one to relate the Weyl tensor to the Riemann curvature of spacetime. The first-order system for the Weyl tensor can be extended to a first-order system including the first and second Cartan structure equations. A hyperbolic system can be extracted by introducing suitable gauge conditions; see Section 2.2.6.3.

2.2.6.2. Null Condition

Consider the Cauchy problem for the semilinear wave equation on Minkowski space,

$$\nabla^a \nabla_a \psi = Q^{ab} \nabla_a \psi \nabla_b \psi$$

with data $\psi\big|_{t=0} = \epsilon \psi_0$, $\partial_t \psi\big|_{t=0} = \epsilon \psi_1$, where $\epsilon > 0$ and ψ_1, ψ_2 are suitably regular functions. Solutions exist globally for small data (i.e. for sufficiently small $\epsilon > 0$) if and only if Q^{ab} satisfies the null condition, $Q^{ab}\xi_a\xi_b = 0$, for any null vector ξ^a. An example given by Fritz John shows that the equation $\nabla^a\nabla_a\psi = |\partial_t\psi|^2$, for which the null condition fails, can have blowup for small data, cf. [104]. Similar results hold also for quasilinear equations; in particular for quasilinear wave equations satisfying a suitable null condition, one has stability of the trivial solution.

For the vacuum Einstein equation in harmonic coordinates, we have

$$R^{\text{harm}}_{ab} = -\tfrac{1}{2}g^{cd}\partial_c\partial_d g_{ab} + S_{ab}(g, \partial g)$$

where the lower order term S_{ab} contains terms of the form $\partial_a g_{cd}\partial_b g_{ef}g^{ce}g^{df}$, and hence the null condition fails to hold for the Einstein vacuum equation in harmonic coordinates. For this reason the problem of stability of Minkowski space in Einstein gravity is subtle. The stability of Minkowski space was first proved by Christodoulou and Klainerman [37]. Later a proof using harmonic coordinates was given by Lindblad and Rodnianski [76]. This exploits the fact that the equation $R^{\text{harm}}_{ab} = 0$ satisfies a weak form of the null condition. Consider the system

$$\nabla^a\nabla_a\psi = |\partial_t\phi|^2 \tag{2.16a}$$

$$\nabla^a\nabla_a\phi = Q^{ab}\nabla_a\phi\nabla_b\phi \tag{2.16b}$$

on Minkowski space, where Q^{ab} has a null structure. For this system, the null condition fails to hold. However, ϕ satisfies an equation with null structure and therefore has good dispersion. The equation for ψ has a source defined in terms of ϕ but no bad self-interaction. One finds therefore that the solution to (2.16) exists globally for small data, but with slightly slower falloff than a solution of an equation satisfying the null condition.

2.2.6.3. Gauge Source Functions

As has been pointed out by Helmut Friedrich, see [55] for discussion, one may introduce gauge source functions $V^a = F^a(x^b, g^{cd})$ without affecting the reduction procedure. The gauge source functions can be designed to yield damping effects or to control the evolution of the lapse and shift. This has frequently been used in numerical relativity. A related strategy is to add terms involving factors of the constraints C^a. Such terms vanish for a solution of the field equations, but may provide improved behavior for the reduced system.

It is often convenient to introduce a suitably normalized tetrad $e_{\underline{a}}{}^a$. Important examples are orthonormal tetrads, satisfying $e_{\underline{a}}{}^a e_{\underline{b}}{}^b g_{ab} = \text{diag}(+1, -1, -1, -1)$, and the null tetrads $(l^a, n^a, m^a, \bar{m}^a)$ with $l^a n_a = 1$, $m^a\bar{m}_a = -1$, all other inner products being zero. Such tetrads appear naturally when working with spinors; see Section 2.4.

The field equations can be written as a system of equations for tetrad components, connection coefficients, and curvature. Introducing tetrad gauge source functions $V_{\underline{ab}} = (\nabla^c \nabla_c e_{\underline{a}}^a) e_{\underline{b}}^b g_{ab}$ it is possible to extract a first-order symmetric hyperbolic system with $V^a, V_{\underline{ab}}$ taking values involving tetrad, connection coefficients and curvature. This opens up a lot of interesting possibilities, but has not been widely used. The phantom gauge introduced by Chandrasekhar [34, p. 240] was shown in [3] to correspond to a tetrad gauge condition of the above type, and is therefore compatible with a well-posed Cauchy problem.

Let $(\mathcal{M}, g_{ab})$ be a vacuum spacetime. Let $g(s)_{ab}$ be a one-parameter family of vacuum metrics and let

$$h_{ab} = \frac{d}{ds} g(s)_{ab} \bigg|_{s=0}.$$

Then h_{ab} solves the linearized Einstein equation $DR_{ab} = 0$, where DR_{ab} is the Frechet derivative of the Ricci tensor at g_{ab} in the direction h_{ab}. A calculation, cf. [3], shows that if we impose the linearized wave map gauge condition, then h_{ab} satisfies the Lichnerowicz wave equation

$$\nabla^c \nabla_c h_{ab} + 2R_{acbd} h^{cd} = 0.$$

2.2.6.4. Asymptotically Flat Data

The Kerr black hole represents an isolated system, and the appropriate data for the black hole stability problem should therefore be asymptotically flat. To make this precise we suppose there is a compact set K in $\mathcal{M}$ and a map $\Phi : \mathcal{M} \setminus K \to \mathbb{R}^3 \setminus B(R, 0)$, where $B(R, 0)$ is a Euclidean ball. This defines a Cartesian coordinate system on the end $\mathcal{M} \setminus K$ so that $h_{ab} - \delta_{ab}$ falls off to zero at infinity, at a suitable rate. Here δ_{ab} is the Euclidean metric in the Cartesian coordinate system constructed above. Similarly, we require that k_{ab} falls off to zero.

Let x^a be the chosen Euclidean coordinate system and let r be the Euclidean radius $r = (\delta_{ab} x^a x^b)^{1/2}$. Following Regge and Teitelboim [97], see also [25], we assume that $g_{ab} = \delta_{ab} + h_{ab}$ with

$$\begin{aligned} h_{ab} &= O(1/r), \quad \partial_a h_{bc} = O(1/r^2), \\ k_{ab} &= O(1/r^2). \end{aligned}$$

Further, we impose the parity conditions

$$h_{ab}(x) = h_{ab}(-x), \quad k_{ab}(x) = -k_{ab}(-x). \tag{2.17}$$

These falloff and parity conditions guarantee that the ADM 4-momentum and angular momentum are well defined. It was shown in [63] that data satisfying the parity condition conditions (2.17) are dense among data which satisfy an asymptotic flatness condition in terms of weighted Sobolev spaces.

Let ξ^a be an element of the Poincare Lie algebra and assume that $NT^a + X^a$ tends in a suitable sense to ξ^a at infinity. Then the action for Einstein gravity can be written in the form

$$\int_{\mathcal{M}} R d\mu_g = P_a \xi^a + \int \pi^{ij} \dot{h}_{ij} - \int N\mathcal{H} + X^i \mathcal{J}_i.$$

Here we may view P_a as a map to the dual of the Poincare Lie algebra, i.e. a momentum map. Evaluating $P_a\xi^a$ on a particular element of the Poincare Lie algebra gives the corresponding momentum. These can also be viewed as charges at infinity. We have

$$P^0 = \frac{1}{16\pi} \lim_{r\to\infty} \int_{S_r} (\partial_i g_{ji} - \partial_j g_{ii}) d\sigma^i \tag{2.18a}$$

$$P^i = \frac{1}{8\pi} \lim_{r\to\infty} \int_{S_r} \pi_{ij} d\sigma^j \tag{2.18b}$$

where $d\sigma^i$ denotes the hypersurface area element of a family of spheres (which can be taken to be coordinate spheres) S_r foliating a neighborhood of infinity. See [82] and references therein for a recent discussion of the conditions under which these expressions are well-defined.

The energy and linear momentum (P^0, P^i) provide the components of a 4-vector P^a, the ADM 4-momentum. Assuming the dominant energy condition then under the above asymptotic conditions, P^a is future causal, and timelike unless the maximal development $(\mathcal{M}, g_{ab})$ is isometric to Minkowski space. Further, P^a transforms as a Minkowski 4-vector, and the ADM mass is given by $M = \sqrt{P^a P_a}$. The boost theorem [38] implies, given an asymptotically flat Cauchy data set, that one may find a boosted Cauchy surface Σ' in its development such that the data is in the rest frame with respect to Σ', i.e. $P^a = M(\partial_t)^a$.

Since the constraint quantities $\mathcal{H}, \mathcal{J}_i$ vanish for solutions of the Einstein equations, the gravitational Hamiltonian takes the value $P_a\xi^a$, and hence the ADM mass and momenta defined by (2.18) are conserved for an evolution with lapse and shift $(N, X^i) \to (1, 0)$ at infinity. If we consider the analog of the above definitions for a hyperboloidal slice which meets $\mathcal{I}$, then the ADM mass and momentum are replaced by the Bondi mass and momentum. An example of a hyperboloidal slice in Minkowski space is given by a level set of the time function $\mathcal{T}$, cf. (2.5), in the compactification of Minkowski space. For the Bondi 4-momentum, one has the important feature that gravitational energy is radiated through $\mathcal{I}$, which means that it is not conserved. See [40] and references therein for further details.

2.2.6.5. Killing Initial Data

A Killing initial data set is a Cauchy data set (Σ, h_{ab}, k_{ab}) such that the development $(\mathcal{M}, g_{ab})$ is a spacetime with a Killing field ν^a, i.e.

$$\mathcal{L}_\nu g_{ab} = 2\nabla_{(a}\nu_{b)} = 0.$$

Now let ν^a be a solution to the wave equation $\nabla^a\nabla_a\nu_b = 0$, but not necessarily a Killing field. In a vacuum spacetime, we then have

$$\nabla^d\nabla_d(\nabla_{(a}\nu_{b)}) = 2R^c{}_{(ab)}{}^d\nabla_{(c}\nu_{d)}.$$

This implies that the tensor $\mathcal{L}_\nu g_{ab}$ satisfies a wave equation, so if it has trivial Cauchy data on Σ, then ν^a is a Killing field in the domain of dependence of Σ. This allows us to characterize Lie symmetries of a development $(\mathcal{M}, g_{ab})$ purely in terms of the Cauchy data. Another way to formulate this statement is that Lie symmetries propagate. This fact, which is closely related to the global uniqueness for the Cauchy problem, allows one to study symmetry restrictions of the Einstein equations. Much work has been done to study consistent subsystems of the Einstein equation, implied by imposing symmetries on the initial data. Examples include Bianchi (spatially homogenous), T^2 (two commuting spatial Killing fields with torus orbits), U^1 (one spatial Killing field generating a circle action). Note, however, there are also the so-called surface symmetric spacetimes, which arise in a somewhat different manner. In addition, there are consistent subsystems which are not given by symmetry restrictions. Examples are the polarized Gowdy spacetimes and half-polarized T^2-symmetric spacetimes. See [7] and references therein for further details.

The analog of the principle that symmetries propagate is also valid for spinors. This leads to the notion of Killing spinor initial data, which is relevant to the problem of Kerr characterization; see [20] for further details.

2.2.6.6. Komar Integrals

Assume that ν^a is a Killing vector field. Then we have $\nabla_a\nu_b = \nabla_{[a}\nu_{b]}$. A calculation shows

$$\nabla^a(\nabla_a\xi_b - \nabla_b\xi_a) = -2R_{bc}\xi^c.$$

Hence, in a vacuum,

$$\int_S e_{abcd}\nabla^c\xi^d$$

depends only on the homology class of the two-surface S. The analogous fact for the source free Maxwell equation, where we have $\nabla^a F_{ab} = 0$, $\nabla_{[a}F_{bc]} = 0$, is the conservation of the charge integrals $\int_S F_{ab}$, $\int_S \epsilon_{abcd}F^{cd}$, which again depend only on the homology class of S. These statements are immediate consequences of Stokes' theorem.

If we consider asymptotically flat spacetimes, we have in the stationary case, with $\xi^a = (\partial_t)^a$,

$$P^a \xi_a = -\frac{1}{8\pi} \int_S \epsilon_{abcd} \nabla^c \xi^d,$$

where on the left-hand side we have the ADM 4-momentum evaluated at infinity. Similarly, in the axially symmetric case, with $\eta^a = (\partial_\phi)^a$,

$$J = -\frac{1}{16\pi} \int_S \epsilon_{abcd} \nabla^c \eta^d.$$

These integrals again depend only on the homology class of S. See [66, §6] for background to these facts. For a non-symmetric, but asymptotically flat, spacetime, letting S tend to infinity through a sequence of suitably round spheres yields the linkage integrals, which again reproduce the ADM momenta [113].

2.3. Black Holes

2.3.1. The Schwarzschild Solution

Before introducing the Kerr solution, we will discuss the spherically symmetric, static Schwarzschild black hole spacetime. This exhibits some of the features of the Kerr solution and has the advantage that the algebraic form of the line element is much simpler. However, it must be noted that, due to the fact that the Schwarzschild spacetime is static and spherically symmetric, the essential difficulties in analyzing fields on the Kerr background stemming from the complicated trapping (i.e. the fact null geodesics orbiting the black hole fill an open spacetime region, cf. section 2.3.7) and superradiance are not seen in the Schwarzschild case. Therefore, one should be careful in generalizing notions from Schwarzschild to Kerr.

In Schwarzschild coordinates (t, r, θ, ϕ), the Schwarzschild metric takes the form

$$g_{ab}dx^a dx^b = f dt^2 - f^{-1} dr^2 - r^2 d\Omega^2_{S^2} \tag{2.19}$$

with $f = 1 - 2M/r$. Here $d\Omega^2_{S^2} = d\theta^2 + \sin^2\theta d\phi^2$ is the line element on the unit 2-sphere. The coordinate r is the area radius, defined by $4\pi r^2 = A(S(r,t))$, where $S(r,t)$ is the 2-sphere with constant t, r. The line element given in equation (2.19) is valid for $r > 0$, but has a coordinate singularity at $r = 2M$, which is also the location of the event horizon. Historically, this fact caused some confusion, and was only fully cleared up in the 1950s due to the work of Kruskal and Szekeres, see e.g. [84, Chapter 31] and references therein. The metric is in fact regular, and the line element given above is valid also for $0 < r < 2M$. At $r = 0$, there is a curvature singularity, where spacetime curvature diverges as $1/r^3$. The Schwarzschild metric is asymptotically flat and the parameter M coincides with the ADM mass.

We remark that by setting $f = 1 - 2M/r + Q^2/r^2$, the Schwarzschild line element becomes that of Reissner–Nordström, a spherically symmetric solution to the Einstein–Maxwell equations, with field strength of the form $F^{tr} = Q/r^2$. Here $Q = \frac{1}{8\pi}\int_S F^{ab} d\sigma_{ab}$.

In order to get a better understanding of the Schwarzschild spacetime, it is instructive to consider its maximal extension. In order to do this, we first introduce the tortoise coordinate r_*,

$$r_* = r + 2M \log\left(\frac{r}{2M} - 1\right). \tag{2.20}$$

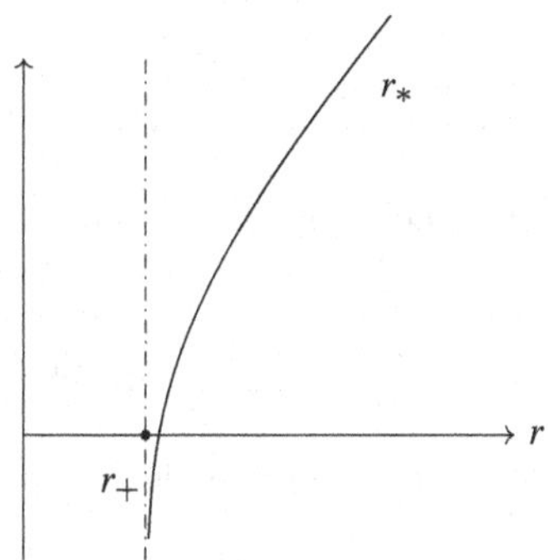

This solves $dr_* = f^{-1}dr$, $r_*(4M) = 4M$. As $r \searrow 2M$, r_* diverges logarithmically to $-\infty$, and for large r, $r_* \sim r$. Inverting (2.20) yields

$$r = 2M\mathrm{W}\left(e^{\frac{r_*}{2M}-1}\right) + 2M \tag{2.21}$$

where W is the principal branch of the Lambert W function.[7] We can now introduce null coordinates

$$u = t - r_*, \quad v = t + r_*.$$

A null tetrad is given by

$$l^a = \sqrt{\frac{2}{f}}\partial_v^a,$$

$$n^a = \sqrt{\frac{2}{f}}\partial_u^a,$$

$$m^a = \frac{1}{\sqrt{2}r}\left(\partial_\theta^a + \frac{i}{\sin\theta}\partial_\phi^a\right).$$

[7] The Lambert W function, or product logarithm, is defined as the solution of $W(x)e^{W(x)} = x$ for $x > 0$. It satisfies $W'(x) = W(x)/((W(x)+1)x)$. The principal branch is analytic at $x = 0$ and is real valued in the range $(-e^{-1}, \infty)$ with values in $(-1, \infty)$. In particular, $W(0) = 0$. See [44].

On the exterior region in Schwarzschild, (u, v) take values in the range $(-\infty, \infty) \times (-\infty, \infty)$. Let $\mathcal{U}, \mathcal{V}$ be a pair of coordinates taking values in $(-\pi/2, \pi/2)$, and related to u, v by

$$u = -4M\log(-\tan\mathcal{U}), \quad \mathcal{U} \in (-\pi/2, 0)$$
$$v = 4M\log(\tan\mathcal{V}), \quad \mathcal{V} \in (0, \pi/2).$$

We have

$$t = \tfrac{1}{2}(v + u) = 4M\log(-\tan\mathcal{V}\tan\mathcal{U})$$
$$r_* = \tfrac{1}{2}(v - u) = 4M\log\left(-\frac{\tan\mathcal{V}}{\tan\mathcal{U}}\right).$$

In terms of $\mathcal{U}, \mathcal{V}$ we have

$$r = 2M\mathrm{W}(-e^{-1}\tan\mathcal{U}\tan\mathcal{V}) + 2M \tag{2.22}$$

and $r > 0$ thus corresponds to $\tan\mathcal{U}\tan\mathcal{V} < 1$. The line element now takes the form

$$g_{ab}dx^a dx^b = \frac{d\mathcal{U}d\mathcal{V}}{\cos^2\mathcal{U}\cos^2\mathcal{V}}\frac{32M^3}{r}e^{-\frac{r}{2M}} - r^2 d\Omega^2_{S^2}. \tag{2.23}$$

The form (2.23) of the Schwarzschild line element is non-degenerate in the range

$$(\mathcal{U}, \mathcal{V}) \in (-\pi/2, \pi/2) \times (-\pi/2, \pi/2) \cap \{-\pi/2 < \mathcal{U} + \mathcal{V} < \pi/2\}. \tag{2.24}$$

In particular, the location $r = 2M$ of the coordinate singularity in the line element (2.19) corresponds to $\mathcal{U}\mathcal{V} = 0$. The line element (2.23) has a coordinate singularity, which is also a curvature singularity, at $r = 0$ (corresponding to $\tan\mathcal{U}\tan\mathcal{V} = 1$), and at $\mathcal{U} = \pm\pi/2$, $\mathcal{V} = \pm\pi/2$ (corresponding to u, v taking unbounded values). Figure 2.3 shows the region given in (2.24), with lines of constant t, r indicated. Using the causal diagram for the extended Schwarzschild solution, one can easily find the null infinities $\mathcal{I}^\pm$, spatial infinity i_0, timelike infinities $i_\pm$, and the horizons $\mathcal{H}^\pm$ at $r = 2M$, which are indicated. Region *I* is the domain of outer communication, i.e. $I^-(\mathcal{I}^+) \cap I^+(\mathcal{I}^-)$, while region *II* is the future trapped (or black hole) region.

The level sets of t hit the bifurcation sphere $\mathcal{B}$ located at $\mathcal{U} = \mathcal{V} = 0$, where $\partial_t = 0$. In particular, we see that the Schwarzschild coordinates are degenerate, since the level sets of t do not foliate the extended Schwarzschild spacetime. On the other hand, a global Cauchy foliation of the maximally extended Schwarzschild spacetime is given by the level sets of the Kruskal time function $\mathcal{T} = \frac{1}{2}(\mathcal{V} + \mathcal{U})$.

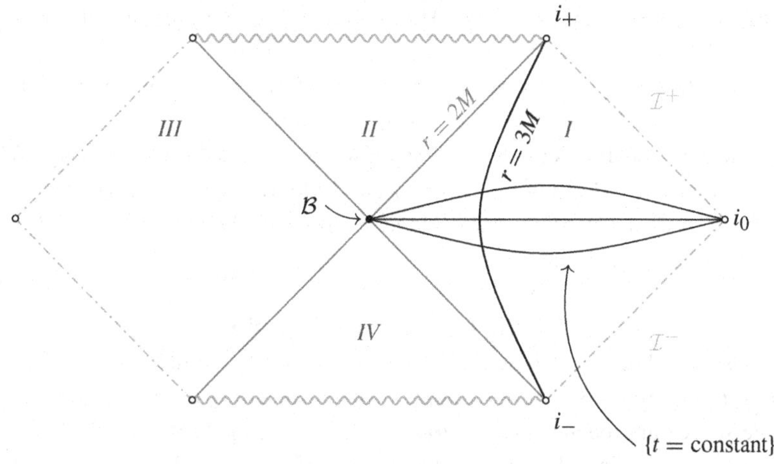

Figure 2.3. Causal diagram for the extended Schwarzschild solution

Given a null vector k^a, perpendicular to a spacelike two-surface S, we may define the null expansion with respect to k^a by

$$\Theta_{k^a} = \tfrac{1}{2}\delta_{k^a} \log(A(S)) \tag{2.25}$$

where δ_{k^a} denotes the variation in the direction k^a.

Then Θ_{k^a} is the expansion of the area element of S, along the null geodesic with velocity k^a. If we let $k^a = (\partial_{\mathcal{V}})^a$, we have

$$\Theta_{k^a} \begin{cases} > 0 & \text{in region } I, \\ = 0 & \text{on } \mathcal{H}_+, \\ < 0 & \text{in region } II. \end{cases}$$

Thus, the area of a bundle of null rays in region I is expanding with respect to a future, outgoing null vector like $\partial_{\mathcal{V}}$, while in region II they are contracting. Actually, in region II, we find that the expansion with respect to any future null vector is negative.

Although null vectors are conventionally drawn at 45° angles, due to the fact that each point in the causal diagram represents a sphere, this does not give a complete description. From the causal diagram it is clear that from each point in the DOC there are null curves which escape through $\mathcal{I}^{\pm}$ or fall in through the horizons $\mathcal{H}^{\pm}$. By continuity, it is clear that there must be null curves which neither escape through $\mathcal{I}$ nor fall in through the horizons $\mathcal{H}$. We refer to these as orbiting or *trapped* null geodesics. In the Schwarzschild spacetime, the trapped null geodesics are located at $r = 3M$; see Figure 2.3. The presence of trapped null geodesics is a robust feature of black hole spacetimes.

Although the region covered by the null coordinates $\mathcal{U}, \mathcal{V}$ is compact, the line element (2.23) is of course isometric to the form given in (2.19). A conformal factor $\Phi = \cos\mathcal{U}\cos\mathcal{V}$ may now be introduced, which brings $\mathcal{I}^{\pm}$ to a finite distance. Letting $\tilde{g}_{ab} = \Phi^2 g_{ab}$, and adding these boundary pieces to $(\mathcal{M}, \tilde{g}_{ab})$, provides a *conformal compactification*[8] of the maximally extended Schwarzschild spacetime.

2.3.1.1. Gravitational Redshift

A robust fact about black hole spacetimes is that radiation emanating from near the event horizon is strongly red shifted before reaching infinity. In the limit as the source approaches the horizon, the redshift tends to infinity. Let $\dot{\gamma}^a$ be a null geodesic. The observed frequency of a plane fronted wave with wave plane perpendicular to $\dot{\gamma}^a$ is

$$\omega = \frac{\xi^a \dot{\gamma}_a}{(\xi^a \xi_a)^{1/2}}$$

where $\xi^a \dot{\gamma}_a$ is conserved along the null geodesic. In Schwarzschild, $\xi^a \xi_a = f = (1 - 2M/r)$. If we let ω_1, ω_2 be the observed frequency at r_1, r_2, we find

$$\frac{\omega_2}{\omega_1} = \frac{1 - 2M/r_1}{1 - 2M/r_2} \searrow 0 \text{ as } r_1 \searrow 2M.$$

2.3.1.2. Orbiting Null Geodesics

Consider a null geodesic γ^a in the Schwarzschild spacetime. Due to the spherical symmetry of the Schwarzschild spacetime, we may assume without loss of generality that $\dot{\theta} = 0$ and set $\theta = \pi/2$, so that γ^a moves in the equatorial plane. Hence the geodesic energy and azimuthal angular momentum $\boldsymbol{e} = -\xi^a \dot{\gamma}_a$ and $\boldsymbol{\ell}_z = \eta^a \dot{\gamma}_a$ are conserved. We have

$$\boldsymbol{\ell}_z = \eta^a \dot{\gamma}^b g_{ab} = r^2 \dot{\phi}$$

In fact the same is true for the momenta corresponding to each of the three rotational Killing fields. Thus, we may consider the total squared angular momentum $\boldsymbol{L}^2$ given by

$$\boldsymbol{L}^2 = 2r^2 m_{(a}\bar{m}_{b)} \dot{\gamma}^a \dot{\gamma}^b = r^4 (g_{S^2})_{ab} \dot{\gamma}^a \dot{\gamma}^b. \tag{2.26}$$

[8] There are subtleties concerning the regularity of the conformal boundary of Schwarzschild; and the naive choice of conformal factor mentioned above does not lead to an analytic compactification. See [60] for recent developments.

For geodesics moving in the equatorial plane, we have $\boldsymbol{L}^2 = \boldsymbol{\ell}_z{}^2$. Rewriting $g_{ab}\dot{\gamma}^a\dot{\gamma}^b = 0$ using (2.4) and these definitions gives

$$\dot{r}^2 + V = \boldsymbol{e}^2 \tag{2.27}$$

where

$$V = \frac{f}{r^2}\boldsymbol{L}^2.$$

Equation 2.27 can be viewed as the equation for a particle moving in a potential V.

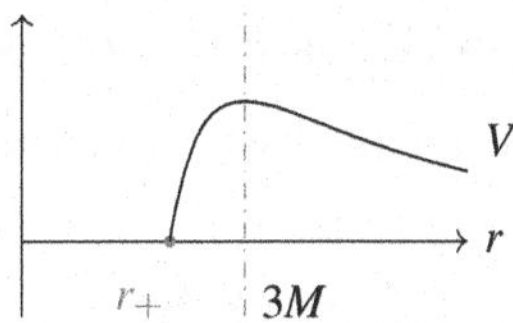

Figure 2.4. The potential for radial motion of null geodesics in Schwarzschild.

An analysis shows that V has a unique critical point at $r = 3M$, and hence a null geodesic with $\dot{r} = 0$ in the Schwarzschild spacetime must orbit at $r = 3M$. We call such null geodesics trapped. The critical point $r = 3M$ is a local maximum for V and hence the orbiting null geodesics are unstable. The sphere $r = 3M$ is called the *photon sphere*. A similar analysis can be performed for massive particles orbiting the Schwarzschild black hole; see [109, Chapter 6] for further details.

The geometric optics correspondence between waves packets and null geodesics indicates that the phenomenon of trapped null geodesics is an obstacle to dispersion, i.e. the tendency for waves to leave every stationary region. For waves of finite energy, the fact that the trapped orbits are unstable can be used to show that such waves in fact disperse. This is a manifestation of the uncertainty principle.

The close relation between the equation for radial motion of null geodesics and the wave equation $\nabla^a\nabla_a\psi = 0$ can be seen as follows. Equation (2.27) can be written in the form

$$r^4\dot{r}^2 + \mathcal{R}(r, \boldsymbol{e}, L) = 0, \tag{2.28}$$

where

$$\mathcal{R} = -r^4\boldsymbol{e}^2 + r^2 fL. \tag{2.29}$$

On the other hand, the wave equation in the Schwarzschild exterior spacetime takes the form

$$r^2\nabla^a\nabla_a\psi = \partial_r(r^2 f)\partial_r + \frac{\mathcal{R}}{r^2 f}.$$

Here $\mathcal{R} = \mathcal{R}(r, \partial_t, \not{\Delta})$ where $\not{\Delta}$ is the spherical Laplacian. This is the same expression as in the equation for the radial motion of null geodesics, but with $\boldsymbol{e}, L^2$ replaced by symmetry operators ∂_t, $\not{\Delta}$, using the correspondence $\boldsymbol{e} \leftrightarrow i\partial_t$, $L^2 \leftrightarrow -\not{\Delta}$. If we perform separation of variables, the angular Laplacian $\not{\Delta}$ is replaced by its eigenvalues $-\ell(\ell+1)$. This relation between the potential for radial motion of null geodesics and the term $\mathcal{R}$ in the d'Alembertian is a curious and interesting fact, and importantly this relation holds also in Kerr (see Figure 2.4).

2.3.2. Raychaudhouri Equation and Comparison Theory

Assume that k^a is a null vector field which generates affinely parametrized geodesics, $k^b\nabla_b k^a = 0$. Let

$$\Theta = \tfrac{1}{2}\nabla_a k^a \tag{2.30}$$

be the divergence, or null expansion,[9] of the null congruence generated by k^a. For any k^a as above, we have

$$k^a\nabla_a\Theta + \Theta^2 + \sigma\bar{\sigma} + \tfrac{1}{2}R_{ab}k^a k^b = 0 \tag{2.31}$$

where $\sigma\bar{\sigma} = \frac{1}{2}(\nabla_{(a}k_{b)}\nabla^{(a}k^{b)} - \frac{1}{2}(\nabla_a k^a)^2$ is the squared shear. Equation (2.31) describes the evolution of the null expansion along null geodesics $\gamma^a(\lambda)$ generated by k^a. Assuming the null energy condition (2.9), we have

$$k^a\nabla_a\Theta + \Theta^2 \leq 0. \tag{2.32}$$

Hence, if $\Theta\big|_S < c_0 < 0$, we find that $\Theta \searrow -\infty$ along γ^a at some finite affine time λ_0.

Recall that a geodesic in a Riemannian manifold ceases to be minimizing at its first conjugate point. This can be shown by "rounding off the corner," which decreases length. In the Lorentzian case, "rounding off the corner," see Figure 2.5, *increases* Lorentzian length, and one finds that points along null geodesic γ^a past $\gamma^a(\lambda_0)$ are timelike related to $\gamma^a(0)$. This means that the

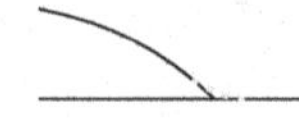

Figure 2.5. Rounding off the corner

[9] The definition of Θ in (2.30) agrees with (2.25); we have dropped the subindex on Θ to avoid clutter. The null expansion is often defined as $\nabla_a k^a$, however we shall here use the normalization as in (2.30).

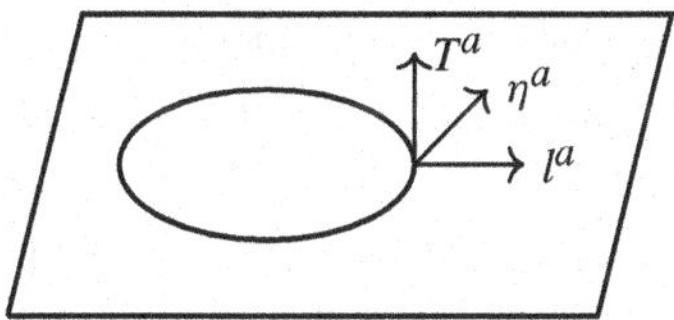

Figure 2.6. The normals to S

geodesic in particular leaves the boundary of the causal future of S. It is known that any $p \in \partial J^+(S)$ is connected to S by a null geodesic without conjugate points. Combining this argument with the inequality (2.32) shows that if $\theta_{k^a} < c_0$ for some $c_0 < 0$, we find that the boundary of the causal future of S can extend only for a finite affine parameter range.

Now let Σ be a spacelike Cauchy surface with future timelike normal T^a. For a two-sided surface $S \subset \Sigma$, we say that a null normal k^a to S is outward pointing if the projection of k^a to Σ points into the exterior of Σ, i.e. the component of $\Sigma \setminus \Sigma$ connected to E. Let η^a be the outward pointing normal to S in Σ. Then $k^a = T^a + \eta^a$ is future directed and outward pointing. Let $H = \nabla_a \eta^a$ be the mean curvature of S in Σ. Then $\Theta_{k^a} = \frac{1}{2}(\mathrm{tr}_S k + H)$, where $\mathrm{tr}_S k = h^{ij} k_{ij} - k_{ij}\eta^i\eta^j$ is the trace of k_{ij} restricted to S. See Figure 2.6. If the outgoing null expansion Θ_{k^a} satisfies $\Theta_{k^a} = 0$ (< 0, > 0), we call S a marginally outer trapped (trapped, untrapped) surface.

Consider the Schwarzschild spacetime; see 2.19. If we designate the null vector $(\partial_V)^a$ as outgoing, then the coordinate spheres $S_{t,r}$ are outer untrapped in regions I, IV, outer trapped in regions II, III, and marginally trapped on $\mathcal{H}$.

Due to their importance, we introduce the acronym MOTS for "marginally outer trapped surface." These are analogs of minimal surfaces in Riemannian geometry. In particular, a MOTS is critical with respect to variation of area along the outgoing null directions. For a stationary black hole spacetime, the event horizon is foliated by MOTS.

As an application of the above remarks, we have the following incompleteness result.

Theorem 2.1 ([15, §7]) *Let $(\mathcal{M}, g_{ab})$ be a globally hyperbolic spacetime satisfying the null energy conditon, and let (Σ, h_{ij}, k_{ij}) be a Cauchy surface in $(\mathcal{M}, g_{ab})$ with a non-compact exterior. Assume that S is outer trapped in the sense that the outgoing null expansion θ of S satisfies $\theta < c_0 < 0$ for some $c_0 < 0$. Then $(\mathcal{M}, g_{ab})$ is causally geodesically incomplete.*

Remark 2.1 Results similar to theorem 2.1 are usually referred to as "singularity theorems," but they actually demonstrate that the spacetime $\mathcal{M}$ has a non-trivial Cauchy horizon $\partial\mathcal{M}$, without giving any information about its properties. Versions of such results were originally proved by Hawking and Penrose, see [62]. Motivated by the strong cosmic censorship conjecture, one expects that for a generic spacetime, the spacetime metric becomes irregular as one approaches $\partial\mathcal{M}$, and hence that a regular extension beyond $\partial\mathcal{M}$ is impossible. For example, in the Schwarzschild spacetime, curvature diverges as $1/r^3$ as one approaches the Cauchy horizon at $r = 0$. This can be seen by looking at the invariantly defined Kretschmann scalar $R_{abcd}R^{abcd} = 48M/r^6$.

The detailed behavior of the geometry at the Cauchy horizon in generic situations is subtle and far from understood; see however [78] and references therein for recent developments. For cosmological singularities, strong cosmic censorship including curvature blowup for generic data has been established in some symmetric situations, see [99, §5.2] and references therein.

By the weak cosmic censorship conjecture, one expects that in a generic asymptotically flat spacetime, $\partial\mathcal{M}$ is hidden from observers at infinity, and hence that the domain of outer communication has a non-trivial boundary, the event horizon. This motivates the idea that MOTS may be viewed as representing the *apparent horizon* of a black hole; see Section 2.3.3. Due to the fact that the MOTS can be understood in terms of Cauchy data, this point of view is important in considering dynamical black holes.

2.3.3. The Apparent Horizon

Consider the Vaidya line element, cf. [96, §5.1.8]

$$ds^2 = fdv^2 - 2dvdr - r^2 d\Omega^s_{S^2} \tag{2.33}$$

with $f = 1 - 2M(v)/r$, where the *mass aspect function* $M(v)$ is an increasing function of the retarded time coordinate v. The matter in the Vaidya spacetime is infalling null dust. Those regions where $dM/dv = 0$ are empty. We see that there is no dr^2 term in (2.33), so r is a null coordinate. Setting $M(v) \equiv M$, gives the Schwarzschild line element in ingoing Eddington–Finkelstein coordinates. A calculation shows that there are MOTS located at $r = 2M(v)$. Hence, if $M(v)$ varies from M_1 to M_2 in an interval (v_1, v_2) and is constant elsewhere (see Figure 2.7) we find that the MOTS move outwards, to the event horizon, located at $r = 2M_2$.

In general, the spacetime tube swept out by the MOTS might, provided it exists, be termed a marginally outer trapped tube (MOTT). By known stability results for MOTS, this exists locally in generic situations, see [16], see also Section 2.3.4. Thus, heuristically the MOTS and MOTT represent the apparent horizon, and the fact that the apparent horizon moves outward corresponds to

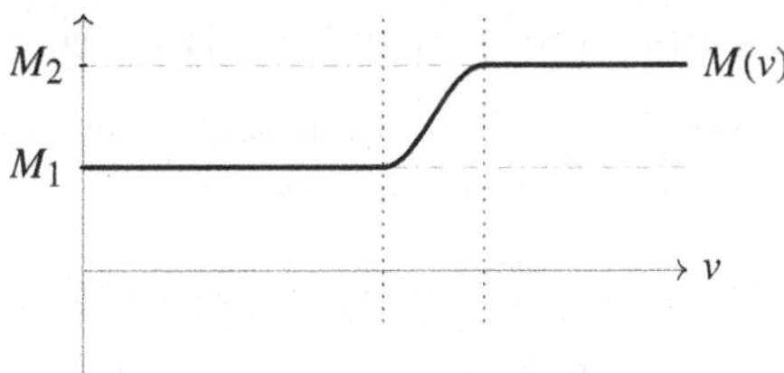

Figure 2.7. Mass aspect function $M(v)$ varying from M_1 to M_2

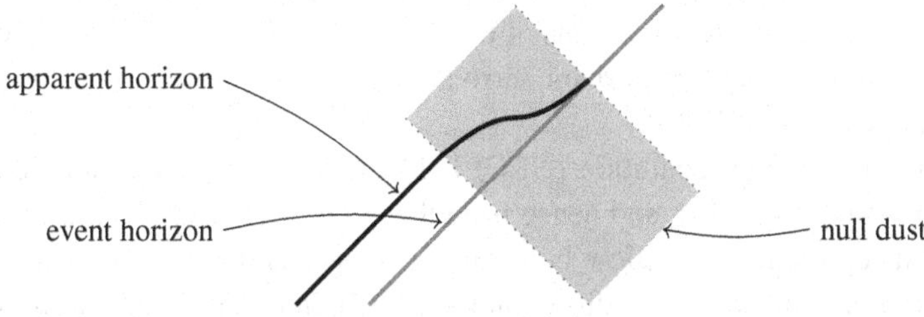

Figure 2.8. Event and apparent horizons in the Vaidya spacetime

the growth of mass of the black hole due to the stress-energy or gravitational energy crossing the horizon; see Figure 2.8.

Remark 2.2 1. The event horizon is teleological, in the sense that determining its location requires complete knowledge of spacetime. In particular, it is not possible to compute its location from Cauchy data without constructing the complete spacetime evolution. On the other hand, the notion of MOTS and apparent horizon are quasilocal notions, which can be determined directly from Cauchy data.

2. The location of MOTSs is not a spacetime concept but depends on the choice of Cauchy slicing. See [26] for results on the region of spacetime containing trapped surfaces. It was shown by Wald and Iyer [111] that there are Cauchy surfaces in the extended Schwarzschild spacetime which approach the singularity arbitrarily closely and such that the past of these Cauchy surfaces does not contain any outer trapped surfaces.
3. The interior of the outermost MOTS is called the trapped region (a notion which depends on the Cauchy slicing). Based on the weak cosmic censorship conjecture, and the above remarks, one expects this to be in the black hole region, which is bounded by the event horizon. See [39, Theorem 6.1] for a result in this direction.

2.3.4. Results on MOTS and the Trapped Region

Several theorems about MOTS have been proved in the last decade. In particular, if a Cauchy surface Σ contains a MOTS, then there is an *outermost* MOTS.

If we conside a Cauchy slicing (Σ_t), then if Σ_{t_0} contains a MOTS, then for $t > t_0$, Σ_t contains a MOTS. However, the location of the MOTS may jump, e.g. due to the formation of a MOTS surrounding the previous one; see Figure 2.9. This phenomenon is seen in numerical simulations of colliding black holes, cf. [86]. There, examples with two merging black holes are considered. When the apparent horizons of the two black holes are sufficiently close together, a new apparent horizon surrounding both is formed, in accordance with the results in [17, 15].

If the null energy condition (NEC) holds, then in a generic situation the MOTT is spacelike [15], and hence from the point of view of the exterior part of $\mathcal{M}$ it represents an outflow boundary. This means that it is not necessary to impose any boundary condition on the MOTT in order to get a well-posed Cauchy problem. This leads to the *exterior Cauchy problem.* As mentioned above, cf. Figure 2.10, in strong field situations, it can happen that the MOTS jumps out. In this case, one must then restart the solving of the exterior Cauchy problem at the jump time. This corresponds closely to what one sees in a

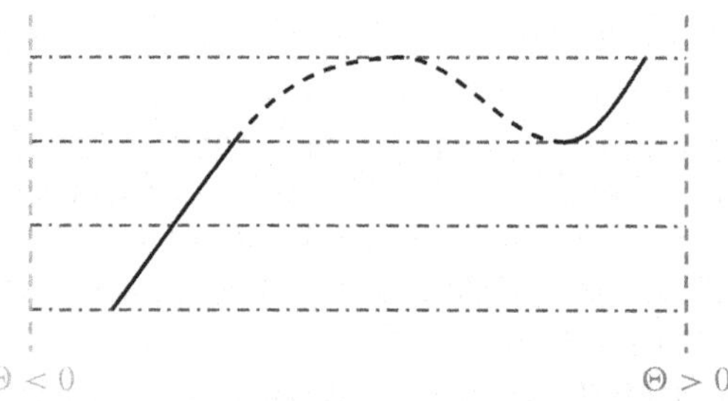

Figure 2.9. The location of the MOTS jumps outward when a new one forms outside the previous one

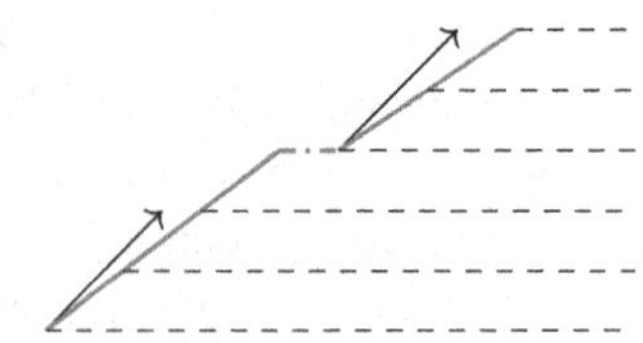

Figure 2.10. The exterior Cauchy problem

numerical evolution of strong field situations, e.g. of merging black holes, when using horizon trackers to determine the location of MOTS.

2.3.5. Formation of Black Holes

The first example of a dynamically forming black hole through the collapse of a cloud of dust was constructed by Oppenheimer and Snyder [91] in 1939. Examples of the formation of a black hole by concentration of gravitational radiation were constructed by Christodoulou [36]. There has been much recent work refining and extending these results, see [70] and references therein.

In order to understand the formation of black holes, it is important to have good conditions for the existence of marginally outer trapped surfaces in a given Cauchy surface. Such results have been proved by Schoen and Yau [102], see also [41]. The result in [102] makes use of Jang's equation to show that MOTS form if a sufficiently dense concentration of matter is present. A related result for the vacuum case is given in [47]; see also [115].

2.3.6. Black Hole Stability

Taking the trapped region as representing a dynamical black hole, the above discussion leads to a picture of the evolution of dynamical black holes, as well as their formation. Based on these general considerations, we can now give a heuristic formulation of the black hole stability problem and related conjectures. Recall that the Kerr black hole spacetime, which we shall study in detail below, is conjectured to be the unique rotating vacuum black hole spacetime, and further to be dynamically stable.

The *black hole stability conjecture* is that Cauchy data sufficiently close, in a suitable sense, to Kerr Cauchy data[10] have a maximal development which is future asymptotic to a Kerr spacetime; see Figure 2.11. In approaching this problem, one may use the results on the evolution of MOTS mentioned above, cf. Section 2.3.4 to consider only the exterior Cauchy problem. It is important to note that the parameters of the "limiting" Kerr spacetime cannot be determined in any effective manner from the initial data.

As discussed above, cf. Section 2.2.6.6, if we are restricted to axial symmetry, then angular momentum is quasilocally conserved. This means that if we further restrict to zero angular momentum, the end state of the evolution must be a Schwarzschild black hole.

[10] See [10], see also e.g. [21, 81] for discussions of the problem of characterizing Cauchy data as Kerr data.

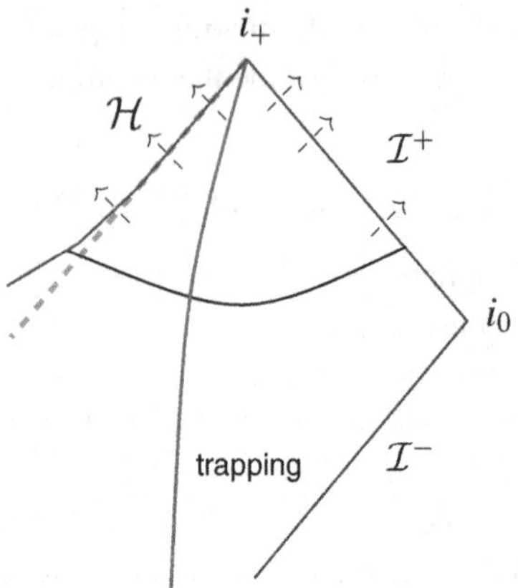

Figure 2.11. Expected causal diagram for the exterior of a solution arising from a perturbation of the Kerr solution

Thus, the *black hole stability conjecture for the axially symmetric case* is that the maximal development of sufficiently small (in a suitable sense), axially symmetric, deformations of Schwarzschild Cauchy data with zero angular momentum is asymptotic in the future to a Schwarzschild spacetime.[11] In this case, due to the loss of energy through $\mathcal{I}^+$, the mass of the "limiting" Schwarzschild black hole cannot be determined directly from the Cauchy data.

A conjecture related to the black hole stability conjecture, but which is even more far reaching, may be termed *the end state conjecture*. Here the idea is that the maximal evolution of generic asymptotically flat vacuum initial data is asymptotic in a suitable sense, to a collection of black holes moving apart, with the near region of each black hole approaching a Kerr geometry. No smallness condition is implied.

The heuristic ideas relating to weak cosmic censorship and Kerr as the final state of the evolution of an isolated system, together with Hawking's area theorem, was used by Penrose to motivate the Penrose inequality,

$$\sqrt{\frac{A_{\text{min}}}{16\pi}} \leq M_{\text{ADM}}$$

where A_{min} is the minimal area of any surface surrounding all past and future trapped regions in a given Cauchy surface, and M_{ADM} is the ADM mass at infinity. The Riemannian version of the Penrose inequality has been proved by Bray [31], and Huisken and Ilmanen [65]. The spacetime version of the

[11] The zero angular momentum, axially symmetric case of the black hole stability problem could also be called the polarized case.

Penrose inequality remains open. It should be stressed that the formulation of the inequality given above may have to be adjusted. Interesting possible approaches to the problem have been developed by Bray and Khuri, see [61] and references therein.

2.3.7. The Kerr Metric

In this section we shall discuss the Kerr metric, which is the main object of our considerations. Although many features of the geometry and analysis on black hole spacetimes are seen in the Schwarzschild case, there are many new and fundamental phenomena present in the Kerr case. Among these are complicated trapping, i.e. the fact that trapped null geodesics fill up an open spacetime region, the fact that the Kerr metric admits only two Killing fields, but a hidden symmetry manifested in the Carter constant, and the fact that the stationary Killing vector field ξ^a fails to be timelike in the whole domain of outer communications, which leads to a lack of a positive conserved energy for waves in the Kerr spacetimes. This fact is the origin of superradiance and the Penrose process. See [108] for a recent survey.

The Kerr metric describes a family of stationary, axisymmetric, asymptotically flat vacuum spacetimes, parametrized by ADM mass M and angular momentum per unit mass a. The expressions for mass and angular momentum introduced in Section 2.2.6 when applied in Kerr geometry yield M and $J = aM$. In Boyer–Lindquist coordinates (t, r, θ, ϕ), the Kerr metric takes the form

$$\begin{aligned} g_{ab} = {} & \frac{(\Delta - a^2 \sin^2\theta) dt_a dt_b}{\Sigma} + \frac{2a \sin^2\theta (a^2 + r^2 - \Delta) dt_{(a} d\phi_{b)}}{\Sigma} \\ & - \frac{\Sigma dr_a dr_b}{\Delta} - \Sigma d\theta_a d\theta_b - \frac{\sin^2\theta \big((a^2+r^2)^2 - a^2 \sin^2\theta \Delta\big) d\phi_a d\phi_b}{\Sigma}, \end{aligned} \tag{2.34}$$

where $\Delta = a^2 - 2Mr + r^2$ and $\Sigma = a^2 \cos^2\theta + r^2$. The volume element is

$$\sqrt{|\det g_{ab}|} = \Sigma \sin\theta \tag{2.35}$$

There is a ring-shaped singularity at $r = 0$, $\theta = \pi/2$. For $|a| \leq M$, the Kerr spacetime contains a black hole, with event horizon at $r = r_+ \equiv M + \sqrt{M^2 - a^2}$, while for $|a| > M$, the singularity is naked in the sense that it is causally connected to observers at infinity. The area of the horizon is $A_{\text{Hor}} = 4\pi(r_+^2 + a^2)$. This achieves its maximum of $16\pi M^2$ when $a = 0$, which provides one of the ingredients in the heuristic argument for the Penrose inequality; see Section 2.3.6. The case $|a| = M$ is called extreme. We shall here be interested only in the subextreme case, $|a| < M$, as this is the only case where we expect black hole stability to hold.

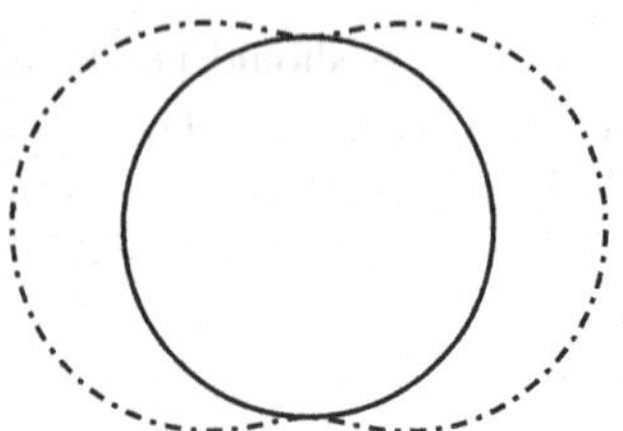

Figure 2.12. The ergoregion

The Boyer–Lindquist coordinates are analogous to the Schwarzschild coordinates in Section 2.3.1 and, upon setting $a = 0$, (2.34) reduces to (2.19). The line element takes a simple form in Boyer-Lindquist coordinates, but, similarly to the Schwarzschild coordinates, they have the drawback that they are not regular at the horizon.

The Kerr metric admits two Killing vector fields $\xi^a = (\partial_t)^a$ (stationary) and $\eta^a = (\partial_\phi)^a$ (axial). Although the stationary Killing field ξ^a is timelike near infinity, since $g_{ab}\xi^a\xi^b \to 1$ as $r \to \infty$, ξ^a becomes spacelike for r sufficiently small, when $1 - 2M/\Sigma < 0$. In the Schwarzschild case $a = 0$, this occurs at the event horizon $r = 2M$. However, for a rotating Kerr black hole with $0 < |a| \leq M$, there is a region, called the ergoregion, outside the event horizon, where ξ^a is spacelike. The ergoregion is bounded by the surface $M + \sqrt{M^2 - a^2 \cos^2\theta}$ which touches the horizon at the poles $\theta = 0, \pi$; see Figure 2.12. In the ergoregion, null and timelike geodesics can have negative energy with respect to ξ^a. The fact that there is no globally timelike vector field in the Kerr exterior is the origin of superradiance, i.e. the fact that waves which scatter off the black hole can leave the ergoregion with more energy (as measured by a stationary observer at infinity) than was sent in. This effect was originally found by an analysis based on separation of variables, but it can be demonstrated rigorously, see [52]. However, it is a subtle effect and not easy to demonstrate numerically, see [75].

If we consider a dynamical spacetime containing a rotating black hole, then the presence of the ergoregion allows for the Penrose process, which extracts rotational energy from the black hole, see [74], see also [48] for a numerical study of superradiance of gravitational waves in a dynamical spacetime.

Let $\omega_H = a/(r_+^2 + a^2)$ be the rotation speed of the black hole. The Killing field $\chi^a = \xi^a + \omega_H \eta^a$ is null on the event horizon in Kerr, which is therefore a Killing horizon. For $|a| < M$, there is a neighborhood of the horizon in the black hole exterior where χ^a is timelike. The surface gravity κ, defined by $\kappa^2 = -\frac{1}{2}(\nabla^a\chi^b)(\nabla_a\chi_b)$ takes the value $\kappa = (r_+ - M)/(r_+^2 + a^2)$, and is positive in the subextreme case $|a| < M$. By general results, a Killing horizon with non-vanishing surface gravity is bifurcate, i.e. there is a cross-section where the null generator vanishes. In the Schwarzschild case, this is the 2-sphere $\mathcal{U} = \mathcal{V} = 0$. See [90, 96] for background on the geometry of the Kerr spacetime; see also [89].

2.4. Spin Geometry

The 2-spinor formalism, and the closely related GHP formalism, are important tools in Lorentzian geometry and the analysis of black hole spacetimes; we will introduce them here. A detailed account of this material is given by Penrose and Rindler [94]. Following the conventions there, we use the abstract index notation with lower case Latin letters $a, b, c, \dots$ for tensor indices, and unprimed and primed upper case Latin letters $A, B, C, \dots, A', B', C', \dots$ for spinor indices. Tetrad and dyad indices are boldface Latin letters following the same scheme, $\mathbf{a}, \mathbf{b}, \mathbf{c}, \dots, \mathbf{A}, \mathbf{B}, \mathbf{C}, \dots, \mathbf{A}', \mathbf{B}', \mathbf{C}', \dots$. For coordinate indices we use Greek letters $\alpha, \beta, \gamma, \dots$.

2.4.1. Spinors on Minkowski Space

Consider Minkowski space $\mathbb{M}$, i.e. $\mathbb{R}^4$ with coordinates $(x^\alpha) = (t, x, y, z)$ and metric

$$g_{\alpha\beta}dx^\alpha dx^\beta = dt^2 - dx^2 - dy^2 - dz^2.$$

Define a complex null tetrad (i.e. frame) $(g_{\mathbf{a}}{}^a)_{\mathbf{a}=0,\cdots,3} = (l^a, n^a, m^a, \bar{m}^a)$, as in (2.3) above, normalized so that $l^a n_a = 1$, $m^a \bar{m}_a = -1$, so that

$$g_{ab} = 2(l_{(a}n_{b)} - m_{(a}\bar{m}_{b)}). \tag{2.36}$$

Similarly, let $\epsilon_{\mathbf{A}}{}^A$ be a dyad (i.e. frame) in $\mathbb{C}^2$, with dual frame $\epsilon_A{}^{\mathbf{A}}$. The complex conjugates will be denoted $\bar{\epsilon}_{\mathbf{A}'}{}^{A'}, \bar{\epsilon}_{A'}{}^{\mathbf{A}'}$ and again form a basis in another two-dimensional complex space denoted $\bar{\mathbb{C}}^2$, and its dual. We can identify the space of complex 2×2 matrices with $\mathbb{C}^2 \otimes \bar{\mathbb{C}}^2$. By construction, the tensor products $\epsilon_{\mathbf{A}}{}^A \bar{\epsilon}_{\mathbf{A}'}{}^{A'}$ and $\epsilon_A{}^{\mathbf{A}} \bar{\epsilon}_{A'}{}^{\mathbf{A}'}$form a basis in $\mathbb{C}^2 \otimes \bar{\mathbb{C}}^2$ and its dual.

Now, with $x^{\mathbf{a}} = x^a g_a{}^{\mathbf{a}}$, writing

$$x^{\mathbf{a}} g_{\mathbf{a}}{}^{\mathbf{A}\mathbf{A}'} \equiv \begin{pmatrix} x^0 & x^2 \\ x^3 & x^1 \end{pmatrix} \tag{2.37}$$

defines the soldering forms, also known as Infeld–van der Waerden symbols $g_a{}^{AA'}$, (and analogously $g_{AA'}{}^a$). By a slight abuse of notation we may write $x^{AA'} = x^a$ instead of $x^{\mathbf{A}\mathbf{A}'} = x^{\mathbf{a}} g_{\mathbf{a}}{}^{\mathbf{A}\mathbf{A}'}$ or, dropping reference to the tetrad,[12] $x^{AA'} = x^a g_a{}^{AA'}$. In particular, $x^a \in \mathbb{M}$ corresponds to a 2×2 complex Hermitian

[12] The Infeld–van der Waerden symbols are sometimes denoted $\sigma_a{}^{AA'}$

matrix $x^{\mathbf{AA}'} \in \mathbb{C}^2 \otimes \bar{\mathbb{C}}^2$. Taking the complex conjugate of both sides of (2.37) gives

$$\bar{x}^a = \bar{x}^{A'A} = (x^{AA'})^*,$$

where $*$ denotes Hermitian conjugation. This extends to a correspondence $\mathbb{C}^4 \leftrightarrow \mathbb{C}^2 \otimes \bar{\mathbb{C}}^2$ with complex conjugation corresponding to Hermitian conjugation.

Note that

$$\det(x^{\mathbf{AA}'}) = x^0x^1 - x^2x^3 = x^a x_a/2. \tag{2.38}$$

We see from the above that the group

$$\mathrm{SL}(2,\mathbb{C}) = \left\{ A = \begin{pmatrix} a & b \\ c & d \end{pmatrix}, \quad a,b,c,d \in \mathbb{C}, \quad ad - bc = 1 \right\}$$

acts on $X \in \mathbb{C}^2 \otimes \bar{\mathbb{C}}^2$ by

$$X \mapsto AXA^*.$$

In view of (2.38) this exhibits $\mathrm{SL}(2,\mathbb{C})$ as a double cover of the identity component of the Lorentz group $\mathrm{SO}_0(1,3)$, the group of linear isometries of $\mathbb{M}$. In particular, $\mathrm{SL}(2,\mathbb{C})$ is the spin group of $\mathbb{M}$. The canonical action

$$(A,v) \in \mathrm{SL}(2,\mathbb{C}) \times \mathbb{C}^2 \mapsto Av \in \mathbb{C}^2$$

of $\mathrm{SL}(2,\mathbb{C})$ on $\mathbb{C}^2$ is the spinor representation. Elements of $\mathbb{C}^2$ are called (Weyl) spinors. The conjugate representation given by

$$(A,v) \in \mathrm{SL}(2,\mathbb{C}) \times \mathbb{C}^2 \mapsto \bar{A}v \in \mathbb{C}^2$$

is denoted $\bar{\mathbb{C}}^2$.

Spinors[13] of the form $x^{AA'} = \alpha^A \beta^{A'}$ correspond to matrices of rank one, and hence to complex null vectors. Denoting $o^A = \epsilon_{\mathbf{0}}{}^A, \iota^A = \epsilon_{\mathbf{1}}{}^A$, we have from the above that

$$l^a = o^A o^{A'}, \quad n^a = \iota^A \iota^{A'}, \quad m^a = o^A \iota^{A'}, \quad \bar{m}^a = \iota^A o^{A'}. \tag{2.39}$$

This gives a correspondence between a null frame in $\mathbb{M}$ and a dyad in $\mathbb{C}^2$.

The action of $\mathrm{SL}(2,\mathbb{C})$ on $\mathbb{C}^2$ leaves invariant a complex area element, a skew-symmetric bispinor. Such a unique spinor ϵ_{AB} is determined by the normalization

$$g_{ab} = \epsilon_{AB}\bar{\epsilon}_{A'B'}.$$

[13] It is conventional to refer to spin-tensors e.g. of the form $x^{AA'}$ or $\psi_{ABA'}$ simply as spinors.

The inverse ϵ^{AB} of ϵ_{AB} is defined by $\epsilon_{AB}\epsilon^{CB} = \delta_A{}^C$, $\epsilon^{AB}\epsilon_{AC} = \delta_C{}^B$. As with g_{ab} and its inverse g^{ab}, the spin-metric ϵ_{AB} and its inverse ϵ^{AB} is used to lower and raise spinor indices,

$$\lambda_B = \lambda^A \epsilon_{AB}, \quad \lambda^A = \epsilon^{AB}\lambda_B.$$

We have

$$\epsilon_{AB} = o_A \iota_B - \iota_A o_B.$$

In particular,

$$o_A \iota^A = 1. \tag{2.40}$$

An element $\phi_{A \cdots D A' \cdots D'}$ of $\bigotimes^k \mathbb{C}^2 \bigotimes^l \bar{\mathbb{C}}^2$ is called a spinor of valence (k, l). The space of totally symmetric[14] spinors $\phi_{A \cdots D A' \cdots D'} = \phi_{(A \cdots D)(A' \cdots D')}$ is denoted $S_{k,l}$. The spaces $S_{k,l}$ for k, l non-negative integers yield all irreducible representations of $\mathrm{SL}(2, \mathbb{C})$. In fact, one can decompose any spinor into "irreducible pieces," i.e. as a linear combination of totally symmetric spinors in $S_{k,l}$ with factors of ϵ_{AB}. The above mentioned correspondence between vectors and spinors extends to tensors of any type, and hence the just mentioned decomposition of spinors into irreducible pieces carries over to tensors as well. Examples are given by $\mathcal{F}_{ab} = \phi_{AB}\epsilon_{A'B'}$, a complex anti-self-dual 2-form, and ${}^{-}C_{abcd} = \Psi_{ABCD}\epsilon_{A'B'}\epsilon_{C'D'}$, a complex anti-self-dual tensor with the symmetries of the Weyl tensor. Here, ϕ_{AB} and Ψ_{ABCD} are symmetric.

2.4.2. Spinors on Spacetime

Now let $(\mathcal{M}, g_{ab})$ be a Lorentzian 3+1 dimensional spin manifold with metric of signature $+---$. The spacetimes we are interested in here are spin ones, in particular any orientable, globally hyperbolic 3+1 dimensional spacetime is spin, cf. [58, page 346]. If $\mathcal{M}$ is spin, then the orthonormal frame bundle $\mathrm{SO}(\mathcal{M})$ admits a lift to $\mathrm{Spin}(\mathcal{M})$, a principal $\mathrm{SL}(2, \mathbb{C})$-bundle. The associated bundle construction now gives vector bundles over $\mathcal{M}$ corresponding to the representations of $\mathrm{SL}(2, \mathbb{C})$, in particular we have bundles of valence (k, l) spinors with sections $\phi_{A \cdots D A' \cdots D'}$. The Levi-Civita connection lifts to act on sections of the spinor bundles,

$$\nabla_{AA'} : \varphi_{B \cdots D B' \cdots D'} \to \nabla_{AA'}\varphi_{B \cdots D B' \cdots D'} \tag{2.41}$$

where we have used the tensor–spinor correspondence to replace the index a by AA'. We shall denote the totally symmetric spinor bundles by $S_{k,l}$ and their spaces of sections by $\mathcal{S}_{k,l}$.

[14] The ordering between primed and unprimed indices is irrelevant.

The above mentioned correspondence between spinors and tensors, and the decomposition into irreducible pieces, can be applied to the Riemann curvature tensor. In this case, the irreducible pieces correspond to the scalar curvature, traceless Ricci tensor, and the Weyl tensor, denoted by R, S_{ab}, and C_{abcd}, respectively. The Riemann tensor then takes the form

$$\begin{aligned} R_{abcd} = & -\tfrac{1}{12}g_{ad}g_{bc}R + \tfrac{1}{12}g_{ac}g_{bd}R + \tfrac{1}{2}g_{bd}S_{ac} - \tfrac{1}{2}g_{bc}S_{ad} \\ & - \tfrac{1}{2}g_{ad}S_{bc} + \tfrac{1}{2}g_{ac}S_{bd} + C_{abcd}. \end{aligned} \tag{2.42}$$

The spinor equivalents of these tensors are

$$C_{abcd} = \Psi_{ABCD}\bar{\epsilon}_{A'B'}\bar{\epsilon}_{C'D'} + \bar{\Psi}_{A'B'C'D'}\epsilon_{AB}\epsilon_{CD}, \tag{2.43a}$$

$$S_{ab} = -2\Phi_{ABA'B'}, \tag{2.43b}$$

$$R = 24\Lambda. \tag{2.43c}$$

2.4.3. Fundamental Operators

Projecting (2.41) on its irreducible pieces gives the following four *fundamental operators*, introduced in [9].

Definition 2.2 The differential operators

$$\mathscr{D}_{k,l} : S_{k,l} \to S_{k-1,l-1}, \quad \mathscr{C}_{k,l} : S_{k,l} \to S_{k+1,l-1},$$

$$\mathscr{C}^{\dagger}_{k,l} : S_{k,l} \to S_{k-1,l+1}, \quad \mathscr{T}_{k,l} : S_{k,l} \to S_{k+1,l+1}$$

are defined as

$$(\mathscr{D}_{k,l}\varphi)_{A_1\dots A_{k-1}}{}^{A'_1\dots A'_{l-1}} \equiv \nabla^{BB'}\varphi_{A_1\dots A_{k-1}B}{}^{A'_1\dots A'_{l-1}}{}_{B'}, \tag{2.44a}$$

$$(\mathscr{C}_{k,l}\varphi)_{A_1\dots A_{k+1}}{}^{A'_1\dots A'_{l-1}} \equiv \nabla_{(A_1}{}^{B'}\varphi_{A_2\dots A_{k+1})}{}^{A'_1\dots A'_{l-1}}{}_{B'}, \tag{2.44b}$$

$$(\mathscr{C}^{\dagger}_{k,l}\varphi)_{A_1\dots A_{k-1}}{}^{A'_1\dots A'_{l+1}} \equiv \nabla^{B(A'_1}\varphi_{A_1\dots A_{k-1}B}{}^{A'_2\dots A'_{l+1})}, \tag{2.44c}$$

$$(\mathscr{T}_{k,l}\varphi)_{A_1\dots A_{k+1}}{}^{A'_1\dots A'_{l+1}} \equiv \nabla_{(A_1}{}^{(A'_1}\varphi_{A_2\dots A_{k+1})}{}^{A'_2\dots A'_{l+1})}. \tag{2.44d}$$

The operators are called respectively the divergence, curl, curl-dagger, and twistor operators.

With respect to complex conjugation, the operators $\mathscr{D}, \mathscr{T}$ satisfy $\overline{\mathscr{D}_{k,l}} = \mathscr{D}_{l,k}$, $\overline{\mathscr{T}_{k,l}} = \mathscr{T}_{l,k}$, while $\overline{\mathscr{C}_{k,l}} = \mathscr{C}^{\dagger}_{l,k}$, $\overline{\mathscr{C}^{\dagger}_{k,l}} = \mathscr{C}_{l,k}$.

Denoting the adjoint of an operator by $\mathcal{A}$ with respect to the bilinear pairing

$$(\phi_{A_1\cdots A_kA'_1\cdots A'_l}, \psi_{A_1\cdots A_kA'_1\cdots A'_l}) = \int \phi_{A_1\cdots A_kA'_1\cdots A'_l}\psi^{A_1\cdots A_kA'_1\cdots A'_l}d\mu$$

by $\mathcal{A}^\dagger$, and the adjoint with respect to the sesquilinear pairing

$$\langle \phi_{A_1\cdots A_k A'_1\cdots A'_l}, \psi_{A_1\cdots A_l A'_1\cdots A'_k}\rangle = \int \phi_{A_1\cdots A_k A'_1\cdots A'_l}\bar{\psi}^{A_1\cdots A_k A'_1\cdots A'_l} d\mu$$

by $\mathcal{A}^\star$, we have

$$(\mathscr{D}_{k,l})^\dagger = -\mathscr{T}_{k-1,l-1}, \qquad (\mathscr{T}_{k,l})^\dagger = -\mathscr{D}_{k+1,l+1},$$
$$(\mathscr{C}_{k,l})^\dagger = \mathscr{C}^\dagger_{k+1,l-1}, \qquad (\mathscr{C}^\dagger_{k,l})^\dagger = \mathscr{C}_{k-1,l+1},$$

and

$$(\mathscr{D}_{k,l})^\star = -\mathscr{T}_{l-1,k-1}, \qquad (\mathscr{T}_{k,l})^\star = -\mathscr{D}_{l+1,k+1},$$
$$(\mathscr{C}_{k,l})^\star = \mathscr{C}_{l-1,k+1}, \qquad (\mathscr{C}^\dagger_{k,l})^\star = \mathscr{C}^\dagger_{l+1,k-1}.$$

As we will see in Section 2.4.4, the kernels of $\mathscr{C}^\dagger_{2s,0}$ and $\mathscr{C}_{0,2s}$ are the massless spin-s fields. The kernels of $\mathscr{T}_{k,l}$ are the valence (k,l) Killing spinors, which we will discuss further in Section 2.4.5 and Section 2.4.7. A complete set of commutator properties of these operators can be found in [9].

2.4.4. Massless Spin-*s* Fields

For $s \in \frac{1}{2}\mathbb{N}$, $\varphi_{A\cdots D} \in \ker \mathscr{C}^\dagger_{2s,0}$ is a totally symmetric spinor $\varphi_{A\cdots D} = \varphi_{(A\cdots D)}$ of valence $(2s,0)$ which solves the massless spin-s equation

$$(\mathscr{C}^\dagger_{2s,0}\varphi)_{A\cdots BD'} = 0.$$

For $s = 1/2$, this is the Dirac–Weyl equation $\nabla_{A'}{}^{A}\varphi_A = 0$, for $s = 1$, we have the left and right Maxwell equation $\nabla_{A'}{}^{B}\phi_{AB} = 0$ and $\nabla_{A}{}^{B'}\varphi_{A'B'} = 0$, i.e. $(\mathscr{C}^\dagger_{2,0}\phi)_{AA'} = 0$, $(\mathscr{C}_{0,2}\varphi)_{AA'} = 0$.

An important example is the Coulomb Maxwell field on Kerr,

$$\phi_{AB} = -\frac{2}{(r - ia\cos\theta)^2} o_{(A}\iota_{B)}. \tag{2.45}$$

This is a non-trivial sourceless solution of the Maxwell equation on the Kerr background. We note that the scalars components, see Section 2.4.8 below, of the Coulomb field $\phi_1 = (r - ia\cos\theta)^{-2}$ while $\phi_0 = \phi_2 = 0$.

For $s > 1$, the existence of a non-trivial solution to the spin-s equation implies curvature conditions, a fact known as the Buchdahl constraint [32],

$$0 = \Psi_{(A}{}^{DEF}\phi_{B\ldots C)DEF}. \tag{2.46}$$

This is easily obtained by commuting the operators in

$$0 = (\mathscr{D}_{2s-1,1}\mathscr{C}^\dagger_{2s,0}\phi)_{A\ldots C}. \tag{2.47}$$

For the case $s = 2$, the equation $\nabla_{A'}{}^{D}\Psi_{ABCD} = 0$ is the Bianchi equation, which holds for the Weyl spinor in any vacuum spacetime. Due to the Buchdahl constraint, it holds that in any sufficiently general spacetime, a solution of the spin-2 equation is proportional to the Weyl spinor of the spacetime.

2.4.5. Killing Spinors

Spinors $\varkappa_{A_1\cdots A_k}{}^{A'_1\cdots A'_l} \in \mathcal{S}_{k,l}$ satisfying

$$(\mathscr{T}_{k,l}\varkappa)_{A_1\cdots A_{k+1}}{}^{A'_1\cdots A'_{l+1}} = 0,$$

are called Killing spinors of valence (k,l). We denote the space of Killing spinors of valence (k,l) by $\mathcal{KS}_{k,l}$. The Killing spinor equation is an overdetermined system. The space of Killing spinors is a finite dimensional space, and the existence of Killing spinors imposes strong restrictions on $\mathcal{M}$; see Section 2.4.7 below. Killing spinors $\nu_{AA'} \in \mathcal{KS}_{1,1}$ are simply conformal Killing vector fields, satisfying $\nabla_{(a}\nu_{b)} - \frac{1}{2}\nabla^{c}\nu_{c}g_{ab}$. A Killing spinor $\kappa_{AB} \in \mathcal{KS}_{2,0}$ corresponds to a complex anti-self-dual conformal Killing–Yano 2-form $\mathcal{Y}_{ABA'B'} = \kappa_{AB}\epsilon_{A'B'}$ satisfying the equation

$$\nabla_{(a}\mathcal{Y}_{b)c} - 2\zeta_{c}g_{ab} + \zeta_{(a}g_{b)c} = 0, \tag{2.48}$$

where in the four-dimensional case, $\zeta_a = \frac{1}{3}\nabla_b\mathcal{Y}^{b}{}_{a}$.

In the mathematics literature, Killing spinors of valence $(1,0)$ are known as twistor spinors. The terms "conformal Killing–Yano form" or "twistor form" is used also for the real 2-forms corresponding to Killing spinors of valence $(2,0)$, as well as for forms of higher degree and on higher dimensional manifolds, appearing as kernel elements of an analogous Stein–Weiss operator. Further, we may mention that Killing spinors $L_{ABA'B'} \in \mathcal{KS}_{2,2}$ are traceless symmetric conformal Killing tensors L_{ab}, satisfying the equation

$$\nabla_{(a}L_{bc)} - \tfrac{1}{3}g_{(ab}\nabla^{d}L_{c)d} = 0. \tag{2.49}$$

In particular, any tensor of the form ζg_{ab} for some scalar field ζ is a conformal Killing tensor. If γ^a is a null geodesic and L_{ab} is a conformal Killing tensor, then $L_{ab}\dot{\gamma}^a\dot{\gamma}^b$ is conserved along γ^a. For any $\kappa_{AB} \in \mathcal{KS}_{2,0}$ we have $L_{ABA'B'} = \kappa_{AB}\bar{\kappa}_{A'B'} \in \mathcal{KS}_{2,2}$. See Section 2.4.7 below for further details.

2.4.6. Algebraically Special Spacetimes

Let $\varphi_{A\cdots D} \in \mathcal{S}_{k,0}$. A spinor α_A is a *principal spinor* of $\varphi_{A\cdots D}$ if

$$\varphi_{A\cdots D}\alpha^{A}\cdots\alpha^{D} = 0.$$

Table 2.1. *The Petrov classification*

I	$\{1,1,1,1\}$	$\Psi_{ABCD} = \alpha_{(A}\beta_B\gamma_C\delta_{D)}$	I
II	$\{2,1,1\}$	$\Psi_{ABCD} = \alpha_{(A}\alpha_B\gamma_C\delta_{D)}$	↘ II
D	$\{2,2\}$	$\Psi_{ABCD} = \alpha_{(A}\alpha_B\beta_C\beta_{D)}$	↙ ↘ D III
III	$\{3,1\}$	$\Psi_{ABCD} = \alpha_{(A}\alpha_B\alpha_C\beta_{D)}$	↘ ↙ N
N	$\{4\}$	$\Psi_{ABCD} = \alpha_A\alpha_B\alpha_C\alpha_D$	↙
O	$\{-\}$	$\Psi_{ABCD} = 0$	O

An application of the fundamental theorem of algebra shows that any $\varphi_{A\cdots D} \in \mathcal{S}_{k,0}$ has exactly k principal spinors $\alpha_A, \ldots, \delta_A$, and hence is of the form

$$\varphi_{A\cdots D} = \alpha_{(A} \cdots \delta_{D)}.$$

If $\varphi_{A\cdots D} \in \mathcal{S}_{k,0}$ has n distinct principal spinors $\alpha_A^{(i)}$, repeated m_i times, then $\varphi_{A\cdots D}$ is said to have algebraic type $\{m_1, \ldots, m_n\}$. Applying this to the Weyl tensor leads to the Petrov classification; see Table 2.1.[15]

A principal spinor o_A determines a principal null direction $l_a = o_A\bar{o}_{A'}$. The Goldberg–Sachs theorem states that, in a vacuum spacetime, the congruence generated by a null field l^a is geodetic and shear free[16] if and only if l_a is a repeated principal null direction of the Weyl tensor C_{abcd} (or equivalently o_A is a repeated principal spinor of the Weyl spinor Ψ_{ABCD}).

2.4.6.1. Petrov Type D

The Kerr metric is of Petrov type D, and many of its important properties follow from this fact. The vacuum type *D* spacetimes have been classified by Kinnersley [69], see also Edgar et al. [49]. The family of Petrov type D spacetimes includes the Kerr–NUT family and the boost-rotation symmetric C-metrics. The only Petrov type D vacuum spacetime which is asymptotically flat and has positive mass is the Kerr metric; see theorem 2.3.

A Petrov type D spacetime has two repeated principal spinors o_A, ι_A, and correspondingly there are two repeated principal null directions l^a, n^a, for the Weyl tensor. We can without loss of generality assume that $l^a n_a = 1$ and define a null tetrad by adding complex null vectors $m^a, \bar{m}^a$ normalized such that $m^a \bar{m}_a = -1$. By the Goldberg–Sachs theorem both l^a, n^a are geodetic and shear free, and only one of the five independent complex Weyl scalars is non-zero, namely

[15] The Petrov classification is exclusive, so a spacetime belongs at each point to exactly one Petrov class.

[16] If l^a is geodetic and shear then the spin coefficients σ, κ, cf. (2.61) below, satisfy $\sigma = \kappa = 0$.

$$\Psi_2 = -l^a m^b \bar{m}^d n^c C_{abcd}. \tag{2.50}$$

In this case, the Weyl spinor takes the form

$$\Psi_{ABCD} = \frac{1}{6}\Psi_2 o_{(A}o_B\iota_C\iota_{D)}.$$

See (2.64) below for the explicit form of Ψ_2 in the Kerr spacetime.

The following result is a consequence of the Bianchi identity.

Theorem 2.2 ([112]) *Assume $(\mathcal{M}, g_{ab})$ is a vacuum spacetime of Petrov type* D. *Then $(\mathcal{M}, g_{ab})$ admits a one-dimensional space of Killing spinors κ_{AB} of the form*

$$\kappa_{AB} = -2\kappa_1 o_{(A}\iota_{B)} \tag{2.51}$$

where o_A, ι_A are the principal spinors of Ψ_{ABCD} and $\kappa_1 \propto \Psi_2^{-1/3}$.

Remark 2.3 Since the Petrov classes are exclusive, we have $\Psi_2 \neq 0$ for a Petrov type D space.

2.4.7. Spacetimes Admitting a Killing Spinor

Differentiating the Killing spinor equation $(\mathscr{T}_{k,l}\phi)_{A\cdots DA'\cdots D'} = 0$, and commuting derivatives, yields an algebraic relation between the curvature, Killing spinor, and their covariant derivatives which restrict the curvature spinor, see [9, §2.3], see also [10, §3.2]. In particular, for a Killing spinor $\kappa_{A\cdots D}$ of valence $(k,0)$, $k \geq 1$, the condition

$$\Psi_{(ABC}{}^{F}\kappa_{D\cdots E)F} = 0 \tag{2.52}$$

must hold, which restricts the algebraic type of the Weyl spinor. For a valence $(2,0)$ Killing spinor κ_{AB}, the condition takes the form

$$\Psi_{(ABC}{}^{E}\kappa_{D)E} = 0. \tag{2.53}$$

It follows from (2.53) that a spacetime admitting a valence $(2,0)$ Killing spinor is of type D, N, or O. The space of Killing spinors of valence $(2,0)$ on Minkowski space (or any space of Petrov type O) has complex dimension 10. The explicit form in Cartesian coordinates $x^{AA'}$ is

$$\kappa^{AB} = U^{AB} + 2x^{A'(A}V^{B)}{}_{A'} + x^{AA'}x^{BB'}W_{A'B'},$$

where $U^{AB}, V^{B}{}_{A'}, W^{A'B'}$ are arbitrary constant symmetric spinors, see[2, Eq. (4.5)]. One of these corresponds to the spinor in (2.51), in spheroidal coordinates it takes the form given in (2.65) below.

A further application of the commutation properties of the fundamental operators yields the fact that the 1-form

$$\xi_{AA'} = (\mathscr{C}^{\dagger}_{2,0}\kappa)_{AA'} \tag{2.54}$$

is a Killing field, $\nabla_{(a}\xi_{b)} = 0$, provided $\mathcal{M}$ is a vacuum. Clearly the real and imaginary parts of ξ_a are also Killing fields. If ξ_a is proportional to a real Killing field,[17] we can without loss of generality assume that ξ_a is real. In this case, the 2-form

$$Y_{ab} = \tfrac{3}{2}i(\kappa_{AB}\bar{\epsilon}_{A'B'} - \bar{\kappa}_{A'B'}\epsilon_{AB}) \tag{2.55}$$

is a Killing–Yano tensor, $\nabla_{(a}Y_{b)c} = 0$, and the symmetric 2-tensor

$$K_{ab} = Y_a{}^c Y_{cb} \tag{2.56}$$

is a Killing tensor,

$$\nabla_{(a}K_{bc)} = 0. \tag{2.57}$$

Further, in this case,

$$\zeta_a = \xi^b K_{ab} \tag{2.58}$$

is a Killing field, see [64, 43]. Recall that the quantity $L_{ab}\dot{\gamma}^a\dot{\gamma}^b$ is conserved along null geodesics if L_{ab} is a conformal Killing tensor. For Killing tensors, this fact extends to all geodesics, so that if K_{ab} is a Killing tensor, then $K_{ab}\dot{\gamma}^a\dot{\gamma}^b$ is conserved along a geodesic γ^a. See [10] for further details and references.

2.4.8. GHP Formalism

Taking the point of view that the null tetrad components of tensors are sections of complex line bundles with the action of the non-vanishing complex scalars corresponding to the rescalings of the tetrad, respecting the normalization, leads to the GHP formalism [59].

Given a null tetrad $l^a, n^a, m^a, \bar{m}^a$ we have a spin dyad o_A, ι_A as discussed above. For a spinor $\varphi_{A\cdots D} \in \mathcal{S}_{k,0}$, it is convenient to introduce the Newman–Penrose scalars

$$\varphi_i = \varphi_{A_1\cdots A_i A_{i+1}\cdots A_k}\iota^{A_1}\cdots\iota^{A_i}o^{A_{i+1}}\cdots o^{A_k}. \tag{2.59}$$

In particular, Ψ_{ABCD} corresponds to the five complex Weyl scalars $\Psi_i, i = 0, \ldots 4$. The definition φ_i extends in a natural way to the scalar components of spinors of valence (k, l).

[17] We say that such spacetimes are of the generalized Kerr–NUT class, see [19] and references therein.

The normalization (2.40) is left invariant under rescalings $o_A \to \lambda o_A$, $\iota_A \to \lambda^{-1}\iota_A$ where λ is a non-vanishing complex scalar field on $\mathcal{M}$. Under such rescalings, the scalars defined by projecting on the dyad, such as φ_i given by (2.59), transform as sections of complex line bundles. A scalar φ is said to have type $\{p, q\}$ if $\varphi \to \lambda^p \bar{\lambda}^q \varphi$ under such a rescaling. Such fields are called properly weighted. The lift of the Levi-Civita connection $\nabla_{AA'}$ to these bundles gives a covariant derivative denoted Θ_a. Projecting on the null tetrad $l^a, n^a, m^a, \bar{m}^a$ gives the GHP operators

$$\text{þ} = l^a \Theta_a, \quad \text{þ}' = n^a \Theta_a, \quad \eth = m^a \Theta_a, \quad \eth' = \bar{m}^a \Theta_a.$$

The GHP operators are properly weighted, in the sense that they take properly weighted fields to properly weighted fields, for example if φ has type $\{p, q\}$, then $\text{þ}\varphi$ has type $\{p+1, q+1\}$. This can be seen from the fact that $l^a = o^A \bar{o}^{A'}$ has type $\{1, 1\}$. There are 12 connection coefficients in a null frame, up to complex conjugation. Of these, eight are properly weighted, the GHP spin coefficients. The other connection coefficients enter in the connection 1-form for the connection Θ_a.

The following formal operations take weighted quantities to weighted quantities,

$$\begin{aligned}
{}^{-}(\text{bar}) &: \ l^a \to l^a,\ n^a \to n^a,\ m^a \to \bar{m}^a,\ \bar{m}^a \to m^a, && \{p, q\} \to \{q, p\},\\
{}'(\text{prime}) &: \ l^a \to n^a,\ n^a \to l^a,\ m^a \to \bar{m}^a,\ \bar{m}^a \to m^a, && \{p, q\} \to \{-p, -q\},\\
{}^{*}(\text{star}) &: \ l^a \to m^a,\ n^a \to -\bar{m}^a,\ m^a \to -l^a,\ \bar{m}^a \to n^a, && \{p, q\} \to \{p, -q\}.
\end{aligned} \tag{2.60}$$

The properly weighted spin coefficients can be represented as

$$\kappa = m^b l^a \nabla_a l_b, \quad \sigma = m^b m^a \nabla_a l_b, \quad \rho = m^b \bar{m}^a \nabla_a l_b, \quad \tau = m^b n^a \nabla_a l_b, \tag{2.61}$$

together with their primes $\kappa', \sigma', \rho', \tau'$.

A systematic application of the above formalism allows one to write the tetrad projection of the geometric field equations in a compact form. For example, the Maxwell equation corresponds to the four scalar equations given by

$$(\text{þ} - 2\rho)\phi_1 - (\eth' - \tau')\phi_0 = -\kappa\phi_2, \tag{2.62}$$

with its primed and starred versions.

Working in a spacetime of Petrov type D gives drastic simplifications, in view of the fact that choosing the null tedrad so that l^a, n^a are aligned with principal null directions of the Weyl tensor (or equivalently choosing the spin dyad so that o_A, ι_A are principal spinors of the Weyl spinor); as has already been mentioned, the Weyl scalars are zero with the exception of Ψ_2, and the only non-zero spin coefficients are ρ, τ and their primed versions.

2.5. The Kerr Spacetime

Taking into account the background material given in Section 2.4, we can now state some further properties of the Kerr spacetime. As mentioned above, the Kerr metric is algebraically special, of Petrov type D. An explicit principal null tetrad $(l^a, n^a, m^a, \bar{m}^a)$ is given by the Carter tetrad [116]

$$l^a = \frac{a(\partial_\phi)^a}{\sqrt{2}\Delta^{1/2}\Sigma^{1/2}} + \frac{(a^2+r^2)(\partial_t)^a}{\sqrt{2}\Delta^{1/2}\Sigma^{1/2}} + \frac{\Delta^{1/2}(\partial_r)^a}{\sqrt{2}\Sigma^{1/2}}, \tag{2.63a}$$

$$n^a = \frac{a(\partial_\phi)^a}{\sqrt{2}\Delta^{1/2}\Sigma^{1/2}} + \frac{(a^2+r^2)(\partial_t)^a}{\sqrt{2}\Delta^{1/2}\Sigma^{1/2}} - \frac{\Delta^{1/2}(\partial_r)^a}{\sqrt{2}\Sigma^{1/2}}, \tag{2.63b}$$

$$m^a = \frac{(\partial_\theta)^a}{\sqrt{2}\Sigma^{1/2}} + \frac{i\csc\theta(\partial_\phi)^a}{\sqrt{2}\Sigma^{1/2}} + \frac{ia\sin\theta(\partial_t)^a}{\sqrt{2}\Sigma^{1/2}}. \tag{2.63c}$$

In view of the normalization of the tetrad, the metric takes the form $g_{ab} = 2(l_{(a}n_{b)} - m_{(a}\bar{m}_{b)})$. We may remark that the choice of l^a, n^a to be aligned with the principal null directions of the Weyl tensor, together with the normalization of the tetrad, fixes the tetrad up to rescalings.

We have

$$\Psi_2 = -\frac{M}{(r - ia\cos\theta)^3}. \tag{2.64}$$

$$\kappa_{AB} = \tfrac{2}{3}(r - ia\cos\theta)o_{(A}\iota_{B)}. \tag{2.65}$$

With κ_{AB} as in (2.65), equation (2.54) yields

$$\xi^a = (\partial_t)^a, \tag{2.66}$$

and from (2.55) we get

$$Y_{ab} = a\cos\theta\, l_{[a}n_{b]} - irm_{[a}\bar{m}_{b]}. \tag{2.67}$$

With the normalizations above, the Killing tensor (2.56) takes the form

$$K_{ab} = \tfrac{1}{4}(2\Sigma l_{(a}n_{b)} - r^2 g_{ab}) \tag{2.68}$$

and (2.58) gives

$$\zeta^a = a^2(\partial_t)^a + a(\partial_\phi)^a. \tag{2.69}$$

Recall that for a geodesic γ, the quantity $\boldsymbol{k} = 4K_{ab}\dot{\gamma}^a\dot{\gamma}^b$, known as Carter's constant, is conserved. Explicitly,

$$\boldsymbol{k} = \dot{\gamma}_\theta^2 + a^2\sin^2\theta\boldsymbol{e}^2 + 2a\boldsymbol{e}\boldsymbol{\ell}_z + a^2\cos^2\theta\boldsymbol{\mu}^2 \tag{2.70}$$

where $\dot{\gamma}_\theta = \dot{\gamma}^a(\partial_\theta)_a$. For $a \neq 0$, the tensor K_{ab} cannot be expressed as a tensor product of Killing fields [112], and similarly Carter's constant $\boldsymbol{k}$ cannot be expressed in terms of the constants of motion associated with Killing fields.

In this sense K_{ab} and $\boldsymbol{k}$ manifest a *hidden symmetry* of the Kerr spacetime. As we shall see in Section 2.7, these structures are also related to symmetry operators and separability properties, as well as conservation laws, for field equations on Kerr, and more generally in spacetimes admitting Killing spinors satisfying certain auxiliary conditions.

2.5.1. Characterizations of Kerr

Consider a vacuum Cauchy data set (Σ, h_{ij}, k_{ij}). We say that (Σ, h_{ij}, k_{ij}) is asymptotically flat if Σ has an end $\mathbb{R}^3 \setminus B(0,R)$ with a coordinate system (x^i) such that

$$h_{ij} = \delta_{ij} + O_\infty(r^\alpha), \quad k_{ij} = O_\infty(r^{\alpha-1}) \tag{2.71}$$

for some $\alpha < -1/2$. The Cauchy data set (Σ, h_{ij}, k_{ij}) is asymptotically Schwarzschildean if

$$h_{ij} = -\left(1 + \frac{2A}{r}\right)\delta_{ij} - \frac{\alpha}{r}\left(\frac{2x_i x_j}{r^2} - \delta_{ij}\right) + o_\infty(r^{-3/2}), \tag{2.72a}$$

$$k_{ij} = \frac{\beta}{r^2}\left(\frac{2x_i x_j}{r^2} - \delta_{ij}\right) + o_\infty(r^{-5/2}), \tag{2.72b}$$

where A is a constant, and α, β are functions on S^2, see [20, §6.5] for details. Here, the symbols $o_\infty(r^\alpha)$ are defined in terms of weighted Sobolev spaces, see [20, §6.2] for details.

If $(\mathcal{M}, g_{ab})$ is a vacuum and contains a Cauchy surface (Σ, h_{ij}, k_{ij}) satisfying (2.71) or (2.72), then $(\mathcal{M}, g_{ab})$ is asymptotically flat, respectively asymptotically Schwarzschildean, at spatial infinity. In this case there is a spacetime coordinate system (x^α) such that $g_{\alpha\beta}$ is asymptotic to the Minkowski line element with asymptotic conditions compatible with (2.72). For such spacetimes, the ADM 4-momentum P^μ is well defined. The positive mass theorem states that P^μ is future directed causal, $P^\mu P_\mu \geq 0$ (where the contraction is in the asymptotic Minkowski line element), $P^0 \geq 0$, and gives conditions under which P^μ is strictly timelike. This holds in particular if Σ contains an apparent horizon.

Mars [80] has given a characterization of the Kerr spacetime as an asymptotically flat vacuum spacetime with a Killing field ξ^a asymptotic to a time translation, positive mass, and an additional condition on the Killing form $F_{AB} = (\mathscr{C}_{1,1}\xi)_{AB}$,

$$\Psi_{ABCD}F^{CD} \propto F_{AB}.$$

A characterization in terms of algebraic invariants of the Weyl tensor has been given by Ferrando and Saez [51]. The just mentioned characterizations are in terms of spacetime quantities. As mentioned in Section 2.2.6.5 Killing spinor initial data propagates, which can be used to formulate a characterization of Kerr in terms of Cauchy data, see [19, 20, 21, 22].

We here give a characterization in terms of spacetimes admitting a Killing spinor of valence $(2,0)$.

Theorem 2.3 *Assume that $(\mathcal{M}, g_{ab})$ is vacuum, asymptotically Schwarzschildean at spacelike infinity and contains a Cauchy slice bounded by an apparent horizon. Assume further that $(\mathcal{M}, g_{ab})$ admits a non-vanishing Killing spinor κ_{AB} of valence $(2,0)$. Then $(\mathcal{M}, g_{ab})$ is locally isometric to the Kerr spacetime.*

Proof Let P^μ be the ADM 4-momentum vector for $\mathcal{M}$. By the positive mass theorem, $P^\mu P_\mu \geq 0$. In the case where $\mathcal{M}$ contains a Cauchy surface bounded by an apparent horizon, then $P^\mu P_\mu > 0$ by [23, Remark 11.5].[18]

Recall that a spacetime with a Killing spinor of valence $(2,0)$ is of Petrov type D, N, or O. From asymptotic flatness and the positive mass theorem, we have $C_{abcd}C^{abcd} = O(1/r^6)$, and hence there is a neighborhood of spatial infinity where $\mathcal{M}$ is Petrov type D. It follows that near spatial infinity, $\kappa_{AB} = -2\kappa_1 o_{(A}\iota_{B)}$, with $\kappa_1 \propto \Psi_2^{-1/3} = O(r)$. It follows from our asymptotic conditions that the Killing field $\xi_{AA'} = (\mathscr{C}^\dagger_{2,0}\kappa)_{AB}$ is $O(1)$ and hence asymptotic to a translation, $\xi^\mu \to A^\mu$ as $r \to \infty$, for some constant vector A^μ. It follows from the discussion in [4, §4] that A^μ is non-vanishing. Now, by [24, §*III*], it follows that in the case $P^\mu P_\mu > 0$, then A^μ is proportional to P^μ, see also [25]. We are now in the situation considered in the work by Bäckdahl and Valiente Kroon, see [21, Theorem B.3], and hence we can conclude that $(\mathcal{M}, g_{ab})$ is locally isometric to the Kerr spacetime. □

Remark 2.4 1. This result can be turned into a characterization in terms of Cauchy data along the lines in [20].
2. Theorem 2.3 can be viewed as a variation on the Kerr characterization given in [21, Theorem B.3]. In the version given here, the asymptotic conditions on the Killing spinor have been removed.

2.6. Monotonicity and Dispersion

The dispersive properties of fields, i.e. the tendency of the energy density contained within any stationary region to decrease asymptotically to the future,

[18] Section 11 appears only in the ArXiv version of [23].

is a crucial property for solutions of field equations on spacetimes, and any proof of stability must exploit this phenomenon. In view of the geometric optics approximation, the dispersive property of fields can be seen in an analogous dispersive property of null geodesics, i.e. the fact that null geodesics in the Kerr spacetime which do not orbit the black hole at a fixed radius must leave any stationary region in at least one of the past or future directions. In Section 2.6.1 we give an explanation for this fact using tools which can readily be adapted to the case of field equations, while in Section 2.6.2 we outline how these ideas apply to fields.

We begin by a discussion of conservation laws. For a null geodesic γ^a, we define the energy associated with a vector field X and evaluated on a Cauchy hypersurface Σ to be

$$e_X[\gamma](\Sigma) = g_{ab}X^a\dot{\gamma}^b|_\Sigma.$$

Since $\dot{\gamma}^b\nabla_b\dot{\gamma}^a = 0$ for a geodesic, integrating the derivative of the energy gives

$$e_X[\gamma](\Sigma_2) - e_X[\gamma](\Sigma_1) = \int_{\lambda_1}^{\lambda_2} (\dot{\gamma}_a\dot{\gamma}_b)\nabla^{(a}X^{b)}d\lambda, \tag{2.73}$$

where λ_i is the unique value of λ such that $\gamma(\lambda)$ is the intersection of γ with Σ_i. Formula (2.73) is particularly easy to work with, if one recalls that

$$\nabla^{(a}X^{b)} = -\frac{1}{2}\mathcal{L}_X g^{ab}.$$

The tensor $\nabla^{(a}X^{b)}$ is commonly called the "deformation tensor." In the following, unless there is the possibility of confusion, we will drop reference to γ and Σ in referring to e_X.

Conserved quantities play a crucial role in understanding the behavior of geodesics as well as fields. By (2.73), the energy e_X is conserved if X^a is a Killing field. In the Kerr spacetime we have the Killing fields $\xi^a = (\partial_t)^a$, $\eta^a = (\partial_\phi)^a$ with the corresponding conserved quantities energy $\boldsymbol{e} = (\partial_t)^a\dot{\gamma}_a$ and azimuthal angular momentum $\boldsymbol{\ell}_z = (\partial_\phi)^a\dot{\gamma}_a$. In addition, the squared particle mass $\boldsymbol{\mu}^2 = g_{ab}\dot{\gamma}^a\dot{\gamma}^b$ and the Carter constant $\boldsymbol{k} = K_{ab}\dot{\gamma}^a\dot{\gamma}^b$ are conserved along any geodesic γ^a in the Kerr spacetime. The presence of the extra conserved quantity allows one to integrate the equations of geodesic motion.[19]

For a covariant field equation derived from an action principle which depends on the background geometry only via the metric and its derivatives, the

[19] In general, the geodesic equation in a four-dimensional stationary and axi-symmetric spacetime cannot be integrated, and the dynamics of particles may in fact be chaotic; see [57, 79] and references therein. Note however that the geodesic equations are not *separable* in the Boyer–Lindquist coordinates. On the other hand, the Darboux coordinates have this property, cf. [56].

symmetric stress-energy tensor T_{ab} is conserved. As an example, we consider the wave equation

$$\nabla^a \nabla_a \psi = 0 \tag{2.74}$$

which has the stress-energy tensor

$$T_{ab} = \nabla_{(a}\psi \nabla_{b)}\bar{\psi} - \tfrac{1}{2}\nabla^c \psi \nabla_c \bar{\psi} g_{ab}. \tag{2.75}$$

Let ψ be a solution to (2.74). Then T_{ab} is conserved, $\nabla^a T_{ab} = 0$. For a vector field X^a we have $\nabla^a(T_{ab}X^b)$ given in terms of the deformation tensor,

$$\nabla^a(T_{ab}X^b) = T_{ab}\nabla^{(a}X^{b)}.$$

Let $(J_X)_a = T_{ab}X^b$ be the current corresponding to X^a. By the above, we have conserved currents J_ξ and J_η corresponding to the Killing fields ξ^a, η^a.

An application of Gauss's law gives the analog of (2.73),

$$\int_{\Sigma_2} (J_X)_a d\sigma^a - \int_{\Sigma_1} (J_X)_a d\sigma^a = \int_\Omega T_{ab}\nabla^{(a}X^{b)}$$

where Ω is a spacetime region bounded by Σ_1, Σ_2.

2.6.1. Monotonicity for Null Geodesics

We shall consider only null geodesics, i.e. $\boldsymbol{\mu} = 0$. In this case we have

$$\begin{aligned} \boldsymbol{k} &= K_{ab}\dot{\gamma}^a\dot{\gamma}^b \\ &= 2\Sigma l_{(a}n_{b)}\dot{\gamma}^a\dot{\gamma}^b \\ &= 2\Sigma m_{(a}\bar{m}_{b)}\dot{\gamma}^a\dot{\gamma}^b. \end{aligned} \tag{2.76}$$

We note that the tensors $2\Sigma l_{(a}n_{b)}$ and $2\Sigma m_{(a}\bar{m}_{b)}$ are conformal Killing tensors; see Section 2.4.5. From (2.76) it is clear that $\boldsymbol{k}$ is non-negative. A calculation using (2.63) gives

$$2\Sigma l^{(a}n^{b)}\partial_a\partial_b = \frac{1}{\Delta}[(r^2+a^2)\partial_t + a\partial_\phi]^2 - \Delta\partial_r^2$$

$$2\Sigma m^{(a}\bar{m}^{b)}\partial_a\partial_b = \partial_\theta^2 + \frac{1}{\sin^2\theta}\partial_\phi^2 + a^2\sin^2\theta\,\partial_t^2 + 2a\partial_t\partial_\phi.$$

Let $Z = (r^2+a^2)\boldsymbol{e} + a\boldsymbol{\ell}_z$. Recall that $\dot{r} = \dot{\gamma}^r = g^{rr}\dot{\gamma}_r$ where $g^{rr} = -\Delta/\Sigma$. Now we can write $0 = g_{ab}\dot{\gamma}^a\dot{\gamma}^b$ in the form

$$\Sigma^2\dot{r}^2 + \mathcal{R}(r; \boldsymbol{e}, \boldsymbol{\ell}_z, \boldsymbol{k}) = 0 \tag{2.77}$$

where

$$\mathcal{R} = -Z^2 + \Delta\boldsymbol{k} \tag{2.78}$$

Equation (2.77) is the exact analog of (2.28) for the Schwarzschild case. It is clear from (2.76) that, for null geodesics, $\boldsymbol{k}$ corresponds in the Schwarzschild case with $a = 0$, to $\boldsymbol{L}^2$, the squared total angular momentum. It is possible to derive equations similar to (2.77) for the other coordinates t, θ, ϕ, which allows the solution of the geodesic equations by quadratures, see e.g. [105] for details.

Equation (2.77) allows one to make a qualitative analysis of the motion of null geodesics in the Kerr spacetime. In particular, we find that the location of orbiting null geodesics is determined by $\mathcal{R} = 0$, $\partial_r \mathcal{R} = 0$. Due to the form of $\mathcal{R}$, the location of orbiting null geodesics depends only on the ratios $\boldsymbol{k}/\boldsymbol{\ell}_z{}^2, \boldsymbol{e}/\boldsymbol{\ell}_z$. One finds that orbiting null geodesics exist for a range of radii $r_1 \leq r \leq r_2$, with $r_+ < r_1 < 3M < r_2$. Here r_1, r_2 depend on a, M and as $|a| \nearrow M$, $r_1 \searrow r_+$, and $r_2 \nearrow 4M$. The orbits at r_1, r_2 are restricted to the equatorial plane, those at r_1 are co-rotating, while those at r_2 are counter-rotating. For $r_1 < r < r_2$, the range of θ depends on r (see Figure 2.13). There is $r_3 = r_3(a, M)$, $r_1 < r_3 < r_2$ such that the orbits at r_3 reach the poles, i.e. $\theta = 0, \theta = \pi$; see Figure 2.13. For such geodesics, it holds that $\boldsymbol{\ell}_z = 0$.

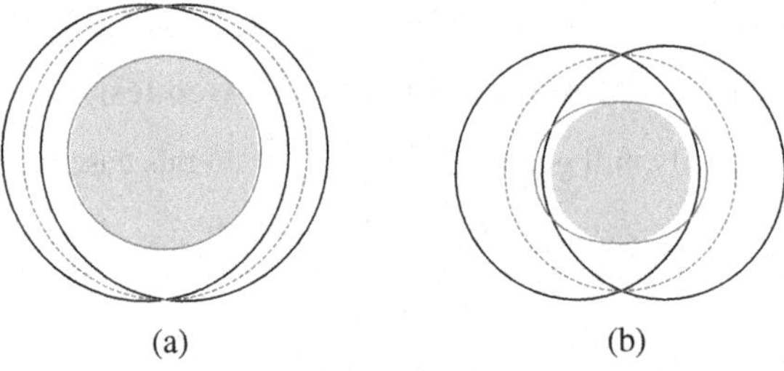

Figure 2.13. The Kerr photon region. In subfigure (a), $|a| \ll M$ and the ergoregion, see Section 2.3.7, is well separated from the photon region (bordered in black). The radius r_3 where geodesics reach the poles is indicated by a gray, dashed line. In subfigure (b), $|a|$ is close to M and the ergoregion overlaps the photon region

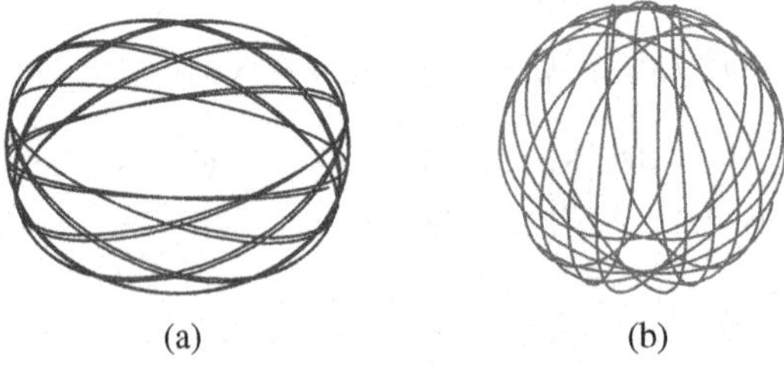

Figure 2.14. Examples of orbiting null geodesics in Kerr with $a = M/2$. In subfigure (a), $\boldsymbol{k}/\boldsymbol{\ell}_z{}^2$ is small, while in subfigure (b), this constant is larger

For the following discussion, it is convenient to introduce

$$\boldsymbol{q} = \boldsymbol{k} - 2a\boldsymbol{e}\boldsymbol{\ell}_z - \boldsymbol{\ell}_z{}^2 = Q^{ab}\dot{\gamma}_a\dot{\gamma}_b,$$

where

$$Q^{ab} = (\partial_\theta)^a(\partial_\theta)^b + \frac{\cos^2\theta}{\sin^2\theta}(\partial_\phi)^a(\partial_\phi)^b + a^2\sin^2\theta(\partial_t)^a(\partial_t)^b. \tag{2.79}$$

By construction, $\boldsymbol{q}$ is a sum of conserved quantities, and is therefore conserved. Further, it is non-negative, since it is a sum of non-negative terms. In the following we use $(\boldsymbol{e}, \boldsymbol{\ell}_z, \boldsymbol{q})$ as parameters for null geodesics. Since we are considering only null geodesics, there is no loss of generality compared to using $(\boldsymbol{e}, \boldsymbol{\ell}_z, \boldsymbol{k})$ as parameters.

For a null geodesic with given parameters $(\boldsymbol{e}, \boldsymbol{\ell}_z, \boldsymbol{q})$, a simple turning point analysis shows that there is a number $r_o \in (r_+, \infty)$ such that the quantity $(r - r_o)\dot{\gamma}^r$ increases overall. This quantity corresponds to the energy e_A for the vector field $A = -(r - r_o)\partial_r$. Following this idea, we may now look for a function $\mathcal{F}$ which will play the role of $-(r - r_o)$, so that for $A = \mathcal{F}\partial_r$, the energy e_A is non-decreasing for all λ and not merely non-decreasing overall. For $a \neq 0$, both r_o and $\mathcal{F}$ will necessarily depend on both the Kerr parameters (M, a) and the constants of motion $(\boldsymbol{e}, \boldsymbol{\ell}_z, \boldsymbol{q})$; the function $\mathcal{F}$ will also depend on r, but no other variables.

We define $A^a = \mathcal{F}(\partial_r)^a$ with

$$\mathcal{F} = \mathcal{F}(r; M, a, \boldsymbol{e}, \boldsymbol{\ell}_z, \boldsymbol{q}).$$

It is important to note that this is a map from the tangent bundle to the tangent bundle, and hence $A^a = \mathcal{F}(\partial_r)^a$ cannot be viewed as a standard vector field, which is a map from the manifold to the tangent bundle.

To derive a monotonicity formula, we wish to choose $\mathcal{F}$ so that e_A has a non-negative derivative. We define the covariant derivative of A by holding the values of $(\boldsymbol{e}, \boldsymbol{\ell}_z, \boldsymbol{q})$ fixed and computing the covariant derivative as if A were a regular vector field. Similarly, we define $\mathcal{L}_A g^{ab}$ by fixing the values of the constants of geodesic motion. Since the constants of motion have zero derivative along null geodesics, equation (2.73) remains valid.

Recall that null geodesics are conformally invariant up to reparameterization. Hence, it is sufficient to work with the conformally rescaled metric Σg^{ab}. Furthermore, since γ is a null geodesic, for any function q_{reduced}, we may subtract $q_{\text{reduced}}\Sigma g^{ab}\dot{\gamma}_a\dot{\gamma}_a$ wherever it is convenient. Thus, the change in e_A is given as the integral of

$$\Sigma\dot{\gamma}_a\dot{\gamma}_b\nabla^{(a}A^{b)} = \left(-\frac{1}{2}\mathcal{L}_A(\Sigma g^{ab}) - q_{\text{reduced}}\Sigma g^{ab}\right)\dot{\gamma}_a\dot{\gamma}_b.$$

The Kerr metric can be written as

$$\Sigma g^{ab} = -\Delta(\partial_r)^a(\partial_r)^b - \frac{1}{\Delta}\mathcal{R}^{ab}, \tag{2.80}$$

where the tensorial form of $\mathcal{R}^{ab}$ can be read off from the earlier definitions. We now calculate $-\mathcal{L}_A g^{ab}\dot{\gamma}_a\dot{\gamma}_b$ using (2.80). Ignoring distracting factors of Σ, Δ, the most important terms are

$$-2(\partial_r\mathcal{F})\dot{\gamma}_r\dot{\gamma}_r + \mathcal{F}(\partial_r\mathcal{R}^{ab})\dot{\gamma}_a\dot{\gamma}_b = -2(\partial_r\mathcal{F})\dot{\gamma}_r\dot{\gamma}_r + \mathcal{F}(\partial_r\mathcal{R}).$$

The second term in this sum will be non-negative if $\mathcal{F} = \partial_r\mathcal{R}(r;M,a;\boldsymbol{e},\boldsymbol{\ell}_z,\boldsymbol{q})$. Recall that the vanishing of $\partial_r\mathcal{R}(r;M,a;\boldsymbol{e},\boldsymbol{\ell}_z,\boldsymbol{q})$ is one of the two conditions for orbiting null geodesics. With this choice of $\mathcal{F}$, the instability of the null geodesic orbits ensures that, for these null geodesics, the coefficient in the first term, $-2(\partial_r\mathcal{F})$, will be positive. These observations motivate the form of $\mathcal{F}$ which yields non-negativity for all null geodesics.

It remains to make explicit the choices of $\mathcal{F}$ and q_{reduced}. Once these choices are made, the necessary calculations are straightforward but rather lengthy. Let z and w be smooth functions of r and the Kerr parameters (M,a). Let $\tilde{\mathcal{R}}'$ denote $\partial_r(\frac{z}{\Delta}\mathcal{R}(r;M,a;\boldsymbol{e},\boldsymbol{\ell}_z,\boldsymbol{q}))$ and choose $\mathcal{F} = zw\tilde{\mathcal{R}}'$ and $q_{\text{reduced}} = (1/2)(\partial_r z)w\tilde{\mathcal{R}}'$. In terms of these functions,

$$\Sigma\dot{\gamma}_a\dot{\gamma}_b\nabla^{(a}A^{b)} = \tfrac{1}{2}w(\tilde{\mathcal{R}}')^2 - z^{1/2}\Delta^{3/2}\left(\partial_r\left(w\frac{z^{1/2}}{\Delta^{1/2}}\tilde{\mathcal{R}}'\right)\right)\dot{\gamma}_r^2. \tag{2.81}$$

If z and w are chosen to be positive, then the first term on the right-hand side of (2.81) which contains a square $(\tilde{\mathcal{R}}')^2$ is non-negative. If we now take $z = z_1 = \Delta(r^2+a^2)^{-2}$ and $w = w_1 = (r^2+a^2)^4/(3r^2-a^2)$, then[20]

$$-\partial_r\left(w\frac{z^{1/2}}{\Delta^{1/2}}\tilde{\mathcal{R}}'\right) = 2\frac{3r^4+a^4}{(3r^2-a^2)^2}\boldsymbol{\ell}_z{}^2 + 2\frac{3r^4-6a^2r^2-a^4}{(3r^2-a^2)^2}\boldsymbol{q}. \tag{2.82}$$

The coefficient of $\boldsymbol{q}$ is positive for $r > r_+$ when $|a| < 3^{1/4}2^{-1/2}M \cong 0.93M$. Since $\boldsymbol{q}$ is non-negative, the right-hand side of (2.82) is non-negative, and hence also the right-hand side of equation (2.81) is non-negative, for this range of a. Since equation (2.81) gives the rate of change, the energy e_A is monotone.

These calculations reveal useful information about the geodesic motion. The positivity of the term on the right-hand side of (2.82) shows that $\tilde{\mathcal{R}}'$ can have at most one root, which must be simple. In turn, this shows that $\mathcal{R}$ can have at

[20] Equation (2.82) corrects a misprint in [12, Eq. (1.15b)].

most two roots. For orbiting null geodesics $\mathcal{R}$ must have a double root, which must coincide with the root of $\tilde{\mathcal{R}}'$. It is convenient to think of the corresponding value of r as being r_o.

The first term in (2.81) vanishes at the root of $\tilde{\mathcal{R}}'$, as it must, so that e_A can be constantly zero on the orbiting null geodesics. When $a = 0$, the quantity $\tilde{\mathcal{R}}'$ reduces to $-2(r - 3M)r^{-4}(\boldsymbol{\ell}_z{}^2 + \boldsymbol{q})$, so that the orbits occur at $r = 3M$. The continuity in a of $\tilde{\mathcal{R}}'$ guarantees that its root converges to $3M$ as $a \to 0$ for fixed $(\boldsymbol{e}, \boldsymbol{\ell}_z, \boldsymbol{q})$.

From the geometrics optics approximation, it is natural to imagine that the monotone quantity constructed in this section for null geodesics might imply the existence of monotone quantities for fields, which would imply some form of dispersion. For the wave equation, this is true. In fact, the above discussion, when carried over to the case of the wave equation, closely parallels the proof of the Morawetz estimate for the wave equation given in [12], see Section 2.6.2 below. The quantity $(\dot{\gamma}_\alpha \dot{\gamma}_\beta)(\nabla^{(\alpha} X^{\beta)})$ corresponds to the Morawetz density, i.e. the divergence of the momentum corresponding to the Morawetz vector field. The role of the conserved quantities $(\boldsymbol{e}, \boldsymbol{\ell}_z, \boldsymbol{q})$ for geodesics is played, in the case of fields, by the energy fluxes defined via second-order symmetry operators corresponding to these conserved quantities. The fact that the quantity $\mathcal{R}$ vanishes quadratically on the trapped orbits is reflected in the Morawetz estimate for fields, by a quadratic degeneracy of the Morawetz density at the trapped orbits.

2.6.2. Dispersive Estimates for Fields

As discussed in Section 2.6.1, one may construct a suitable function of the conserved quantities for null geodesics in the Kerr spacetime which is monotone along the geodesic flow. This function may be viewed as arising from a generalized vector field on phase space. The monotonicity property implies, as discussed there, that non-trapped null geodesics disperse, in the sense that they leave any stationary region in the Kerr spacetime. As mentioned in Section 2.6.1, in view of the geometric optics approximation for the wave equation, such a monotonicity property for null geodesics reflects the tendency for waves in the Kerr spacetime to disperse.

At the level of the wave equation, the analog of the just mentioned monotonicity estimate is called the Morawetz estimate. For the wave equation $\nabla^a \nabla_a \psi = 0$, a Morawetz estimate provides a current J_a defined in terms of ψ and some of its derivatives, with the property that $\nabla^a J_a$ has suitable positivity properties, and that the flux of J_a can be controlled by a suitable energy defined in terms of the field.

Let ψ be a solution of the wave equation $\nabla^a\nabla_a\psi = 0$. Define the current J_a by

$$J_a = T_{ab}A^b + \tfrac{1}{2}q(\bar{\psi}\nabla_a\psi + \psi\nabla_a\bar{\psi}) - \tfrac{1}{2}(\nabla_a q)\psi\bar{\psi},$$

where T_{ab} is the stress-energy tensor given by (2.75). We have

$$\nabla^a J_a = T_{ab}\nabla^{(a}A^{b)} + q\nabla^c\psi\nabla_c\bar{\psi} - \tfrac{1}{2}(\nabla^c\nabla_c q)\psi\bar{\psi}. \qquad (2.83)$$

We now specialize to Minkowski space, with the line element $g_{ab}dx^a dx^b = dt^2 - dr^2 - d\theta^2 - r^2\sin^2\theta d\phi^2$. Let

$$E(\tau) = \int_{\{t=\tau\}} T_{tt} d^3x$$

be the energy of the field at time τ, where T_{tt} is the energy density. The energy is conserved, so that $E(t)$ is independent of t.

Setting $A^a = r(\partial_r)^a$, we have

$$\nabla^{(a}A^{b)} = g^{ab} - (\partial_t)^a(\partial_t)^b. \qquad (2.84)$$

With $q = 1$, we get

$$\nabla^a J_a = -T_{tt}.$$

With the above choices, the bulk term $\nabla^a J_a$ has a sign. This method can be used to prove dispersion for solutions of the wave equation. In particular, by introducing suitable cutoffs, one finds that for any $R_0 > 0$, there is a constant C, so that

$$\int_{t_0}^{t_1}\int_{|r|\le R_0} T_{tt} d^3x dt \le C(E(t_0) + E(t_1)) \le 2CE(t_0), \qquad (2.85)$$

see [85]. The local energy, $\int_{|r|\le R_0} T_{tt} d^3x$, is a function of time. By (2.85) it is integrable in t, and hence it must decay to zero as $t \to \infty$, at least sequentially. This shows that the field disperses. Estimates of this type are called Morawetz or integrated local energy decay estimates.

For a solution ϕ_{AB} of the Maxwell equation $(\mathscr{C}^\dagger_{2,0}\phi)_{AA'} = 0$, the stress-energy tensor T_{ab} given by

$$T_{ab} = \phi_{AB}\bar{\phi}_{A'B'}$$

is conserved, $\nabla^a T_{ab} = 0$. Further, T_{ab} has trace zero, with $T^a{}_a = 0$.

Restricting to Minkowski space and setting $J_a = T_{ab}A^b$, with $A^a = r(\partial_r)^a$ we have

$$\nabla^a J_a = -T_{tt}$$

which again gives local energy decay for the Maxwell field on Minkowski space.

For the wave equation on Schwarzschild we can choose

$$A^a = \frac{(r-3M)(r-2M)}{3r^2}(\partial_r)^a, \tag{2.86a}$$

$$q = \frac{6M^2 - 7Mr + 2r^2}{6r^3}. \tag{2.86b}$$

This gives

$$\begin{aligned} -\nabla^{(a}A^{b)} = & -\frac{Mg^{ab}(r-3M)}{3r^3} + \frac{M(r-2M)^2(\partial_r)^a(\partial_r)^b}{r^4} \\ & + \frac{(r-3M)^2((\partial_\theta)^a(\partial_\theta)^b + \csc^2\theta(\partial_\phi)^a(\partial_\phi)^b)}{3r^5}, \end{aligned} \tag{2.87a}$$

$$\begin{aligned} -\nabla_a J^a = & \frac{M|\partial_r\psi|^2(r-2M)^2}{r^4} + \frac{\left(|\partial_\theta\psi|^2 + |\partial_\phi\psi|^2\csc^2\theta\right)(r-3M)^2}{3r^5} \\ & + \frac{M|\psi|^2(54M^2 - 46Mr + 9r^2)}{6r^6}. \end{aligned} \tag{2.87b}$$

Here, A^a was chosen so that the last two terms (2.87a) have good signs. The form of q given here was chosen to eliminate the $|\partial_t\psi|^2$ term in (2.87b). The first terms in (2.87b) are clearly non-negative, while the last is of a lower-order and can be estimated using a Hardy estimate [12]. The effect of trapping in Schwarzschild at $r = 3M$ is manifested in the fact that the angular derivative term vanishes at $r = 3M$.

In the case of the wave equation on Kerr, the above argument using a classical vector field cannot work due to the complicated structure of the trapping. However, making use of higher-order currents constructed using second-order symmetry operators for the wave equation, we can construct a generalized Morawetz vector field analogous to the vector field A^a. This is discussed in Section 2.6.1 and has been carried out in detail in [12].

If we apply the same idea for the Maxwell field on Schwarzschild, there is no reason to expect that local energy decay should hold, in view of the fact that the Coulomb solution is a time-independent solution of the Maxwell equation which does not disperse. In fact, with

$$A^a = \mathcal{F}(r)\Big(1 - \frac{2M}{r}\Big)(\partial_r)^a, \tag{2.88}$$

we have

$$-T_{ab}\nabla^{(a}A^{b)} = -\phi^{AB}\bar{\phi}^{A'B'}(\mathscr{T}_{1,1}A)_{ABA'B'} \tag{2.89}$$

$$\begin{aligned} &= \left(|\phi_0|^2 + |\phi_2|^2\right)\frac{(r-2M)}{2r}\mathcal{F}'(r) \\ &\quad - \frac{|\phi_1|^2\left(r(r-2M)\mathcal{F}'(r) - 2\mathcal{F}(r)(r-3M)\right)}{r^2}. \end{aligned} \tag{2.90}$$

If $\mathcal{F}'$ is chosen to be positive, then the coefficient of the extreme components in (2.90) is positive. However, at $r = 3M$, the coefficient of the middle component is necessarily of the opposite sign. It is possible to show that no choice of $\mathcal{F}$ will give positive coefficients for all components in (2.90).

The dominant energy condition, that $T_{ab}V^aW^b \geq 0$ for all causal vectors V^a, W^a, is a common and important condition on stress energy tensors. In Riemannian geometry, a natural condition on a symmetric 2-tensor T_{ab} would be non-negativity, i.e. the condition that for all X^a, one has $T_{ab}X^aX^b \geq 0$.

However, in order to prove dispersive estimates for null geodesics and the wave equation, the dominant energy condition on its own is not sufficient and non-negativity cannot be expected for stress energy tensors. Instead, a useful condition to consider is non-negativity modulo trace terms, i.e. the condition that for every X^a there is a q such that $T_{ab}X^aX^b + qT^a{}_a \geq 0$. For null geodesics and the wave equation, the tensors $\dot{\gamma}_a\dot{\gamma}_b$ and $\nabla_a u \nabla_b u = T_{ab} + T^\gamma{}_\gamma g_{ab}$ are both non-negative, so $\dot{\gamma}_a\dot{\gamma}_b$ and T_{ab} are non-negative modulo trace terms.

From equation (2.87a), we see that $-\nabla^{(a}A^{b)}$ is of the form $f_1 g^{ab} + f_2 \partial_r^a \partial_r^b + f_3 \partial_\theta^a \partial_\theta^b + f_4 \partial_\phi^a \partial_\phi^b$ where f_2, f_3, and f_4 are non-negative functions. That is $-\nabla^{(a}A^{b)}$ is a sum of a multiple of the metric plus a sum of terms of the form of a non-negative coefficient times a vector tensored with itself. Thus, from the non-negativity modulo trace terms, for null geodesics and the wave equation respectively, there are functions q such that $\dot{\gamma}_a\dot{\gamma}_b\nabla^aA^b = \dot{\gamma}_a\dot{\gamma}_b\nabla^aA^b + qg^{ab}\dot{\gamma}_a\dot{\gamma}_b \leq 0$ and $T_{ab}\nabla^aA^b + qT^a{}_a \leq 0$. For null geodesics, since $g^{ab}\dot{\gamma}_a\dot{\gamma}_b = 0$, the q term can be ignored. For the wave equation, one can use the terms involving q in equations (2.83), to cancel the $T^a{}_a$ term in $\nabla^a J_a$. For the wave equation, this gives non-negativity for the first-order terms in $-\nabla^a J_a$, and one can then hope to use a Hardy estimate to control the zeroth-order terms.

If we now consider the Maxwell equation, we have the fact that the Maxwell stress energy tensor is traceless, $T^a{}_a = 0$ and does not satisfy the non-negativity condition. Therefore it also does not satisfy the condition of non-negativity modulo trace. This appears to be the fundamental underlying obstruction to proving a Morawetz estimate using T_{ab}. This can be seen as a manifestation of the fact that the Coulomb solution does not disperse. In fact, it is immediately clear that the Maxwell stress energy cannot be used directly to prove dispersive estimates since it does not vanish for the Coulomb field (2.45) on the Kerr spacetime. We may remark that the existence of the Coulomb solution on the Kerr spacetime is a consequence of the fact that the exterior of the black hole contains non-trivial 2-spheres and the existence of two conserved charge integrals $\int_S F_{ab} d\sigma^{ab}$, $\int_S (*F)_{ab} d\sigma^{ab}$. Hence this is valid also for dynamical black hole spacetimes.

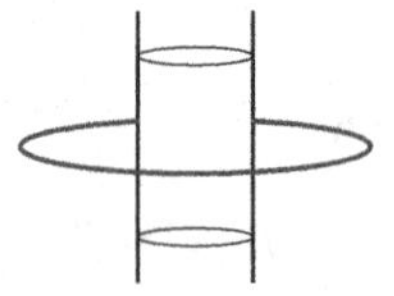

2.7. Symmetry Operators

A symmetry operator for a field equation is an operator which takes solutions to solutions. In order to analyze higher spin fields on the Kerr spacetime, it is important to gain an understanding of the symmetry operators for this case. In the paper [9] we have given a complete characterization of those spacetimes admitting symmetry operators of a second order for the field equations of spins $0, 1/2, 1$, i.e. the conformal wave equation, the Dirac–Weyl equation and the Maxwell equation, respectively, and given the general form of the symmetry operators, up to equivalence. In order to simplify the presentation here, we shall discuss only the spin-1 case, and restrict ourselves to spacetimes admitting a valence $(2,0)$ Killing spinor κ_{AB}. We first give some background on the wave equation.

2.7.1. Symmetry Operators for the Kerr Wave Equation

As shown by Carter [33], if K_{ab} is a Killing tensor in a Ricci flat spacetime, the operator

$$K = \nabla_a K^{ab} \nabla_b \tag{2.91}$$

is a commuting symmetry operator for the d'Alembertian,

$$[\nabla^a \nabla_a, K] = 0.$$

In particular there is a second-order symmetry operator for the wave equation, i.e. an operator which maps solutions to solutions,

$$\nabla^a \nabla_a \psi = 0 \quad \Rightarrow \quad \nabla^a \nabla_a K \psi = 0.$$

Due to the form of the Carter Killing tensor, K_{ab}, cf. (2.68), the operator K defined by (2.91) contains derivatives with respect to all coordinates.

Recall that $\nabla^a \nabla_a = \frac{1}{\mu_g} \partial_a \mu_g g^{ab} \partial_b$, where $\mu_g = \sqrt{\det(g_{ab})}$ is the volume element. For Kerr in Boyer–Lindquist coordinates, we have from (2.35) $\mu_g = \Sigma \upmu$, with $\upmu = \sin\theta$. After rescaling the d'Alembertian by Σ, and using the just mentioned facts, one finds

$$\Sigma \nabla^a \nabla_a = -\partial_r \Delta \partial_r + \frac{\mathcal{R}(r; \partial_t, \partial_\phi, Q)}{\Delta} \tag{2.92}$$

where

$$Q = \frac{1}{\upmu} \partial_a \upmu Q^{ab} \partial_b.$$

In view of the form of Q^{ab} given in (2.79), we see that Q contains derivatives only with respect to θ, ϕ, t, but not with respect to r. Thus, it is clear from

(2.92) that Q is a commuting symmetry operator for the rescaled d'Alembertian $\Sigma\nabla^a\nabla_a$,

$$[\Sigma\nabla^a\nabla_a, Q] = 0.$$

In addition to the symmetry operator Q related to the Carter constant, we have the second-order symmetry operators generated by the Killing fields $\xi^a\nabla_a = \partial_t$, $\eta^a\nabla_a = \partial_\phi$. The operator Q can be termed a hidden symmetry, since it cannot be represented in terms of operators generated by the Killing fields.

The above shows that we can write

$$\Sigma\nabla^a\nabla_a = \mathrm{R} + \mathrm{S}$$

where the operators R, S commute, $[\mathrm{R}, \mathrm{S}] = 0$, and R contains derivatives with respect to the non-symmetry coordinate r, and the two symmetry coordinates t, ϕ, while S contains derivatives with respect to the non-symmetry coordinate θ, and with respect to t, ϕ.

By making a separated ansatz

$$\psi_{\omega,\ell,m}(t, r, \theta, \phi) = e^{-i\omega t}e^{im\phi}R_{\omega,\ell,m}(r)S_{\omega,\ell,m}(\theta)$$

the equation $\nabla^a\nabla_a\psi = 0$ becomes a pair of scalar ordinary differential equations

$$\mathrm{R}R + \lambda R = 0 \tag{2.93a}$$

$$\mathrm{S}S = \lambda S \tag{2.93b}$$

where $\lambda = \lambda_{\omega,\ell,m}$. Here it should be noted that equation (2.93b) is to be considered as a boundary value problem on $[0, \pi]$ with boundary conditions determined by the requirement that ϕ be smooth. In the Schwarzschild case $a = 0$, we can take $\mathrm{S} = \not{\Delta}$, the angular Laplacian. The eigenfunctions of $\not{\Delta}$ are the spherical harmonics $Y_{\ell,m}(\theta, \phi) = e^{im\phi}Y_\ell(\theta)$. The eigenvalues of $\not{\Delta}$ are $\lambda_{\ell,m} = -\ell(\ell + 1)$.

The solutions to the eigenproblem $\mathrm{S}S = \lambda S$ are the spheroidal harmonics; the eigenvalues in this case are not known in closed form, but depend on the time frequency ω, and are indexed by ℓ, m. For real ω, it is known that the eigensystem is complete, but for general ω this is not known.

One may now apply a Fourier transform and represent a typical solution ψ to the wave equation in the form

$$\psi = \int d\omega \sum_{\ell,m} e^{-i\omega t}e^{im\phi}R_{\omega,\ell,m}S_{\omega,\ell,m},$$

analyze the behavior of the separated modes $\psi_{\omega,\ell,m}$, and recoved estimates for ψ after inverting the Fourier transform. In order to do this, one must show a priori that the Fourier transform can be applied. This can be done by applying cutoffs, and removing these after estimates have been proved using Fourier techniques. This approach has been followed in e.g. [45, 14, 13]. In recent work by Dafermos, Rodnianski, and Shlapentokh-Rothman, see [46], proving boundedness and decay for the wave equation on Kerr for the whole range $|a| < M$, makes use of the technical condition of time integrability, i.e. that the solution to the wave equation and its derivatives to a sufficiently high order is bounded in L^2 on time lines,

$$\int_{-\infty}^{\infty} dt |\partial^\alpha \psi(t,r,\theta,\phi)|^2.$$

This condition is consistent with integrated local energy decay and is removed at the end of the argument.

However, by working directly with currents defined in terms of second-order symmetry operators, one may prove a Morawetz estimate directly for the wave equation on the Kerr spacetime. This was carried out for the case $|a| \ll M$ in [12]. This involves introducing a generalization of the vector field method to allow for currents defined in terms of generalized, operator valued, vector fields. These are operator analogs of the generalized vector field A^a introduced in Section 2.6.1.

Fundamental for either of the above mentioned approaches, is that the analysis of the wave equation on the Kerr spacetime is based on the hidden symmetry manifested in the existence of the Carter constant, or the conserved quantity $\boldsymbol{q}$, and its corresponding symmetry operator Q.

2.7.2. Symmetry Operators for the Maxwell Field

There are two spin-1 equations (left and right) depending on the helicity of the spinor. These are

$$(\mathscr{C}^\dagger_{2,0}\phi)_{AA'} = 0 \quad \text{(left)}, \quad \text{and} \quad (\mathscr{C}_{0,2}\varphi)_{AA'} = 0 \quad \text{(right)}.$$

The real Maxwell equation $\nabla^a F_{ab} = 0$, $\nabla_{[a}F_{bc]} = 0$ for a real two form $F_{ab} = F_{[ab]}$ is equivalent to either the right or the left Maxwell equations. Henceforth we will always assume that ϕ_{AB} solves the left Maxwell equation.

Given a conformal Killing vector $\nu^{AA'}$, we follow [6, Equations (2) and (15)], see also [5], and define a conformally weighted Lie derivative acting on a symmetric valence $(2s, 0)$ spinor field as follows.

Definition 2.3 For $\nu^{AA'} \in \ker \mathscr{T}_{1,1}$, and $\varphi_{A_1\dots A_{2s}} \in \mathcal{S}_{2\mathfrak{s},0}$, we define

$$\hat{\mathcal{L}}_\nu \varphi_{A_1\dots A_{2\mathfrak{s}}} \equiv \nu^{BB'} \nabla_{BB'} \varphi_{A_1\dots A_{2\mathfrak{s}}} + s \varphi_{B(A_2\dots A_{2\mathfrak{s}}} \nabla_{A_1)B'} \nu^{BB'} \\ + \tfrac{1-s}{4} \varphi_{A_1\dots A_{2\mathfrak{s}}} \nabla^{CC'} \nu_{CC'}. \tag{2.94}$$

If ν^a is a conformal Killing field, then $(\mathscr{C}^\dagger_{2,0} \hat{\mathcal{L}}_\nu \varphi)_{AA'} = \hat{\mathcal{L}}_\nu (\mathscr{C}^\dagger_{2,0} \varphi)_{AA'}$. It follows that the first-order operator $\varphi \to \hat{\mathcal{L}}_\nu \varphi$ defines a symmetry operator of the first order, which is also of the first kind. For the equations of spins 0 and 1, the only first-order symmetry operators are given by conformal Killing fields. For the spin-1 equation, we may have symmetry operators of the first kind, taking left fields to left, i.e. $\ker \mathscr{C}^\dagger \mapsto \ker \mathscr{C}^\dagger$ and of the second kind, taking left fields to right, $\ker \mathscr{C}^\dagger \mapsto \ker \mathscr{C}$. Observe that symmetry operators of the first kind are linear symmetry operators in the usual sense, while symmetry operators of the second kind followed by complex conjugation gives anti-linear symmetry operators in the usual sense.

Recall that the Kerr spacetime admits a constant of motion for geodesics $\boldsymbol{q}$ which is not reducible to the conserved quantities defined in terms of Killing fields, but rather is defined in terms of a Killing tensor. Similarly, in a spacetime with Killing spinors, the geometric field equations may admit symmetry operators of order greater than one, not expressible in terms of the symmetry operators defined in terms of (conformal) Killing fields. We refer to such symmetry operators as "hidden symmetries."

In general, the existence of symmetry operators of the second order implies the existence of Killing spinors (of valence $(2,2)$ for the conformal wave equation and for Maxwell symmetry operators of the first kind for Maxwell, or $(4,0)$ for Maxwell symmetry operators of the second kind) satisfying certain auxiliary conditions. The conditions given in [9] are valid in arbitrary four-dimensional spacetimes, with no additional conditions on the curvature. As shown in [9], the existence of a valence $(2,0)$ Killing spinor is a sufficient condition for the existence of second-order symmetry operators for the spin-$\mathfrak{s}$ equations, for $\mathfrak{s} = 0, 1/2, 1$.

Remark 2.5 1. If κ_{AB} is a Killing spinor of valence $(2,0)$, then $L_{ABA'B'} = \kappa_{AB}\bar{\kappa}_{A'B'}$ and $L_{ABCD} = \kappa_{(AB}\kappa_{CD)}$ are Killing spinors of valence $(2,2)$ and $(4,0)$, respectively, satisfying the auxiliary conditions given in [9].

2. In the case of aligned matter with respect to Ψ_{ABCD}, any valence $(4,0)$ Killing spinor L_{ABCD} factorizes, i.e. $L_{ABCD} = \kappa_{(AB}\kappa_{CD)}$ for some Killing spinor κ_{AB} of valence $(2,0)$ [9, Theorem 8]. An example of a spacetime with aligned matter which admits a valence $(2,2)$ Killing spinor that does not factorize is given in [9, §6.3], see also [83].

Proposition 2.1 ([9]) 1. *The general symmetry operator of the first kind for the Maxwell field, of order at most two, is of the form*

$$\chi_{AB} = Q\phi_{AB} + (\mathscr{C}_{1,1}A)_{AB}, \tag{2.95}$$

where ϕ_{AB} is a Maxwell field, and $A_{AA'}$ is a linear concomitant[21] *of the first order, such that $A_{AA'} \in \ker \mathscr{C}^{\dagger}_{1,1}$ and $Q \in \ker \mathscr{T}_{0,0}$, i.e. is locally constant.*

2. *The general symmetry operator of the second kind for the Maxwell field is of the form*

$$\omega_{A'B'} = (\mathscr{C}^{\dagger}_{1,1}B)_{A'B'}, \tag{2.96}$$

where $B_{AA'}$ is a first-order linear concomitant of ϕ_{AB} such that $B_{AA'} \in \ker \mathscr{C}_{1,1}$.

Remark 2.6 The operators $\mathscr{C}^{\dagger}_{1,1}$ and $\mathscr{C}_{1,1}$ are the adjoints of the left and right Maxwell operators $\mathscr{C}^{\dagger}_{2,0}$ and $\mathscr{C}_{0,2}$. The conserved currents for the Maxwell field can be characterized in terms of solutions of the adjoint Maxwell equations

$$(\mathscr{C}^{\dagger}_{1,1}A)_{A'B'} = 0 \tag{2.97a}$$

$$(\mathscr{C}_{1,1}B)_{AB} = 0. \tag{2.97b}$$

Definition 2.4 Given a spinor $\kappa_{AB} \in \mathcal{S}_{2,0}$ we define the operators $\mathscr{E}_{2,0} : \mathcal{S}_{2,0} \to \mathcal{S}_{2,0}$ and $\bar{\mathscr{E}}_{0,2} : \mathcal{S}_{0,2} \to \mathcal{S}_{0,2}$ by

$$(\mathscr{E}_{2,0}\varphi)_{AB} = -2\kappa_{(A}{}^{C}\varphi_{B)C}, \tag{2.98a}$$

$$(\bar{\mathscr{E}}_{0,2}\phi)_{A'B'} = -2\bar{\kappa}_{(A'}{}^{C'}\phi_{B')C'}. \tag{2.98b}$$

Let κ_i be the Newman–Penrose scalars for κ_{AB}. If κ_{AB} is of algebraic type $\{1,1\}$ then $\kappa_0 = \kappa_2 = 0$, in which case $\kappa_{AB} = -2\kappa_1 o_{(A}\iota_{B)}$. A direct calculation gives the following result.

Lemma 2.1 *Let $\kappa_{AB} \in \mathcal{S}_{2,0}$ and assume that κ_{AB} is of algebraic type $\{1,1\}$. Then the operators $\mathscr{E}_{2,0}, \bar{\mathscr{E}}_{2,0}$ remove the middle component and rescale the extreme components as*

$$(\mathscr{E}_{2,0}\varphi)_0 = -2\kappa_1\varphi_0, \quad (\mathscr{E}_{2,0}\varphi)_1 = 0, \quad (\mathscr{E}_{2,0}\varphi)_2 = 2\kappa_1\varphi_2, \tag{2.99a}$$

$$(\bar{\mathscr{E}}_{0,2}\phi)_{0'} = -2\bar{\kappa}_{1'}\phi_{0'}, \quad (\bar{\mathscr{E}}_{0,2}\phi)_{1'} = 0, \quad (\bar{\mathscr{E}}_{0,2}\phi)_{2'} = 2\bar{\kappa}_{1'}\phi_{2'}. \tag{2.99b}$$

Remark 2.7 If κ_{AB} is a Killing spinor in a Petrov type D spacetime, then κ_{AB} is of algebraic type $\{1,1\}$.

[21] A concomitant is a covariant, local partial differential operator.

Definition 2.5 Define the first-order 1-form linear concomitants $A_{AA'}, B_{AA'}$ by

$$A_{AA'}[\kappa_{AB}, \phi_{AB}] = -\tfrac{1}{3}(\mathscr{E}_{2,0}\phi)_{AB}(\mathscr{C}_{0,2}\bar{\kappa})^{B}{}_{A'} + \bar{\kappa}_{A'B'}(\mathscr{C}^{\dagger}_{2,0}\mathscr{E}_{2,0}\phi)_{A}{}^{B'}, \quad (2.100a)$$

$$A_{AA'}[\nu_{AA'}, \phi_{AB}] = \nu_{BA'}\phi_{A}{}^{B}, \quad (2.100b)$$

$$B_{AA'}[\kappa_{AB}, \phi_{AB}] = \kappa_{AB}(\mathscr{C}^{\dagger}_{2,0}\mathscr{E}_{2,0}\phi)^{B}{}_{A'} + \tfrac{1}{3}(\mathscr{E}_{2,0}\phi)_{AB}(\mathscr{C}^{\dagger}_{2,0}\kappa)^{B}{}_{A'}. \quad (2.100c)$$

When there is no room for confusion, we suppress the arguments and write simply $A_{AA'}, B_{AA'}$. The following result shows that $A_{AA'}, B_{AA'}$ solves the adjoint Maxwell equations, provided ϕ_{AB} solves the Maxwell equation.

Lemma 2.2 ([9, §7]) *Assume that κ_{AB} is a Killing spinor of valence $(2,0)$, that $\nu_{AA'}$ is a conformal Killing field, and that ϕ_{AB} is a Maxwell field. Then, with $A_{AA'}, B_{AA'}$ given by* (2.100) *it holds that $A_{AA'}[\kappa_{AB}, \phi_{AB}]$ and $A_{AA'}[\nu_{AA'}, \phi_{AB}]$ satisfy $(\mathscr{C}^{\dagger}_{1,1}A)_{A'B'} = 0$, and $B_{AA'}[\kappa_{AB}, \phi_{AB}]$ satisfies $(\mathscr{C}_{1,1}B)_{AB} = 0$.*

Remark 2.8 Proposition 2.1 together with Lemma 2.2 show that the existence of a valence $(2,0)$ Killing spinor implies that there are non-trivial second-order symmetry operators of the first and second kind for the Maxwell equation.

2.8. Conservation Laws for the Teukolsky System

Recall that the operators $\mathscr{C}$ and $\mathscr{C}^{\dagger}$ are adjoints, and hence their composition yields a wave operator. We have the identities (valid in a general spacetime)

$$\square\varphi_{AB} + 8\Lambda\varphi_{AB} - 2\Psi_{ABCD}\varphi^{CD} = -2(\mathscr{C}_{1,1}\mathscr{C}^{\dagger}_{2,0}\varphi)_{AB}, \quad (2.101a)$$

$$\square\varphi_{ABCD} - 6\Psi_{(AB}{}^{FH}\varphi_{CD)FH} = -2(\mathscr{C}_{3,1}\mathscr{C}^{\dagger}_{4,0}\varphi)_{ABCD}. \quad (2.101b)$$

Here φ_{AB} and φ_{ABCD} are elements of $\mathcal{S}_{2,0}$ and $\mathcal{S}_{4,0}$, respectively. This means that the Maxwell equation $(\mathscr{C}^{\dagger}_{2,0}\phi)_{AA'} = 0$ in a vacuum spacetime implies the wave equation

$$\square\phi_{AB} - 2\Psi_{ABCD}\phi^{CD} = 0. \quad (2.102)$$

Similarly, in a vacuum spacetime, the Bianchi system $(\mathscr{C}^{\dagger}_{4,0}\Psi)_{A'ABC} = 0$ holds for the Weyl spinor, and we arrive at the Penrose wave equation

$$\square\Psi_{ABCD} - 6\Psi_{(AB}{}^{FH}\Psi_{CD)FH} = 0. \quad (2.103)$$

Restricting to a vacuum type D spacetime, and projecting the Maxwell wave equation (2.102) and the linearized Penrose wave equation (2.103) on the principal spin dyad, one obtains wave equations for the extreme Maxwell scalars ϕ_0, ϕ_2 and the extreme linearized Weyl scalars $\dot{\Psi}_0, \dot{\Psi}_4$.

Letting $\psi^{(s)}$ denote $\phi_0, \Psi_2^{-2/3}\phi_2$ for s $=$ $1, -1$, respectively, and $\dot{\Psi}_0, \Psi_2^{-4/3}\dot{\Psi}_4$ for s $= 2, -2$, respectively, one finds that these fields satisfy the system

$$[\boxdot_{2s} - 4s^2\Psi_2]\psi^{(s)} = 0, \tag{2.104}$$

see [3, §3], where, in GHP notation

$$\boxdot_p = 2(\text{þ} - p\rho - \bar{\rho})(\text{þ}' - \rho') - 2(\eth - p\tau - \bar{\tau}')(\eth' - \tau') + (3p-2)\Psi_2. \tag{2.105}$$

Equation (2.104) was first derived by Teukolsky [106, 107] for massless spin-$\mathfrak{s}$ fields and linearized gravity on Kerr, and is referred to as the Teukolsky Master Equation (TME). It was shown by Ryan [100] that the tetrad projection of the linearized Penrose wave equation yields the TME, see also Bini et al. [29, 30]. In the Kerr case, The TME admits a commuting symmetry operator, and hence allows separation of variables; the equation is valid for fields of all half-integer spins between 0 and 2.

As discussed above, the TME is a wave equation for the weighted field $\psi^{(s)}$. It is derived from the spin-$\mathfrak{s}$ field equation by applying a first-order operator and hence is valid for the extreme scalar components of the field, rescaled as explained above. It is important to emphasize that there is a loss of information in deriving the TME from the spin-$\mathfrak{s}$ equation. For example, if we consider two independent solutions of the TME with spin weights s $= \pm 1$, these will not in general be components of a single Maxwell field. If indeed this is the case, the Teukolsky–Starobinsky identities (TSI) (also referred to as Teukolsky–Press relations), see [67] and references therein, hold.

The TME admits commuting symmetry operators S_s, R_s, so that

$$\boxdot_{2s} - 4s^2\Psi_2 = R_s + S_s$$

with $[R_s, S_s] = 0$, and such that, as in the case for the wave equation discussed in Section 2.7.1, the operators R_s, S_s involve derivatives with respect to r and θ, respectively, in addition to derivatives in the symmetry directions t, ϕ. This shows that one may make a consistent separated ansatz

$$\psi^{(s)}(t, r, \theta, \phi) = e^{-i\omega t}e^{im\phi}R^{(s)}(r)S^{(s)}(\theta)$$

where $R^{(s)}$ solves the radial TME

$$(R_s + \lambda_{s,\omega,\ell,m})R^{(s)} = 0$$

where $\lambda_{s,\omega,\ell,m}$ is an eigenvalue for the angular Teukolsky equation $S_s S^{(s)} = \lambda S^{(s)}$, which is the equation for a spin-weighted spheroidal harmonic.

Although the TSI are usually discussed in terms of separated forms of $\psi^{(s)}$, we are here interested in the TSI as differential relations between the scalars of extreme spin weights. From this point of view, the TSI expresses the fact that the Debye potential construction starting from the different Maxwell scalars for a given Maxwell field ϕ_{AB} yields scalars of *the same* Maxwell field. The equations for the Maxwell scalars in terms of Debye potentials can be found in Newman–Penrose notation in [42]. These expressions correspond to the components of a symmetry operator of the second kind. See [2, §5.4.2] for further discussion, where also the GHP version of the formulas can be found. An analogous situation obtains for the case of linearized gravity, see [77]. In this case, the TSI are of the fourth order. Thus, for a Maxwell field, or a solution of the linearized Einstein equations on a Kerr, or more generally a vacuum type D background, the pair of Newman–Penrose scalars of extreme spin weights for the field satisfy a system of differential equations consisting of both the TME and the TSI.

Although the TME is derived from an equation governed by a variational principle, it has been argued by Anco, see the discussion in [95], that the Teukolsky system admits no *real* variational principle, due to the fact that the operator $\boxminus_p$ defined by the above fails to be formally self-adjoint. Hence, the issue of real conserved currents for the Teukolsky system, which appear to be necessary for estimates of the solutions, appears to be open. However, as we shall demonstrate here, if we consider the *combined* TME and TSI in the spin-1 or Maxwell case, as a system of equations for both of the extreme Maxwell scalars ϕ_0, ϕ_2, this system does admit both a conserved current and a conserved stress-energy-like tensor.

2.8.1. A New Conserved Tensor for Maxwell

Let $\phi_{AB} \to \chi_{AB}$ be the second order symmetry operator of the first kind given by (2.95) with $Q = 0$ and $A_{AA'}$ given by (2.100a), and let $\xi_{AA'}$ be given by (2.54). Then the current

$$\Psi_{AA'} = \tfrac{1}{2}\xi^{BB'}\chi_{AB}\bar{\phi}_{A'B'} + \tfrac{1}{2}\xi^{BB'}\phi_{AB}\bar{\chi}_{A'B'},$$

is conserved. In fact, as discussed in [10, section 6], $-\Psi_{AA'}$ is equivalent to a current $V_{ab}\xi^b$ defined in terms of a symmetric tensor V_{ab} which we shall now introduce. Let

$$\eta_{AA'} \equiv (\mathscr{C}^{\dagger}_{2,0}\mathscr{E}_{2,0}\phi)_{AA'}. \tag{2.106}$$

where $(\mathscr{E}_{2,0}\phi)_{AB}$ is given by (2.98a) and define the symmetric tensor V_{ab} by

$$V_{ABA'B'} \equiv \tfrac{1}{2}\eta_{AB'}\bar{\eta}_{A'B} + \tfrac{1}{2}\eta_{BA'}\bar{\eta}_{B'A} + \tfrac{1}{3}(\mathscr{E}_{2,0}\phi)_{AB}(\hat{\mathcal{L}}_{\bar{\xi}}\bar{\phi})_{A'B'} \\ + \tfrac{1}{3}(\bar{\mathscr{E}}_{2,0}\bar{\phi})_{A'B'}(\hat{\mathcal{L}}_{\xi}\phi)_{AB}. \tag{2.107}$$

Then, as we shall now show, V_{ab} is itself conserved,

$$\nabla^a V_{ab} = 0,$$

and hence may be viewed as a higher-order stress-energy tensor for the Maxwell field. The tensor V_{ab} has several important properties. First of all, if $\mathcal{M}$ is of Petrov type D, it depends only on the extreme Maxwell scalars ϕ_0, ϕ_2, and hence cancels the static Coulomb Maxwell field (2.45) on Kerr which has only the middle scalar non-vanishing. This can be proved using Lemma 2.1, cf. [10, Corollary 6.2]. Further, the tensor

$$U_{AA'BB'} = \tfrac{1}{2}\eta_{AB'}\bar{\eta}_{A'B} + \tfrac{1}{2}\eta_{BA'}\bar{\eta}_{B'A}$$

is a superenergy tensor for the 1-form field $\eta_{AA'}$, and hence satisfies the dominant energy condition, cf. [27, 103]. Note that the notion of a superenergy tensor extends to spinors of arbitrary valence. Similarly to the wave equation stress energy, V_{ab} has a non-vanishing trace, $V^a{}_a = U^a{}_a = -\bar{\eta}^a\eta_a$.

In order to analyze V_{ab}, we first collect some properties of the one-form $\eta_{AA'}$ as defined in (2.106).

Lemma 2.3 ([8, Lemma 2.4]) *Let $\kappa_{AB} \in \mathcal{KS}_{2,0}$, and assume the aligned matter condition holds with respect to κ_{AB}. Let $\xi_{AA'}$ be given by* (2.54). *Further, let ϕ_{AB} be a Maxwell field, and let $\eta_{AA'}$ be given by* (2.106). *Then we have*

$$(\mathscr{D}_{1,1}\eta) = 0, \tag{2.108a}$$

$$(\mathscr{C}_{1,1}\eta)_{AB} = \tfrac{2}{3}(\hat{\mathcal{L}}_\xi\phi)_{AB}, \tag{2.108b}$$

$$(\mathscr{C}^\dagger_{1,1}\eta)_{A'B'} = 0, \tag{2.108c}$$

$$\eta_{AA'}\xi^{AA'} = \kappa^{AB}(\hat{\mathcal{L}}_\xi\phi)_{AB}. \tag{2.108d}$$

The following lemma gives a general condition, without assumptions on the spacetime geometry, for a tensor constructed along the lines of V_{ab} to be conserved. The proof is a straightforward computation.

Lemma 2.4 ([8, Lemma 3.1]) *Assume that $\varphi_{AB} \in \mathcal{S}_{2,0}$ satisfies the system*

$$(\mathscr{C}^\dagger_{1,1}\mathscr{C}^\dagger_{2,0}\varphi)_{A'B'} = 0, \tag{2.109a}$$

$$(\mathscr{C}_{1,1}\mathscr{C}^\dagger_{2,0}\varphi)_{AB} = \varpi_{AB}, \tag{2.109b}$$

for some $\varpi_{AB} \in \mathcal{S}_{2,0}$. Let

$$\varsigma_{AA'} = (\mathscr{C}^\dagger_{2,0}\varphi)_{AA'}, \tag{2.110}$$

and define the symmetric tensor $X_{ABA'B'}$ by

$$X_{ABA'B'} = \tfrac{1}{2}\varsigma_{AB'}\bar{\varsigma}_{A'B} + \tfrac{1}{2}\varsigma_{BA'}\bar{\varsigma}_{B'A} + \tfrac{1}{2}\bar{\varpi}_{A'B'}\varphi_{AB} + \tfrac{1}{2}\varpi_{AB}\bar{\varphi}_{A'B'}. \tag{2.111}$$

Then

$$\nabla^{BB'}X_{ABA'B'} = 0. \tag{2.112}$$

We now have the following result, which follows directly from Lemma 2.4 and the identities for $\eta_{AA'}$ given in Lemma 2.3 together with the above remarks.

Theorem 2.4 ([8, Theorem 1.1]) *Assume that* $(\mathcal{M}, g_{ab})$ *admits a valence* $(2,0)$ *Killing spinor* κ_{AB} *and assume that the aligned matter condition holds with respect to* κ_{AB}. *Let* ϕ_{AB} *be a solution of the Maxwell equation. Then the tensor* $V_{ABA'B'}$ *given by* (2.107) *is conserved, i.e.*

$$\nabla^{AA'} V_{ABA'B'} = 0$$

If in addition $(\mathcal{M}, g_{ab})$ *is of Petrov type* D, *then* V_{ab} *depends only on the extreme components of* ϕ_{AB}.

The properties of V_{ab} indicate that V_{ab}, rather than the Maxwell stress-energy T_{ab}, may be used in proving dispersive estimates for the Maxwell field. In Section 2.9 we shall outline the proof of a Morawetz estimate for the Maxwell field on the Schwarzschild background, making use of a related approach.

2.8.2. Teukolsky Equation and Conservation Laws

We end this section by pointing out the relation between the fact that V_{ab} is conserved and the TME and TSI which follow from the Maxwell equation in a Petrov type D spacetime.

A computation shows that the identities

$$(\mathscr{C}^{\dagger}_{1,1}\mathscr{C}^{\dagger}_{2,0}\mathscr{E}_{2,0}\phi)_{A'B'} = 0 \tag{2.113a}$$

$$(\mathscr{E}_{2,0}\mathscr{C}_{1,1}\mathscr{C}^{\dagger}_{2,0}\mathscr{E}_{2,0}\phi)_{AB} = \tfrac{2}{3}(\hat{\mathcal{L}}_{\xi}\mathscr{E}_{2,0}\phi)_{AB} \tag{2.113b}$$

follow from the Maxwell equations, cf. [8, Eq. (3.5)]. We see that this system is equivalent to (2.109a), (2.109b), with $\varphi_{AB} = \mathscr{E}_{2,0}\phi$ and $\varpi_{AB} = \frac{2}{3}(\hat{\mathcal{L}}_{\xi}\phi)_{AB}$. This shows that the fact that V_{ab} is conserved is a direct consequence of (2.113), which in fact are the covariant versions of the TME and TSI. In order to make this clear for the case of the TME, we project (2.113b)on the dyad. A calculation shows that

$$0 = -\text{þ}\,\text{þ}'\varphi_0 + \rho\,\text{þ}'\varphi_0 + \bar{\rho}\,\text{þ}'\varphi_0 + \eth\,\eth'\varphi_0 - \tau\,\eth'\varphi_0 - \overline{\tau'}\,\eth'\varphi_0, \tag{2.114a}$$

$$0 = -\rho'\,\text{þ}\varphi_2 - \overline{\rho'}\,\text{þ}\varphi_2 + \text{þ}'\,\text{þ}\varphi_2 + \bar{\tau}\,\eth\varphi_2 + \tau'\,\eth\varphi_2 - \eth'\,\eth\varphi_2 \tag{2.114b}$$

where $\varphi_0 = -2\kappa_1\phi_0$ and $\varphi_2 = 2\kappa_1\phi_2$. We see from this that (2.113b) is equivalent to the scalar form of TME for Maxwell given in (2.104) above. Further, one can show along the same lines that (2.113a) is equivalent to the TSI for Maxwell given in scalar form in [2, §5.4.2], cf. [8, §3.1].

2.9. A Morawetz Estimate for the Maxwell Field on Schwarzschild

In this section, we shall outline the proof of the Morawetz estimate for the Maxwell field on the Schwarzschild spacetime given recently in [11].

Assume that κ_{AB} is a valence $(2,0)$ Killing spinor, such that $\kappa_{CD}\kappa^{CD} \neq 0$. In the case of the Schwarzschild spacetime, κ_{AB} is given by (2.65). Define the Killing fields $\xi^{AA'}$, $\eta^{AA'}$ in terms of κ_{AB} by (2.54) and (2.106), respectively.

Now let ϕ_{AB} be a solution to the source-free Maxwell equation $(\mathscr{C}^\dagger_{2,0}\phi)_{AA'} = 0$ and define

$$U_{AA'} = -\tfrac{1}{2}\nabla_{AA'}\log(-\kappa_{CD}\kappa^{CD}), \tag{2.115a}$$

$$\Upsilon = \kappa^{AB}\phi_{AB}, \tag{2.115b}$$

$$\Theta_{AB} = (\mathscr{E}_{2,0}\phi)_{AB}, \tag{2.115c}$$

$$\beta_{AA'} = \eta_{AA'} + U^{B}{}_{A'}\Theta_{AB}. \tag{2.115d}$$

In the Schwarzschild case, we have

$$\xi^{AA'} = (\partial_t)^{AA'}, \qquad U_{AA'} = -r^{-1}\nabla_{AA'}r. \tag{2.116}$$

Analogously to Lemma 2.3, we have

Lemma 2.5 ([11, Lemma 8])

$$\beta_{AA'} = -U_{AA'}\Upsilon + (\mathscr{T}_{0,0}\Upsilon)_{AA'}, \tag{2.117a}$$

$$(\mathscr{D}_{1,1}\beta) = -U^{AA'}\beta_{AA'}, \tag{2.117b}$$

$$(\mathscr{C}_{1,1}\beta)_{AB} = U_{(A}{}^{A'}\beta_{B)A'}, \tag{2.117c}$$

$$(\mathscr{C}^\dagger_{1,1}\beta)_{A'B'} = U^{A}{}_{(A'}\beta_{|A|B')}. \tag{2.117d}$$

The superenergy tensors for $\beta_{AA'}$ and Θ_{AB} are given by

$$\mathbf{H}_{ABA'B'} = \tfrac{1}{2}\beta_{AB'}\overline{\beta}_{A'B} + \tfrac{1}{2}\beta_{BA'}\overline{\beta}_{B'A}, \tag{2.118a}$$

$$\mathbf{W}_{ABA'B'} = \Theta_{AB}\overline{\Theta}_{A'B'}. \tag{2.118b}$$

Choosing the principal tetrad in Schwarzschild given by specializing (2.63) to $a = 0$ gives in a standard manner an orthonormal frame,

$$\widehat{T}^{AA'} \equiv \tfrac{1}{\sqrt{2}}(o^{A}\bar{o}^{A'} + \iota^{A}\bar{\iota}^{A'}), \qquad \widehat{X}^{AA'} \equiv \tfrac{1}{\sqrt{2}}(\bar{o}^{A'}\iota^{A} + o^{A}\bar{\iota}^{A'}),$$

$$\widehat{Y}^{AA'} \equiv \tfrac{i}{\sqrt{2}}(-\bar{o}^{A'}\iota^{A} + o^{A}\bar{\iota}^{A'}), \qquad \widehat{Z}^{AA'} \equiv \tfrac{1}{\sqrt{2}}(o^{A}\bar{o}^{A'} - \iota^{A}\bar{\iota}^{A'}).$$

The tensor $\mathbf{H}_{ABA'B'}$, which agrees up to lower order terms with the conserved tensor $V_{ABA'B'}$ introduced in Section 2.8.1, is not itself conserved, it yields a conserved energy current.

Lemma 2.6 ([11, Lemma 11]) *For the Schwarzschild spacetime we have*

$$\nabla^{BB'}\mathbf{H}_{ABA'B'} = -U_{AA'}\beta^{BB'}\bar{\beta}_{B'B}, \tag{2.119a}$$

$$\xi^{AA'}\nabla^{BB'}\mathbf{H}_{ABA'B'} = 0. \tag{2.119b}$$

In particular, $\xi^{BB'}\mathbf{H}_{ABA'B'}$ *is a future causal conserved current.*

This result makes use of the fact that the Schwarzschild spacetime is non-rotating. For the Kerr spacetime with non-vanishing angular momentum, the 1-form $U_{AA'}$ fails to be real and the current $\mathbf{H}_{ab}\xi^b$ is not conserved.

For a vector field A^a and a scalar q, define the Morawetz current $\mathbf{P}_a$ by

$$\begin{aligned}\mathbf{P}_{AA'} = {} & \mathbf{H}_{ABA'B'}A^{BB'} - \tfrac{1}{2}q\bar{\beta}_{A'}{}^{B}\Theta_{AB} - \tfrac{1}{2}q\beta_{A}{}^{B'}\overline{\Theta}_{A'B'} \\ & + \tfrac{1}{2}\Theta_{A}{}^{B}\overline{\Theta}_{A'}{}^{B'}(\mathscr{T}_{0,0}q)_{BB'}.\end{aligned} \tag{2.120}$$

For any spacelike hypersurface Σ, we define the energy integrals

$$E_{\xi}(\Sigma) = \int_{\Sigma}\mathbf{H}_{ab}\xi^{b}N^{a}\mathrm{d}\mu_{\Sigma}, \tag{2.121}$$

$$E_{\xi+A,q}(\Sigma) = \int_{\Sigma}\left(\mathbf{H}_{ab}\xi^{b} + \mathbf{P}_{a}\right)N^{a}\mathrm{d}\mu_{\Sigma}. \tag{2.122}$$

In view of Lemma 2.6, $E_{\xi}(\Sigma)$ is non-negative and conserved.

We shall make the following explicit choices of the A^a and q,

$$A^{a} = \frac{(r-3M)(r-2M)}{2r^{2}}(\partial_{r})^{a}, \tag{2.123a}$$

$$q = \frac{9M^{2}(r-2M)(2r-3M)}{4r^{5}}. \tag{2.123b}$$

2.9.1. Positive Energy

Before proving the integrated decay estimate, we shall verify that the energy (2.122) is non-negative and uniformly equivalent to the energy (2.121).

From the properties of spin-weighted spherical harmonics, one derives the inequalities

$$\int_{S_r}|\varphi_0|^2 d\mu_{S_r} \le r^2\int_{S_r}|\eth'\varphi_0|^2\mathrm{d}\mu_{S_r}, \tag{2.124a}$$

$$\int_{S_r}|\varphi_2|^2 d\mu_{S_r} \le r^2\int_{S_r}|\eth\varphi_2|^2\mathrm{d}\mu_{S_r} \tag{2.124b}$$

for the extreme scalars φ_0, φ_2 of a smooth symmetric spinor field φ_{AB}, cf. [11, Lemma 6]. Here S_r is a sphere with constant t, r in the Schwarzschild spacetime.

By making use of the Cauchy–Schwarz inequality, and the Hardy type inequalities (2.124), we get

Theorem 2.5 ([11, Theorem 13, Corollary 14]) *Let $A^{AA'}$ and q be given by* (2.123a) *and* (2.123b).

1. *For any constant $|c_1| \leq 10/9$ and any spherically symmetric slice Σ with future pointing timelike normal $N^{AA'}$ such that $N^{AA'}N_{AA'} = 1$ we have a positive energy*

$$\int_\Sigma N^{AA'}(\mathbf{H}_{ABA'B'}\xi^{BB'} + c_1 \mathbf{P}_{AA'})d\mu_{\Sigma_i} \geq 0. \tag{2.125}$$

2. *For any spherically symmetric slice Σ with future pointing timelike normal $N^{AA'}$ such that $N^{AA'}N_{AA'} = 1$ the energies $E_\xi(\Sigma)$ and $E_{\xi+A,q}(\Sigma)$ are uniformly equivalent,*

$$\tfrac{1}{10}E_\xi(\Sigma) \leq E_{\xi+A,q}(\Sigma) \leq \tfrac{19}{10}E_\xi(\Sigma). \tag{2.126}$$

In particular, we find that using Theorem 2.5, we can dominate the integral of the bulk term for the Morawetz current over a spacetime domain bounded by Cauchy surfaces Σ_1, Σ_2, in terms of the energies $E_\xi(\Sigma_1)$, $E_\xi(\Sigma_2)$. This is the essential step in the proof of an integrated energy decay (or Morawetz) estimate.

We shall apply (2.124) to Θ_{AB}. We have

$$|\beta_{\widehat{Z}}|^2 + |\beta_{\bar{\widehat{Z}}}|^2 = |\eth\, \Theta_2|^2 + |\eth'\, \Theta_0|^2, \tag{2.127a}$$

and

$$\mathbf{W}_{\widehat{T}\widehat{T}} = \widehat{T}^{AA'}\widehat{T}^{BB'}\Theta_{AB}\overline{\Theta}_{A'B'} = \tfrac{1}{2}|\Theta_0|^2 + \tfrac{1}{2}|\Theta_2|^2. \tag{2.127b}$$

Equations (2.127a), (2.127b) and the inequalities (2.124) combine to give the estimate

$$\int_{S_r} \mathbf{W}_{\widehat{T}\widehat{T}} d\mu_{S_r} \leq \frac{r^2}{2}\int_{S_r} |\beta_{\widehat{T}}|^2 + |\beta_{\widehat{Z}}|^2 \mathrm{d}\mu_{S_r}, \tag{2.128}$$

cf. [11, Lemma 15]. From the form (2.120) of the Morawetz current $\mathbf{P}_a$, the definition of $\beta_{AA'}$ and the properties of β_a given in Lemma 2.5 we get

$$\begin{aligned} -(\mathscr{D}_{1,1}\mathbf{P}) = {} & -\beta^{AA'}\bar{\beta}^{B'B}(\mathscr{T}_{1,1}A)_{ABA'B'} \\ & + \beta^{AA'}\bar{\beta}_{A'A}\big(\tfrac{1}{4}(\mathscr{D}_{1,1}A) + A^{BB'}U_{BB'} - q\big) \\ & + \Theta_{AB}\overline{\Theta}_{A'B'}\big(U^{AA'}(\mathscr{T}_{0,0}q)^{BB'} - \tfrac{1}{2}(\mathscr{T}_{1,1}\mathscr{T}_{0,0}q)^{ABA'B'}\big). \end{aligned} \tag{2.129}$$

With the explicit choices (2.123a) and (2.123b) for the Morawetz vector field A^a and the scalar q, respectively, the above estimates now yield

$$\int_\Omega -(\mathscr{D}_{1,1}\mathbf{P})d\mu_\Omega \geq \int_\Omega \frac{1}{8}|\beta_{AA'}|^2_{1,\text{deg}} + \frac{M}{100r^4}|\Theta_{AB}|^2_2 d\mu_\Omega, \tag{2.130}$$

for any spherically symmetric spacetime region Ω of the Schwarzschild spacetime.

We now make use of Gauss's formula to evaluate the left-hand side of (2.130). Theorem 2.5 and the estimates just proved then yield the following energy bound and Morawetz estimate for the Maxwell field on the Schwarzschild spacetime.

Theorem 2.6 ([11, Theorem 2]) *Let Σ_1 and Σ_2 be spherically symmetric spacelike hypersurfaces in the exterior region of the Schwarzschild spacetime such that Σ_2 lies in the future of Σ_1 and $\Sigma_2 \cup -\Sigma_1$ is the oriented boundary of a spacetime region Ω.*

If ϕ_{AB} is a solution of the Maxwell equations on the Schwarzschild exterior, and Θ_{AB} and $\beta_{AA'}$ are defined by equations (2.115d) *to* (2.115c)*, then*

$$E_\xi(\Sigma_2) = E_\xi(\Sigma_1), \tag{2.131}$$

$$\int_\Omega |\beta_{AA'}|^2_{1,deg} + \frac{2M}{25r^4}|\Theta_{AB}|^2_2 d\mu_\Omega \leq \frac{72}{5}E_\xi(\Sigma_1), \tag{2.132}$$

where $E_\xi(\Sigma_i)$ is the energy associated with ξ^a, evaluated on Σ_i, and $|\beta_{AA'}|_{1,deg}$ and $|\Theta_{AB}|_2$ are, respectively, the degenerate norm of $\beta_{AA'}$ and the norm of Θ_{AB} defined by

$$\begin{aligned}
|\beta_{AA'}|^2_{1,deg} &= \frac{(r-3M)^2}{r^3}\left(|\beta_{\widehat{X}}|^2 + |\beta_{\widehat{Y}}|^2\right) + \frac{M(r-2M)}{r^3}|\beta_{\widehat{Z}}|^2 \\
&\quad + \frac{M(r-3M)^2(r-2M)}{r^5}|\beta_{\widehat{T}}|^2, \\
|\Theta_{AB}|^2_2 &= \frac{(r-2M)}{r}\mathbf{W}_{\widehat{T}\widehat{T}}.
\end{aligned}$$

□

Acknowledgments

We are grateful to Steffen Aksteiner, Siyuan Ma, Marc Mars, and Claudio Paganini for helpful remarks. Pieter Blue and Thomas Bäckdahl were supported by EPSRC grant EP/J011142/1. Lars Andersson thanks the Institut Henri Poincaré, Paris, for hospitality and support during part of the work on this chapter.

References

[1] B. P. Abbott *et al.* Observation of gravitational waves from a binary black hole merger. *Physical Review Letters*, 116:061102, Feb. 2016.

[2] S. Aksteiner. Geometry and analysis in black hole spacetimes. PhD thesis, Gottfried Wilhelm Leibniz Universität Hannover, 2014. http://d-nb.info/1057896721/34.

[3] S. Aksteiner and L. Andersson. Linearized gravity and gauge conditions. *Classical and Quantum Gravity*, 28(6):065001, Mar. 2011. arXiv.org:1009.5647.

[4] S. Aksteiner and L. Andersson. Charges for linearized gravity. *Classical and Quantum Gravity*, 30(15):155016, Aug. 2013. arXiv.org:1301.2674.

[5] S. C. Anco and J. Pohjanpelto. Conserved currents of massless fields of spin $s \geq \frac{1}{2}$. *Proceedings of the Royal Society of London. Series A, Mathematical and Physical Sciences*, 459(2033):1215–1239, 2003.

[6] S. C. Anco and J. Pohjanpelto. Symmetries and currents of massless neutrino fields, electromagnetic and graviton fields. In *Symmetry in physics*, vol. 34 of *CRM Proceedings and Lecture Notes*, pp. 1–12. *American Mathematical Society*, Providence, RI, 2004. math-ph/0306072.

[7] L. Andersson. The global existence problem in general relativity. In Piotr T. Chrusciel and Helmut Friedrich, editors, *The Einstein equations and the large scale behavior of gravitational fields*, pp. 71–120. Birkhäuser, Basel, 2004.

[8] L. Andersson, T. Bäckdahl, and P. Blue. A new tensorial conservation law for Maxwell fields on the Kerr background. Dec. 2014. arXiv.org:1412.2960, *to appear in the Journal of Differential Geometry.*

[9] L. Andersson, T. Bäckdahl, and P. Blue. Second order symmetry operators. *Classical and Quantum Gravity*, 31(13):135015, Jul. 2014. arXiv.org:1402.6252.

[10] L. Andersson, T. Bäckdahl, and P. Blue. Spin geometry and conservation laws in the Kerr spacetime. In L. Bieri and S.-T. Yau, editors, *One hundred years of general relativity*, pp. 183–226. International Press, Boston, 2015. arXiv.org:1504.02069.

[11] L. Andersson, T. Bäckdahl, and P. Blue. Decay of solutions to the Maxwell equation on the Schwarzschild background. *Classical and Quantum Gravity*, 33(8):085010, 2016. arXiv.org:1501.04641.

[12] L. Andersson and P. Blue. Hidden symmetries and decay for the wave equation on the Kerr spacetime. *Annals of Mathematics (2)*, 182(3):787–853, 2015.

[13] L. Andersson and P. Blue. Uniform energy bound and asymptotics for the Maxwell field on a slowly rotating Kerr black hole exterior. *Journal of Hyperbolic Differential Equations*, 12(04):689–743, Oct. 2015.

[14] L. Andersson, P. Blue, and J.-P. Nicolas. A decay estimate for a wave equation with trapping and a complex potential. *International Mathematics Research Notices*, (3):548–561, 2013.

[15] L. Andersson, M. Mars, J. Metzger, and W. Simon. The time evolution of marginally trapped surfaces. *Classical and Quantum Gravity*, 26(8):085018, Apr. 2009.

[16] L. Andersson, M. Mars, and W. Simon. Local Existence of Dynamical and Trapping Horizons. *Physical Review Letters*, 95(11):111102, Sep. 2005.

[17] L. Andersson and J. Metzger. The area of horizons and the trapped region. *Communications in Mathematical Physics*, 290(3):941–972, 2009.

[18] L. Andersson and V. Moncrief. Einstein spaces as attractors for the Einstein flow. *Journal of Differential Geometry*, 89(1):1–47, 2011.

[19] T. Bäckdahl and J. A. Valiente Kroon. Geometric invariant measuring the deviation from Kerr Data. *Physical Review Letters*, 104(23):231102, Jun. 2010.

[20] T. Bäckdahl and J. A. Valiente Kroon. On the construction of a geometric invariant measuring the deviation from Kerr data. *Annals Henri Poincaré*, 11(7):1225–1271, 2010.

[21] T. Bäckdahl and J. A. Valiente Kroon. The "non-Kerrness" of domains of outer communication of black holes and exteriors of stars. *Royal Society of London. Series A, Mathematical and Physical Sciences*, 467:1701–1718, Jun. 2011. arXiv.org:1010.2421.

[22] T. Bäckdahl and J. A. Valiente Kroon. Constructing "non-Kerrness" on compact domains. *Journal of Mathematical Physics*, 53(4):042503, Apr. 2012.

[23] R. A. Bartnik and P. T. Chruściel. Boundary value problems for Dirac-type equations. *Journal fur die reine und angewandte Mathematik*, 579:13–73, 2005. arXiv.org:math/0307278.

[24] R. Beig and P. T. Chruściel. Killing vectors in asymptotically flat space–times. I. Asymptotically translational Killing vectors and the rigid positive energy theorem. *Journal of Mathematical Physics*, 37:1939–1961, Apr. 1996.

[25] R. Beig and N. Ó Murchadha. The Poincaré group as the symmetry group of canonical general relativity. *Annals of Physics*, 174(2):463–498, 1987.

[26] I. Bengtsson and J. M. M. Senovilla. Region with trapped surfaces in spherical symmetry, its core, and their boundaries. *Physical Review*, 83(4):044012, Feb. 2011.

[27] G. Bergqvist. Positivity of General Superenergy Tensors. *Communications in Mathematical Physics*, 207:467–479, 1999.

[28] A. N. Bernal and M. Sánchez. Further results on the smoothability of Cauchy hypersurfaces and Cauchy time functions. *Letters in Mathematical Physics*, 77:183–197, Aug. 2006.

[29] D. Bini, C. Cherubini, R. T. Jantzen, and R. Ruffini. Teukolsky master equation–de Rham wave equation for the gravitational and electromagnetic fields in vacuum. *Progress of Theoretical Physics*, 107:967–992, May 2002. arXiv.org:gr-qc/0203069.

[30] D. Bini, C. Cherubini, R. T. Jantzen, and R. Ruffini. De Rham wave equation for tensor valued p-forms. *International Journal of Modern Physics D*, 12:1363–1384, 2003.

[31] H. L. Bray. Proof of the Riemannian Penrose inequality using the positive mass theorem. *Journal of Differential Geometry*, 59(2):177–267, 2001.

[32] H. Buchdahl. On the compatibility of relativistic wave equations for particles of higher spin in the presence of a gravitational field. *Il Nuovo Cimento*, 10(1):96–103, 1958.

[33] B. Carter. Killing tensor quantum numbers and conserved currents in curved space. *Physical Review*, 16:3395–3414, Dec. 1977.

[34] S. Chandrasekhar. *The mathematical theory of black holes*, vol. 69 of *International Series of Monographs on Physics*. The Clarendon Press, Oxford University Press, New York, 1992. Revised reprint of the 1983 original, Oxford Science Publications.

[35] Y. Choquet-Bruhat and R. Geroch. Global aspects of the Cauchy problem in general relativity. *Communications in Mathematical Physics*, 14:329–335, Dec. 1969.

[36] D. Christodoulou. *The Formation of Black Holes in General Relativity*. European Mathematical Society, Zurich, 2009.

[37] D. Christodoulou and S. Klainerman. *The global nonlinear stability of the Minkowski space*, vol. 41 of *Princeton Mathematical Series*. Princeton University Press, Princeton, NJ, 1993.

[38] D. Christodoulou and N. O'Murchadha. The boost problem in general relativity. *Communications in Mathematical Physics*, 80:271–300, Jun. 1981.

[39] P. T. Chruściel, G. J. Galloway, and D. Solis. Topological censorship for Kaluza–Klein space–times. *Annales Henri Poincaré*, 10:893–912, Jul. 2009.

[40] P. T. Chruściel, J. Jezierski, and J. Kijowski. *Hamiltonian field theory in the radiating regime*, vol. 70 of *Lecture Notes in Physics. Monographs*. Springer-Verlag, Berlin, 2002.

[41] C. J. S. Clarke. A condition for forming trapped surfaces. *Classical Quantum Gravity*, 5(7):1029–1032, 1988.

[42] J. M. Cohen and L. S. Kegeles. Electromagnetic fields in curved spaces: A constructive procedure. *Physical Review*, 10:1070–1084, Aug. 1974.

[43] C. D. Collinson and P. N. Smith. A comment on the symmetries of Kerr black holes. *Communications in Mathematical Physics*, 56:277–279, Oct. 1977.

[44] R. M. Corless, G. H. Gonnet, D. E. G. Hare, D. J. Jeffrey, and D. E. Knuth. On the LambertW function. *Advances in Computational Mathematics*, 5(1):329–359, 1996.

[45] M. Dafermos and I. Rodnianski. A proof of the uniform boundedness of solutions to the wave equation on slowly rotating Kerr backgrounds. *Inventiones Mathematicae*, 185(3):467–559, 2011.

[46] M. Dafermos, I. Rodnianski, and Y. Shlapentokh-Rothman. Decay for solutions of the wave equation on Kerr exterior spacetimes III: The full subextremal case $|a| < M$. Feb. 2014. arXiv.org:1402.7034.

[47] D. M. Eardley. Gravitational collapse of vacuum gravitational field configurations. *Journal of Mathematical Physics*, 36(6):3004–3011, 1995.

[48] W. E. East, F. M. Ramazanoğlu, and F. Pretorius. Black hole superradiance in dynamical spacetime. *Physical Review*, 89(6):061503, Mar. 2014.

[49] S. B. Edgar, A. G.-P. Gómez-Lobo, and J. M. Martín-García. Petrov D vacuum spaces revisited: Identities and invariant classification. *Classical and Quantum Gravity*, 26(10):105022, May 2009. arXiv.org:0812.1232.

[50] A. Einstein. Kosmologische Betrachtungen zur allgemeinen Relativitätstheorie. *Sitzungsberichte der Königlich Preußischen Akademie der Wissenschaften (Berlin), Seite 142–152.*, 1917.

[51] J. J. Ferrando and J. A. Sáez. An intrinsic characterization of the Kerr metric. *Classical and Quantum Gravity*, 26(7):075013, Apr. 2009. arXiv.org:0812.3310.

[52] F. Finster, N. Kamran, J. Smoller, and S.-T. Yau. A Rigorous Treatment of Energy Extraction from a Rotating Black Hole. *Communications in Mathematical Physics*, 287:829–847, May 2009.

[53] A. E. Fischer and V. Moncrief. The Einstein flow, the σ-constant and the geometrization of 3-manifolds. *Classical Quantum Gravity*, 16(11):L79–L87, 1999.

[54] Y. Fourès-Bruhat. Théorème d'existence pour certains systèmes d'équations aux dérivées partielles non linéaires. *Acta Mathematica*, 88(1): 141–225, 1952.

[55] H. Friedrich. Hyperbolic reductions for Einstein's equations. *Classical and Quantum Gravity*, 13:1451–1469, Jun. 1996.

[56] V. Frolov and A. Zelnikov. *Introduction to Black Hole Physics*. Oxford University Press Oxford, 2011.

[57] J. R. Gair, C. Li, and I. Mandel. Observable properties of orbits in exact bumpy spacetimes. *Physical Review*, 77(2):024035, Jan. 2008. arXiv.org:0708.0628.

[58] R. Geroch. Spinor structure of space–times in general relativity. II. *Journal of Mathematical Physics*, 11(1):343–348, 1970.

[59] R. Geroch, A. Held, and R. Penrose. A space–time calculus based on pairs of null directions. *Journal of Mathematical Physics*, 14:874–881, Jul. 1973.

[60] J. Haláček and T. Ledvinka. The analytic conformal compactification of the Schwarzschild spacetime. *Classical and Quantum Gravity*, 31(1):015007, Jan. 2014.

[61] Q. Han and M. Khuri. The Conformal Flow of Metrics and the General Penrose Inequality. Aug. 2014. arXiv.org:1409.0067.

[62] S. W. Hawking and G. F. R. Ellis. *The large scale structure of space–time*. Cambridge University Press, London and New York, 1973. Cambridge Monographs on Mathematical Physics, No. 1.

[63] L.-H. Huang. On the center of mass of isolated systems with general asymptotics. *Classical and Quantum Gravity*, 26(1):015012, Jan. 2009.

[64] L. P. Hughston and P. Sommers. The symmetries of Kerr black holes. *Communications in Mathematical Physics*, 33:129–133, Jun. 1973.

[65] G. Huisken and T. Ilmanen. The inverse mean curvature flow and the Riemannian Penrose inequality. *Journal of Differential Geometry*, 59(3):353–437, 2001.

[66] V. Iyer and R. M. Wald. Some properties of the Noether charge and a proposal for dynamical black hole entropy. *Physical Review*, 50:846–864, Jul. 1994.

[67] E. G. Kalnins, W. Miller, Jr., and G. C. Williams. Teukolsky-Starobinsky identities for arbitrary spin. *Journal of Mathematical Physics*, 30:2925–2929, Dec. 1989.

[68] R. P. Kerr. Gravitational field of a spinning mass as an example of algebraically special metrics. *Physical Review Letters*, 11:237–238, Sep. 1963.

[69] W. Kinnersley. Type D Vacuum Metrics. *Journal of Mathematical Physics*, 10:1195–1203, July 1969.

[70] S. Klainerman, J. Luk, and I. Rodnianski. A fully anisotropic mechanism for formation of trapped surfaces in vacuum. Feb. 2013. arXiv.org:1302.5951.

[71] S. Klainerman and I. Rodnianski. Rough solutions of the Einstein-vacuum equations. *Annals of Mathematics (2)*, 161(3):1143–1193, 2005.

[72] S. Klainerman, I. Rodnianski, and J. Szeftel. Overview of the proof of the Bounded L^2 Curvature Conjecture. Apr. 2012. arXiv.org:1204.1772.

[73] S. Klainerman, I. Rodnianski, and J. Szeftel. The bounded L^2 curvature conjecture. *Inventiones Mathematicae*, 202(1):91–216, 2015.

[74] J.-P. Lasota, E. Gourgoulhon, M. Abramowicz, A. Tchekhovskoy, and R. Narayan. Extracting black-hole rotational energy: The generalized Penrose process. *Physical Review*, 89(2):024041, Jan. 2014.

[75] A. László and I. Rácz. Superradiance or total reflection? *Springer Proceedings in Physics*, 157:119–127, 2014. arXiv.org:1212.4847.

[76] H. Lindblad and I. Rodnianski. Global existence for the Einstein vacuum equations in wave coordinates. *Communications in Mathematical Physics*, 256:43–110, May 2005. arXiv.org:math/0312479.

[77] C. O. Lousto and B. F. Whiting. Reconstruction of black hole metric perturbations from the Weyl curvature. *Physical Review*, 66(2):024026, July 2002.

[78] J. Luk. Weak null singularities in general relativity. Nov. 2013. arXiv.org:1311.4970.

[79] G. Lukes-Gerakopoulos, T. A. Apostolatos, and G. Contopoulos. Observable signature of a background deviating from the Kerr metric. *Physical Review*, 81(12):124005, Jun. 2010. arXiv.org:1003.3120.

[80] M. Mars. Uniqueness properties of the Kerr metric. *Classical and Quantum Gravity*, 17:3353–3373, Aug. 2000.

[81] M. Mars, T.-T. Paetz, J. M. M. Senovilla, and W. Simon. Characterization of (asymptotically) Kerr–de Sitter-like spacetimes at null infinity. Mar. 2016. arXiv.org:1603.05839.

[82] B. Michel. Geometric invariance of mass-like asymptotic invariants. *Journal of Mathematical Physics*, 52(5):052504–052504, May 2011.

[83] J.-P. Michel, F. Radoux, and J. Šilhan. Second Order Symmetries of the Conformal Laplacian. *SIGMA*, 10:016, Feb. 2014.

[84] C. W. Misner, K. S. Thorne, and J. A. Wheeler. *Gravitation*. W. H. Freeman and Co., San Francisco, CA, 1973.

[85] C. S. Morawetz. Time decay for the nonlinear Klein–Gordon equations. *Proceedings of the Royal Society of London. Series A, Mathematical and Physical Sciences*, 306:291–296, 1968.

[86] P. Mösta, L. Andersson, J. Metzger, B. Szilágyi, and J. Winicour. The merger of small and large black holes. *Classical and Quantum Gravity*, 32(23):235003, Dec. 2015.

[87] E. T. Newman, E. Couch, K. Chinnapared, A. Exton, A. Prakash, and R. Torrence. Metric of a rotating, charged mass. *Journal of Mathematical Physics*, 6:918–919, Jun. 1965.

[88] E. T. Newman and A. I. Janis. Note on the Kerr spinning-particle metric. *Journal of Mathematical Physics*, 6:915–917, Jun. 1965.

[89] I. D. Novikov and V. P. Frolov. *Physics of black holes*. (Fizika chernykh dyr, Moscow, Izdatel'stvo Nauka, 1986) Dordrecht, Netherlands, Kluwer Academic Publishers, 1989. Translation. Previously cited in issue 19, p. 3128, Accession no. A87-44677., 1989.

[90] B. O'Neill. *The geometry of Kerr black holes*. A K Peters Ltd., Wellesley, MA, 1995.

[91] J. R. Oppenheimer and H. Snyder. On continued gravitational contraction. *Physical Review*, 56:455–459, Sep. 1939.

[92] D. Parlongue. Geometric uniqueness for non-vacuum Einstein equations and applications. Sep. 2011. arXiv.org:1109.0644.

[93] R. Penrose. Zero rest-mass fields including gravitation: Asymptotic behaviour. *Proceedings of the Royal Society of London. Series A, Mathematical and Physical Sciences*, 284:159–203, 1965.

[94] R. Penrose and W. Rindler. *Spinors and space–time I & II*. Cambridge Monographs on Mathematical Physics. Cambridge University Press, Cambridge, 1986.

[95] Z. Perjés and Á. Lukács. Canonical quantization and black hole perturbations. In J. Lukierski and D. Sorokin, editors, *Fundamental interactions and twistor-Like methods*, volume 767 of *American Institute of Physics Conference Series*, pp. 306–315, Apr. 2005.

[96] E. Poisson. *A relativist's toolkit*. Cambridge University Press, Cambridge, 2004. The mathematics of black-hole mechanics.

[97] T. Regge and C. Teitelboim. Role of surface integrals in the Hamiltonian formulation of general relativity. *Annals of Physics*, 88:286–318, Nov. 1974.

[98] H. Ringström. Cosmic censorship for Gowdy spacetimes. *Living Reviews in Relativity*, 13(2), 2010.

[99] H. Ringström. Origins and development of the Cauchy problem in general relativity. *Classical and Quantum Gravity*, 32(12):124003, Jun. 2015.

[100] M. P. Ryan. Teukolsky equation and Penrose wave equation. *Physical Review*, 10:1736–1740, Sep. 1974.

[101] J. Sbierski. On the existence of a maximal Cauchy development for the Einstein equations: A dezornification. *Annales Henri Poincaré*, 17(2):301–329, 2016.

[102] R. Schoen and S. T. Yau. The existence of a black hole due to condensation of matter. *Communications in Mathematical Physics*, 90(4):575–579, 1983.

[103] J. M. M. Senovilla. Super-energy tensors. *Classical and Quantum Gravity*, 17:2799–2841, Jul. 2000.

[104] C. D. Sogge. *Lectures on non-linear wave equations*. International Press, Boston, MA, second edition, 2008.

[105] E. Teo. Spherical photon orbits around a Kerr black hole. *General Relativity Gravitation*, 35(11):1909–1926, 2003.

[106] S. A. Teukolsky. Rotating black holes: separable wave equations for gravitational and electromagnetic perturbations. *Physical Review Letters*, 29:1114–1118, Oct. 1972.

[107] S. A. Teukolsky. Perturbations of a rotating black hole. I. Fundamental equations for gravitational, electromagnetic, and neutrino-field perturbations. *Astrophysical Journal*, 185:635–648, Oct. 1973.

[108] S. A. Teukolsky. The Kerr metric. *Classical and Quantum Gravity*, 32(12):124006, Jun. 2015.

[109] R. M. Wald. *General relativity*. University of Chicago Press, Chicago, IL, 1984.

[110] R. M. Wald. Gravitational collapse and cosmic censorship. Oct. 1997. arXiv.org:gr-qc/9710068.

[111] R. M. Wald and V. Iyer. Trapped surfaces in the Schwarzschild geometry and cosmic censorship. *Physical Review*, 44:R3719–R3722, Dec. 1991.

[112] M. Walker and R. Penrose. On quadratic first integrals of the geodesic equations for type {2,2} spacetimes. *Communications in Mathematical Physics*, 18:265–274, Dec. 1970.

[113] J. Winicour and L. Tamburino. Lorentz-covariant gravitational energy-momentum linkages. *Physical Review Letters*, 15:601–605, Oct. 1965.

[114] W. W.-Y. Wong. A comment on the construction of the maximal globally hyperbolic Cauchy development. *Journal of Mathematical Physics*, 54(11):113511–113511, Nov. 2013.

[115] S. T. Yau. Geometry of three manifolds and existence of black hole due to boundary effect. *Advances in Theoretical and Mathematical Physics*, 5(4):755–767, 2001.

[116] R. L. Znajek. Black hole electrodynamics and the Carter tetrad. *Monthly Notices of the Royal Astronomical Society*, 179:457–472, May 1977.

[117] P. Hintz and A. Vasy. *The global non-linear stability of the Kerr–de Sitter family of black holes. Preprint*, arXiv:1606.04014, 2016.

Albert Einstein Institute, Am Mühlenberg 1, D-14476 Potsdam, Germany
E-mail address: laan@aei.mpg.de

Mathematical Sciences, Chalmers University of Technology and University of Gothenburg, SE-412 96 Gothenburg, Sweden and The School of Mathematics, University of Edinburgh, James Clerk Maxwell Building, Peter Guthrie Tait Road, Edinburgh EH9 3FD, UK
E-mail address: thobac@chalmers.se

The School of Mathematics and the Maxwell Institute, University of Edinburgh, James Clerk Maxwell Building, Peter Guthrie Tait Road, Edinburgh EH9 3FD,UK
E-mail address: P.Blue@ed.ac.uk

3

An Introduction to Conformal Geometry and Tractor Calculus, with a view to Applications in General Relativity

Sean N. Curry and A. Rod Gover

Abstract. The chapter consists of expanded notes for the course of eight one-hour lectures given by the second author at the 2014 summer school entitled Asymptotic Analysis in General Relativity held in Grenoble by the Institut Fourier. The first four lectures deal with conformal geometry and the conformal tractor calculus, taking as primary motivation the search for conformally invariant tensors and differential operators. The final four lectures apply the conformal tractor calculus to the study of conformally compactified geometries, motivated by the conformal treatment of infinity in general relativity.

3.1. Introduction

Definition 3.1 A conformal n-manifold ($n \geq 3$) is the structure $(M, \boldsymbol{c})$ where

- M is an n-manifold,
- $\boldsymbol{c}$ is a conformal equivalence class of signature (p,q) metrics, that is $g, \widehat{g} \in \boldsymbol{c} \overset{\text{def.}}{\Longleftrightarrow} \widehat{g} = \Omega^2 g$ and $C^\infty(M) \ni \Omega > 0$.

To any pseudo-Riemannian n-manifold (M,g) with $n \geq 3$ there is an associated conformal manifold $(M,[g])$ where $[g]$ is the set of all metrics $\hat{g}$ which are smooth positive multiples of the metric g. In Riemannian signature $(p,q) = (n,0)$ passing to the conformal manifold means geometrically that we are forgetting the notion of lengths (of tangent vectors and of curves) and retaining only the notion of angles (between tangent vectors and curves) and of ratios of lengths (of tangent vectors at a fixed point) associated to

2010 *Mathematics Subject Classification.* Primary 53A30, 35Q75, 53B15, 53C25; Secondary 83C05, 35Q76, 53C29.
Key words and phrases. Einstein metrics, conformal differential geometry, conformal infinity.
ARG gratefully acknowledges support from the Royal Society of New Zealand via Marsden Grant 13-UOA-018.

the metric g. In Lorentzian signature $(n-1,1)$ passing to the conformal manifold means forgetting the 'spacetime interval' (analogous to length in Riemannian signature) and retaining only the light cone structure of the Lorentzian manifold. On a conformal Lorentzian manifold one also has the notion of angles between intersecting spacelike curves and the notion of orthogonality of tangent vectors at a point, but the conformal structure itself is determined by the light cone structure, justifying the use of the word 'only' in the previous sentence.

The significance of conformal geometry for general relativity largely stems from the fact that the light cone structure determines the *causal structure* of spacetime (and, under some mild assumptions, the causal structure determines the light cone structure). On top of this we shall see in the lectures that the Einstein field equations admit a very nice interpretation in terms of conformal geometry.

In these notes we will develop the natural invariant calculus on conformal manifolds, the (conformal) tractor calculus, and apply it to the study of conformal invariants and conformally compactified geometries. The course is divided into two parts consisting of four lectures each. The first four lectures deal with conformal geometry and the conformal tractor calculus, taking as primary motivation the problem of constructing conformally invariant tensors and differential operators. The tools developed for this problem however allow us to tackle much more than our original problem of invariants and invariant operators. In the final four lectures they will be applied in particular to the study of conformally compactified geometries, motivated by the conformal treatment of infinity in general relativity. Along the way we establish the connection between the conformal tractor calculus and Helmut Friedrich's conformal field equations. We also digress for one lecture, discussing conformal hypersurface geometry, in order to facilitate the study of the relationship of the geometry of conformal infinity to that of the interior. Finally we show how the tractor calculus may be applied to treat aspects of the asymptotic analysis of boundary problems on conformally compact manifolds. For completeness an appendix has been added which covers further aspects of the conformal tractor calculus as well as discussing briefly the canonical conformal Cartan bundle and connection.

The broad philosophy behind our ensuing discussion is that conformal geometry is important not only for understanding conformal manifolds, or conformally invariant aspects of pseudo-Riemannian geometry (such as conformally invariant field equations), but that it is highly profitable to think of a pseudo-Riemannian manifold as a kind of symmetry breaking (linked to holonomy reduction) of a conformal manifold whenever there are any (even remotely) conformal geometry related aspects of the problem being considered. Our discussion of the conformal tractor calculus will lead us naturally to the notions of almost Einstein and almost pseudo-Riemannian geometries, which include Einstein and pseudo-Riemannian manifolds (respectively) as well as their respective conformal compactifications (should they admit one). The general theory of Cartan holonomy reductions then enables us to put constraints on the smooth structure and the geometry of conformal infinities of Einstein manifolds, and the tractor calculus enables us to generalise partially these results to pseudo-Riemannian manifolds.

We deal exclusively with conformal manifolds of at least three dimensions in these notes. That is not to say that two-dimensional conformal manifolds cannot be fitted into the framework which we describe. However, in order to have a canonical tractor calculus on two-dimensional conformal manifolds the conformal manifold needs to be equipped with an extra structure, weaker than a Riemannian structure but stronger than a conformal structure, called a *Möbius structure*. In higher dimensions a conformal structure determines a canonical Möbius structure via the construction of the canonical conformal Cartan bundle and connection (outlined in Section A.1). In two dimensions there is no canonical Cartan bundle and connection associated with a conformal manifold, so this (Möbius) structure must be imposed as an additional assumption if we wish to work with the tractor calculus. We note that to any Riemannian 2-manifold, or to any (non-degenerate) two-dimensional submanifold of a higher dimensional conformal manifold, there is associated a canonical Möbius structure and corresponding tractor calculus.

Also left out in these notes is any discussion of conformal spin geometry. In this case there is again a canonical tractor calculus, known as *spin tractor calculus* or *local twistor calculus*, which is a refinement of the usual conformal tractor calculus in the same way that spinor calculus is a refinement of the usual tensor calculus on pseudo-Riemannian spin manifolds. The interested reader is referred to [4, 51].

3.1.1. Notation and Conventions

We may use abstract indices, or no indices, or frame indices according to convenience. However, we will make particularly heavy use of the abstract index notation. For instance if L is a linear endomorphism of a finite dimensional vector space V then we may choose to write L using abstract indices as $L^a{}_b$ (or $L^b{}_c$, or $L^{a'}{}_{b'}$, it makes no difference as the indices are just place holders meant to indicate tensor type, and contractions). In this case we would write a vector $v \in V$ as v^a (or v^b, or $v^{a'}$, ...) and we would write the action of L on

v as $L^a{}_b v^b$ (repeated indices denote tensor contraction so $L^a{}_b v^b$ simply means $L(v)$). Similarly if $w \in V^*$ then using abstract indices we would write w as w_a (or w_b, or $w_{a'}$, ...) and $w(v)$ as $w_a v^a$, whereas the outer product $v \otimes w \in \mathrm{End}(V)$ would be written as $v^a w_b$ (or $v^b w_c$, or $w_b v^a$, ...). A covariant 2-tensor $T \in \otimes^2 V^*$ may be written using abstract indices as T_{ab}, the symmetric part of T is then denoted by $T_{(ab)}$ and the antisymmetric part by $T_{[ab]}$, that is

$$T_{(ab)} = \frac{1}{2}\left(T_{ab} + T_{ba}\right) \quad \text{and} \quad T_{[ab]} = \frac{1}{2}\left(T_{ab} - T_{ba}\right).$$

Note that swapping the indices a and b in T_{ab} amounts to swapping the 'slots' of the covariant 2-tensor (so that b becomes the label for the first slot and a for the second), which gives in general a different covariant 2-tensor from T_{ab} whose matrix with respect to any basis for V would be the transpose of that of T_{ab}. We can similarly define the symmetric or antisymmetric part of any covariant tensor $T_{ab\cdots e}$, and these are denoted by $T_{(ab\cdots e)}$ and $T_{[ab\cdots e]}$ respectively. We use the same bracket notation for the symmetric and antisymmetric parts of contravariant tensors. Note that we do not have to symmetrise or skew-symmetrise over all indices, for instance $T^{a[bc]}{}_d$ denotes

$$\frac{1}{2}\left(T^{abc}{}_d - T^{acb}{}_d\right).$$

The abstract index notation carries over in the obvious way to vector and tensor fields on a manifold. The virtue of using abstract index notation on manifolds is that it makes immediately apparent the type of tensorial object one is dealing with and its symmetries without having to bring in extraneous vector fields or 1-forms. We will commonly denote the tangent bundle of M by $\mathcal{E}^a$, and the cotangent bundle of M by $\mathcal{E}_a$. We then denote the bundle of covariant 2-tensors by $\mathcal{E}_{ab}$, its subbundle of symmetric 2-tensors by $\mathcal{E}_{(ab)}$, and so on. In order to avoid confusion, when working on the tangent bundle of a manifold M we will always use lower case Latin abstract indices taken from the beginning of the alphabet (a, b, etc.) whereas we will take our frame indices from a later part of the alphabet (starting from i, j, etc.).

It is common when working with abstract index notation to use the same notation $\mathcal{E}^a$ for the tangent bundle and its space of smooth sections. Here however we have used the notation $\Gamma(\mathcal{V})$ for the space of smooth sections of a vector bundle $\mathcal{V}$ consistently throughout, with the one exception that for a differential operator D taking sections of a vector bundle $\mathcal{U}$ to sections of $\mathcal{V}$ we have written

$$D : \mathcal{U} \to \mathcal{V}$$

in order to simplify the notation.

Consistent with our use of $\mathcal{E}^a$ for TM we will often denote by $\mathcal{E}$ the trivial $\mathbb{R}$-bundle over our manifold M, so that $\Gamma(\mathcal{E}) = C^\infty(M)$. When using index free notation we denote the space of vector fields on M by $\mathfrak{X}(M)$, and we use the shorthand Λ^k for $\Lambda^k T^*M$ (when the underlying manifold M is understood). Unless otherwise indicated $[\,\cdot\,,\,\cdot\,]$ is the commutator bracket acting on pairs of endomorphisms. Note that the Lie bracket arises in this way when we consider vector fields as derivations of the algebra of smooth functions. In all of the following all structures (manifolds, bundles, tensor fields, etc.) will be assumed smooth, meaning C^∞.

3.1.1.1. Coupled Connections

We assume that the reader is familiar with the notion of a linear connection on a vector bundle $\mathcal{V} \to M$ and the special case of an affine connection (being a connection on the tangent bundle of a manifold). Given a pair of vector bundles $\mathcal{V}$ and $\mathcal{V}'$ over the same manifold M and linear connections ∇ and ∇' defined on $\mathcal{V}$ and $\mathcal{V}'$ respectively, then there is a natural way to define a connection on the tensor product bundle $\mathcal{V} \otimes \mathcal{V}' \to M$. The *coupled connection* $\nabla^\otimes$ on $\mathcal{V} \otimes \mathcal{V}'$ is given on a simple section $v \otimes v'$ of $\mathcal{V} \otimes \mathcal{V}'$ by the Leibniz formula

$$\nabla^\otimes_X(v \otimes v') = (\nabla_X v) \otimes v' + v \otimes (\nabla'_X v')$$

for any $X \in \mathfrak{X}(M)$. Since $\Gamma(\mathcal{V} \otimes \mathcal{V}')$ is (locally) generated by simple sections, this formula determines the connection $\nabla^\otimes$ uniquely. In order to avoid clumsy notation we will often simply write all of our connections as ∇ when it is clear from the context which (possibly coupled) connection is being used.

3.1.1.2. Associated Bundles

We assume also that the reader is familiar with the notion of a vector bundle on a smooth manifold M as well as that of an H-principal bundle over M, where H is a Lie group. If $\pi : \mathcal{G} \to M$ is a (right) H-principal bundle and $\mathbb{V}$ is a (finite dimensional) representation of H then the *associated vector bundle* $\mathcal{G} \times_H \mathbb{V} \to M$ is the vector bundle with total space defined by

$$\mathcal{G} \times_H \mathbb{V} = \mathcal{G} \times \mathbb{V} / \sim$$

where $\sim$ is the equivalence relation

$$(u, v) \sim (u \cdot h, h^{-1} \cdot v), \quad h \in H$$

on $\mathcal{G} \times \mathbb{V}$; the projection of $\mathcal{G} \times_H \mathbb{V}$ to M is simply defined by taking $[(u, v)]$ to $\pi(u)$. For example if $\mathcal{F}$ is the linear frame bundle of a smooth n-manifold M then

$$TM = \mathcal{F} \times_{\mathrm{GL}(n)} \mathbb{R}^n \quad \text{and} \quad T^*M = \mathcal{F} \times_{\mathrm{GL}(n)} (\mathbb{R}^n)^*.$$

Similarly if H is contained in a larger Lie group G then one can extend any principal H-bundle $\mathcal{G} \to M$ to a principal G-bundle $\tilde{\mathcal{G}} \to M$ with total space

$$\tilde{\mathcal{G}} = \mathcal{G} \times_H G = \mathcal{G} \times G/\sim \quad \text{where} \quad (u,g) \sim (u \cdot h, h^{-1} \cdot g).$$

For example if (M,g) is a Riemannian n-manifold and $\mathcal{O}$ denotes its orthonormal frame bundle then

$$\mathcal{F} = \mathcal{O} \times_{\mathrm{O}(n)} \mathrm{GL}(n)$$

is the linear frame bundle of M.

3.2. Lecture 1: Riemannian Invariants and Invariant Operators

Recall that if ∇ is an affine connection then its *torsion* is the tensor field $T^\nabla \in \Gamma(TM \otimes \Lambda^2 T^*M)$ defined by

$$T^\nabla(u,v) = \nabla_u v - \nabla_v u - [u,v] \quad \text{for all} \quad u,v \in \mathfrak{X}(M).$$

It is interesting that this is a tensor; by its construction one might expect a differential operator. Importantly torsion is an *invariant* of connections. On a smooth manifold the map

$$\nabla \mapsto T^\nabla$$

taking connections to their torsion depends only the smooth structure. By its construction here it is clear that it is independent of any choice of coordinates.

For any connection ∇ on a vector bundle $\mathcal{V}$ its curvature is defined by

$$R^\nabla(u,v)W := [\nabla_u, \nabla_v]W - \nabla_{[u,v]}W \quad \text{for all} \quad u,v \in \mathfrak{X}(M), \text{ and } W \in \Gamma(\mathcal{V}),$$

and $R^\nabla \in \Gamma(\Lambda^2 T^*M \otimes \mathrm{End}(\mathcal{V}))$. Again the map taking connections to their curvatures

$$\nabla \mapsto R^\nabla$$

clearly depends only on the (smooth) vector bundle structure of $\mathcal{V}$ over the smooth manifold M. The curvature is an invariant of linear connections. In particular this applies to affine connections, that is when $\mathcal{V} = TM$.

3.2.1. Ricci Calculus and Weyl's Invariant Theory

The most familiar setting for these objects is pseudo-Riemannian geometry. In this case we obtain a beautiful local calculus that is sometimes called the Ricci calculus. Let us briefly recall how this works. Recall that a pseudo-Riemannian manifold consists of an n-manifold M equipped with a metric g of signature (p,q), that is a section $g \in \Gamma(S^2T^*M)$ such that pointwise g is non-degenerate and of signature (p,q). Then g canonically determines a distinguished affine connection called the Levi-Civita connection. This is the unique connection ∇ satisfying:

- $\nabla g = 0$ (metric compatibility) and
- $T^\nabla = 0$ (torsion freeness).

Thus on a smooth manifold we have a canonical map from each metric to its Levi-Civita connection

$$g \mapsto \nabla^g$$

and, as above, a canonical map which takes each Levi-Civita connection to its curvature $\nabla^g \mapsto R^{\nabla^g}$, called the *Riemannian curvature*. Composing these we get a canonical map that takes each metric to its curvature

$$g \longmapsto R^g,$$

and this map depends only on the smooth structure of M. So we say that R^g is an invariant of the pseudo-Riemannian manifold (M,g). How can we construct more such invariants? Or 'all' invariants, in some perhaps restricted sense?

The first, and perhaps most important, observation is that using the Levi-Civita connection and Riemannian curvature one can proliferate Riemannian invariants. To simplify the explanation let's fix a pseudo-Riemannian manifold (M,g) and use abstract index notation (when convenient). Then the metric is written g_{ab} and we write $R_{ab}{}^c{}_d$ for the Riemannian curvature. So if u, v, w are tangent vector fields then so is $R(u,v)w$ and this is written

$$u^a v^b R_{ab}{}^c{}_d w^d.$$

From curvature we can form the *Ricci* and *Scalar* curvatures, respectively:

$$\mathrm{Ric}_{ab} := R_{ca}{}^c{}_b \quad \text{and} \quad \mathrm{Sc} := g^{ab}\,\mathrm{Ric}_{ab},$$

and these are invariants of (M,g). As also are:

$$\nabla_b R_{cdef},\ \nabla_a \nabla_b R_{cdef},$$

which are *tensor valued invariants* and

$$\mathrm{Ric}^{ab}\,\mathrm{Ric}_{ab}\,\mathrm{Sc} = |\,\mathrm{Ric}\,|^2\,\mathrm{Sc},\ \ R_{abcd}R^{abcd} = |R|^2,\ \ (\nabla_a R_{bcde})\nabla^a R^{bcde} = |\nabla R|^2,$$

which are some *scalar valued invariants*. Here we have used the metric (and its inverse) to raise and lower indices, and contractions are indicated by repeated indices.

Since this is a practical and efficient way to construct invariants, it would be useful to know: Do *all local Riemannian invariants* arise in this way? That is from partial or complete contractions of expressions made using g, R and its covariant derivatives $\nabla \cdots \nabla R$ (and the metric volume form vol^g, if M is oriented). We shall term invariants constructed this way *Weyl invariants*.

Before answering this, one first needs to be careful about what is meant by a local invariant. For example, the following is a reasonable definition for scalar invariants.

Definition 3.2 A *scalar Riemannian invariant* P is a function which assigns to each pseudo-Riemannian n-manifold (M, g) a function $P(g)$ such that:

(i) $P(g)$ is natural, in the sense that for any diffeomorphism $\phi : M \to M$ we have $P(\phi^* g) = \phi^* P(g)$.
(ii) P is given by a universal polynomial expression of the nature that, given a local coordinate system (x^i) on (M, g), $P(g)$ is given by a polynomial in the variables $g_{mn}, \partial_{i_1} g_{mn}, \cdots, \partial_{i_1}\partial_{i_2}\cdots\partial_{i_k} g_{mn}, (\det g)^{-1}$, for some positive integer k.

Then with this definition, and a corresponding definition for tensor-valued invariants, it is true that all local invariants arise as *Weyl invariants*, and the result goes by the name of *Weyl's classical invariant theory*, see for example [1, 53]. Given this result, in the following when we mention pseudo-Riemannian invariants we will mean Weyl invariants.

3.2.2. Invariant Operators, and Analysis

In a similar way we can use the Ricci calculus to construct invariant differential operators on pseudo-Riemannian manifolds. For example the (Bochner) Laplacian is given by the formula

$$\nabla^a \nabla_a = \Delta : \mathcal{E} \longrightarrow \mathcal{E},$$

in terms of the Levi-Civita connection ∇. There are also obvious ways to make operators with curvature in coefficients, for example

$$R_a{}^c{}_b{}^d \nabla_c \nabla_d : \mathcal{E} \longrightarrow \mathcal{E}_{(ab)}.$$

With suitable restrictions imposed, in analogy with the case of invariants, one can make the statement that all local natural invariant differential operators arise in this way. It is beyond our current scope to make this precise; suffice to say that when we discuss invariant differential operators on pseudo-Riemannian manifolds we will again mean operators constructed in this way.

Remark 3.1 If a manifold has a spin structure, then essentially the above is still true but there is a further ingredient involved, namely the Clifford product. This allows the construction of important operators such as the Dirac operator.

The main point here is that the 'Ricci calculus' provides an effective and geometrically transparent route to the construction of invariants and invariant operators.

Invariants and invariant operators are the basic objects underlying the first steps (and often significantly more than just the first steps) of treating problems in general relativity and, more generally, in:

- The global analysis of manifolds;
- The study and application of geometric PDE;
- Riemannian spectral theory;
- Physics and mathematical physics.

Furthermore from a purely theoretical point of view, we cannot claim to understand a geometry if we do not have a good theory of local invariants and invariant operators.

3.3. Lecture 2: Conformal Transformations and Conformal Covariance

A good theory of conformal geometry should provide some hope of treating the following closely related problems.

Problem 1 Describe a practical way to generate/construct (possibly all) local natural invariants of a conformal structure.

Problem 2 Describe a practical way to generate/construct (possibly all) natural linear differential operators that are canonical and well-defined on (i.e. are invariants of) a conformal structure.

We have not attempted to be precise in these statements, since here they are mainly for the purpose of motivation. Let us first approach these naïvely.

3.3.1. Conformal Transformations

Recall that for any metric g we can associate its Levi-Civita connection ∇. Let $e_i = \frac{\partial}{\partial x^i} = \partial_i$ be a local coordinate frame and E^i its dual. Locally, any connection is determined by how it acts on a frame field. For the Levi-Civita connection the resulting *connection coefficients*

$$\Gamma^i{}_{jk} := E^i(\nabla_k e_j), \quad \text{where} \quad \nabla_k := \nabla_{e_k},$$

are often called the *Christoffel symbols*, and are given by the Koszul formula:

$$\Gamma^i{}_{jk} := \frac{1}{2} g^{il} \left(g_{lj,k} + g_{lk,j} - g_{jk,l} \right)$$

where $g_{ij} = g(e_i, e_j)$ and $g_{lj,k} = \partial_k g_{lj}$.

Using this formula for the Christoffel symbols we can easily compute the transformation formula for ∇ under a *conformal transformation* $g \mapsto \widehat{g} = \Omega^2 g$. Let $\Upsilon_a := \Omega^{-1} \nabla_a \Omega$, $v^a \in \Gamma(\mathcal{E}^a)$ and $\omega_b \in \Gamma(\mathcal{E}_b)$. Then we have:

$$\nabla^{\widehat{g}}_a v^b = \nabla_a v^b + \Upsilon_a v^b - \Upsilon^b v_a + \Upsilon^c v_c \delta^b_a, \tag{3.1}$$

$$\nabla^{\widehat{g}}_a \omega_b = \nabla_a \omega_b - \Upsilon_a \omega_b - \Upsilon_b \omega_a + \Upsilon^c \omega_c g_{ab}. \tag{3.2}$$

For $\omega_b \in \Gamma(\mathcal{E}_b)$ we have from (3.2) that

$$\omega_b \mapsto \nabla_a \omega_b - \nabla_b \omega_a$$

is conformally invariant. But this is just the exterior derivative $\omega \mapsto d\omega$, its conformal invariance is better seen from the fact that it is defined on a smooth manifold without further structure: for $u, v \in \mathfrak{X}(M)$

$$d\omega(u, v) = u\omega(v) - v\omega(u) - \omega([u, v]).$$

Inspecting (3.2) it is evident that this is the only first-order conformally invariant linear differential operator on T^*M that takes values in an irreducible bundle.

The Levi-Civita connection acts also on other tensor bundles. We can use the formulae (3.1) and (3.2) along with the product rule to compute the conformal transformation of the result. For example for a simple covariant 2-tensor

$$u_b \otimes w_c \quad \text{we have} \quad \widehat{\nabla}_a (u_b \otimes w_c) = (\widehat{\nabla}_a u_b) \otimes w_c + u_b \otimes (\widehat{\nabla}_a w_c).$$

Thus we can compute $\widehat{\nabla}_a (u_b \otimes w_c)$ by using (3.2) for each term on the right hand side. But locally any covariant 2-tensor is a linear combination of simple 2-tensors and so we conclude that for a covariant 2-tensor F_{bc}

$$\widehat{\nabla}_a F_{bc} = \nabla_a F_{bc} - 2\Upsilon_a F_{bc} - \Upsilon_b F_{ac} - \Upsilon_c F_{ba} + \Upsilon^d F_{dc} g_{ab} + \Upsilon^d F_{bd} g_{ac}. \tag{3.3}$$

By the obvious extension of this idea one quickly calculates the formula for the conformal transformation for the Levi-Civita covariant derivative of an (r, s)-tensor.

From the formula (3.3) we see that the completely skew part $\nabla_{[a} F_{bc]}$ is conformally invariant. In the case F is skew, in that $F_{bc} = -F_{cb}$, this recovers that dF is conformally invariant. A more interesting observation arises with the divergence $\nabla^b F_{bc}$. We have

$$\widehat{\nabla}^b F_{bc} = \widehat{g}^{ab} \widehat{\nabla}_a F_{bc},$$

and $\widehat{g}^{ab} = \Omega^{-2} g^{ab}$. Thus we obtain

$$\begin{aligned}\widehat{\nabla}^b F_{bc} &= \Omega^{-2} g^{ab} (\nabla_a F_{bc} - 2\Upsilon_a F_{bc} - \Upsilon_b F_{ac} - \Upsilon_c F_{ba} + \Upsilon^d F_{dc} g_{ab} + \Upsilon^d F_{bd} g_{ac}) \\ &= \Omega^{-2} \big(\nabla^b F_{bc} + (n-3)\Upsilon^d F_{dc} + \Upsilon^d F_{cd} - \Upsilon_c F_b{}^b\big).\end{aligned}$$

In particular then, if F is skew then $F_b{}^b = 0$ and we have simply

$$\widehat{\nabla}^b F_{bc} = \Omega^{-2} \big(\nabla^b F_{bc} + (n-4)\Upsilon^d F_{dc}\big). \tag{3.4}$$

So we see that something special happens in dimension 4. Combining with our earlier observation we have the following result.

Proposition 3.1 *In dimension 4 the differential operators*

$$\mathrm{Div} : \Lambda^2 \to \Lambda^1 \quad \textit{and} \quad \mathrm{Max} : \Lambda^1 \to \Lambda^1$$

given by

$$F_{bc} \mapsto \nabla^b F_{bc} \quad \textit{and} \quad u_c \mapsto \nabla^b \nabla_{[b} u_{c]}$$

respectively, are conformally covariant, in that

$$\widehat{\nabla}^b F_{bc} = \Omega^{-2} \nabla^b F_{bc} \quad \textit{and} \quad \widehat{\nabla}^b \widehat{\nabla}_{[b} u_{c]} = \Omega^{-2} \nabla^b \nabla_{[b} u_{c]}. \tag{3.5}$$

The non-zero powers of Ω (precisely Ω^{-2}) appearing in (3.5) mean that these objects are only *covariant* rather than invariant. Conformal covariance is still a strong symmetry property however, as we shall see. Before we discuss that in more detail note that, for the equations, these factors of Ω make little difference:

$$\nabla^b F_{bc} = 0 \quad \Leftrightarrow \quad \widehat{\nabla}^b F_{bc} = 0$$

and

$$\nabla^b \nabla_{[b} u_{c]} = 0 \quad \Leftrightarrow \quad \widehat{\nabla}^b \widehat{\nabla}_{[b} u_{c]} = 0.$$

In this sense these equations are conformally invariant.

Remark 3.2 In fact these equations are rather important. If we add the condition that F is closed then on Lorentzian signature 4-manifolds the system

$$\mathrm{d}F = 0 \quad \text{and} \quad \mathrm{Div}(F) = 0$$

is the field strength formulation of the source-free *Maxwell equations* of electromagnetism. The locally equivalent equations $\mathrm{Div}(du) = 0$ give the potential formulation of the (source-free) Maxwell equations. The conformal invariance of these has been important in physics. ∎

Returning to our search for conformally covariant operators and equations, our preliminary investigation suggests that such things might be rather rare. From (3.3) we see that the divergence of a 2-form is not conformally covariant except in dimension 4. In fact, in contrast to what this might suggest, there is a rich theory of conformally covariant operators. However there are some subtleties involved. Before we come to this it will be useful to examine how conformal rescaling affects the curvature.

3.3.2. Conformal Rescaling and Curvature

Using (3.1), (3.2) and the observations following these we can compute, for example, the conformal transformation formulae for the Riemannian curvature and its covariant derivatives and so forth. At the very lowest orders this provides a tractable approach to finding conformal invariants.

Let us fix a metric g. With respect to metric traces, we can decompose the curvature tensor of g into a trace-free part and a trace part in the following way:

$$R_{abcd} = \underbrace{W_{abcd}}_{\text{trace-free}} + \underbrace{2g_{c[a}P_{b]d} + 2g_{d[b}P_{a]c}}_{\text{trace part}} .$$

Here P_{ab}, so defined, is called the Schouten tensor, while the tensor $W_{ab}{}^{c}{}_{d}$ is called the Weyl tensor. In dimensions $n \geq 3$ we have $\mathrm{Ric}_{ab} = (n-2)P_{ab} + Jg_{ab}$, where $J := g^{ab}P_{ab}$. So the Schouten tensor P_{ab} is a trace modification of the Ricci tensor.

Exercise 1 Prove using (3.1) that under a conformal transformation $g \mapsto \widehat{g} = \Omega^2 g$, as above, the Weyl and Schouten tensors transform as follows:

$$W^{\widehat{g}}{}_{ab}{}^{c}{}_{d} = W^{g}{}_{ab}{}^{c}{}_{d}$$

and

$$P^{\widehat{g}}{}_{ab} = P_{ab} - \nabla_a \Upsilon_b + \Upsilon_a \Upsilon_b - \frac{1}{2} g_{ab} \Upsilon_c \Upsilon^c. \tag{3.6}$$

Thus the Weyl curvature is a conformal invariant, while objects such as $|W|^2 := W_{abcd}W^{abcd}$ may be called *conformal covariants* because under the conformal change they pick up a power of the conformal factor $|\widehat{W}|^2 = \Omega^{-4}|W|^2$ (where for simplicity we are hatting the symbol for the object rather than the metric). We will see shortly that such objects correspond to invariants.

Here we are defining *conformal invariants* to be Riemannian invariants that have the additional property of being unchanged under conformal transformation.

Exercise 2 Consider computing the conformal transformation of the derivatives of the curvature: ∇R, $\nabla\nabla R$, and so on then possibly using this and an undetermined coefficient approach to finding conformal invariants or conformal covariants. This rapidly gets intractable.

3.3.3. Conformally Invariant Linear Differential Operators

We may try the corresponding approach for constructing further conformally invariant linear differential operators:

$$D^g : \mathcal{U} \to \mathcal{V} \tag{3.7}$$

such that $D^{\widehat{g}} = D^g$. Namely, first consider the possible Riemannian invariant linear differential operators between the bundles concerned and satisfying some order constraint. Next compute their transformation under a conformal change. Finally seek a linear combination of these that forms a conformally invariant operator. For our purposes this also defines what we mean by conformally invariant linear differential operator. For many applications we require that the domain and target bundles $\mathcal{U}$ and $\mathcal{V}$ are irreducible.

It turns out that irreducible tensor (or even spinor) bundles are not sufficient to deal with conformal operators. Let us see a first glimpse of this by recalling the construction of one of the most well known conformally invariant differential operators.

3.3.3.1. The Conformal Wave Operator

For analysis on pseudo-Riemannian manifolds (M, g) the Laplacian is an extremely important operator. The *Laplacian* Δ on functions, which is also called the *Laplace–Beltrami operator*, is given by

$$\Delta := \nabla^a \nabla_a : \mathcal{E} \to \mathcal{E},$$

where ∇ is the Levi-Civita connection for g.

Let us see how this behaves under conformal rescaling. For a function f, $\nabla_a f$ is simply the exterior derivative of f and so is conformally invariant. So to compute the Laplacian for $\widehat{g} = \Omega^2 g$ we need only use (3.2) with $u := df$:

$$\begin{aligned}
\Delta^{\widehat{g}} f &= \widehat{\nabla}^a \nabla_a f \\
&= \widehat{g}^{ab} \widehat{\nabla}_b \nabla_a f \\
&= \Omega^{-2} g^{ab} \left(\nabla_b \nabla_a f - \Upsilon_b \nabla_a f - \Upsilon_a \nabla_b f + g_{ab} \Upsilon^c \nabla_c f \right).
\end{aligned}$$

So

$$\Delta^{\hat{g}} f = \Omega^{-2} \left(\Delta f + (n-2) \Upsilon^c \nabla_c f \right). \tag{3.8}$$

By inspecting this formula we learn two things. Let us summarise with a proposition.

Proposition 3.2 *The Laplacian on functions is conformally covariant in dimension 2, but not in other dimensions.*

This is reminiscent of our observations surrounding the expression (3.4) and the Maxwell system (cf. Proposition 3.1).

We need to modify the Laplacian to have any hope of obtaining an invariant operator in other dimensions. A key idea will be to introduce a curvature into the formula for a new Laplacian. However, inspecting the formulae for curvature transformation in Section 3.3.2 it is easily seen that this manoeuvre alone will not deal with the term $\Upsilon^c\nabla_c f$ in (3.8).

To eliminate the term $(n-2)\Upsilon^c\nabla_c f$ we will make what seems like a strange move (and we will explain later the mathematics behind this). We will allow the domain 'function' to depend on the choice of metric in the following way. We decree that the function f on (M,g) corresponds to $\widehat{f} := \Omega^{1-\frac{n}{2}}f$ on $(M,\widehat{g})$, where $\widehat{g}=\Omega^2 g$ as above. Now let us calculate $\Delta^{\hat{g}}\hat{f}$. First we have:

$$\begin{aligned}\nabla_a(\Omega^{1-\frac{n}{2}}f) &= \left(1-\frac{n}{2}\right)\Omega^{1-\frac{n}{2}}\Upsilon_a f + \Omega^{1-\frac{n}{2}}df\\ &= \Omega^{1-\frac{n}{2}}\left(\nabla_a f + \left(1-\frac{n}{2}\right)\Upsilon_a f\right).\end{aligned}$$

We use this in the next step:

$$\begin{aligned}\Delta^{\hat{g}}\widehat{f} &= \Omega^{-2}g^{ab}\widehat{\nabla}_b\left[\Omega^{1-\frac{n}{2}}\left(\nabla_a f + \left(1-\frac{n}{2}\right)\Upsilon_a f\right)\right].\\ &= \Omega^{-1-\frac{n}{2}}g^{ab}\left[\nabla_b\nabla_a f + \left(1-\frac{n}{2}\right)\Upsilon_a\nabla_b f + \left(1-\frac{n}{2}\right)\Upsilon_b\nabla_a f\right.\\ &\qquad + \left(1-\frac{n}{2}\right)^2\Upsilon_a\Upsilon_b f + \left(1-\frac{n}{2}\right)f\nabla_b\Upsilon_a - \Upsilon_b\nabla_a f\\ &\qquad - \left(1-\frac{n}{2}\right)\Upsilon_b\Upsilon_a f - \Upsilon_a\nabla_b f - \left(1-\frac{n}{2}\right)\Upsilon_a\Upsilon_b f\\ &\qquad \left. + g_{ab}\left(\Upsilon^c\nabla_c f + \left(1-\frac{n}{2}\right)\Upsilon^2 f\right)\right]\\ &= \Omega^{-1-\frac{n}{2}}\left[\Delta f + \left(1-\frac{n}{2}\right)\left(\nabla^a\Upsilon_a + \Upsilon^2\left(\frac{n}{2}-1\right)\right)f\right],\end{aligned}$$

where we have used (3.2) once again and have written Υ^2 as a shorthand for $\Upsilon^a\Upsilon_a$.

Now in the last line of the formula for $\Delta^{\hat{g}}\widehat{f}$ the terms involving Υ do not involve derivatives of f. Thus there is hope of matching this with a curvature transformation. Indeed contracting (3.6) with $\hat{g}^{-1}$ gives (with $\mathsf{J}^g := g^{ab}P^g_{ab}$)

$$\mathsf{J}^{\widehat{g}} = \Omega^{-2}\left(\mathsf{J} - \nabla^a\Upsilon_a + \left(1-\frac{n}{2}\right)\Upsilon^2\right), \qquad (3.9)$$

and so

$$\left(\Delta^{\widehat{g}} + \left(1 - \frac{n}{2}\right) \mathsf{J}^{\widehat{g}}\right)\widehat{f} = \Omega^{-1-\frac{n}{2}} \left(\Delta + \left(1 - \frac{n}{2}\right) \mathsf{J}\right) f, \tag{3.10}$$

and we have found a Laplacian operator with a symmetry under conformal rescaling. Using the relation between J and the scalar curvature this is written as in the definition here.

Definition 3.3 On a pseudo-Riemannian manifold (M, g) the operator

$$Y^g : \mathcal{E} \to \mathcal{E} \quad \text{defined by} \quad Y^g := \Delta^g - \frac{n-2}{4(n-1)} \mathrm{Sc}^g$$

is called the *conformal Laplacian* or, in Lorentzian signature, the *conformal wave operator.*

Remark 3.3 On Lorentzian signature manifolds it is often called the *conformal wave operator* because the leading term agrees with the operator giving the usual wave equation. It seems that it was in this setting that the operator was first discovered and applied [54, 16]. On the other hand in the setting of Riemannian signature Y is often called the *Yamabe operator* because of its role in the Yamabe problem of scalar curvature prescription. ∎

According to our calculations above this has the following remarkable symmetry property with respect to conformal rescaling.

Proposition 3.3 *The conformal Laplacian is conformally covariant in the sense that*

$$Y^{\widehat{g}} \circ \Omega^{1-\frac{n}{2}} = \Omega^{-1-\frac{n}{2}} \circ Y^g.$$

This property of the conformal Laplacian motivates a definition.

Definition 3.4 On pseudo-Riemannian manifolds a natural linear differential operator P^g, on a function or tensor/spinor field, is said to be a *conformally covariant operator* if for all positive functions Ω

$$P^{\widehat{g}} \circ \Omega^{w_1} = \Omega^{w_2} \circ P^g,$$

where $\widehat{g} = \Omega^2 g$, $(w_1, w_2) \in \mathbb{R} \times \mathbb{R}$, and where we view the powers of Ω as multiplication operators.

In this definition it is not meant that the domain and target bundles are necessarily the same. The example in our next exercise will be important for our later discussions.

Exercise 3 On pseudo-Riemannian manifolds $(M^{n\geq 3}, g)$ show that

$$\begin{aligned} A^g_{ab} : \mathcal{E} &\longrightarrow \mathcal{E}_{(ab)_0} \text{ given by} \\ f &\longmapsto \nabla_{(a}\nabla_{b)_0} f + \mathsf{P}_{(ab)_0} f \end{aligned} \tag{3.11}$$

is conformally covariant with $(w_1, w_2) = (1, 1)$. That is if $\widehat{g} = \Omega^2 g$, for some positive function Ω, then

$$A^{\widehat{g}}(\Omega f) = \Omega(A^g f).$$

3.3.4. Conformal Geometry

Recall that we defined a conformal manifold as a manifold M equipped only with an equivalence class of conformally related metrics (see Definition 3.1). Conformally covariant operators, as in Definition 3.4, have good conformal properties but (in general) fail to be invariant in the sense of (3.7). This is not just an aesthetic shortcoming, it means that they are not well-defined on conformal manifolds $(M, \boldsymbol{c})$. To construct operators on (M, g) that do descend to the corresponding conformal structure $(M, \boldsymbol{c} = [g])$ we need the notion of conformal densities.

3.3.4.1. Conformal Densities and the Conformal Metric

Let $(M, \boldsymbol{c})$ be a *conformal manifold* of signature (p, q) (with $p + q = n$). For a point $x \in M$, and two metrics g and $\hat{g}$ from the conformal class, there is an element $s \in \mathbb{R}_+$ such that $\hat{g}_x = s^2 g_x$ (where the squaring of s is a convenient convention). Thus, we may view the conformal class as being given by a smooth ray subbundle $\mathcal{Q} \subset S^2T^*M$, whose fibre at x is formed by the values of g_x for all metrics g in the conformal class. By construction, $\mathcal{Q}$ has fibre $\mathbb{R}_+$ and the metrics in the conformal class are in bijective correspondence with smooth sections of $\mathcal{Q}$.

Denoting by $\pi : \mathcal{Q} \to M$ the restriction to $\mathcal{Q}$ of the canonical projection $S^2T^*M \to M$, we can view this as a principal bundle with structure group $\mathbb{R}_+$. The usual convention is to rescale a metric g to $\hat{g} = \Omega^2 g$. This corresponds to a principal action given by $\rho^s(g_x) = s^2 g_x$ for $s \in \mathbb{R}_+$ and $g_x \in \mathcal{Q}_x$, the fibre of $\mathcal{Q}$ over $x \in M$.

With this, we immediately have a family of basic real line bundles $\mathcal{E}[w] \to M$ for $w \in \mathbb{R}$ by defining $\mathcal{E}[w]$ to be the associated bundle to $\mathcal{Q}$ with respect to the action of $\mathbb{R}_+$ on $\mathbb{R}$ given by $s \cdot t := s^{-w}t$. The usual correspondence between sections of an associated bundle and equivariant functions on the total space of a principal bundle then identifies the space $\Gamma(\mathcal{E}[w])$ of smooth sections of $\mathcal{E}[w]$ with the space of all smooth functions $f : \mathcal{Q} \to \mathbb{R}$ such that $f(\rho^s(g_x)) = s^w f(g_x)$ for all $s \in \mathbb{R}_+$. We shall call $\mathcal{E}[w]$ the bundle of *conformal densities of weight w*. Note that each bundle $\mathcal{E}[w]$ is trivial and inherits an orientation from that on $\mathbb{R}$. We write $\mathcal{E}_+[w]$ for the ray subbundle consisting of positive elements.

If $\widehat{g} = \Omega^2 g \in c$ then the conformally related metrics $\widehat{g}$ and g each determine sections of $\mathcal{Q}$. We may pull back f via these sections and we obtain functions on M related by

$$f^{\widehat{g}} = \Omega^w f^g.$$

With $w = 1 - \frac{n}{2}$ this explains the 'strange move' in Section 3.3.3.1 for the domain function of the conformal wave operator, which is really an operator on the bundle $\mathcal{E}[1 - \frac{n}{2}]$.

Although the bundle $\mathcal{E}[w]$ as we defined it depends on the choice of the conformal structure, it is naturally isomorphic to a density bundle (which is independent of the conformal structure). Recall that the bundle of α–densities is associated with the full linear frame bundle of M with respect to the one-dimensional representation $A \mapsto |\det(A)|^{-\alpha}$ of the group $GL(n, \mathbb{R})$. In particular, 2-densities may be canonically identified with the oriented bundle $(\Lambda^n T^* M)^2$, and one-densities are exactly the geometric objects on manifolds that may be integrated (in a coordinate-independent way).

To obtain the link with conformal densities, as defined above, recall that any metric g on M determines a nowhere vanishing 1-density, the volume density $\mathrm{vol}(g)$. In a local frame, this density is given by $\sqrt{|\det(g_{ij})|}$, which implies that for a positive function Ω we get $\mathrm{vol}(\Omega^2 g) = \Omega^n \,\mathrm{vol}(g)$. So there is bijectively a map from 1-densities to functions $\mathcal{Q} \to \mathbb{R}$ that are homogeneous of degree $-n$ given by the map

$$\phi \mapsto \phi(x)/\,\mathrm{vol}(g)(x),$$

and this gives an identification of the 1-density bundle with $\mathcal{E}[-n]$ and thus an identification of $\mathcal{E}[w]$ with the bundle of $(-\frac{w}{n})$-densities on M.

So we may think of conformal density bundles as those bundles associated with the frame bundle via one-dimensional representations, just as tensor bundles are associated with higher rank representations. Given any vector bundle $\mathcal{B}$ we will use the notation

$$\mathcal{B}[w] := \mathcal{B} \otimes \mathcal{E}[w],$$

and say the bundle is *weighted* of weight w. Note that $\mathcal{E}[w] \otimes \mathcal{E}[w'] = \mathcal{E}[w+w']$ and that in the above we assume that $\mathcal{B}$ is not a density bundle itself and is unweighted (weight zero).

Clearly, sections of such weighted bundles may be viewed as homogeneous (along the fibres of $\mathcal{Q}$) sections of the pull-back along $\pi : \mathcal{Q} \to M$. Now the tautological inclusion of $\tilde{\boldsymbol{g}} : \mathcal{Q} \to \pi^* S^2 T^* M$ is evidently homogeneous of degree 2, as for $(s^2 g_x, x) \in \mathcal{Q}$, we have $\tilde{\boldsymbol{g}}(s^2 g_x, x) = (s^2 g_x, x) \in \pi^* S^2 T^* M$. So $\tilde{\boldsymbol{g}}$ may be identified with a canonical section of $\boldsymbol{g} \in \Gamma(S^2 T^* M[2])$ which provides another description of the conformal class. We call $\boldsymbol{g}$ the *conformal metric*.

Another way to recover $\boldsymbol{g}$ is to observe that any metric $g \in \boldsymbol{c}$ is a section of $\mathcal{Q}$, and hence determines a section $\sigma_g \in \Gamma(\mathcal{E}_+[1])$ with the characterising property that the corresponding homogeneous function $\tilde{\sigma}_g$ on $\mathcal{Q}$ takes the value 1 along the section g. Then

$$\boldsymbol{g} = (\sigma_g)^2 g \tag{3.12}$$

and it is easily verified that this is independent of the choice of $g \in c$. Conversely it is clear that any section $\sigma \in \Gamma(\mathcal{E}_+[1])$ determines a metric via

$$g := \sigma^{-2}\boldsymbol{g}.$$

On a conformal manifold we call $\sigma \in \Gamma(\mathcal{E}_+[1])$, or equivalently the corresponding $g \in \boldsymbol{c}$, a *choice of scale*.

A nice application of $\boldsymbol{g}$ is that we can use it to raise, lower and contract tensor indices on a conformal manifold, for example

$$\boldsymbol{g}_{ab} : \mathcal{E}^a \to \mathcal{E}_b[2] \quad \text{by} \quad v^a \mapsto \boldsymbol{g}_{ab}v^a,$$

just as we use the metric in pseudo-Riemannian geometry. Also $\boldsymbol{g}$ gives the isomorphism

$$\otimes^n \boldsymbol{g} : \left(\Lambda^n TM\right)^2 \xrightarrow{\simeq} \mathcal{E}[2n]. \tag{3.13}$$

3.3.4.2. Some Calculus with Conformal Densities

Observe that a choice of scale $g \in c$ determines a connection on $\mathcal{E}[w]$ via the formula

$$\nabla^g \tau := \sigma^w\big(\mathrm{d}(\sigma^{-w}\tau)\big), \quad \tau \in \Gamma(\mathcal{E}[w]), \tag{3.14}$$

where d is the exterior derivative and $\sigma \in \Gamma(\mathcal{E}_+[1])$ satisfies $\boldsymbol{g} = \sigma^2 g$, as $\sigma^{-w}\tau$ is a function (i.e. is a section of $\mathcal{E}[0] = \mathcal{E}$). Coupling this to the Levi-Civita connection for g (and denoting both ∇^g) we have at once that $\nabla^g\sigma = 0$ and hence

$$\nabla^g \boldsymbol{g} = 0. \tag{3.15}$$

On the other hand the Levi-Civita connection directly determines a linear connection on $\mathcal{E}[w]$ since the latter is associated with the frame bundle, as mentioned above. But (3.15), with (3.13), shows that this agrees with (3.14). That is, (3.14) is the Levi-Civita connection on $\mathcal{E}[w]$. Thus we have:

Proposition 3.4 *On a conformal manifold* $(M, \boldsymbol{c})$ *the conformal metric* $\boldsymbol{g}$ *is preserved by the Levi-Civita connection* ∇^g *for any* $g \in \boldsymbol{c}$.

In view of Proposition 3.4 it is reasonable to use the conformal metric to raise and lower tensor indices even when working in a scale! We shall henceforth do this unless we state otherwise. This enables us to give formulae for natural conformally invariant operators acting between density bundles. For example, now choosing $g \in \boldsymbol{c}$ and forming the Laplacian, and the trace of the Schouten tensor,

$$\Delta := \boldsymbol{g}^{ab}\nabla_a\nabla_b \quad \text{and} \quad \mathsf{J} := \boldsymbol{g}^{ab}P_{ab},$$

we have the following. The conformal Laplacian can be interpreted as a differential operator

$$Y : \mathcal{E}\left[1 - \frac{n}{2}\right] \to \mathcal{E}\left[-1 - \frac{n}{2}\right], \quad \text{given by} \quad \Delta + \left(1 - \frac{n}{2}\right)\mathsf{J},$$

that is *conformally invariant*, meaning that it is well-defined on conformal manifolds.

Similarly the operator A^g_{ab} of (3.11) is equivalent to a conformally invariant operator,

$$A_{ab} : \mathcal{E}[1] \to \mathcal{E}_{(ab)_0}[1].$$

3.3.4.3. Conformal Transformations

The above constructions enable us to understand how conformally covariant objects may be reinterpreted as objects that descend to invariant operators on conformal manifolds. A similar result applies to curvature covariants. However, we have not advanced the problem of constructing these.

It follows at once from the formula (3.14) that under conformal transformations $g \mapsto \widehat{g} = \Omega^2 g$ the Levi-Civita connection on $\mathcal{E}[w]$ transforms by

$$\nabla^{\widehat{g}}_a \tau = \nabla^g_a \tau + w\Upsilon_a \tau, \tag{3.16}$$

since $\sigma_{\widehat{g}} = \Omega^{-1}\sigma_g$. We can combine this with the transformation formulae (3.1), (3.2) and (3.6) to compute the conformal transformations of weighted tensors and Riemannian invariants. However this remains a hopeless approach to finding conformal invariants.

3.4. Lecture 3: Prolongation and the Tractor Connection

If we are going to be successful at calculating conformal invariants we are going to need a better way to calculate, one which builds in conformal invariance from the start. In the next two lectures we develop such a calculus, the *conformal tractor calculus*, which can be used to proliferate conformally invariant tensor (or tractor) expressions.

For treating conformal geometry it would be clearly desirable to find an analogue of the Ricci calculus available in the pseudo-Riemannian setting.

On the tangent bundle a conformal manifold $(M, \boldsymbol{c})$ has a distinguished equivalence class of connections but no distinguished connection from this class. So at first the situation does not look promising. However we will see that if we pass from the tangent bundle to a vector bundle with two more dimensions (the standard tractor bundle), then there is indeed a distinguished connection.

There are many ways to see how the tractor calculus arises on a conformal manifold; we will give a very explicit construction which facilitates calculation, but first it will be very helpful to examine how the tractor calculus arises on the flat model space of (Riemannian signature) conformal geometry, the conformal sphere.

Remark 3.4 Recall that inverse stereographic projection maps Euclidean space conformally into the sphere as a one point (conformal) compactification, so that it makes sense to think of the sphere as the conformally 'flat' model of Riemannian signature conformal geometry. The real reason however is that the conformal sphere arises naturally as the geometry of a homogeneous space of Lie groups and that conformal geometry can be thought of as the geometry of curved analogues of this homogeneous geometry in the sense of Élie Cartan (see, e.g. [12]). ∎

3.4.1. The Model of Conformal Geometry: The conformal Sphere

We now look at the conformal sphere, which is an extremely important example. We shall see that the sphere can be viewed as a homogeneous space on which is naturally endowed a conformal structure, and that the conformal tractor calculus arises naturally from this picture. This is the model for Riemannian signature conformal geometry, but a minor variation of this applies to other signatures.

First some notation. Consider an $(n+2)$-dimensional real vector space $\mathbb{V} \cong \mathbb{R}^{n+2}$. Considering the equivalence relation on $\mathbb{V} \setminus \{0\}$ given by

$$v \sim v', \quad \text{if and only if} \quad v' = rv \quad \text{for some} \quad r > 0$$

we now write

$$\mathbb{P}_+(\mathbb{V}) := \{[v] \mid v \in \mathbb{V} \setminus \{0\}\}$$

where $[v]$ denotes the equivalence class of v. We view this as a smooth manifold by the identification with one, equivalently any, round sphere in $\mathbb{V}$.

Suppose now that $\mathbb{V}$ is equipped with a non-degenerate bilinear form $\mathcal{H}$ of signature $(n+1, 1)$. The *null cone* $\mathcal{N}$ of zero-length vectors forms a quadratic

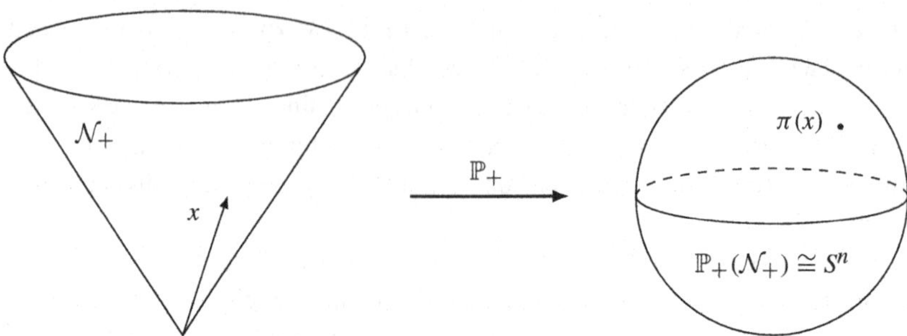

Figure 3.1. The conformal sphere $\mathbb{S}^n$ as the ray projectivisation of the forward null cone $\mathcal{N}_+$

variety in $\mathbb{V}$. Choosing a time orientation, let us write $\mathcal{N}_+$ for the forward part of $\mathcal{N}\setminus\{0\}$. Under the $\mathbb{R}_+$-ray projectivisation of $\mathbb{V}$, meaning the natural map to equivalence classes $\mathbb{V} \to \mathbb{P}_+(\mathbb{V})$, the forward cone $\mathcal{N}_+$ is mapped to a quadric in $\mathbb{P}_+(\mathbb{V})$. This image is topologically a sphere S^n and we will write π for the submersion $\mathcal{N}_+ \to S^n = \mathbb{P}_+(\mathcal{N}_+)$ (see Figure 3.1).

Each point $x \in \mathcal{N}_+$ determines a positive definite inner product g_x on $T_{\pi(x)}S^n$ by $g_x(u, v) = \mathcal{H}_x(u', v')$ where $u', v' \in T_x\mathcal{N}_+$ are *lifts* of $u, v \in T_{\pi(x)}S^n$, meaning that $\pi(u') = u$, $\pi(v') = v$. For a given vector $u \in T_{\pi(x)}S^n$ two lifts to $x \in \mathcal{N}_+$ differ by a vertical vector (i.e. a vector in the kernel of $d\pi$). By differentiating the defining equation for the cone we see that any vertical vector is normal to the cone with respect to $\mathcal{H}$ (null tangent vectors to hypersurfaces are normal), and so it follows that g_x is independent of the choices of lifts. Clearly then, each section of π determines a metric on S^n and by construction this is smooth if the section is. Evidently the metric agrees with the pull-back of $\mathcal{H}$ via the section concerned. We may choose coordinates X^A, $A = 0, \cdots, n+1$, for $\mathbb{V}$ so that $\mathcal{N}$ is the zero locus of the form $-(X^0)^2 + (X^1)^2 + \cdots + (X^{n+1})^2$, in which terms the usual round sphere arises as the section $X^0 = 1$ of π.

Now, viewed as a metric on $T\mathbb{V}$, $\mathcal{H}$ is homogeneous of degree 2 with respect to the standard Euler (or position) vector field E on $\mathbb{V}$, that is $\mathcal{L}_E\mathcal{H} = 2\mathcal{H}$, where $\mathcal{L}$ denotes the Lie derivative. In particular this holds on the cone, which we note is generated by E. Write $\boldsymbol{g}$ for the restriction of $\mathcal{H}$ to vector fields in $T\mathcal{N}_+$ which are the lifts of vector fields on S^n. Note that u' is the lift of a vector field u on S^n, which means that for all $x \in \mathcal{N}_+$, $d\pi(u'(x)) = u(\pi(x))$, and so $\mathcal{L}_E u' = 0$ (modulo vertical vector fields) on $\mathcal{N}_+$. Thus for any pair $u, v \in \Gamma(TS^n)$, with lifts to vector fields u', v' on $\mathcal{N}_+$, $\boldsymbol{g}(u', v')$ is a function on $\mathcal{N}_+$ homogeneous of degree 2, and which is independent of how the vector fields were lifted. It follows that if $s > 0$ then $g_{sx} = s^2 g_x$, for all $x \in \mathcal{N}_+$. Evidently

$\mathcal{N}_+$ may be identified with the total space $\mathcal{Q}$ of a bundle of conformally related metrics on $\mathbb{P}_+(\mathcal{N}_+)$. Thus $\boldsymbol{g}(u', v')$ may be identified with a conformal density of weight 2 on S^n. That is, this construction canonically determines a section of $\mathcal{E}_{(ab)}[2]$ that we shall also denote by $\boldsymbol{g}$. This has the property that if σ is any section of $\mathcal{E}_+[1]$ then $\sigma^{-2}\boldsymbol{g}$ is a metric g^σ. Obviously different sections of $\mathcal{E}_+[1]$ determine conformally related metrics and, by the last observation in the previous paragraph, there is a section σ_o of $\mathcal{E}_+[1]$ so that g^{σ_o} is the round metric.

Thus we see that $\mathbb{P}_+(\mathcal{N}_+)$ is canonically equipped with the standard conformal structure on the sphere, but with no preferred metric from this class. Furthermore $\boldsymbol{g}$, which arises here from $\mathcal{H}$ by restriction, is the conformal metric on $\mathbb{P}_+(\mathcal{N}_+)$. In summary then we have the following.

Lemma 3.1 *Let* $\boldsymbol{c}$ *be the conformal class of* $S^n = \mathbb{P}_+(\mathcal{N}_+)$ *determined canonically by* $\mathcal{H}$*. This includes the round metric. The map* $\mathcal{N}_+ \ni (\pi(x), g_x) \in \mathcal{Q}$ *gives an identification of* $\mathcal{N}_+$ *with* $\mathcal{Q}$*, the bundle of conformally related metrics on* $(S^n, \boldsymbol{c})$*.*

Via this identification: functions homogeneous of degree w *on* $\mathcal{N}_+$ *are equivalent to functions homogeneous of degree* w *on* $\mathcal{Q}$ *and hence correspond to conformal densities of weight* w *on* $(S^n, \boldsymbol{c})$*; the conformal metric* $\boldsymbol{g}$ *on* $(S^n, \boldsymbol{c})$ *agrees with, and is determined by, the restriction of* $\mathcal{H}$ *to the lifts of vector fields on* $\mathbb{P}_+(\mathcal{N}_+)$*.*

The conformal sphere, as constructed here, is acted on transitively by $G = \mathrm{O}_+(\mathcal{H}) \cong \mathrm{O}_+(n+1, 1)$, where this is the time orientation preserving subgroup of the orthogonal group preserving $\mathcal{H}$, $\mathrm{O}(\mathcal{H}) \cong \mathrm{O}(n+1, 1)$. Thus, as a homogeneous space, $\mathbb{P}_+(\mathcal{N}_+)$ may be identified with G/P, where P is the (parabolic) Lie subgroup of G preserving a nominated null ray in $\mathcal{N}_+$.

3.4.1.1. Canonical Calculus on the Model

Here we sketch briefly one way to see the tractor connection on the model $\mathbb{P}_+(\mathcal{N}_+)$.

Note that as a manifold $\mathbb{V}$ has some special structures that we have already used. In particular an origin and, from the vector space structure of $\mathbb{V}$, the Euler vector field E which assigns to each point $X \in \mathbb{V}$ the vector $X \in T_X\mathbb{V}$, via the canonical identification of $T_X\mathbb{V}$ with $\mathbb{V}$.

The vector space $\mathbb{V}$ has an affine structure and this induces a global parallelism: the tangent space $T_x\mathbb{V}$ at any point $x \in \mathbb{V}$ may be canonically identified with $\mathbb{V}$. Thus, in particular, for any parameterised curve in $\mathbb{V}$ there is a canonical notion of parallel transport along the given curve. This exactly means that, viewing $\mathbb{V}$ as a manifold, it is equipped with a canonical affine connection $\nabla^{\mathbb{V}}$. The affine structure gives more than this of course. It is isomorphic to $\mathbb{R}^{n+2}$ with its usual affine structure, and so the tangent bundle to $\mathbb{V}$ is trivialised by everywhere parallel tangent fields. It follows that the canonical connection $\nabla^{\mathbb{V}}$ is flat and has trivial holonomy.

Next observe that $\mathcal{H}$ determines a signature $(n+1,1)$ metric on $\mathbb{V}$, where the latter is viewed as an affine manifold. By the definition of its promotion from bilinear form to metric, one sees at once that for any vector fields U, V that are parallel on $\mathbb{V}$ the quantity $\mathcal{H}(U, V)$ is constant. This means that $\mathcal{H}$ is itself parallel since for any vector field W we have

$$(\nabla_W \mathcal{H})(U, V) = W \cdot \mathcal{H}(U, V) - \mathcal{H}(\nabla_W U, V) - \mathcal{H}(U, \nabla_W V) = 0.$$

The second key observation is that a restriction of these structures descends to the conformal n-sphere. We observed above that $\mathcal{N}_+$ is an $\mathbb{R}_+$-ray bundle over S^n. We may identify S^n with $\mathcal{N}_+/\sim$ where the equivalence relation is that $x \sim y$ if and only if x and y are points of the same fibre $\pi^{-1}(x')$ for some $x' \in S^n$. The restriction $T\mathbb{V}|_{\mathcal{N}_+}$ is a rank $n+2$ vector bundle over $\mathcal{N}_+$. Now we may define an equivalence relation on $T\mathbb{V}|_{\mathcal{N}_+}$ that covers the relation on $\mathcal{N}_+$. Namely we decree $U_x \sim V_y$ if and only if $x, y \in \pi^{-1}(x')$ for some $x' \in S^n$, *and* U_x and V_y are parallel. Considering parallel transport up the fibres of π, it follows that $T\mathbb{V}|_{\mathcal{N}_+}/\sim$ is isomorphic to the restriction $T\mathbb{V}|_{\mathrm{im}(S)}$ where S is any section of π (that is $S : S^n \to \mathcal{N}_+$ is a smooth map such that $\pi \circ S = \mathrm{id}_{S^n}$). But $\mathrm{im}(S)$ is identified with S^n via π and it follows that $T\mathbb{V}|_{\mathcal{N}_+}/\sim$ may be viewed as a vector bundle $\mathcal{T}$ on S^n. Furthermore it is clear from the definition of the equivalence relation on $T\mathbb{V}|_{\mathcal{N}_+}$ that $\mathcal{T}$ is independent of S. The vector bundle $\mathcal{T}$ on S^n is the *(standard) tractor bundle* of $(S^n, \boldsymbol{c})$.

By restriction, $\mathcal{H}$ and $\nabla^{\mathbb{V}}$ determine, respectively, a (signature $(n+1,1)$) metric and connection on the bundle $T\mathbb{V}|_{\mathcal{N}_+}$ that we shall denote with the same notation. Since a vector field which is parallel along a curve γ in $\mathcal{N}_+$ may be

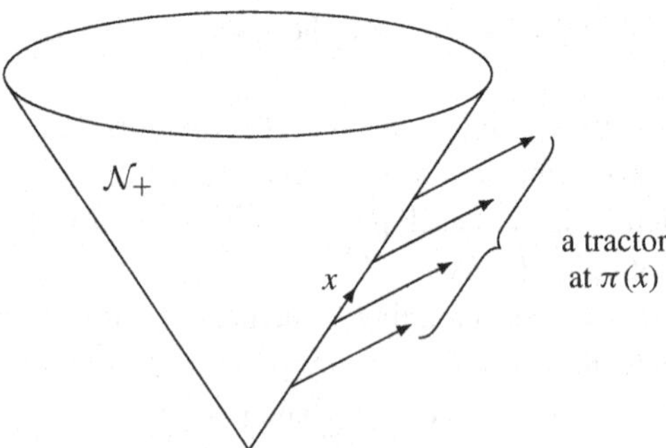

Figure 3.2. An element of $\mathcal{T}_{\pi(x)}$ corresponds to a homogeneous of degree zero vector field along the ray generated by x

uniquely extended to a vector field which is also parallel along every fibre of π through the curve γ, it is clear that $\nabla^{\mathbb{V}}$ canonically determines a connection on $\mathcal{T}$ that we shall denote $\nabla^{\mathcal{T}}$. Sections $U, V \in \Gamma(\mathcal{T})$ are represented on $\mathcal{N}_+$ by vector fields $\tilde{U}, \tilde{V}$ that are parallel in the direction of the fibres of $\pi : \mathcal{N}_+ \to S^n$ (see Figure 3.2). On the other hand $\mathcal{H}$ is also parallel along each fibre of π and so $\mathcal{H}(\tilde{U}, \tilde{V})$ is constant on each fibre. Thus $\mathcal{H}$ determines a signature $(n+1, 1)$ metric h on $\mathcal{T}$ satisfying $h(U, V) = \mathcal{H}(\tilde{U}, \tilde{V})$. What is more, since $\nabla^{\mathbb{V}}\mathcal{H} = 0$ on $\mathcal{N}_+$, it follows that h is preserved by $\nabla^{\mathcal{T}}$, that is

$$\nabla^{\mathcal{T}} h = 0.$$

Summarising the situation thus far we have the following.

Theorem 3.1 *The model $(S^n, \boldsymbol{c})$ is canonically equipped with the following: a canonical rank $n+2$ bundle $\mathcal{T}$; a signature $(n+1, 1)$ metric h on this; and a connection $\nabla^{\mathcal{T}}$ on $\mathcal{T}$ that preserves h.*

Although we shall not go into details here it is straightforward to show the following:

Proposition 3.5 *The tractor bundle $\mathcal{T}$ of the model $(S^n, \boldsymbol{c})$ has a composition structure*

$$\mathcal{T} = \mathcal{E}[1] \mathbin{\rlap{+}{\supset}} TS^n[-1] \mathbin{\rlap{+}{\supset}} \mathcal{E}[-1]. \tag{3.17}$$

The restriction of h to the subbundle $\mathcal{T}^0 = TS^n[-1] \mathbin{\rlap{+}{\supset}} \mathcal{E}[-1]$ induces the conformal metric $\boldsymbol{g} : TS^n[-1] \times TS^n[-1] \to \mathcal{E}$. Any null $Y \in \Gamma\mathcal{T}[-1]$ satisfying $h(X, Y) = 1$ determines a splitting $\mathcal{T} \xrightarrow{\cong} \mathcal{E}[1] \oplus TS^n[-1] \oplus \mathcal{E}[-1]$ such that the metric h is given by $(\sigma,\ \mu,\ \rho) \mapsto 2\sigma\rho + \boldsymbol{g}(\mu, \mu)$ as a quadratic form.

It is easily seen how this composition structure arises geometrically. The subbundle $\mathcal{T}^0$ of $\mathcal{T}$ corresponds to the fact that $T\mathcal{N}_+$ is naturally identified with a subbundle of $T\mathbb{V}|_{\mathcal{N}_+}$. The vertical directions in there correspond to the fact that $\mathcal{E}[-1]$ is a subbundle of $\mathcal{T}^0$, and the *semidirect sum* symbols $\mathbin{\rlap{+}{\supset}}$ record this structure.

3.4.1.2. The Abstract Approach to the Tractor Connection

An alternative way of seeing how the tractor connection arises on the flat model is via the group picture. If G/H is a homogeneous space of Lie groups, then the canonical projection $G \to G/H$ gives rise to a principal H-bundle over G/H (with total space G). If $\mathbb{V}$ is a representation of H then one obtains a homogeneous vector bundle $\mathcal{V} := G \times_H \mathbb{V}$ over G/H whose total space is defined to be the quotient of $G \times \mathbb{V}$ by the equivalence relation

$$(g, v) \sim (gh, h^{-1} \cdot v).$$

If $\mathbb{V}$ is in fact a representation of G, then the bundle $\mathcal{V} := G \times_H \mathbb{V}$ is trivial. The isomorphism

$$G \times_H \mathbb{V} \cong (G/H) \times \mathbb{V}$$

is given by

$$[g, v] \mapsto (gH, g \cdot v)$$

which is easily checked to be well defined. This trivialisation gives rise to a flat connection on $G \times_H \mathbb{V}$. In the case where we have the conformal sphere $G/P \cong S^n$ and $\mathbb{V}$ is the defining representation of G (i.e. $\mathbb{R}^{n+2}$) then $G \times_P \mathbb{V}$ is (naturally identified with) the tractor bundle $\mathcal{T}$ and the connection induced by the trivialisation $G \times_P \mathbb{V} \cong (G/P) \times \mathbb{V}$ is the tractor connection. The trivialisation $\mathcal{T} \cong (G/P) \times \mathbb{V}$ also immediately gives the existence of a bundle metric preserved by the tractor connection induced by the bilinear form on $\mathbb{V}$.

3.4.1.3. The Conformal Model in Other Signatures

With only a little more effort one can treat the case of general signature (p, q). This again begins with a real vector space $\mathbb{V} \cong \mathbb{R}^{n+2}$, but now we equip it with a non-degenerate bilinear form $\mathcal{H}$ of signature $(p+1, q+1)$, where $p+q = n$.

Writing again $\mathcal{N}$ for the quadratic variety of vectors which have zero-length with respect to $\mathcal{H}$, we see that the space of null rays in $\mathcal{N} \setminus \{0\}$ has the topology of $S^p \times S^q$. This is connected, unless p or q is zero in either, in which case we get two copies of S^n. In any case an easy adaption of the earlier discussion shows that $\mathbb{P}_+(\mathcal{N})$ is equipped canonically with a conformal structure of signature (p, q) and on this a tractor connection preserving now a tractor metric of signature $(p+1, q+1)$. (One can easily check that the conformal class $\boldsymbol{c}$ of $\mathbb{P}_+(\mathcal{N})$ contains a metric g which is of the form $g_{S^p} - g_{S^q}$ for some identification of $\mathbb{P}_+(\mathcal{N})$ with $S^p \times S^q$.)

As constructed here the conformal space $\mathbb{P}_+(\mathcal{N})$ is acted on transitively by $G = \mathrm{O}(\mathcal{H}) \cong \mathrm{O}(p+1, q+1)$, the orthogonal group preserving $\mathcal{H}$. As a homogeneous space, $\mathbb{P}_+(\mathcal{N})$ may be identified with G/P, where P is the Lie subgroup of G preserving a nominated null ray in $\mathcal{N}$. We may think of this group picture as a good model for general conformal manifolds. Of course there are other possible choices of G/P (which may result in models which are only locally equivalent to the ones here), as we have already seen in the Riemannian case. See [37] for a discussion of the possible choices of (G, P) and the connection with global aspects of the conformal tractor calculus.

Remark 3.5 Note that the model space $S^1 \times S^{n-1}$ of Lorentzian signature conformal geometry is simply the quotient of the (conformal) Einstein universe $\mathbb{R} \times S^{n-1}$ by integer times 2π translations. Thus the usual embeddings of

Minkowski and de Sitter space into the Einstein universe can be seen (by passing to the quotient) as conformal embeddings into the flat model space. In fact, $S^1 \times S^{n-1}$ can be seen as two copies of Minkowski space glued together along a null boundary with two cone points, or as two copies of de Sitter space glued together along a spacelike boundary with two connected components which are $(n-1)$-spheres. The significance of this will become clearer as we continue to develop the tractor calculus and then move on to study conformally compactified geometries. ∎

3.4.2. Prolongation and the Tractor Connection

Here we construct the tractor bundle, connection and metric on a conformal manifold of dimension at least three. We will see that the 'conformal to Einstein' condition plays an important role. The tractor bundle and connection are obtained by 'prolonging' the 'almost Einstein equation' (3.11).

First we state what is meant by Einstein here.

Definition 3.5 In dimensions $n \geq 3$, a metric will be said to be *Einstein* if

$$\mathrm{Ric}^g = \lambda g$$

for some function λ.

The Bianchi identities imply that, for any metric satisfying this equation, λ is constant (as we assume M is connected). Throughout the following we shall assume that $n \geq 3$.

Recalling the model above, note that a parallel co-tractor I corresponds to a homogeneous polynomial which, in standard coordinates X^A on $\mathbb{V} \xrightarrow{\simeq} \mathbb{R}^{n+2}$, is given by $\tilde{\sigma} = I_A X^A$. Then $\tilde{\sigma} = 1$ is a section of $\mathcal{N}_+$ that corresponds to the intersection of $\mathcal{N}_+$ with the hyperplane $I_A X^A = 1$, in $\mathbb{V}$. For at least some of these distinguished (conic) sections the resulting metric on S^n, $g = \tilde{\sigma}^{-2}\boldsymbol{g}$ (on the open set where the corresponding density $\sigma \in \Gamma(\mathcal{E}[1])$ is nowhere vanishing) is obviously Einstein. For example the round metric has already been discussed. It is in fact true that all metrics obtained on open regions of S^n in this way are Einstein, as we shall shortly see.

3.4.3. The Almost Einstein Equation

Let $(M, \boldsymbol{c})$ be a conformal manifold with $\dim(M) \geq 3$. It is clear that $g^o \in \boldsymbol{c}$ is Einstein if and only if $\mathsf{P}^{g^o}_{(ab)_0} = 0$, and this is the link with the equation (3.11). As a conformally invariant equation this takes the form

$$\nabla^g_{(a}\nabla^g_{b)_0}\sigma + \mathsf{P}^g_{(ab)_0}\sigma = 0, \tag{AE}$$

where $\sigma \in \mathcal{E}_+[1]$ encodes $g^o = \sigma^{-2}\boldsymbol{g}$ and we have used the superscripts to emphasise that we have picked some metric $g \in \boldsymbol{c}$ in order to write the equation.

Suppose that σ is a solution with the property that it is nowhere zero. Then, without loss of generality, we may assume that σ is positive, that is $\sigma \in \mathcal{E}_+[1]$. So σ is a scale, and $g^o = \sigma^{-2}\boldsymbol{g}$ is a well-defined metric. Since (3.11) is conformal invariant, there is no loss if we calculate the equation (AE) in this scale. But then $\nabla^{g^o}\sigma = 0$. Thus we conclude that

$$\mathsf{P}_{(ab)_0} = 0.$$

Conversely suppose that $\mathsf{P}^{g^o}_{(ab)_0} = 0$ for some $g^o \in c$. Then $g^o = \sigma^{-2}\boldsymbol{g}$ for some $\sigma \in \mathcal{E}_+[1]$. Therefore σ solves (AE) by the reverse of the same argument. Thus in summary we have the following, cf. [40].

Proposition 3.6 *$(M, \boldsymbol{c})$ is conformally Einstein (i.e. there is an Einstein metric g^o in $\boldsymbol{c}$) if and only if there exists $\sigma \in \mathcal{E}_+[1]$ that solves* (AE)*. If $\sigma \in \mathcal{E}_+[1]$ solves* (AE) *then $g^o := \sigma^{-2}\boldsymbol{g}$ is the corresponding Einstein metric.*

There are some important points to make here.

Remark 3.6 Equation (AE) is equivalent to a system of $\frac{(n+2)(n-1)}{2}$ scalar equations on one scalar variable. So it is overdetermined and we do not expect it to have solutions in general. ∎

3.4.4. The Connection Determined by a Conformal Structure

Proposition 3.6 has the artificial feature that it is a statement about nowhere vanishing sections of $\mathcal{E}[1]$. Let us rectify this.

Definition 3.6 Let $(M, \boldsymbol{c})$ be a conformal manifold, of any signature, and $\sigma \in \mathcal{E}[1]$. We say that $(M, \boldsymbol{c}, \sigma)$ is an *almost Einstein structure* if $\sigma \in \mathcal{E}[1]$ solves equation (AE).

We shall term (AE) the (conformal) almost Einstein equation.

Since the almost Einstein equation is conformally invariant it is natural to seek integrability conditions that are also conformally invariant. There is a systematic approach to this using a procedure known as prolongation that goes as follows. We fix a metric $g \in \boldsymbol{c}$ to facilitate the calculations. With this understood, for the most part in the following we will omit the decoration by g of the natural objects it determines; for example we shall write ∇_a rather than ∇^g_a.

As a first step observe that the equation (AE) is equivalent to the equation

$$\nabla_a \nabla_b \sigma + \mathsf{P}_{ab}\sigma + \boldsymbol{g}_{ab}\rho = 0 \tag{3.18}$$

where we have introduced the new variable $\rho \in \mathcal{E}[-1]$ to absorb the trace terms. The key idea is to attempt to construct an equivalent first-order closed system. We introduce $\mu_a \in \mathcal{E}_a[1]$, so our equation is replaced by the equivalent system

$$\nabla_a \sigma - \mu_a = 0, \quad \text{and} \quad \nabla_a \mu_b + \mathsf{P}_{ab}\sigma + \boldsymbol{g}_{ab}\rho = 0\,. \tag{3.19}$$

This system is almost closed in the sense that the derivatives of σ and μ_b are given algebraically in terms of the unknowns σ, μ_b and ρ. However to obtain a similar result for ρ we must differentiate the system; by definition *(differential) prolongation* is precisely concerned with this process of producing higher order systems, and their consequences. Here we use notation from earlier.

The Levi-Civita covariant derivative of (3.18) gives

$$\nabla_a \nabla_b \nabla_c \sigma + \boldsymbol{g}_{bc}\nabla_a \rho + (\nabla_a \mathsf{P}_{bc})\sigma + \mathsf{P}_{bc}\nabla_a \sigma = 0.$$

Contracting this using, respectively, $\boldsymbol{g}^{ab}$ and $\boldsymbol{g}^{bc}$ yields

$$\begin{aligned} \boldsymbol{g}^{ab}: \qquad & \Delta \nabla_c \sigma + \nabla_c \rho + (\nabla^a \mathsf{P}_{ac})\sigma + \mathsf{P}^a{}_c \nabla_a \sigma = 0 \quad (1) \\ \boldsymbol{g}^{bc}: \qquad & \nabla_c \Delta \sigma + n\nabla_c \rho + (\nabla_c J)\sigma + J\nabla_c \sigma = 0 \quad (2). \end{aligned}$$

Then the difference (2) – (1) is simply

$$(n-1)\nabla_c \rho + J\nabla_c \sigma - \mathsf{P}^a{}_c \nabla_a \sigma + R_{cb}{}^b{}_d \nabla^d \sigma = 0$$

where we have used the contracted Bianchi identity $\nabla^a \mathsf{P}_{ac} = \nabla_c \mathsf{J}$ and computed the commutator $[\nabla_c, \Delta]$ acting on σ. But

$$R_{cb}{}^b{}_a \nabla^a \sigma = -R_{ca}\nabla^a \sigma = (2-n)\mathsf{P}_c{}^a \nabla_a \sigma - J\nabla_c \sigma,$$

and using this we obtain

$$\nabla_c \rho - \mathsf{P}_c{}^a \mu_a = 0, \tag{3.20}$$

after dividing by the overall factor $(n-1)$. So we have our closed system and, what is more, this system yields a linear connection. We discuss this now.

On a conformal manifold $(M, \boldsymbol{c})$ let us write $[\mathcal{T}]_g$ to mean the pair consisting of a direct sum bundle and $g \in \boldsymbol{c}$, as follows:

$$[\mathcal{T}]_g := \big(\mathcal{E}[1] \oplus \mathcal{E}_a[1] \oplus \mathcal{E}[-1], g\big). \tag{3.21}$$

Proposition 3.7 *On a conformal manifold $(M,\boldsymbol{c})$, fix any metric $g \in \boldsymbol{c}$. There is a linear connection $\nabla^{\mathcal{T}}$ on the bundle*

$$[\mathcal{T}]_g \cong \begin{array}{c} \mathcal{E}[1] \\ \oplus \\ \mathcal{E}_a[1] \\ \oplus \\ \mathcal{E}[-1] \end{array}$$

given by

$$\nabla_a^{\mathcal{T}} \begin{pmatrix} \sigma \\ \mu_b \\ \rho \end{pmatrix} := \begin{pmatrix} \nabla_a \sigma - \mu_a \\ \nabla_a \mu_b + g_{ab}\rho + \mathsf{P}_{ab}\sigma \\ \nabla_a \rho - \mathsf{P}_{ab}\mu^b \end{pmatrix}. \tag{3.22}$$

Solutions of the almost Einstein equation (AE) *are in one-to-one correspondence with sections of the bundle $[\mathcal{T}]_g$ that are parallel for the connection $\nabla^{\mathcal{T}}$.*

Proof It remains only to prove that $\nabla^{\mathcal{T}}$ is a linear connection. But this is an immediate consequence of its explicit formula which we see takes the form $\nabla + \Phi$ where ∇ is the Levi-Civita connection on the direct sum bundle $[\mathcal{T}]_g = \mathcal{E}[1] \oplus \mathcal{E}_a[1] \oplus \mathcal{E}[-1]$ and Φ is a section of $\mathrm{End}([\mathcal{T}]_g)$. □

We shall call $\nabla^{\mathcal{T}}$, of (3.22), the *(conformal) tractor connection*. Interpreted naïvely the statement in the proposition might appear to be not very strong: we have already remarked that on a particular manifold it can be that the equation (AE) has no non-trivial solutions. However this is a universal result, and so it in fact gives an extremely useful tool for investigating the existence of solutions. Note that the connection (3.22) is well defined on any pseudo-Riemannian manifold. The point is that in using (3.22) for any application, we have immediately available the powerful theory of linear connections (e.g. parallel transport and curvature).

It is an immediate consequence of Proposition 3.7 that the almost Einstein equation (AE) can have at most $n+2$ linearly independent solutions. However we shall see that far stronger results are available after we refine our understanding of the tractor connection (3.22). Furthermore this connection will be seen to have a role that extends far beyond the almost Einstein equation.

3.4.5. Conformal Properties of the Tractor Connection

Although derived from a conformally invariant equation, the bundle and connection described in (3.21) and (3.22) are expressed in a way depending a priori on a choice of $g \in c$, so we wish to study their conformal properties.

Let us again fix some choice $g \in c$, before investigating the consequences of changing this conformally.

Given the data of $(\sigma, \mu_b, \rho) \in [\mathcal{T}]_g$ at $x \in M$, it follows from the general properties of linear connections that we may always solve

$$\nabla^{\mathcal{T}}(\sigma, \mu_b, \rho) = 0, \quad \text{at} \quad x \in M. \tag{3.23}$$

This imposes no restriction on either the conformal class c or the choice $g \in c$. Examining the formula (3.22) for the tractor connection we see that if (3.23) holds then, at the point x, we necessarily have

$$\mu_b = \nabla_b \sigma, \quad \rho = -\frac{1}{n}(\Delta\sigma + \mathrm{J}\sigma), \tag{3.24}$$

where the second equation follows by taking a $\boldsymbol{g}^{ab}$ trace of the middle entry on the right hand side of (3.22). Thus canonically associated to the tractor connection there is the second-order differential $\mathcal{E}[1] \to [\mathcal{T}]_g$ given by

$$[D\sigma]_g = \begin{pmatrix} n\sigma \\ n\nabla_b\sigma \\ -(\Delta\sigma + \mathrm{J}\sigma) \end{pmatrix}, \tag{3.25}$$

where we have included the normalising factor n (=dim(M)) for later convenience.

Recall that we know how the Levi-Civita connection, and hence also Δ and J here, transform conformally. Thus it follows that $[D\sigma]_g$, or equivalently (3.24), determines how the variables σ, μ_b and ρ of the prolonged system must transform if they are to remain compatible with $\nabla^{\mathcal{T}}$ under conformal changes. If $\widehat{g} = \Omega^2 g$, for some positive function Ω, then a brief calculation reveals

$$\nabla_b^{\widehat{g}}\sigma = \nabla_b^{g}\sigma + \Upsilon_b\sigma, \text{ and } -\frac{1}{n}(\Delta^{\widehat{g}}\sigma + \mathrm{J}^{\widehat{g}}\sigma) = -\frac{1}{n}(\Delta^{g}\sigma + \mathrm{J}^{g}\sigma) - \Upsilon^b\nabla_b\sigma - \frac{1}{2}\sigma\Upsilon^b\Upsilon_b,$$

where as usual Υ denotes $d\Omega$. Thus we decree

$$\widehat{\sigma} := \sigma, \quad \widehat{\mu}_b := \mu_b + \Upsilon_b\sigma, \quad \widehat{\rho} := \rho - \Upsilon^b\mu_b - \frac{1}{2}\sigma\Upsilon^b\Upsilon_b,$$

or, writing $\Upsilon^2 := \Upsilon^a\Upsilon_a$, this may be otherwise written using an obvious matrix notation:

$$[\mathcal{T}]_{\widehat{g}} \ni \begin{pmatrix} \widehat{\sigma} \\ \widehat{\mu}_b \\ \widehat{\rho} \end{pmatrix} = \begin{pmatrix} 1 & 0 & 0 \\ \Upsilon_b & \delta_b^c & 0 \\ -\frac{1}{2}\Upsilon^2 & -\Upsilon^c & 1 \end{pmatrix} \begin{pmatrix} \sigma \\ \mu_c \\ \rho \end{pmatrix} \sim \begin{pmatrix} \sigma \\ \mu_b \\ \rho \end{pmatrix} \in [\mathcal{T}]_g. \tag{3.26}$$

Note that at each point $x \in M$ the transformation here is manifestly by a group action. Thus in the obvious way this defines an equivalence relation among the direct sum bundles $[\mathcal{T}]_g$ (of (3.21)) that covers the conformal equivalence of metrics in $\boldsymbol{c}$, and the quotient by this defines what we shall call the conformal standard tractor bundle $\mathcal{T}$ on $(M, \boldsymbol{c})$. More precisely we have the following definition.

Definition 3.7 On a conformal manifold $(M, \boldsymbol{c})$ the *standard tractor bundle* is

$$\mathcal{T} := \bigsqcup_{g \in c} [\mathcal{T}]_g / \sim,$$

meaning the disjoint union of the $[\mathcal{T}]_g$ (parameterised by $g \in c$) modulo equivalence relation given by (3.26). We shall also use the abstract index notation $\mathcal{E}_A$ for $\mathcal{T}$.

We shall carry many conventions from tensor calculus over to bundles and tractor fields. For example we shall write $\mathcal{E}_{(AB)}[w]$ to mean $S^2\mathcal{T} \otimes \mathcal{E}[w]$, and so forth.

There are some immediate consequences of Definition 3.7 that we should observe. First, from this definition, the next statement follows tautologically.

Proposition 3.8 *The formula* (3.25) *determines a conformally invariant differential operator*

$$D : \mathcal{E}[1] \to \mathcal{T}.$$

This operator is evidently intimately connected with the very definition of the standard tractor bundle. Because of its fundamental role we make the following definition.

Definition 3.8 Given a section of $\sigma \in \mathcal{E}[1]$ we shall call

$$I := \frac{1}{n} D\sigma$$

the *scale tractor* corresponding to σ.

Next observe also that from (3.26) it is clear that $\mathcal{T}$ is a filtered bundle; we summarise this using a semi-direct sum notation

$$\mathcal{T} = \mathcal{E}[1] \oplus\!\!\!\!+ \ \mathcal{E}_a[1] \oplus\!\!\!\!+ \ \mathcal{E}[-1] \tag{3.27}$$

meaning that $\mathcal{T}$ has a subbundle $\mathcal{T}^1 \subset \mathcal{T}$ isomorphic to $\mathcal{E}[-1]$, $\mathcal{E}_a[1]$ is isomorphic to a subbundle of the quotient $\mathcal{T}/\mathcal{T}^1$ bundle, and $\mathcal{E}[1]$ is the final factor. We use X to denote the bundle surjection $X : \mathcal{T} \to \mathcal{E}[1]$ or in abstract indices:

$$X^A : \mathcal{E}_A \to \mathcal{E}[1]. \tag{3.28}$$

Note that given any metric $g \in \boldsymbol{c}$ we may interpret X as the map $[\mathcal{T}]_g \to \mathcal{E}[1]$ given by

$$\begin{pmatrix} \sigma \\ \mu_b \\ \rho \end{pmatrix} \mapsto \sigma.$$

For our current purposes the critical result at this point is that the tractor connection $\nabla^{\mathcal{T}}$ 'intertwines' with the transformation (3.26) in the following sense.

Exercise 4 Let $V = (\sigma, \mu_b, \rho)$, a section of $[\mathcal{T}]_g$, and $\widehat{V} = (\widehat{\sigma}, \widehat{\mu}_b, \widehat{\rho})$, a section of $[\mathcal{T}]_{\widehat{g}}$, be related by (3.26), where $\widehat{g} = \Omega^2 g$. Show that then

$$\begin{pmatrix} \widehat{\nabla_a\sigma - \mu_a} \\ \widehat{\nabla_a\mu_b + g_{ab}\rho + \mathsf{P}_{ab}\sigma} \\ \widehat{\nabla_a\rho - \mathsf{P}_{ab}\mu^b} \end{pmatrix} = \begin{pmatrix} 1 & 0 & 0 \\ \Upsilon_b & \delta_b^c & 0 \\ -\frac{1}{2}\Upsilon^2 & -\Upsilon^c & 1 \end{pmatrix} \begin{pmatrix} \nabla_a\sigma - \mu_a \\ \nabla_a\mu_b + g_{ac}\rho + \mathsf{P}_{ac}\sigma \\ \nabla_a\rho - \mathsf{P}_{ac}\mu^c \end{pmatrix}.$$

Here $\Upsilon = d\Omega$, as usual, and for example $\widehat{\nabla_a\sigma - \mu_a}$ means $\widehat{\nabla}_a\widehat{\sigma} - \widehat{\mu}_a$.

Given a tangent vector field v^a, the exercise shows that $v^a\nabla_a^{\mathcal{T}} V$ *transforms conformally as a standard tractor field*, that is by (3.26). Whence $\nabla^{\mathcal{T}}$ descends to a well defined connection on $\mathcal{T}$. Let us summarise as follows.

Theorem 3.2 *Let* $(M, \boldsymbol{c})$ *be a conformal manifold of dimension at least 3. The formula* (3.22) *determines a conformally invariant connection*

$$\nabla^{\mathcal{T}} : \mathcal{T} \to \Lambda^1 \otimes \mathcal{T}.$$

For obvious reasons this will also be called the *conformal tractor connection* (on the standard tractor bundle $\mathcal{T}$); the formula (3.22) is henceforth regarded as the incarnation of this conformally invariant object on the realisation $[\mathcal{T}]_g$ of $\mathcal{T}$, as determined by the choice $g \in \boldsymbol{c}$.

It is important to realise that the conformal tractor connection exists canonically on any conformal manifold (of dimension at least 3). (One also has the tractor connection on two-dimensional *Möbius* conformal manifolds.) In particular its existence does not rely on solutions to the equation (AE). Nevertheless by its construction in Section 3.4.4 (as a prolongation of the equation (AE)), and using also equation (3.24), we have at once the following important property.

Theorem 3.3 *On a conformal manifold* $(M, \boldsymbol{c})$ *we have the following. There is a one-to-one correspondence between sections* $\sigma \in \mathcal{E}[1]$, *satisfying the conformal equation*

$$\nabla_{(a}\nabla_{b)_0}\sigma + \mathrm{P}_{(ab)_0}\sigma = 0, \tag{AE}$$

and parallel standard tractors I. The mapping from almost Einstein scales to parallel tractors is given by $\sigma \mapsto \frac{1}{n}D_A\sigma$ *while the inverse map is* $I_A \mapsto X^A I_A$.

So a parallel tractor is necessarily a scale tractor, as in Definition 3.8, but in general the converse does not hold.

3.4.6. The Tractor Metric

It turns out that the tractor bundle has a beautiful and important structure that is perhaps not expected from its origins via prolongation in Section 3.4.4.

Proposition 3.9 *Let* $(M, \boldsymbol{c})$ *be a conformal manifold of signature* (p,q). *The formula*

$$[V_A]_g = (\sigma, \mu_a, \rho) \mapsto 2\sigma\rho + \boldsymbol{g}^{ab}\mu_a\mu_b =: h(V,V)$$

defines, by polarisation, a signature $(p+1, q+1)$ *metric on* $\mathcal{T}$.

Proof As a symmetric bilinear form field on the bundle $[\mathcal{T}]_g$, h takes the form

$$h(V',V) \stackrel{g}{=} \begin{pmatrix} \sigma' & \mu' & \rho' \end{pmatrix} \begin{pmatrix} 0 & 0 & 1 \\ 0 & \boldsymbol{g}^{-1} & 0 \\ 1 & 0 & 0 \end{pmatrix} \begin{pmatrix} \sigma \\ \mu \\ \rho \end{pmatrix}, \tag{3.29}$$

where $\stackrel{g}{=}$ should be read as 'equals, calculating in the scale g'. So we see that the signature is as claimed. By construction $h(V,V')$ has weight 0. It remains to check the conformal invariance. Here we use the notation from (3.26):

$$\begin{aligned} 2\widehat{\sigma}\widehat{\rho} + \widehat{\mu}^a\widehat{\mu}_a &= 2\sigma(\rho - \Upsilon_c\mu^c - \frac{1}{2}\Upsilon^2\sigma) + (\mu^a + \Upsilon^a\sigma)(\mu_a + \Upsilon_a\sigma) \\ &= 2\sigma\rho + \mu^a\mu_a - 2\sigma\Upsilon\mu - \sigma^2\Upsilon^2 + 2\Upsilon\mu\sigma + \Upsilon^2\sigma^2 \\ &= 2\sigma\rho + \mu^a\mu_a. \end{aligned}$$

□

In the abstract index notation the tractor metric is $h_{AB} \in \Gamma(\mathcal{E}_{(AB)})$, and its inverse h^{BC}.

The standard tractor bundle, as introduced in Sections 3.4.4 and 3.4.5, would more naturally have been defined as the dual tractor bundle. But Proposition (3.9) shows that we have not damaged our development; the tractor bundle is canonically isomorphic to its dual and normally we do not distinguish these, except by the raising and lowering of abstract indices using h_{AB}.

From these considerations we see that there is no ambiguity in viewing X, of (3.28), as a section of $\mathcal{T} \otimes \mathcal{E}[1] \cong \mathcal{T}[1]$. In this spirit we refer to X as the *canonical tractor*. At this point it is useful to note that in view of this canonical self duality $\mathcal{T} \cong \mathcal{T}^*$, and the formula (3.29), we have the following result.

Proposition 3.10 *The canonical tractor X^A is null, in that*

$$h_{AB}X^AX^B = 0.$$

Furthermore $X_A = h_{AB}X^B$ gives the canonical inclusion of $\mathcal{E}[-1]$ into $\mathcal{E}_A$:

$$X_A : \mathcal{E}[-1] \to \mathcal{E}_A.$$

In terms of the decomposition of $\mathcal{E}_A$ given by a choice of metric this is simply

$$\rho \mapsto \begin{pmatrix} 0 \\ 0 \\ \rho \end{pmatrix}.$$

As another immediate application of the metric we observe the following. Given a choice of scale $\sigma \in \mathcal{E}_+[1]$, consider the corresponding scale tractor I, and in particular it squared length $h(I,I)$. This evidently has conformal weight zero and so is a function on $(M,\boldsymbol{c})$ determined only by the choice of scale σ. Explicitly we have

$$h(I,I) \stackrel{g}{=} \boldsymbol{g}^{ab}(\nabla_a\sigma)(\nabla_b\sigma) - \frac{2}{n}\sigma(\mathsf{J}+\Delta)\sigma \tag{3.30}$$

from Definition 3.8 with (3.25) and (3.29). Here we have calculated the right hand side in terms of some metric g in the conformal class (hence the notation $\stackrel{g}{=}$). But since σ is a scale we may, in particular, use $g := \sigma^{-2}\boldsymbol{g}$. Then $\nabla^g\sigma = 0$ and we find the following result.

Proposition 3.11 *On a conformal manifold $(M,\boldsymbol{c})$, let $\sigma \in \mathcal{E}_+[1]$, and $I = \frac{1}{n}D\sigma$ the corresponding scale tractor. Then*

$$h^{AB}I_BI_C = -\frac{2}{n}\mathsf{J}^\sigma,$$

where $\mathsf{J}^\sigma := g^{ab}\mathsf{P}_{ab}$, and $g_{ab} = \sigma^{-2}\boldsymbol{g}_{ab}$.

Note we usually write J to mean the density $\boldsymbol{g}^{ab}\mathsf{P}_{ab}$, as calculated in the scale g. So here $\mathsf{J}^\sigma = \sigma^2\mathsf{J}$. In a nutshell the conformal meaning of scalar curvature is that it is the length squared of the scale tractor (up to a negative constant factor).

Next we come to the main reason the tractor metric is important, namely because it is preserved by the tractor connection. With V as in Proposition 3.9 we have

$$\begin{aligned}\nabla_a h(V,V) &\stackrel{g}{=} 2[\rho\nabla_a\sigma + \sigma\nabla_a\rho + \boldsymbol{g}^{bc}\mu_b\nabla_a\mu_c] \\ &= 2[\rho(\nabla_a\sigma - \mu_a) + \sigma(\nabla_a\rho - P_{ab}\mu^b) + \boldsymbol{g}^{bc}\mu_b(\nabla_a\mu_c + g_{ac}\rho + P_{ac}\sigma)] \\ &= 2h(V, \nabla_a^{\mathcal{T}} V).\end{aligned}$$

We summarise this with the previous result.

Theorem 3.4 *On a conformal manifold $(M, \boldsymbol{c})$ of signature (p,q), the tractor bundle carries a canonical conformally invariant metric h of signature $(p+1, q+1)$. This is preserved by the tractor connection.*

We see at this point that the tractor calculus is beginning to look like an analogue for conformal geometry of the Ricci calculus of (pseudo-)Riemannian geometry: a metric on a manifold canonically determines a unique Levi-Civita connection on the tangent bundle preserving the metric. The analogue here is that a conformal structure of any signature (and dimension at least 3) determines canonically the standard tractor bundle $\mathcal{T}$ equipped with the connection $\nabla^{\mathcal{T}}$ and a metric h preserved by $\nabla^{\mathcal{T}}$.

Note also that the tractor bundle, metric and connection seem to be analogues of the corresponding structures found for the model in Theorem 3.1. Especially in view of the matching filtration structures; (3.27) should be compared with that found on the model in Proposition 3.5 (noting that $TS^n[-1] \cong T^*S^n[1]$). In fact the tractor connection here of Theorem 3.2, and the related structures, generalise the corresponding objects on the model. This follows by more general results in [7], or alternatively may be verified directly by computing the formula for the connection of Theorem 3.1 in terms of the Levi-Civita connection on the round sphere.

3.5. Lecture 4: The Tractor Curvature, Conformal Invariants and Invariant Operators

Let us return briefly to our motivating problems: the construction of invariants and invariant operators.

3.5.1. Tractor Curvature

Since the tractor connection $\nabla^{\mathcal{T}}$ is well defined on a conformal manifold its curvature κ on $\mathcal{E}^A$ depends only on the conformal structure; by construction it is an invariant of that. If we couple the tractor connection with any torsion free connection (in particular with the Levi-Civita connection of any metric in the conformal calss) then, according to our conventions from Lecture 1, the curvature of the tractor connection is given by

$$(\nabla_a \nabla_b - \nabla_b \nabla_a) U^C = \kappa_{ab}{}^C{}_D U^D \quad \text{for all} \quad U^C \in \Gamma(\mathcal{E}^C),$$

where ∇ denotes the coupled connection. Now using the fact that the tractor connection preserves the inverse metric h^{CD} we have

$$0 = (\nabla_a \nabla_b - \nabla_b \nabla_a) h^{CD} = \kappa_{ab}{}^C{}_E h^{ED} + \kappa_{ab}{}^D{}_E h^{CE}.$$

In other words, raising an index with h^{CD}

$$\kappa_{ab}{}^{CD} = -\kappa_{ab}{}^{DC}. \tag{3.31}$$

It is straightforward to compute κ in a scale g. We have

$$(\nabla_a \nabla_b - \nabla_b \nabla_a) \begin{pmatrix} \sigma \\ \mu^c \\ \rho \end{pmatrix} = \begin{pmatrix} 0 & 0 & 0 \\ C_{ab}{}^c & W_{ab}{}^c{}_d & 0 \\ 0 & -C_{abd} & 0 \end{pmatrix} \begin{pmatrix} \sigma \\ \mu^d \\ \rho \end{pmatrix} \tag{3.32}$$

where, recall W is the Weyl curvature and C is the Cotton tensor,

$$C_{abc} := 2\nabla_{[a} \mathrm{P}_{b]c}.$$

In the expression for κ, the zero in the top right follows from the skew symmetry (3.31), while the remaining zeros of the right column show that the canonical tractor X^D annihilates the curvature,

$$\kappa_{ab}{}^C{}_D X^D = 0.$$

This with the skew symmetry determines the top row of the curvature matrix. It follows from the conformal transformation properties of the tractor splittings that the central entry of the matrix is conformally invariant, and this is consistent with the appearance there of the Weyl curvature $W_{ab}{}^c{}_d$. Note that in dimension 3 this necessarily vanishes, and so it follows that then the tractor curvature is fully captured by and equivalent to the Cotton curvature C_{abc}. Again this is consistent with the well known conformal invariance of that quantity in dimension 3. Thus we have the following result.

Proposition 3.12 *The normal conformal tractor connection $\nabla^{\mathcal{T}}$ is flat if and only if the conformal manifold is locally equivalent to the flat model.*

So we shall say a conformal manifold $(M, \boldsymbol{c})$ is conformally flat if $\kappa = 0$.

3.5.1.1. Application: Conformally Invariant Obstructions to Metrics Being Conformal-to-Einstein

Note that as an immediate application we can use the tractor curvature to manufacture easily obstructions to the existence of an Einstein metric in the conformal class $\boldsymbol{c}$.

From Proposition 3.6 and Theorem 3.3 a metric g is Einstein if and only if the corresponding scale tractor $I_A = \frac{1}{n}D_A\sigma$ is parallel, where $g = \sigma^{-2}\boldsymbol{g}$. Thus if g is Einstein then we have

$$\kappa_{ab}{}^{C}{}_{D}I^{D} = 0. \tag{3.33}$$

Recall that $\kappa_{ab}{}^{C}{}_{D}X^{D} = 0$. If at any point p the kernel of $\kappa_{ab}{}^{C}{}_{D} : \mathcal{E}^{D} \to \mathcal{E}_{ab}{}^{C}$ is exactly one dimensional then we say that tractor curvature has maximal rank. We have:

Proposition 3.13 *Let $(M, \boldsymbol{c})$ be a conformal manifold. If at any point $p \in M$ the tractor curvature has maximal rank then there is no Einstein metric in the conformal equivalence class* $\boldsymbol{c}$.

From this observation it is easy to manufacture conformal invariants that must vanish on an Einstein manifold, see [29].

Proof If a metric g is Einstein then σ is a true scale and hence nowhere zero. Thus I_p^D is not parallel to X_p^D, and the result follows from (3.33). □

3.5.2. Toward Tractor Calculus

Although the tractor connection and its curvature are conformally invariant it is still not evident how to manufacture easily conformal invariants. The tractor curvature takes values in $\Lambda^2(T^*M) \otimes \mathrm{End}(\mathcal{T})$ and this is not a bundle on which the tractor connection acts.

To deal with this and the related problem of constructing differential invariants of tensors and densities we need additional tools.

The simplest of these is the Thomas-D operator. Recall that in Proposition 3.8 (cf. (3.25)) we constructed a differential operator $D : \mathcal{E}[1] \to \mathcal{T}$ that was, by construction, tautologically conformally invariant. This generalises. Let us write $\mathcal{E}^{\Phi}[w]$ to denote any tractor bundle of weight w. Then:

Proposition 3.14 *There is a conformally invariant differential operator*

$$D_A : \mathcal{E}^{\Phi}[w] \to \mathcal{E}_A \otimes \mathcal{E}^{\Phi}[w-1],$$

defined in a scale g by

$$V \mapsto [D_A V]_g := \begin{pmatrix} (n+2w-2)wV \\ (n+2w-2)\nabla_a V \\ -(\Delta V + wJV) \end{pmatrix}$$

where ∇ denotes the coupled tractor–Levi-Civita connection.

Proof Under a conformal transformation $g \mapsto \widehat{g} = \Omega^2 g$, $[D_A f]_g$ transforms by (3.26). □

Note that this result is not as trivial as the written proof suggests since V (with any indices suppressed) is a section of any tractor bundle, and the operator is second order. In fact there are nice ways to construct the Thomas-D operator from more elementary invariant operators [8, 25].

3.5.2.1. Application: Differential Invariants of Densities and Weighted Tractors

A key point about the Thomas-D operator is that it can be iterated. For $V \in \Gamma(\mathcal{E}^{\Phi}[w])$ we may form (suppressing all indices):

$$V \mapsto (V,\ DV,\ D \circ DV,\ D \circ D \circ DV, \cdots)$$

and, for generic weights $w \in \mathbb{R}$, in a conformally invariant way this encodes the jets of the section V entirely into weighted tractor bundles. Thus we can proliferate invariants of V.

For example for $f \in \Gamma(\mathcal{E}[w])$ we can form

$$(D^A f) D_A f = -2w(n + 2w - 2) f(\Delta f + w \mathrm{J} f) + (n + 2w - 2)^2 (\nabla^a f)\nabla_a f.$$

By construction this is conformally invariant, for *any weight* w. So in fact it is a family of invariants (of densities). Similarly we may form

$$(D^A D^B f) D_A B_B f = -2(n + 2w - 4)(n + 2w - 2)w(w - 1) f \Delta^2 f \\ + \textit{lower order terms},$$

and so forth.

3.5.2.2. Conformal Laplacian-Type Linear Operators

One might hope that the tractor-D operator is also effective for the construction of conformally invariant linear differential operators. In particular the construction of Laplacian type operators is important. Certainly $D^A D_A$ is by construction conformally invariant but, on any weighted tractor bundle:

$$D^A D_A = 0.$$

This is verified by a straightforward calculation, but it should not be surprising as by construction it would be invariant on, for example, densities of any weight. On the standard conformal sphere it is well-known that there is no non-trivial operator with this property.

From our earlier work we know the domain bundle of the conformal Laplacian is $\mathcal{E}[1 - \frac{n}{2}]$. Observe that for $V \in \Gamma(\mathcal{E}^{\Phi}[1 - \frac{n}{2}])$ we have $(n + 2(1 - \frac{n}{2}) - 2) = 0$, and

$$D_A f \stackrel{g}{=} \begin{pmatrix} 0 \\ 0 \\ -(\Delta + \frac{2-n}{2}\mathsf{J})V \end{pmatrix}, \quad \text{that is} \quad D_A V = -X_A \Box V$$

where $\Box$ is the (tractor-twisted) conformal Laplacian. In particular, the proof of Proposition 3.14 was also a proof of this result:

Lemma 3.2 *The operator* $\Box \stackrel{g}{=} (\Delta + \frac{2-n}{2}\mathsf{J})$ *is a conformally invariant differential operator*

$$\Box : \mathcal{E}^{\Phi}\left[1 - \frac{n}{2}\right] \to \mathcal{E}^{\Phi}\left[-1 - \frac{n}{2}\right],$$

where $\mathcal{E}^{\Phi}$ *is any tractor bundle.*

This is already quite useful, as the next exercise shows.

Exercise 5 Show that if $f \in \mathcal{E}[2 - \frac{n}{2}]$ then

$$\Box D_A f = -X_A P_4 f. \tag{3.34}$$

Thus there is a conformally invariant differential operator

$$P_4 : \mathcal{E}\left[2 - \frac{n}{2}\right] \to \mathcal{E}\left[-2 - \frac{n}{2}\right] \quad \text{where} \quad P_4 \stackrel{g}{=} \Delta^2 + \textit{lower order terms}.$$

In fact, this is the celebrated *Paneitz operator* discovered by Stephen Paneitz in 1983, see [46] for a reproduction of his preprint from the time. See [25, 30] for further discussion and generalisations.

An important point is that (3.34) does *not* hold if we replace f with a tractor field of the same weight! Those who complete the exercise will observe the first hint of this subtlety, in that during the calculation derivatives will need to be commuted.

3.5.3. Splitting Tractors

Although the importance of the tractor connection stems from its conformal invariance we need efficient ways to handle the decomposition of the tractor bundle corresponding to a choice of scale.

Recall that a metric $g \in \boldsymbol{c}$ determines an isomorphism

$$\begin{matrix} \mathcal{E}[1] \\ \oplus \\ \mathcal{E}_a[1] \\ \oplus \\ \mathcal{E}[-1] \end{matrix} \stackrel{\simeq}{\longrightarrow} \mathcal{E}^A, \quad \text{mapping} \quad \begin{pmatrix} \sigma \\ \mu_a \\ \rho \end{pmatrix} = [U^A]_g \mapsto U^A \in \mathcal{E}^A. \tag{3.35}$$

	Y^A	Z^{Ac}	X^A
Y_A	0	0	1
Z_{Ab}	0	$\delta_b{}^c$	0
X_A	1	0	0

Figure 3.3. Tractor inner product

The inclusion of $\mathcal{E}[-1]$ into the direct sum followed by this map is just $X^A : \mathcal{E}[-1] \to \mathcal{E}^A$, as observed earlier in Proposition (3.10). This is conformally invariant. However let us now fix the notation

$$Z^{Aa} : \mathcal{E}_a[1] \to \mathcal{E}^A, \quad \text{and} \quad Y^A : \mathcal{E}[1] \to \mathcal{E}^A, \tag{3.36}$$

for the other two bundle maps determined by (3.35). We call these (along with X^A) the *tractor projectors* and view them as bundle sections $Z^{Aa} \in \Gamma(\mathcal{E}^{Aa})[-1]$, and $Y^A \in \mathcal{E}^A[-1]$. So in summary $[U^A]_g = (\sigma, \mu_a, \rho)$ is equivalent to

$$U^A = Y^A\sigma + Z^{Aa}\mu_a + X^A\rho. \tag{3.37}$$

Using the formula (3.29) for the tractor metric it follows at once that $X^A Y_A = 1$, $Z^{Aa}Z_{Ab} = \delta^a_b$ and all other quadratic combinations of the X, Y and Z are zero as summarised in Figure 3.3. Thus we also have $Y_A U^A = \rho, X_A U^A = \sigma$, $Z_{Ab}U^A = \mu_b$ and the metric may be decomposed into a sum of projections, $h_{AB} = Z_A{}^c Z_{Bc} + X_A Y_B + Y_A X_B$.

The projectors Y and Z depend on the metric $g \in \boldsymbol{c}$. If $\hat{Y}^A$ and $\hat{Z}^A{}_b$ are the corresponding quantities in terms of the metric $\hat{g} = \Omega^2 g$ then altogether we have

$$\widehat{X}^A = X^A,\ \widehat{Z}^{Ab} = Z^{Ab} + \Upsilon^b X^A,\ \widehat{Y}^A = Y^A - \Upsilon_b Z^{Ab} - \tfrac{1}{2}\Upsilon_b\Upsilon^b X^A \tag{3.38}$$

as follows immediately from (3.37) and (3.26).

Remark 3.7 In the notation $Z^A{}_a$ the tractor and tensor indices are both abstract. If we move to a concrete frame field for TM, $e_1, \cdots, e_n$ and then write $Z^A{}_i$, $i = 1, \cdots, n$, for the contraction $Z^A{}_a e^a_i$, we come to the (weighted) tractor frame field:

$$X^A, Z^A{}_1, \cdots, Z^A{}_n, Y^A.$$

This is a frame for the tractor bundle adapted to the filtration (3.27) and metric, as reflected in the conformal and inner product properties described in (3.38) and Figure 3.3. ∎

Of course tensor products of the tractor bundle are also decomposed by the isomorphism (3.35) and this is described by the tensor products of the projectors in an obvious way. To illustrate consider the case of the bundle of *tractor k-forms* $\Lambda^k\mathcal{T}$ (which note is non-zero for $k = 0, \cdots, n+2$). The composition series $\mathcal{T} = \mathcal{E}[1] \oplus T^*M[-1] \oplus \mathcal{E}[-1]$ determines the composition series for $\Lambda^k\mathcal{T}$,

$$\Lambda^k\mathcal{T} \cong \Lambda^{k-1}[k] \oplus \begin{array}{c}\Lambda^{k-2}[k-2]\\ \oplus \\ \Lambda^k[k]\end{array} \oplus \Lambda^{k-1}[k-2]. \tag{3.39}$$

Given a choice of metric g from the conformal class there is a splitting of this composition series corresponding to the splitting (3.35) of $\mathcal{T}$, and this is easily computed and dealt with using the 'projectors' X, Y and Z, see for example [26].

3.5.4. The Connection

Just as connections are often described in terms of their action on a suitable frame field it is useful to give the tractor connection in terms of its action on the projectors X, Y and Z. The tractor covariant derivative of a field $U^A \in \Gamma(\mathcal{E}^A)$, as in (3.37), is given by (3.22). Using the isomorphism (3.35) this is written

$$\nabla_a U^B = Y^B(\nabla_a\sigma - \mu_a) + Z^{Bb}(\nabla_a\mu_b + g_{ab}\rho + \mathsf{P}_{ab}\sigma) + X^B(\nabla_a\rho - \mathsf{P}_{ab}\mu^b).$$

On the other hand applying the connection directly to U^A expanded as in (3.37) we have

$$\nabla_a U^B = Y^B\nabla_a\sigma + \sigma\nabla_a Y^B + Z^{Bb}\nabla_a\mu_b + \mu_b\nabla_a Z^{Bb} + X^B\nabla_a\rho + \rho\nabla_a X^B,$$

where we have used the Leibniz rule for ∇_a, viewed as the coupled tractor-Levi-Civita connection. Comparing these we obtain:

$$\nabla_a X^B = Z^B{}_a\,, \quad \nabla_a Z^B{}_b = -\mathsf{P}_{ab}X^B - g_{ab}Y^B\,, \quad \nabla_a Y^B = \mathsf{P}_{ab}Z^{Bb}. \tag{3.40}$$

This gives the transport equations for the projectors (and determines these for the adapted frame as in Remark 3.7). From a practical point of view the formulae (3.40) here enable the easy computation of the connection acting on a tractor field of any valence.

3.5.4.1. Application: Computing and Conformal Laplacian Operators

The formulae are effective for reducing most tractor calculations to a routine task. For example suppose we want to compute $D^AX_Af = D^A(X_Af)$ for

$f \in \Gamma(\mathcal{E}^{\Phi}[w])$, that is a section of any weighted tractor bundle. We calculate in some scale g. First note that

$$\begin{aligned}
X^A(\Delta + (w+1)\mathrm{J})X_A f &= X^A \Delta X_A f, \\
&= X^A[\Delta, X_A] f, \\
&= X^A \nabla^b [\nabla_b, X_A] f + X^A [\nabla_b, X_A] \nabla^b f \\
&= X^A (\nabla^b Z_{bA}) f + 2X^A Z_{bA} \nabla^b f \\
&= X^A(-\mathrm{J} X_A - n Y_A) f \\
&= -nf,
\end{aligned}$$

where we used $X^A X_A = 0$, $X^A Z_{bA} = 0$ and $X^A Y_A = 1$. Thus we have

$$\begin{aligned}
D^A X_A f &= [(w+1)(n+2w)Y^A + (n+2w)Z^{Aa}\nabla_a - X^A \Delta] X_A f \\
&= (w+1)(n+2w)f + (n+2w)Z^{aA} Z_{aA} f + nf \\
&= (w+1)(n+2w)f + (n+2w)nf + nf,
\end{aligned}$$

and collecting terms we come to

$$D^A X_A f = (n + 2w + 2)(n + w) f. \tag{3.41}$$

A straightforward induction using (3.41) and (3.40) then enables us to show that:

Theorem 3.5 *On a conformal manifold $(M^n, \boldsymbol{c})$, for any tractor bundle $\mathcal{E}^{\Phi}$, and for each $k \in \mathbb{Z}_{\geq 1}$ if n is odd (or $k \in \mathbb{Z}_{\geq 1}$ and $2k < n$ if n is even) there is a conformally invariant differential operator*

$$\square_{2k} : \mathcal{E}^{\Phi}\left[k - \frac{n}{2}\right] \to \mathcal{E}^{\Phi}\left[-k - \frac{n}{2}\right]$$

of the form Δ^k + lower order terms (up to a non-zero constant factor). These are given by

$$\square_{2k} := D^{A_1} \cdots D^{A_{k-1}} \square D_{A_{k-1}} \cdots D_{A_1}.$$

Remark 3.8 The operators in the theorem were first reported in [25], (and with a different proof) as part of joint work of the second author with M. G. Eastwood.

Acting on densities of weight $(k-n/2)$ there are the Graham–Jenne–Mason–Sparling (GJMS) operators of [36]. For $k \geq 3$ these operators of Theorem 3.5 differ from the GJMS operators, as follows easily from the discussion in [30]. ∎

3.5.5. Constructing Invariants

We can now put together the above tools and proliferate curvature invariants. As a first step, following [27] we may form

$$W_{AB}{}^{K}{}_{L} := \frac{3}{n-2} D^P X_{[P} Z_A{}^a Z^b_{B]} \kappa_{ab}{}^{K}{}_{L}.$$

This is conformally invariant by construction, as it is immediate from (3.38) that $X_{[P} Z_A{}^a Z^b_{B]}$ is conformally invariant. It is exactly the object $\mathbb{X}^3$ which, for example, gives the conformally invariant injection

$$\mathbb{X}^3 : \Lambda^2[1] \to \Lambda^3 \mathcal{T}.$$

It turns out that the W-tractor W_{ABCD} has the symmetries of an algebraic Weyl tensor. In fact in a choice of conformal scale, W_{ABCE} is given by

$$\begin{aligned}(n-4)\big(&Z_A{}^a Z_B{}^b Z_C{}^c Z_E{}^e W_{abce} - 2Z_A{}^a Z_B{}^b X_{[C} Z_{E]}{}^e C_{abe} \\ &-2X_{[A} Z_{B]}{}^b Z_C{}^c Z_E{}^e C_{ceb}\big) + 4X_{[A} Z_{B]}{}^b X_{[C} Z_{E]}{}^e B_{eb},\end{aligned} \tag{3.42}$$

where C_{abc} is the Cotton tensor, and

$$B_{ab} := \nabla^c C_{cba} + \mathsf{P}^{dc} W_{dacb}, \tag{3.43}$$

see [31]. Note that from (3.42) it follows that, in dimension 4, B_{eb} is conformally invariant. This is the *Bach tensor*.

Since W_{ABCD} takes values in a weighted tractor bundle we may apply the Thomas-D to this to capture jets of structure invariantly

$$W_{ABCD} \mapsto (W,\ DW,\ D \circ DW,\ D \circ D \circ DW, \cdots)$$

following the idea of Section 3.5.2.1. Contractions of these terms then yield invariants. On odd dimensional manifolds this idea with some minor additional input produces a generating set of scalar conformal invariants [27]. An alternative (but closely related [7]) approach to the construction of conformal invariants uses the Fefferman–Graham ambient metric [3, 17]. With either approach, in even dimensions the situation is far more subtle, deeper ideas are needed, and even for the construction of scalar conformal invariants open problems remain (as mentioned in the sources referenced).

3.6. Lecture 5: Conformal Compactification of Pseudo-Riemannian Manifolds

At this point we change directions from developing the theory of tractor calculus for conformal geometries to applying this calculus to various problems inspired by asymptotic analysis in general relativity. We will see that the conformal treatment of infinity in general relativity fits very nicely with the conformal tractor calculus, and our basic motivation will be to produce results which are useful in this setting. For the most part we will work fairly generally however, and much of what is presented will be applicable in other situations involving hypersurfaces or boundaries (such as Cauchy surfaces or various kinds of horizons in general relativity, or to the study of Poincaré–Einstein metrics in differential geometry). From this point on we will adopt the convention that $d = n + 1$ (rather than n as before) is the dimension of our manifold M, so that n will be the dimension of ∂M or of a hypersurface in M.

3.6.1. Asymptotic Flatness and Conformal Infinity in General Relativity

It is natural in seeking to understand and describe the physics of general relativity to want to study isolated systems. In particular we want to be able to understand the mass and/or energy of the system (as well as other physical quantities) and how the system radiates gravitational energy (or how the system interacts with gravitational radiation coming in from infinity, i.e. from outside the system). Clearly it is quite unnatural to try to isolate a physical system from a spacetime by simply considering the inside of a timelike tube containing the system (events inside the tube would depend causally on events outside, and there would be no natural choice of such a tube anyway). From early on in the history of general relativity then, physicists have sought to define isolated systems in terms of spacetimes which are *asymptotically flat*, that is which approach the geometry of Minkowski space in a suitable way as you approach 'infinity'.

Definitions of asymptotic flatness typically involve the existence of special coordinate systems in which the metric components and other physical fields fall off sufficiently quickly as you approach infinity (infinity being defined by the coordinate system). Once a definition of asymptotically flat spacetimes is established one can talk about the *asymptotic symmetry group* of such spacetimes (which is not in general the Poincare group) and the corresponding physical quantities such as mass and energy. One must be careful in defining the asymptotic flatness condition not to require the fields to fall off too fast (thereby excluding massive spacetimes or gravitationally radiating ones) or too slow (so that one loses the asymptotic symmetries needed to define physical quantities).

A condition on asymptotically flat spacetimes which came to be seen as important is the *Sachs peeling-off property*, which is the condition that along a future directed null geodesic which goes out to infinity with affine parameter λ the Weyl curvature $W^a{}_{bcd}$ satisfies

$$W = \frac{W^{(4)}}{\lambda} + \frac{W^{(3)}}{\lambda^2} + \frac{W^{(2)}}{\lambda^3} + \frac{W^{(1)}}{\lambda^4} + O\left(\frac{1}{\lambda^5}\right)$$

where each tensor $W^{(k)}$ is of special algebraic type with the tangent vector to the null geodesic as a k-fold repeated principal null direction. Typically asymptotically flat spacetimes were required to be Ricci flat near infinity, so the Weyl curvature was in fact the full curvature tensor in this region; the term $W^{(4)}$ in the asymptotic expansion was interpreted as the gravitational radiation reaching infinity.

The mass associated with this approach to isolated systems is called the *Bondi mass*, which is not a conserved quantity but satisfies a *mass loss formula* as energy is radiated away from the spacetime. The asymptotic symmetry group (which is the same abstract group for any spacetime) is known as the *BMS* (Bondi–Metzner–Sachs) *group*. There is an alternative way of defining asymptotic flatness using the 3+1 formalism. In this case one talks about the spacetime being *asymptotically flat at spatial infinity*, the corresponding mass is the *ADM* (Arnowitt–Deser–Misner) *mass*, and the asymptotic symmetry group is known as the *spatial infinity (Spi) group*. Questions of how these two notions of asymptotic flatness are related to each other can be subtle and tricky. For a good introductory survey of these issues see the chapter on asymptotic flatness in [56].

After a great deal of work in these areas, a new approach was suggested by Roger Penrose in the 1960s [47, 48, 50]. Penrose required of an asymptotically flat spacetime that the conformal structure of spacetime extend to a pair of null hypersurfaces called future and past null infinity. This mirrored the conformal compactification of Minkowski space obtained by adding a lightcone at infinity (see Figure 3.4). It was quickly shown that (i) Penrose's notion of conformal infinity satisfied the appropriate uniqueness property [22], (ii) this form of asymptotic flatness implied the peeling property [47, 50], and (iii) the corresponding group of asymptotic symmetries was the usual BMS group [23]. It is not totally surprising that this idea worked out so well: the causal (or light-cone) structure of spacetime, which is encoded by the conformal structure, had played an important role in the analysis of gravitational radiation up to that point. The conformal invariance of the zero rest mass equations for arbitrary spin particles and their peeling-off properties also fit nicely with Penrose's proposal, providing further motivation. What is really nice about this approach however is that it is both natural and coordinate free.

Let us now give the formal definition(s), following [19].

Definition 3.9 A smooth (time and space-orientable) spacetime (M_+, g_+) is called *asymptotically simple* if there exists another smooth Lorentzian manifold (M, g) such that

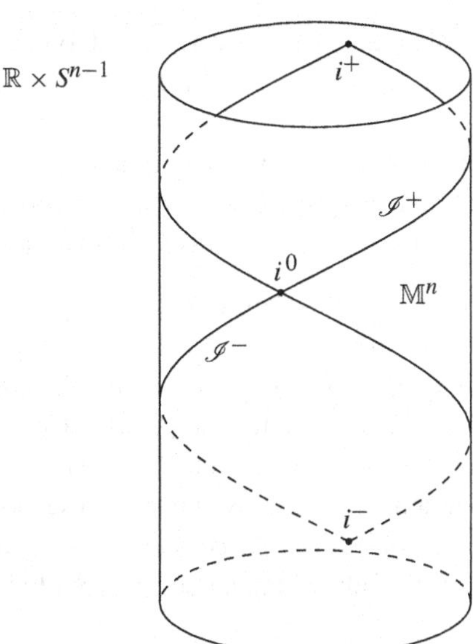

Figure 3.4. The standard conformal embedding of n-dimensional Minkowski space $\mathbb{M}^n$ into the Einstein cylinder

(i) M_+ is an open submanifold of M with smooth boundary $\partial M_+ = \mathscr{I}$;
(ii) There exists a smooth scalar field Ω on M, such that $g = \Omega^2 g_+$ on M_+, and so that $\Omega = 0$, $\mathrm{d}\Omega \neq 0$ on $\mathscr{I}$;
(iii) Every null geodesic in M_+ acquires a future and a past endpoint on $\mathscr{I}$.

An asymptotically simple spacetime is called *asymptotically flat* if in addition $\mathrm{Ric}^{g_+} = 0$ in a neighbourhood of $\mathscr{I}$.

The Lorentz manifold (M, g) is commonly referred to as the *unphysical spacetime*, and g is called the *unphysical metric*. One can easily see that $\mathscr{I}$ must have two connected components (assuming M_+ is connected) $\mathscr{I}^+$ and $\mathscr{I}^-$ consisting of the future and past endpoints of null geodesics respectively. It is also easy to see that $\mathscr{I}^+$ and $\mathscr{I}^-$ are both smooth null hypersurfaces. Usually (in the four-dimensional setting) $\mathscr{I}^+$ and $\mathscr{I}^-$ will have topology $\mathbb{S}^2 \times \mathbb{R}$, however this is not necessarily the case.

Remark 3.9 The third condition is in some cases too strong a requirement, for instance in a Schwarzschild black hole spacetime there are null geodesics which circle about the singularity forever. To include Schwarzschild and other such spacetimes one must talk about *weakly asymptotically flat* spacetimes (see [49]). ∎

Remark 3.10 (For those wondering 'whatever happened to the cosmological constant?') One can also talk about *asymptotically de Sitter* spacetimes in which case the boundary hypersurface(s) will be spacelike and one asks for the spacetime to be Einstein, rather than Ricci flat, in a neighbourhood of the conformal infinity. Similarly one can talk about *asymptotically anti-de Sitter* spacetimes, which have a timelike hypersurface as conformal infinity. ∎

Just how many spacetimes satisfy this definition of asymptotic flatness, and how do we get our hands on them? These questions lead us to the discussion of Friedrich's conformal field equations. We will see that the conformal field equations arise very naturally from the tractor picture. However, first we will discuss more generally the mathematics of conformal compactification (inspired by the conformal treatment of infinity above) for pseudo-Riemannian manifolds, and the geometric constraints placed on such 'compactifications'. Again we shall see that this fits nicely within a tractor point of view.

3.6.2. Conformal Compactification

A pseudo-Riemannian manifold (M_+, g_+) is said to be *conformally compact* if M_+ can be identified with the interior of a smooth compact manifold with boundary M and there is a defining function r for the boundary $\Sigma = \partial M_+$ so that

$$g_+ = r^{-2} g \quad \text{on } M_+$$

where g is a smooth metric on M. In calling r a *defining function* for Σ we mean that r is a smooth real valued function on M such that Σ is exactly the zero locus $\mathcal{Z}(r)$ of r and furthermore that $\mathrm{d}r$ is non-zero at every point of Σ.

In the case that $dr|_\Sigma$ is nowhere null, such a g induces a metric $\overline{g}$ on Σ, but the defining function r may be replaced by $r' = f \cdot r$ where f is any non-vanishing function. This changes $\overline{g}$ conformally and so canonically the boundary has a conformal structure determined by g_+, but no metric. The metric g_+ is then complete and $(\Sigma, \overline{c})$ is sometimes called the conformal infinity of M_+. Actually for our discussion here we are mainly interested in the structure and geometry of the boundary and the asymptotics of M_+ near this, so it is not really important to us that M is compact. An important problem is how to link the conformal geometry and conformal field theory of Σ to the corresponding pseudo-Riemannian objects on M_+.

We will see that certainly this kind of 'compactification' is not always possible, but it is useful in a number of settings. Clearly conformal geometry is involved. In a sense it arises in two ways (that are linked). Most obviously the boundary has a conformal structure. Secondly the interior was conformally rescaled to obtain the metric g that extends to the boundary. This suggests a strong role for conformal geometry – and it is this that we want to discuss.

Toward our subsequent discussion let us first make a first step by linking this notion of compactification to our conformal tools and notations. Let us fix a choice of r and hence g above. If $\tau \in \Gamma(\mathcal{E}[1])$ is any non-vanishing scale on M then $\sigma := r\tau$ is also a section of $\mathcal{E}[1]$ but now with zero locus $\mathcal{Z}(\sigma) = \Sigma$. This satisfies that $\nabla^g \sigma$ is nowhere zero along Σ and so we say that σ is a *defining density* for Σ. Clearly we can choose τ so that

$$g = \tau^{-2}\boldsymbol{g}$$

where $\boldsymbol{g}$ is the conformal metric on M. Then

$$g_+ = \sigma^{-2}\boldsymbol{g}.$$

Thus we may think of a conformally compact manifold, as defined above, as a conformal manifold with boundary $(M, \boldsymbol{c})$ equipped with a section $\sigma \in \Gamma(\mathcal{E}[1])$ that is a defining density for the boundary ∂M.

The key examples of conformally compactified manifolds fit nicely within this framework. Indeed, consider the Poincare ball model of hyperbolic space $\mathbb{H}^d$. This conformal compactification can be realised by considering the extra structure induced on the conformal sphere $(S^d, \boldsymbol{c})$ by a choice of constant spacelike tractor field I on S^d (i.e. a constant vector I in $\mathbb{V}$). This choice gives a symmetry reduction of the conformal group $G = \mathrm{SO}_+(\mathcal{H})$ to the subgroup $H \cong \mathrm{SO}_+(d, 1)$ which stabilises I. On each of the two open orbits of S^d under the action of H there is induced a hyperbolic metric, whereas the closed orbit (an n-sphere) receives only the conformal structure induced from $(S^d, \boldsymbol{c})$. The two open orbits correspond to the two regions where the scale σ of I is positive and negative respectively, and the closed orbit is the zero locus of σ. If we choose coordinates X^A for $\mathbb{V}$ as before, then the 1-density σ corresponds to the homogeneous degree one polynomial $\tilde{\sigma} = I_A X^A$ (dualising I using $\mathcal{H}$); that $\sigma^{-2}\boldsymbol{g}$ gives a hyperbolic metric on each of the open orbits can be seen by noting that the hyperplanes $I_A X^A = \pm 1$ intersect the future light cone $\mathcal{N}_+$ in hyperbolic sections, and the fact that $\mathcal{Z}(\sigma)$ receives only a conformal structure can be seen from the fact that the hyperplane $I_A X^A = 0$ intersects $\mathcal{N}_+$ in a subcone. Thus we see that the scale corresponding to a constant spacelike tractor I on $(S^d, \boldsymbol{c})$ gives rise to a decomposition of the conformal d-sphere into two copies of conformally compactified hyperbolic space glued along their boundaries (see Figure 3.5).

One can repeat the construction above in the case of the Lorentzian signature model space $S^n \times S^1$ in which case one obtains a decomposition of $S^n \times S^1$ into two copies of conformally compactified $AdS^d/\mathbb{Z}$ glued together along their

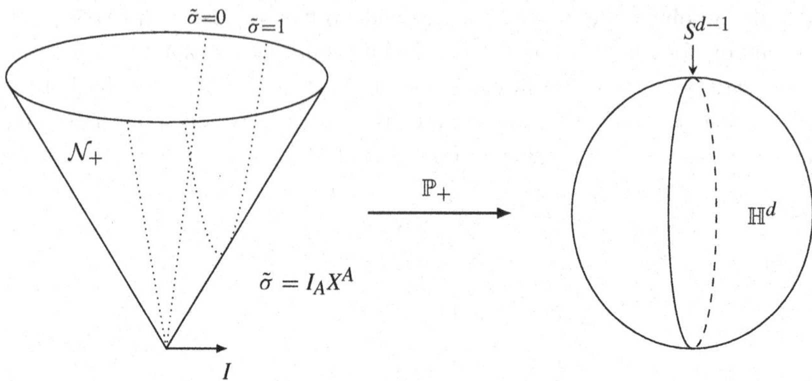

Figure 3.5. The orbit decomposition of the conformal sphere corresponding to the subgroup H of the conformal group G preserving a fixed spacelike vector I. An open orbit may be thought of as $\mathbb{H}^d$ conformally embedded into $\mathbb{S}^d$

conformal infinities (where the $\mathbb{Z}$-action on anti-de Sitter space is given by integer times 2π time translations); this conformal compactification of $AdS^d/\mathbb{Z}$ corresponds to the usual embedding of AdS^d into (half of) the Einstein universe after we take the quotient by $\mathbb{Z}$. If one instead takes a constant null tractor I on $S^n \times S^1$ then the corresponding scale σ gives a decomposition of $S^n \times S^1$ into two copies of conformally compactified Minkowski space glued along their null boundaries (plus two isolated points where $\mathcal{Z}(\sigma)$ has a double cone-like singularity).

3.6.3. Geometry of Scale

There is an interesting perspective which emerges from the above correspondence between constant (parallel) tractors and conformal compactifications. This may be motivated by Proposition 3.7 and Theorem 3.3 which linked almost Einstein scales to parallel tractors. Together they imply:

Proposition 3.15 *An Einstein manifold (M,g) is the same as a conformal manifold $(M,\boldsymbol{c})$ equipped with a parallel standard tractor I such that $\sigma := X^A I_A$ is nowhere zero: Given $\boldsymbol{c}$ and I the metric is recovered by*

$$g = \sigma^{-2}\boldsymbol{g}. \tag{3.44}$$

Conversely a metric g determines $\boldsymbol{c} := [g]$, and $\sigma \in \mathcal{E}_+[1]$ by (3.44) *again, now used as an expression for this variable. Then $I = \frac{1}{n}D\sigma$.*

So this suggests that if we wish to draw on conformal geometry then using the package $(M, \boldsymbol{c}, I)$ may give perspectives not easily seen via the (M, g) framework. Some questions arise:

(1) First the restriction that $\sigma := X^A I_A$ is nowhere zero seems rather unnatural from this point of view. So what happens if we drop that? Then, if I is a parallel standard tractor, recall we say that $(M, \boldsymbol{c}, I)$ is an almost Einstein structure. What, for example, does the zero locus $\mathcal{Z}(\sigma)$ of $\sigma = X^A I_A$ look like in this case?
(2) Is there a sensible way to drop the Einstein condition and use this approach on general pseudo-Riemannian manifolds?

3.6.3.1. The Zero Locus of Almost Einstein and Almost Scalar Constant (ASC) Manifolds

In fact there is now known a way to answer the first question via a very general theory. Using 'the package $(M, \boldsymbol{c}, I)$' amounts to recovering the underlying pseudo-Riemannian structure (and its generalisations as below) as a type of *structure group reduction* of a conformal Cartan geometry. If I is parallel then this a *holonomy reduction*. On an extension of the conformal Cartan bundle to a principal bundle with fibre group $G = O(p+1, q+1)$, the parallel tractor I gives a bundle reduction to a principal bundle with fibre H where this is: (i) $O(p, q+1)$, if I is spacelike; (ii) $O(p+1, q)$, if I is timelike; and (iii) a pseudo-Euclidean group $\mathbb{R}^d \rtimes O(p, q)$, if I is null.

By the general theory of Cartan holonomy reductions [11] any such reduction yields a canonical stratification of the underlying manifold into a disjoint union of *curved orbits*. These are parametrised by $H \backslash G/P$, as are the orbits of H on G/P. Each curved orbit is an initial submanifold carrying a canonically induced Cartan geometry of the same type as that of the corresponding orbit in the model. Furthermore this curved orbit decomposition must look locally like the decomposition of the model G/P into H-orbits. This means that in an open neighbourhood U of any point on M there is a diffeomorphism from U to an open set $U_\circ$ in the model that maps each curved orbit (intersected with U) diffeomorphically to the corresponding H-orbit (intersected with $U_\circ$) of G/P.

In our current setting P is the stabiliser in G of a null ray in $\mathbb{R}^{p+1,q+1}$ and the model G/P is isomorphic to conformal $S^d \times \{1, -1\}$, if $\boldsymbol{c}$ is Riemannian, and conformal $S^p \times S^q$ otherwise. The curved orbits arise here because as we move around the manifold the algebraic relationship between the parallel object I and the canonical tractor X changes. (For general holonomy reductions of Cartan geometries the situation is a simple generalisation of this.) In particular, in Riemannian signature (or if $I^2 \neq 0$), it is easily verified using these tools that the curved orbits (and the H-orbits on the model) are distinguished by the strict sign of $\sigma = X^A I_A$, see [11, Section 3.5]. By examining these sets on the model we conclude.

Theorem 3.6 *The curved orbit decomposition of an almost Einstein manifold* $(M, \boldsymbol{c}, I)$ *is according to the strict sign of* $\sigma = I_A X^A$. *The zero locus satisfies:*

- *If* $I^2 \neq 0$ *(i.e.* g^o *Einstein and not Ricci flat) then* $\mathcal{Z}(\sigma)$ *is either empty or is a smooth embedded hypersurface.*
- *If* $I^2 = 0$ *(i.e.* g^o *Ricci flat) then* $\mathcal{Z}(\sigma)$ *is, after possibly excluding isolated points from* $\mathcal{Z}(\sigma)$*, either empty or a smooth embedded hypersurface.*

Here g^o means the metric $\sigma^{-2}\boldsymbol{g}$ on the open orbits (where σ is nowhere zero).

Remark 3.11 Much more can be said using the tools mentioned. For example in the case of Riemannian signature it is easily shown that:

- If $I^2 < 0$ (i.e. g^o Einstein with positive scalar curvature) then $\mathcal{Z}(\sigma)$ is empty.
- If $I^2 = 0$ (i.e. g^o Ricci flat) then $\mathcal{Z}(\sigma)$ is either empty or consists of isolated points.
- If $I^2 > 0$ (i.e. g^o Einstein with negative scalar curvature) then $\mathcal{Z}(\sigma)$ is either empty or is a smooth embedded separating hypersurface.

This holds because this is how things are locally on the flat model [24, 11]; if $I^2 < 0$ then on the model we have a round metric induced on the whole conformal sphere, the one open orbit is the whole space; if $I^2 = 0$ then on the model we are looking at the one point conformal compactification of Euclidean space given by inverse stereographic projection, and there are two orbits, one open and one an isolated point, so in the curved case $\mathcal{Z}(\sigma)$ is either empty or consists of isolated points; if $I^2 > 0$ then on the model we are looking at two copies of conformally compactified hyperbolic space glued along their boundaries as discussed earlier, in this case the closed orbit is a separating hypersurface, so in the curved case $\mathcal{Z}(\sigma)$ is either empty or is a smooth embedded separating hypersurface (see Figure 3.6). In the Lorentzian setting one can obtain a similar improvement of Theorem 3.6 by considering the model cases. ∎

In fact similar results hold in greater generality (related to question (2) above) and this is easily seen using the earlier tractor calculus and elementary considerations. We learn from the Einstein case above that an important role is played by the scale tractor

$$I_A = \frac{1}{d} D_A \sigma.$$

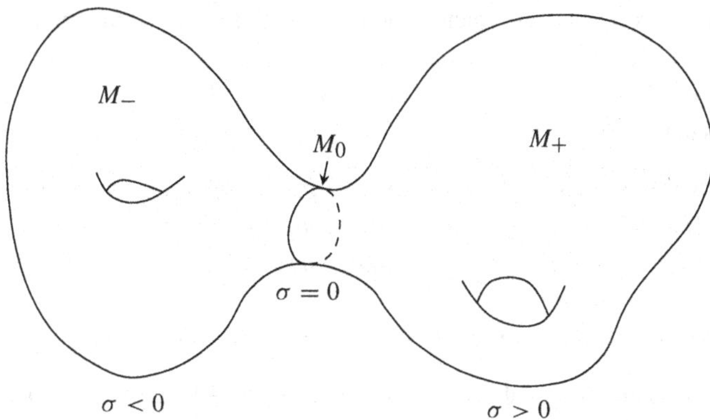

Figure 3.6. The curved orbit decomposition of an almost Einstein manifold with $\mathcal{Z}(\sigma)$ an embedded separating hypersurface

In the Einstein case this is parallel and hence non-zero everywhere. Let us drop the condition that I is parallel and for convenience say that a structure

$$(M^d, \boldsymbol{c}, \sigma) \quad \text{where} \quad \sigma \in \Gamma(\mathcal{E}[1])$$

is *almost pseudo-Riemannian* if the scale tractor $I_A := \frac{1}{d} D_A \sigma$ is nowhere zero. Note then that σ is non-zero on an open dense set, since $D_A\sigma$ encodes part of the 2-jet of σ. So on an almost pseudo-Riemannian manifold there is the pseudo-Riemannian metric $g^o = \sigma^{-2}\boldsymbol{g}$ on the same open dense set. In the following the notation I will always refer to a scale tractor, so $I = \frac{1}{d}D\sigma$, for some $\sigma \in \Gamma(\mathcal{E}[1])$. Then we often mention I instead of σ and refer to $(M, \boldsymbol{c}, I)$ as an almost pseudo-Riemannian manifold.

Now recall from (3.30) that

$$I^2 \stackrel{g}{=} \boldsymbol{g}^{ab}(\nabla_a\sigma)(\nabla_b\sigma) - \frac{2}{d}\sigma(\mathrm{J} + \Delta)\sigma \tag{3.45}$$

where g is any metric from $\boldsymbol{c}$ and ∇ its Levi-Civita connection. This is well-defined everywhere on an almost pseudo-Riemannian manifold, while according to Proposition 3.11, where σ is non-zero, it computes

$$I^2 = -\frac{2}{d}\mathrm{J}^{g^o} = -\frac{\mathrm{Sc}^{g^o}}{d(d-1)} \quad \text{where} \quad g^o = \sigma^{-2}\boldsymbol{g}.$$

Thus I^2 gives a generalisation of the scalar curvature (up to a constant factor $-1/d(d-1)$); it is canonical and smoothly extends the scalar curvature to include the zero set of σ. We shall use the term *ASC manifold* to mean an almost pseudo-Riemannian manifold with I^2 = *constant*. Since the tractor connection preserves the tractor metric, an almost Einstein manifold is a special case, just as Einstein manifolds have constant scalar curvature.

Much of the previous theorem still holds in the almost pseudo-Riemannian setting when $I^2 \neq 0$:

Theorem 3.7 *Let $(M, \boldsymbol{c}, I)$ be an almost pseudo-Riemannian manifold with $I^2 \neq 0$. Then $\mathcal{Z}(\sigma)$, if not empty, is a smooth embedded separating hypersurface. This has a spacelike (resp. timelike) normal if g has negative scalar (resp. positive) scalar curvature.*

If $\boldsymbol{c}$ has Riemannian signature and $I^2 < 0$ then $\mathcal{Z}(\sigma)$ is empty.

Proof This is an immediate consequence of (3.45). Along the zero locus $\mathcal{Z}(\sigma)$ of σ (assuming it is non-empty) we have

$$I^2 = \boldsymbol{g}^{ab}(\nabla_a \sigma)(\nabla_b \sigma),$$

in particular $\nabla\sigma$ is nowhere zero on $\mathcal{Z}(\sigma)$, and so σ is a defining density. Thus $\mathcal{Z}(\sigma)$ is a smoothly embedded hypersurface by the implicit function theorem. Evidently $\mathcal{Z}(\sigma)$ separates M according to the sign of σ. Also $\nabla\sigma$ is a (weight 1) conormal field along $\mathcal{Z}(\sigma)$ so the claimed signs for the normal also follow from the display.

Finally if $\boldsymbol{c}$ (and hence $\boldsymbol{g}$) has Riemannian signature, then the display shows that at any point of $Z(\sigma)$ the constant I^2 must be positive. This is a contradiction if $I^2 < 0$ and so $Z(\sigma) = \emptyset$. □

Remark 3.12 Note that if M is compact then Theorem 3.7 gives a decomposition of M into conformally compact manifolds glued along their conformal infinities. Note also that if M is allowed to have a boundary then we only mean that $\mathcal{Z}(\sigma)$ is separating if it is not a boundary component of M. ■

What we can conclude from all this is that almost pseudo-Riemannian manifolds $(M, \boldsymbol{c}, I)$ naturally give rise to nicely conformally compactified metrics (at least in the case where $I^2 \neq 0$). If the scalar curvature of the manifold (M_+, g_+) admitting a conformal compactification (M, c) is bounded away from zero then the conformal compactification must arise from an almost pseudo-Riemannian manifold $(M, \boldsymbol{c}, I)$ in this way because I^2 is then bounded away from zero on M_+ and hence I extends to be nowhere zero on M (all that we really need is that $I^2 \neq 0$ on ∂M_+). Thus we can apply the results above to the conformal compactification of (M_+, g_+). This tells us for instance that Riemannian manifolds with negative scalar curvature bounded away from zero must have a nice smooth boundary as conformal infinity if they can be conformally compactified. In the next section we develop this kind of idea.

3.6.4. Constraints on Possible Conformal Compactifications

Here we show that elementary geometric considerations restrict the topological (and geometric) possibilities for a conformal infinity.

Theorem 3.8 *Let (M_+, g_+) be a geodesically complete pseudo-Riemannian manifold and $i : M_+ \to M$ an embedding of M_+ as an open submanifold in a closed conformal manifold $(M, \boldsymbol{c})$ and σ a smooth section of $\mathcal{E}[1]$ on M so that on the image $i(M_+)$, which we identify with M_+, we have that*

$$g_+ = \sigma^{-2}\boldsymbol{g}$$

where $\boldsymbol{g}$ is the conformal metric on M. If the scalar curvature of g_+ is bounded away from zero then either $M_+ = M$ or the boundary points of M_+ in M form a smooth embedded hypersurface in M. This has a timelike normal field if $\mathrm{Sc}^{g_+} > 0$ and a spacelike normal field if $\mathrm{Sc}^{g_+} < 0$.

Proof We can form the scale tractor

$$I_A = \frac{1}{d} D_A \sigma.$$

Then I^2 is a smooth function on M that is bounded away from zero on the open set $M_+ \subset M$. Thus $I^2 \neq 0$ on the topological closure $\overline{M}_+$ of M_+ in M. Now since $I^2 \neq 0$ on $\overline{M}_+$ we have at any point $p \in \mathcal{Z}(\sigma) \cap \overline{M}_+$ that $\nabla\sigma(p) \neq 0$. Thus the zero locus of σ (intersected with $\overline{M}_+$) is a regular embedded submanifold of $\overline{M}_+$. By construction this lies in the set ∂M_+ of boundary points to M_+.

On the other hand any boundary point is in the zero locus, as otherwise for any $p \in \partial M_+$ such that $\sigma(p) \neq 0$ there is an open neighbourhood of p in $\overline{M}_+$ such that σ is nowhere zero and so g extends as a metric to this neighbourhood. So then there is a geodesic from a point in M_+ that reaches p in finite time, which contradicts that M_+ is geodesically complete. □

In particular if g_+ is Einstein with non-zero cosmological constant then I is parallel, and hence I^2 is constant and non-zero. So the above theorem applies in this case. In fact in the Einstein case we can also get results which in some aspects are stronger. For example:

Theorem 3.9 *Suppose that M_+ is an open dense submanifold in a compact connected conformal manifold $(M^d, \boldsymbol{c})$, possibly with boundary, and that one of the following two possibilities holds: either M is a manifold with boundary ∂M and $M \setminus M_+ = \partial M$, or M is closed and $M \setminus M_+$ is contained in a smoothly embedded submanifold of M of codimension at least 2. Suppose also that g_+ is a geodesically complete Einstein, but not Ricci flat, pseudo-Riemannian* *metric on M_+ such that on M_+ the conformal structure $[g_+]$ coincides with the restriction of* $\boldsymbol{c}$*. Then, either*

- *M_+ is closed and $M_+ = M$; or*
- *we are in the first setting with $M \setminus M_+$ the smooth n-dimensional boundary for M.*

Proof M_+ is canonically equipped with a parallel standard tractor I such that $I^2 = c \neq 0$, where c is constant.

Now $[g_+]$ on M_+ is the restriction of a smooth conformal structure $\boldsymbol{c}$ on M. Thus the conformal tractor connection on M_+ is the restriction of the smooth tractor connection on M. Working locally it is straightforward to use parallel transport along a congruence of curves to give a smooth extension of I to a sufficiently small open neighbourhood of any point in $M \setminus M_+$, and since M_+ is dense in M the extension is parallel and unique. (See [15] for this extension result in general for the case where $M \setminus M_+$ lies in a submanifold of dimension at least 2.) It follows that I extends as a parallel field to all of M.

It follows that $\sigma := I_A X^A$ also extends smoothly to all of M. This puts us back in the setting of Theorem 3.8, so that $M \setminus M_+$ is either empty or is an embedded hypersurface. If $M \setminus M_+$ is empty then M is closed and $M = N$. If $M \setminus M_+$ is an embedded hypersurface then, by our assumptions, we must be in the case where M is a manifold with boundary and $M \setminus M_+ = \partial M$ (since an embedded hypersurface cannot be contained in a submanifold of codimension two). □

Remark 3.13 In Theorem 3.8 we had to assume that there was a smooth globally defined conformal 1 density σ which glued the metric on M_+ smoothly to the conformal structure of M. In Theorem 3.9 we are able to drop this condition because we are requiring that the metric g_+ on M_+ is Einstein with conformal structure agreeing with that on M, allowing us to recover the smooth global 1 density σ by parallel extension of the scale tractor for g_+ to $M \setminus M_+$. However, it is not hard to see that one must place some restriction on the submanifold $M \setminus M_+$ in order for a parallel extension to exist; this is where the condition that $M \setminus M_+$ must lie in a submanifold of codimension two comes in for the case where M is closed. ■

3.6.5. Friedrich's Conformal Field Equations and Tractors

In this section we briefly discuss Friedrich's conformal field equations and their applications. We will see that the equations can be very easily arrived at using tractor calculus, and that the tractor point of view is also nicely compatible with the way the equations are used in applications.

3.6.5.1. The Equations Derived

Suppose that (M_+, g_+) is an asymptotically flat spacetime with corresponding unphysical spacetime (M, g) and conformal factor Ω. (We may assume that $M = M_+ \cup \mathscr{I}$.) Then since (M_+, g_+) is Ricci flat near the conformal infinity $\mathscr{I}$, the conformal scale tractor I defined on M_+ corresponding to the metric g_+ is parallel (and null) near $\mathscr{I}$ and thus has a natural extension (via parallel transport or simply by taking the limit) to all of M. Let us assume for simplicity that (M_+, g_+) is globally Ricci flat, then the naturally extended tractor I is globally parallel on the non-physical spacetime (M, g). If we write out the equation $\nabla_a I^B = 0$ on M in slots using the decomposition of the standard tractor bundle of $(M, [g])$ induced by g then we get the system of equations

$$\nabla_a \sigma = \mu_a$$
$$\nabla_a \mu_b = -\sigma \mathsf{P}_{ab} - \rho \boldsymbol{g}_{ab}$$
$$\nabla_a \rho = \mathsf{P}_{ab} \mu^b$$

where $I \stackrel{g}{=} (\sigma, \mu_a, \rho)$. If we also trivialise the conformal density bundles using g then σ becomes Ω, $\boldsymbol{g}$ becomes g, and we recognise the above equations as the first three equations in what are commonly known as *Friedrich's conformal field equations*. A fourth member of the conformal field equations can be obtained by writing $I_A I^A = 0$ out as

$$2\sigma\rho + \mu_a \mu^a = 0.$$

Next we observe that $\nabla_a I^B = 0$ clearly implies $\kappa_{ab}{}^C{}_D I^D = 0$, and the projecting part of this equation gives the *conformal C-space equation*

$$\nabla_b \mathsf{P}_{ac} - \nabla_a \mathsf{P}_{bc} = W_{abcd} \sigma^{-1} \mu^d$$

after multiplication by σ^{-1}. (The 'bottom slot' of $\kappa_{ab}{}^C{}_D I^D = 0$ taken w.r.t. g is $(\nabla_b \mathsf{P}_{ac} - \nabla_a \mathsf{P}_{bc})\mu^c = 0$ which also follows from contracting the above displayed equation with μ^c.) The contracted Bianchi identity states that

$$\nabla^d W_{abcd} = \nabla_b \mathsf{P}_{ac} - \nabla_a \mathsf{P}_{bc}$$

and from this and the previous display it follows that

$$\nabla^d (\sigma^{-1} W_{abcd}) = 0$$

(where we have used that $\nabla_a \sigma = \mu_a$). It can be shown [19] that the field $\sigma^{-1} W_{ab}{}^c{}_d$ is regular at $\mathscr{I}$ (recall the Sachs peeling property), and we will write this field as $K_{ab}{}^c{}_d$. If we substitute $K_{abcd} = \boldsymbol{g}_{ce} K_{ab}{}^e{}_d$ for $\sigma^{-1} W_{abcd}$ in the C-space equation and in the last equation displayed above we get the remaining two equations from Friedrich's conformal system

$$\nabla_b \mathsf{P}_{ac} - \nabla_a \mathsf{P}_{bc} = K_{abcd}\mu^d$$

and

$$\nabla^a K_{abcd} = 0,$$

which hold not only on M_+ but on all of M. This pair of equations encode not only the fact that (M_+, g_+) is Cotton flat (both equations being in some sense conformal C-space equations), but also by their difference they encode the contracted Bianchi identity (and hence the full Bianchi identity if M_+ is four-dimensional [21]).

The equations from Friedrich's conformal system are thus seen to be elementary consequences of the system of four tractor equations

$$\nabla_a I^B = 0$$
$$I^A I_A = 0$$
$$\kappa_{ab}{}^C{}_D I^D = 0$$
$$\nabla_{[a}\kappa_{bc]DE} = 0.$$

Remark 3.14 There are in fact more equations that could be considered as a part of Friedrich's conformal system, however these simply define the connection and the curvature terms used in the equations we have given. These equations would be the expressions $\nabla g = 0$ and $T^\nabla = 0$ satisfied by the Levi-Civita connection of g, and the decomposition of the Riemannian curvature tensor of ∇ as

$$R_{abcd} = \sigma K_{abcd} + 2\mathsf{P}_{a[c}g_{d]b} - 2\mathsf{P}_{b[c}g_{d]a}$$

which serves to define the curvature tensors K_{abcd} and P_{ab} (along with the appropriate symmetry conditions on K_{abcd} and P_{ab} as well as the condition that $\boldsymbol{g}^{ac}K_{abcd} = 0$).

Note that we can easily allow for a non-zero cosmological constant in our tractor system by taking $I^A I_A$ to be a constant rather than simply zero. ∎

3.6.5.2. The Purpose of the Equations

What are the conformal field equations for? The answer is that one seeks solutions to them in the same way that one seeks solutions to Einstein's field equations. A global solution to the conformal field equations gives a conformally compactified solution of Einstein's field equations. (In fact solutions of the conformal field equations may extend to regions 'beyond infinity' where Ω becomes negative.) We should note one can very easily allow for a non-zero cosmological constant and even for non-zero matter fields (especially ones with nice conformal behaviour) in the conformal field equations. The conformal field equations are then a tool for obtaining and investigating isolated systems in general relativity.

Like Einstein's field equations, the conformal field equations have an initial value formulation where initial data is specified on a Cauchy hypersurface. But when working with the conformal field equations it is also natural to prescribe data on the conformal infinity; when initial data is prescribed on (part of) $\mathscr{I}^-$ as well as on an ingoing null hypersurface which meets $\mathscr{I}^-$ transversally we have the *characteristic initial value problem.* When initial data is specified on a spacelike hypersurface which meets conformal infinity transversally we have the *hyperboloidal initial value problem.* (The name comes from the fact that spacelike hyperboloids in Minkowski space are the prime examples of such initial data hypersurfaces.) If the hyperboloidal initial data hypersurface meets $\mathscr{I}^-$ (rather than $\mathscr{I}^+$) then one should also prescribe data on the part of $\mathscr{I}^-$ to the future of the hypersurface, giving rise to an initial-boundary value problem.

What is being sought in the study of these various geometric PDE problems? Firstly information about when spacetimes will admit a conformal infinity and what kind of smoothness it might have. Secondly information about gravitational radiation produced by various gravitational systems as well as the way that such systems interact with gravitational radiation (scattering properties). Thirdly, one is obviously interested in the end in having a general understanding of the solutions of the conformal field equations (though this is a very hard problem). These problems have been studied a good deal, both analytically and numerically, however there are many questions left to be answered. For helpful overviews of this work consult [19, 21].

3.6.5.3. Regularity

An important feature of the conformal field equations is that they are *regular* at infinity. From the tractor calculus point of view this is obvious since the conformal structure extends to the conformal infinity so the tractor system displayed above cannot break down there, but one can also easily see this from the usual form of the equations. The significance of this is highlighted when one considers what field equations one might naively expect to use in this setting: the most obvious guess would be to rewrite $\mathrm{Ric}^{g_+} = 0$ in terms of the Ricci tensor of the unphysical metric g and its Levi-Civita connection yielding

$$\mathrm{Ric}_{ab} + \frac{d-2}{\Omega}\nabla_a\nabla_b\Omega - \boldsymbol{g}_{ab}\left(\frac{1}{\Omega}\nabla^c\nabla_c\Omega - \frac{d-1}{\Omega^2}(\nabla^c\Omega)\nabla_c\Omega\right) = 0$$

which degenerates as $\Omega \to 0$.

It is worth noting the significance of the regularity of the conformal field equations in the area of numerical relativity. The appeal of 'conformal compactification' to numerical relativists should be obvious: it means that one is dealing with finite domains.

3.6.5.4. Different Reductions and Different Forms of the Equations

The conformal field equations are a system of geometric partial differential equations. In order to study them analytically or numerically they need to be reduced to a classical system of PDE; this involves introducing coordinates and adding conditions on various fields to pin down the natural gauge freedom in the equations (gauge fixing). There is a significant amount of freedom in how one reduces the system. If this process is done carefully with the conformal field equations one can obtain (in four dimensions) a symmetric hyperbolic system of evolution equations together with an elliptic system of constraint equations (reflecting the fact that the conformal field equations are overdetermined). The constraint equations can be taken as conditions on initial data for the Cauchy problem for the conformal field equations, whereas the evolution equations can be taken as prescribing how such data will evolve off the Cauchy hypersurface. For further discussion of the reduction process and a demonstration of how the conformal field equations can be reduced see [19].

One may employ different reductions of the conformal field equations in different settings and for different purposes. Indeed, the form of Friedrich's conformal field equations which we presented above is by no means the only form of the equations that is used, and the significance of the other forms is that they allow for a still broader range of different reductions. As an example of this, note that one could instead have employed the splitting of the tractor bundle induced by a Weyl connection [8] to obtain a system of conformal field equations in a similar way to what we did above; this would result in a different form of the equations which have different gauge freedoms and from which we can obtain different reduced systems; the equations which one would obtain this way are what Friedrich calls the *general conformal field equations* whereas the equations we presented above are referred to as the *metric conformal field equations* [21]. Friedrich also frequently casts the equations in spinor form, and we have now seen that the conformal field equations take a very simple tractor form

$$
\begin{aligned}
\nabla_a I^B &= 0 \\
I^A I_A &= 0 \\
\kappa_{ab}{}^C{}_D I^D &= 0 \\
\nabla_{[a}\kappa_{bc]DE} &= 0
\end{aligned}
$$

(where we also need to impose the appropriate conditions on ∇ and κ as variables). It may indeed prove profitable to see the various reductions of the conformal field equations as being reductions of this tractor system. The gauge freedom(s) in Friedrich's field equations can be seen as coming from the splitting of the tractor equations into tensor equations (along with freedom to choose coordinates). Friedrich's use of conformal geodesics [21] in constructing coordinates as part of the reduction process also fits quite nicely with the tractor picture (see, e.g., [42, 44]).

Remark 3.15 In [20] Fraundiener and Sparling observed that in the four-dimensional (spin) case Friedrich's conformal field equations could be recast in terms of local twistors involving the so-called 'infinity twistor' $I^{\alpha\beta}$. In this case the bundle of real bitwistors is canonically isomorphic to the standard tractor bundle, and the 'infinity twistor' $I^{\alpha\beta}$ corresponds (up to a constant factor) to the scale tractor I^A, so that the local twistor system in [20] is closely related to the tractor system above. Indeed the connection is so close (the modern approach to conformal tractors having developed out of study of the local twistor calculus [2]) that we may consider the local twistor formulation to be the origin of the above tractor system. The conformal field equations have also been presented in terms of tractors by Christian Lübbe [42]. Lübbe and Tod have applied the tractor calculus to the study of conformal gauge singularities in general relativity (see, e.g., [44]). ∎

3.7. Lecture 6: Conformal Hypersurfaces

In order to progress in our study of conformal compactification we first need to spend some time considering the geometry of embedded hypersurfaces in conformal manifolds (or of boundaries to conformal manifolds if you like). In particular we will need to examine how the standard tractor bundle (and connection, etc.) of the hypersurface with its induced conformal structure is related to the standard tractor bundle (and connection, etc.) of the ambient space. Although our main motivation is the study of conformal compactification for this lecture we will consider conformal hypersurfaces more generally since there are many other important kinds of hypersurface which turn up in general relativity. We will however restrict ourselves to the case of non-degenerate hypersurfaces; in particular we do not give a treatment of null hypersurfaces here.

3.7.1. Conformal Hypersurfaces

By a *hypersurface* Σ in a manifold M we mean a smoothly embedded codimension 1 submanifold of M. We recall some facts concerning hypersurfaces in a conformal manifold (M^d, c), $d \geq 3$. In fact we wish to include the case that Σ might be a boundary component. In the case of a pseudo-Riemannian (or conformal pseudo-Riemannian) manifold with boundary then, without further comment, we will assume that the conformal structure extends smoothly to the boundary.

Here we shall restrict ourselves to hypersurfaces Σ with the property that any conormal field along Σ is nowhere null (i.e. to non-degenerate hypersurfaces). In this case the restriction of any metric $g \in \boldsymbol{c}$ gives a metric $\bar{g}$ on Σ. Different metrics, among the metrics so obtained, are related conformally, and so the conformal class $\boldsymbol{c}$ determines a conformal structure $\bar{\boldsymbol{c}}$ on Σ. To distinguish from the ambient objects, we shall overline the corresponding objects intrinsic to this conformal structure. For example $\bar{\boldsymbol{g}}$ denotes the conformal metric on Σ.

Then by working locally we may assume that there is a section $n_a \in \Gamma(\mathcal{E}_a[1])$ on M such that, along Σ, n_a is a conormal satisfying $|n|^2_{\boldsymbol{g}} := \boldsymbol{g}^{ab} n_a n_b = \pm 1$. This means that if $g = \sigma^{-2}\boldsymbol{g}$ is any metric in $\boldsymbol{c}$ then $n^g_a = \sigma^{-1} n_a$ is an extension to M of a unit conormal to Σ in (M, g). So $n_a|_\Sigma$ is the conformally invariant version of a unit conormal in pseudo-Riemannian geometry; n_a must have conformal weight 1 since $\boldsymbol{g}^{-1}$ has conformal weight -2.

We choose to work with the weighted (extended) conormal field n_a even in the presence of a metric g from the conformal class. Thus we end up with a weight 1 *second fundamental form* L_{ab} by restricting, along Σ, $\nabla_a n_b$ to $T\Sigma \times T\Sigma \subset (TM \times TM)|_\Sigma$, where $\nabla = \nabla^g$. (Here we are viewing $T^*\Sigma$ as the subbundle of $T^*M|_\Sigma$ orthogonal to n^a.) Explicitly L_{ab} is given by

$$L_{ab} := \nabla_a n_b \mp n_a n^c \nabla_c n_b \quad \text{along } \Sigma,$$

since $|n|^2_{\boldsymbol{g}}$ is constant along Σ. From this formula, it is easily verified that L_{ab} is independent of how n_a is extended off Σ. It is timely to note that L_{ab} harbours a hypersurface conformal invariant: using the formulae (3.2) and (3.16) we compute that under a conformal rescaling, $g \mapsto \widehat{g} = e^{2\omega} g$, L_{ab} transforms according to

$$L^{\widehat{g}}_{ab} = L^g_{ab} + \bar{\boldsymbol{g}}_{ab} \Upsilon_c n^c,$$

where as usual Υ is the exterior derivative of ω (which is equal to the log exterior derivative $\Omega^{-1} d\Omega$ of $\Omega = e^\omega$) and we use this notation below without further mention. Thus we see easily the following well-known result:

Proposition 3.16 *The trace-free part of the second fundamental form*

$$\mathring{L}_{ab} = L_{ab} - H\bar{\mathbf{g}}_{ab}, \quad \textit{where} \quad H := \frac{1}{d-1}\bar{\mathbf{g}}^{cd}L_{cd}$$

is conformally invariant.

The averaged trace of L ($= L^g$), denoted H above, is the *mean curvature* of Σ. Evidently this a conformal -1-density and under a conformal rescaling, $g \mapsto \widehat{g} = e^{2\omega}g$, H^g transforms to $H^{\widehat{g}} = H^g + n^a\Upsilon_a$.

Thus we obtain a conformally invariant section N of $\mathcal{T}|_\Sigma$

$$N_A \stackrel{g}{=} \begin{pmatrix} 0 \\ n_a \\ -H^g \end{pmatrix},$$

and from (3.29) $h(N,N) = \pm 1$ along Σ; the conformal invariance follows because the right hand side of the display transforms according to the 'tractor characterising' transformation (3.26). Obviously N is independent of any choices in the extension of n_a off Σ. This is the *normal tractor* of [2] and may be viewed as a tractor bundle analogue of the unit conormal field from the theory of pseudo-Riemannian hypersurfaces.

3.7.2. Umbilicity

A point p in a hypersurface is said to be an *umbilic point* if, at that point, the trace-free part $\mathring{L}$ of the second fundamental form is zero. Evidently this is a conformally invariant condition. A hypersurface is *totally umbilic* if this holds at all points. As an easy first application of the normal tractor we recall that it leads to a nice characterisation of the umbilicity condition.

Differentiating N tangentially along Σ using $\nabla^{\mathcal{T}}$, we obtain the following result.

Lemma 3.3

$$\mathbb{L}_{aB} := \underline{\nabla}_a N_B \stackrel{g_{cb}}{=} \begin{pmatrix} 0 \\ \mathring{L}_{ab} \\ -\frac{1}{d-2}\nabla^b\mathring{L}_{ab} \end{pmatrix} \tag{3.46}$$

where $\underline{\nabla}$ is the pull-back to Σ of the ambient tractor connection.

Proof Using the formula (3.22) for the tractor connection, we have

$$\nabla_c N_B \stackrel{g}{=} \begin{pmatrix} -n_c \\ \nabla_c n_b - \mathbf{g}_{cb}H \\ -\nabla_c H - P_{cb}n^b \end{pmatrix}.$$

Thus applying the orthogonal $TM|_\Sigma \to T\Sigma$ projector $\Pi_a^c := (\delta_a^c \mp n^c n_a)$ we obtain immediately the top two terms on the right hand side of (3.46). The remaining term follows after using the hypersurface Codazzi equation for pseudo-Riemannian geometry

$$\overline{\nabla}_a L_{bc} - \overline{\nabla}_b L_{ac} = \Pi_a^{a'} \Pi_b^{b'} R_{a'b'cd} n^d,$$

where $\overline{\nabla}$ denotes the Levi-Civita connection for the metric $\overline{g}$ induced by g (for full details see [55, 32]). □

Thus we recover the following result.

Proposition 3.17 [2] *Along a conformal hypersurface Σ, the normal tractor N is parallel, with respect to $\nabla^{\mathcal{T}}$, if and only if the hypersurface Σ is totally umbilic.*

3.7.3. Conformal Calculus for Hypersurfaces

The local calculus for hypersurfaces in Riemannian geometry is to a large extent straightforward because there is a particularly simple formula, known as the *Gauss formula*, which relates the ambient Levi-Civita connection to the Levi-Civita connection of the induced metric. It is natural to ask if we have the same with our conformal tractor calculus.

Given an ambient metric g we write $\underline{\nabla}$ to denote the pull-back of the ambient Levi-Civita connection along the embedding of the hypersurface Σ, that is the ambient connection differentiating sections of $TM|_\Sigma$ in directions tangent to the hypersurface. Then the Gauss formula may be expressed as

$$\underline{\nabla}_a v^b = \overline{\nabla}_a v^b \mp n^b L_{ac} v^c$$

for any tangent vector field v to Σ (thought of as a section v^a of $\mathcal{E}^a|_\Sigma$ satisfying $v^a n_a = 0$). We now turn towards deriving a tractor analogue of this.

Note that by dint of its conformal structure the hypersurface $(\Sigma, \overline{c})$ has its own *intrinsic* tractor calculus, and in particular a rank $d+1$ standard tractor bundle $\overline{\mathcal{T}}$. Before we could hope to address the question above we need to relate $\overline{\mathcal{T}}$ to the ambient tractor bundle $\mathcal{T}$.

First observe that, along Σ, $\mathcal{T}$ has a natural rank $(d+1)$-subbundle, namely $N^\perp$ the orthogonal complement to N_B. As noted in [5, 38], there is a canonical (conformally invariant) isomorphism

$$N^\perp \xrightarrow{\simeq} \overline{\mathcal{T}}. \tag{3.47}$$

Calculating in a scale g on M, the tractor bundle $\mathcal{T}$, and hence also $N^\perp$, decomposes into a triple. Then the mapping of the isomorphism is

$$[N^\perp]_g \ni \begin{pmatrix} \sigma \\ \mu_b \\ \rho \end{pmatrix} \mapsto \begin{pmatrix} \sigma \\ \mu_b \mp H n_b \sigma \\ \rho \pm \frac{1}{2} H^2 \sigma \end{pmatrix} \in [\overline{\mathcal{T}}]_{\overline{g}} \tag{3.48}$$

where, as usual, H denotes the mean curvature of Σ in the scale g and $\overline{g}$ is the pull-back of g to Σ. Since (σ, μ_b, ρ) is a section of $[N^\perp]_g$ we have $n^a \mu_a = H\sigma$. Using this one easily verifies that the mapping is conformally invariant: if we transform to $\widehat{g} = e^{2\omega} g$, $\omega \in \mathcal{E}$, then (σ, μ_b, ρ) transforms according to (3.26). Using that $\widehat{H} = H + n^a \Upsilon_a$ one calculates that the image of (σ, μ_b, ρ) (under the map displayed) transforms by the intrinsic version of (3.26), that is by (3.26) except where Υ_a is replaced by $\overline{\Upsilon}_a = \Upsilon_a \mp n_a n^b \Upsilon_b$ (which on Σ agrees with $\overline{\mathrm{d}}\omega$, the tangential derivative of ω). This signals that the explicit map displayed in (3.48) descends to a conformally invariant map (3.47). We henceforth use this to identify $N^\perp$ with $\overline{\mathcal{T}}$, and write $\mathrm{Proj}_\Sigma : \mathcal{T}|_\Sigma \to \overline{\mathcal{T}}$ for the orthogonal projection afforded by N (or using abstract indices $\Pi^A_B = \delta^A_B \mp N^A N_B$).

It follows easily from (3.48) that the tractor metric $\overline{h}$ on $\overline{\mathcal{T}}$ agrees with the restriction of the ambient tractor metric h to $N^\perp$. In summary we have:

Theorem 3.10 *Let $(M^d, \boldsymbol{c})$ be a conformal manifold of dimension $d \geq 4$ and Σ a regular hypersurface in M. Then, with $\overline{\mathcal{T}}$ deonting the intrinsic tractor bundle of the induced conformal structure $\boldsymbol{c}_\Sigma$, there is a canonical isomorphism*

$$\overline{\mathcal{T}} \to N^\perp.$$

Furthermore the tractor metric of $\boldsymbol{c}_\Sigma$ coincides with the pull-back of the ambient tractor metric, under this map.

Henceforth we shall simply identify $\overline{\mathcal{T}}$ and $N^\perp$.

Remark 3.16 Note that if Σ is minimal in the scale g, that is if $H^g = 0$, then the isomorphism (3.48) is simply

$$[N^\perp]_g \ni \begin{pmatrix} \sigma \\ \mu_b \\ \rho \end{pmatrix} \mapsto \begin{pmatrix} \sigma \\ \mu_b \\ \rho \end{pmatrix} \in [\overline{\mathcal{T}}]_{\overline{g}}. \tag{3.49}$$

Moreover, it is easy to see that one can always find such a minimal scale g for Σ. Let $g = \sigma^{-2}\boldsymbol{g}$ be any metric in $\boldsymbol{c}$ and let $\omega := \mp s\sigma H^g$, where s is a normalised defining function for Σ (at least in a neighbourhood of Σ) and H^g has been extended off Σ arbitrarily. Then if $\hat{g} = e^{2\omega} g$ we have that, along Σ,

$$H^{\hat{g}} = H^g + n^a \Upsilon_a = H^g + n^a \nabla_a \omega = 0$$

since by assumption $\nabla_a s = \sigma^{-1} n_a$ along Σ and $s|_\Sigma = 0$. ∎

Now we have two connections on $\overline{\mathcal{T}}$ that we may compare, namely the intrinsic tractor connection $\overline{\nabla}^{\overline{\mathcal{T}}}$, meaning the normal tractor connection determined by conformal structure $(\Sigma, \boldsymbol{c})$, and the *projected ambient tractor connection* $\check{\nabla}$. On $U \in \Gamma(\overline{\mathcal{T}})$ the latter is defined by

$$\check{\nabla}_a U^B := \Pi^B_C(\Pi^c_a \nabla_c U^C) \quad \text{along } \Sigma,$$

where we view $U \in \Gamma(\overline{\mathcal{T}})$ as a section of $N^\perp$, and make an arbitrary smooth extension of this to a section of $\mathcal{T}$ in a neighbourhood (in M) of Σ. It is then easily verified that $\check{\nabla}$ is a connection on $\overline{\mathcal{T}}$. By construction it is conformally invariant. Thus the difference between this and the intrinsic tractor connection is some canonical conformally invariant section of $T^*\Sigma \otimes \mathrm{End}(\overline{\mathcal{T}})$.

The difference between the projected ambient and the intrinsic tractor connections can be expressed using the tractor contorsion $\mathrm{S}_a{}^B{}_C$ defined by the equation

$$\check{\nabla}_a V^B = \overline{\nabla}_a V^B \mp \mathrm{S}_a{}^B{}_C V^C \tag{3.50}$$

where $V^B \in \Gamma(\overline{\mathcal{E}}^B)$ is an intrinsic tractor. The intrinsic tractor contorsion can be computed explicitly in any scale g to take the form

$$\mathrm{S}_a{}^B{}_C \stackrel{g}{=} \bar{X}^B \bar{Z}_C{}^c \mathcal{F}_{ac} - \bar{Z}^B{}_b \bar{X}_C \mathcal{F}_a{}^b \tag{3.51}$$

or in other words

$$\mathrm{S}_{aBC} = \overline{\mathbb{X}}_{BC}{}^c \mathcal{F}_{ac}, \tag{3.52}$$

where evidently $\mathcal{F}_{ac}$ must be some conformal invariant of hypersurfaces (here $\bar{X}^B$ and $\bar{Z}^B{}_b$ are standard tractor projectors for Σ and $\overline{\mathbb{X}}_{BC}{}^c = 2\bar{X}_{[B}\bar{Z}_{C]}{}^c$, which is conformally invariant). In fact the details of the computation (see [52, 55]) reveal this is the *Fialkow tensor* (cf. [18])

$$\mathcal{F}_{ab} = \frac{1}{n-2}\Big(W_{acbd}n^c n^d + \mathring{L}^2_{ab} - \frac{|\mathring{L}|^2}{2(n-1)}\bar{g}_{ab}\Big), \tag{3.53}$$

where $\mathring{L}^2_{ab} := \mathring{L}_a{}^c \mathring{L}_{cb}$ and $n = d - 1$. Altogether we have,

$$\begin{aligned} \underline{\nabla}_a V^B &= \Pi^B{}_C \underline{\nabla}_a V^C \pm \mathrm{N}^B \mathrm{N}_C \underline{\nabla}_a V^C \\ &= \check{\nabla}_a V^B \mp \mathrm{N}^B \mathbb{L}_{aC} V^C \\ &= \overline{\nabla}_a V^B - S_a{}^B{}_C V^C \mp \mathrm{N}^B \mathbb{L}_{aC} V^C \end{aligned} \tag{3.54}$$

for an intrinsic tractor $V^B \in \overline{\mathcal{T}}^B$. The tractor Gauss formula is therefore

$$\underline{\nabla}_a V^B = \overline{\nabla}_a V^B \mp S_a{}^B{}_C V^C \mp \mathrm{N}^B \mathbb{L}_{aC} V^C \tag{3.55}$$

for any $V^B \in \overline{\mathcal{T}}^B$. Here we see that the object

$$\mathbb{L}_{aB} := \underline{\nabla}_a \mathrm{N}_B$$

is a tractor analogue of the second fundamental form; we shall therefore call it the *tractor shape form*.

These results provide the first steps in a calculus for conformal hypersurfaces that is somewhat analogous to the local invariant calculus for Riemannian hypersurfaces. In particular it can be used to proliferate hypersurface conformal invariants and conformally invariant operators [55, 32]. We will apply this calculus to the study of conformal infinities in the final two lectures.

3.8. Lecture 7: Geometry of Conformal Infinity

3.8.1. Geometry of Conformal Infinity and its Embedding

We return now to the study of conformally compact geometries. We will consider in particular those which near the conformal infinity are asymptotically of constant non-zero scalar curvature. By imposing a constant dilation we may assume that I^2 approaches ± 1.

We begin by observing that the normal tractor is linked, in an essential way, to the ambient geometry off the hypersurface $\Sigma := \mathcal{Z}(\sigma)$.

Proposition 3.18 *Let $(M^d, \boldsymbol{c}, I)$ be an almost pseudo-Riemannian structure with scale singularity set $\Sigma \neq \emptyset$ and $I^2 = \pm 1 + \sigma^2 f$ for some smooth (weight -2) density f. Then by Theorem 3.7 Σ is a smoothly embedded hypersurface and, with N denoting the normal tractor for Σ, we have $N = I|_\Sigma$.*

Proof For simplicity let us first assume $I^2 = \pm 1$ (so $f = 0$ and the structure is ASC). As usual let us write $\sigma := h(X, I)$. By definition

$$I_A = \frac{1}{d} D_A \sigma \stackrel{g}{=} \begin{pmatrix} \sigma \\ \nabla_a \sigma \\ -\frac{1}{d}(\Delta\sigma + J\sigma) \end{pmatrix},$$

where $g \in \boldsymbol{c}$ and ∇ denotes its Levi-Civita connection. Let us write $n_a := \nabla_a \sigma$. Along Σ we have $\sigma = 0$, therefore

$$I|_\Sigma \stackrel{g}{=} \begin{pmatrix} 0 \\ n_a \\ -\frac{1}{d}\Delta\sigma \end{pmatrix},$$

along Σ. Clearly then $|n|_g^2 = \pm 1$, along Σ, since $I^2 = \pm 1$. So $n_a|_\Sigma$ is a conformal weight 1 conormal field for Σ.

Next we calculate the mean curvature $H = H^g$ in terms of σ. Recall $(d-1)H = \nabla^a n_a \mp n^a n^b \nabla_b n_a$, on Σ. We calculate the right hand side in a neighbourhood of Σ. Since $n_a = \nabla_a \sigma$, we have $\nabla^a n_a = \Delta\sigma$. On the other hand

$$n^a n^b \nabla_b n_a = \frac{1}{2} n^b \nabla_b (n^a n_a) = \frac{1}{2} n^b \nabla_b \left(\pm 1 + \frac{2}{d}\sigma\Delta\sigma + \frac{2}{d} J\sigma^2 \right),$$

where we used that $|\frac{1}{d}D\sigma|^2 = \pm 1$ and so $n^a n_a = \pm 1 + \frac{2}{d}\sigma\Delta\sigma + \frac{2}{d}J\sigma^2$. Now along Σ we have $\pm 1 = n^a n_a = n^a \nabla_a \sigma$, and so there this simplifies to

$$n^a n^b \nabla_b n_a = \pm \frac{1}{d} \Delta\sigma.$$

Putting these results together, we have

$$(d-1)H = \frac{1}{d}(d-1)\Delta\sigma \quad \Rightarrow \quad H = \frac{1}{d}\Delta\sigma \quad \text{along } \Sigma.$$

Thus

$$I|_\Sigma \stackrel{g}{=} \begin{pmatrix} 0 \\ n_a \\ -H \end{pmatrix},$$

as claimed. Now note that if we repeat the calculation with $I^2 = \pm 1 + \sigma^2 f$ then the result still holds, as in the calculation this relation was differentiated just once. □

Corollary 3.1 *Let $(M^d, \boldsymbol{c}, I)$ be an almost pseudo-Riemannian structure with scale singularity set $\Sigma \neq \emptyset$, and that is asymptotically Einstein in the sense that $I^2|_\Sigma = \pm 1$, and $\nabla_a I_B = \sigma f_{aB}$ for some smooth (weight -1) tractor valued 1-form f_{aB}. Then Σ is a totally umbilic hypersurface.*

Proof The assumptions on the scale tractor I imply that $I^2 = \pm 1 + \sigma^2 f$ for some smooth function f (since we assume $\boldsymbol{c}$ and σ smooth). Thus it follows from Proposition 3.18 that, along the singularity hypersurface, I agrees with the normal tractor N. Thus N is parallel along Σ and so, from Proposition 3.17, Σ is totally umbilic. □

Note that a hypersurface is totally umbilic if and only if it is conformally totally geodesic: if Σ is totally umbilic then locally it is straightforward to find a metric $g \in \boldsymbol{c}$ so that $H^g = 0$ whence $L^g_{ab} = 0$ (this was demonstrated in Remark 3.16). In this scale any geodesic on the submanifold Σ, with its induced metric $\overline{g}$, is also a geodesic of the ambient (M, g). So the condition of being totally umbilic is a strong matching of the conformal structures.

In fact for a conformally compact metric that is asymptotically Einstein, as in Corollary 3.1, there is an even stronger compatibility (involving a higher order of contact) between the geometry of $\boldsymbol{c}$ and $\overline{\boldsymbol{c}}$. First a preliminary result.

Proposition 3.19 *Let $(M^{d\geq 4}, \boldsymbol{c}, I)$ be an almost pseudo-Riemannian structure with scale singularity set $\Sigma \neq \emptyset$, and that is asymptotically Einstein in the sense that $I^2|_\Sigma = \pm 1$, and $\nabla_a I_B = \sigma^2 f_{aB}$ for some smooth (weight -2) tractor valued 1-form field f_{aB}. Then the Weyl curvature $W_{ab}{}^c{}_d$ satisfies*

$$W_{ab}{}^c{}_d n^d = 0, \quad \textit{along } \Sigma,$$

where n^d is the normal field.

Proof Since, along Σ, I_B is parallel to the given order we have that the tractor curvature satisfies

$$\kappa_{ab}{}^C{}_D I^D = \kappa_{ab}{}^C{}_D N^D = 0 \quad \text{along } \Sigma,$$

and from the formulae for κ and N the result is immediate. □

Theorem 3.11 *Let $(M^{d\geq 4}, \boldsymbol{c}, I)$ be an almost pseudo-Riemannian structure with scale singularity set $\Sigma \neq \emptyset$, and that is asymptotically Einstein in the sense that $I^2|_\Sigma = \pm 1$, and $\nabla_a I_B = \sigma^2 f_{aB}$ for some smooth (weight -2) tractor valued 1-form field f_{aB}. Then the tractor connection of $(M, \boldsymbol{c})$ preserves the intrinsic tractor bundle of Σ, where the latter is viewed as a subbundle of the ambient tractors: $\mathcal{T}_\Sigma \subset \mathcal{T}$. Furthermore the restriction of the parallel transport of $\nabla^{\mathcal{T}}$ coincides with the intrinsic tractor parallel transport of $\nabla^{\mathcal{T}_\Sigma}$.*

Proof From Theorem 3.10 the tractor bundle $\overline{\mathcal{T}}$ of $(\Sigma, \boldsymbol{c})$ may be identified with the subbundle $N^\perp$ in $\mathcal{T}|_\Sigma$, consisting of standard tractors orthogonal to the normal tractor N. Since $N = I|_\Sigma$ is parallel along Σ this subbundle is preserved by the ambient tractor connection and the projected ambient tractor connection $\check{\nabla}$ coincides with restriction to $\overline{\mathcal{T}}$ of the pull-back of the ambient tractor connection:

$$\check{\nabla}_a = \Pi_a^b \nabla_b^{\mathcal{T}} \quad \text{on} \quad \overline{\mathcal{T}} \subset \mathcal{T}.$$

Thus the result follows from the tractor Gauss formula if the Fialkow tensor

$$\mathcal{F}_{ab} = \frac{1}{n-2}\Big(W_{acbd} n^c n^d + \mathring{L}^2_{ab} - \frac{|\mathring{L}|^2}{2(n-1)} \overline{g}_{ab}\Big),$$

of (3.53) vanishes. Now from Corollary 3.1 Σ is totally umbilic, and so $\mathring{L}_{ab} = 0$, while Proposition 3.19 states that $W_{acbd} n^d = 0$ along Σ. Thus $\mathcal{F}_{ab} = 0$. □

Remark 3.17 In the case of an almost Einstein manifold Theorem 3.11 can also be seen to follow directly from the general theory of [11], see Theorem 3.5. ■

3.9. Lecture 8: Boundary Calculus and Asymptotic Analysis

In the previous lectures we have seen that for conformally compact manifolds with a non-degenerate conformal infinity there are useful tools for studying the link between the conformal structure on the conformal infinity and its relation to the ambient pseudo-Riemannian structure. These showed for example that conditions like asymptotically ASC and then asymptotically Einstein lead to a higher order of contact between the ambient and boundary conformal structures.

Here we show that the same tools, which are canonical to the geometry of almost-pseudo Riemannian manifolds, lead first to an interesting canonical calculus along the boundary and then to canonical boundary problems that can be partly solved by these tools. Further details of the boundary calculus for conformally compactified manifolds presented in this lecture, and on applications, can be found in [28, 33, 34, 35].

3.9.1. The Canonical Degenerate Laplacian

On an almost pseudo-Riemannian manifold $(M, \boldsymbol{c}, I)$ there is a canonical degenerate Laplacian type differential operator, namely

$$I\cdot D := I^A D_A.$$

This acts on any weighted tractor bundle, preserving its tensor type but lowering the weight:

$$I\cdot D : \mathcal{E}^{\Phi}[w] \to \mathcal{E}^{\Phi}[w-1].$$

In the following it will be useful to define the *weight operator* $\mathbf{w}$. On sections of a conformal density bundle this is just the linear operator that returns the weight. So if $\tau \in \Gamma(\mathcal{E}[w_0])$ then

$$\mathbf{w}\,\tau = w_0 \tau.$$

Then this is extended in an obvious way to weighted tensor or tractor bundles. So if $\mathcal{B}$ is some vector bundle of conformal weight zero then $\mathbf{w}$ acts as the zero operator on its sections and then if $\beta \in \Gamma(\mathcal{B}[w_0])$ we have

$$\mathbf{w}\,\beta = w_0 \beta.$$

Now expanding $I\cdot D$ in terms of some background metric $g \in \boldsymbol{c}$, we have

$$I\cdot D \stackrel{g}{=} \begin{pmatrix} -\frac{1}{d}(\Delta\sigma + \mathrm{J}\sigma) & \nabla^a\sigma & \sigma \end{pmatrix} \begin{pmatrix} \mathbf{w}(d+2\mathbf{w}-2) \\ \nabla_a(d+2\mathbf{w}-2) \\ -(\Delta + \mathrm{J}\mathbf{w}) \end{pmatrix}.$$

As an operator on any density or weighted tractor bundle of weight w each occurrence of $\mathbf{w}$ evaluates to w. So then

$$I\,D \stackrel{g}{=} -\sigma\Delta + (d+2w-2)\Big[(\nabla^a\sigma)\nabla_a - \frac{w}{d}(\Delta\sigma)\Big] - \frac{2w}{d}(d+w-1)\sigma \mathsf{J} \quad (3.56)$$

on $\mathcal{E}^\Phi[w]$. Now if we calculate in the metric $g_+ = \sigma^{-2}\boldsymbol{g}$, away from the zero locus of σ, and trivialise the densities accordingly, then σ is represented by 1 in the trivialisation, and we have

$$I{\cdot}D \stackrel{g_+}{=} -\left(\Delta^{g_+} + \frac{2w(d+w-1)}{d}\mathsf{J}^{g_+}\right).$$

In particular if g_+ satisfies $\mathsf{J}^{g_+} = \mp\frac{d}{2}$ (i.e. $\mathrm{Sc}^{g_+} = \mp d(d-1)$ or equivalently $I^2 = \pm 1$) then, relabeling $d+w-1 =: s$ and $d-1 =: n$, we have

$$I{\cdot}D \stackrel{g_+}{=} -\big(\Delta^{g_+} \pm s(n-s)\big). \quad (3.57)$$

On the other hand, looking again to (3.56), we see that $I{\cdot}D$ degenerates along the conformal infinity $\Sigma = \mathcal{Z}(\sigma)$ (assumed to be non-empty), and there the operator is first order. In particular if the structure is asymptotically ȦSC in the sense that $I^2 = \pm 1 + \sigma f$, for some smooth (weight -1) density f, then along Σ

$$I{\cdot}D = (d+2w-2)\delta_n\,,$$

on $\mathcal{E}^\Phi[w]$, where δ_n is the conformal Robin operator,

$$\delta_n \stackrel{g}{=} n^a\nabla^g_a - wH^g,$$

of [13, 5] (twisted with the tractor connection); here n^a is a length ± 1 normal and H^g the mean curvature, as measured in the metric g.

3.9.2. A Boundary Calculus for the Degenerate Laplacian

Let $(M, \boldsymbol{c})$ be a conformal structure of dimension $d \geq 3$ and of any signature. Given σ a section of $\mathcal{E}[1]$, write I_A for the corresponding scale tractor as usual. That is $I_A = \frac{1}{d}D_A\sigma$. Then $\sigma = X^A I_A$.

3.9.2.1. The $\mathfrak{sl}(2)$-algebra

Suppose that $f \in \mathcal{E}^\Phi[w]$, where $\mathcal{E}^\Phi$ denotes any tractor bundle. Select $g \in \boldsymbol{c}$ for the purpose of calculation, and write $I_A \stackrel{g}{=} (\sigma,\ \nu_a,\ \rho)$ to simplify the notation. Then using $\nu_a = \nabla_a\sigma$, we have

$$I{\cdot}D(\sigma f) = (d+2w)\big((w+1)\sigma\rho f + \sigma\nu_a\nabla^a f + f\nu_a\nu^a\big)$$
$$-\sigma\big(\sigma\Delta f + 2\nu_a\nabla^a f + f\Delta\sigma + (w+1)\mathsf{J}\sigma f\big),$$

while

$$-\sigma\, I{\cdot}Df = -\sigma(d+2w-2)\big(w\rho f + \nu_a\nabla^a f\big) + \sigma^2(\Delta f + w\mathrm{J}f)\,.$$

So, by virtue of the fact that $\rho = -\frac{1}{d}(\Delta\sigma + \mathrm{J}\sigma)$, we have

$$[I{\cdot}D,\sigma]f = (d+2w)(2\sigma\rho + \nu_a\nu^a)f.$$

Now $I^A I_A = I^2 \stackrel{g}{=} 2\sigma\rho + \nu_a\nu^a$, whence the last display simplifies to

$$[I{\cdot}D,\sigma]f = (d+2w)I^2 f.$$

Denoting by $\mathbf{w}$ the weight operator on tractors, we have the following.

Lemma 3.4 *Acting on any section of a weighted tractor bundle we have*

$$[I{\cdot}D,\sigma] = I^2(d+2\mathbf{w}),$$

where $\mathbf{w}$ *is the weight operator.*

Remark 3.18 A similar computation to above shows that, more generally,

$$I\cdot D\big(\sigma^\alpha f\big) - \sigma^\alpha I\cdot Df = \sigma^{\alpha-1}\alpha\, I^2(d+2\mathbf{w}+\alpha-1)f\,,$$

for any constant α. ∎

The operator $I{\cdot}D$ lowers conformal weight by 1. On the other hand, as an operator (by tensor product) σ raises conformal weight by 1. We can record this by the commutator relations

$$[\mathbf{w}, I{\cdot}D] = -I{\cdot}D \quad \text{and} \quad [\mathbf{w},\sigma] = \sigma,$$

so with the lemma we see that the operators σ, $I{\cdot}D$ and $\mathbf{w}$, acting on weighted scalar or tractor fields, generate an $\mathfrak{sl}(2)$ Lie algebra, provided I^2 is nowhere vanishing. It is convenient to fix a normalisation of the generators; we record this and our observations as follows.

Proposition 3.20 *Suppose that* (M,c,σ) *is such that* I^2 *is nowhere vanishing. Setting* $x := \sigma$, $y := -\frac{1}{I^2}I{\cdot}D$ *and* $h := d+2\mathbf{w}$ *we obtain the commutation relations*

$$[h,x] = 2x, \quad [h,y] = -2y, \quad [x,y] = h,$$

of standard $\mathfrak{sl}(2)$*-algebra generators.*

In the case of $I^2 = 0$ *the result is an Inönü–Wigner contraction of the* $\mathfrak{sl}(2)$*-algebra:*

$$[h,x] = 2x, \quad [h,y] = -2y, \quad [x,y] = 0,$$

where h *and* x *are as before, but now* $y = -I{\cdot}D$.

Subsequently $\mathfrak{g}$ will be used to denote this ($\mathfrak{sl}(2)$) Lie algebra of operators. From Proposition 3.20 (and in concordance with Remark 3.18) follow some useful identities in the universal enveloping algebra $\mathcal{U}(\mathfrak{g})$.

Corollary 3.2

$$
\begin{gathered}
[x^k, y] = x^{k-1}k(d + 2\mathbf{w} + k - 1) = x^{k-1}k(h + k - 1) \\
\text{and} \\
[x, y^k] = y^{k-1}k(d + 2\mathbf{w} - k + 1) = y^{k-1}k(h - k + 1)\,.
\end{gathered}
\tag{3.58}
$$

3.9.3. Tangential Operators and Holographic Formulæ

Suppose that $\sigma \in \Gamma(\mathcal{E}[1])$ is such that $I_A = \frac{1}{d} D_A \sigma$ satisfies that $I^A I_A = I^2$ is nowhere zero. As explained in Section 3.6.3, the zero locus $\mathcal{Z}(\sigma)$ of σ is then either empty or forms a smooth hypersurface.

Conversely if Σ is any smooth oriented hypersurface then, at least in a neighbourhood of Σ, there is a smooth defining function s. Now take $\sigma \in \Gamma(\mathcal{E}[1])$ to be the unique density which gives s in the trivialisation of $\mathcal{E}[1]$ determined by some $g \in \boldsymbol{c}$. It follows then that $\Sigma = \mathcal{Z}(\sigma)$ and $\nabla^g \sigma$ is non-zero at all points of Σ. If $\nabla^g \sigma$ is nowhere null along Σ then Σ is non-degenerate and I^2 is nowhere vanishing in a neighbourhood of Σ, and we are in the situation of the previous paragraph. We call such a σ a *defining density* for Σ (recall our definition from Lecture 6), and to simplify the discussion we shall take M to agree with this neighbourhood of Σ. Until further notice σ will mean such a section of $\mathcal{E}[1]$ with $\Sigma = \mathcal{Z}(\sigma)$ non-empty and non-degenerate. Note that Σ has a conformal structure $\bar{\boldsymbol{c}}$ induced in the obvious way from $(M, \boldsymbol{c})$ and is a conformal infinity for the metric $g_+ := \sigma^{-2}\boldsymbol{g}$ on $M \setminus \Sigma$.

3.9.3.1. Tangential Operators

Suppose that σ is a defining density for a hypersurface Σ. Let $P : \mathcal{E}^\Phi[w_1] \to \mathcal{E}^\Phi[w_2]$ be some linear differential operator defined in a neighbourhood of Σ. We shall say that P *acts tangentially (along* Σ*)* if $P \circ \sigma = \sigma \circ \widetilde{P}$, where $\widetilde{P} : \mathcal{E}^\Phi[w_1 - 1] \to \mathcal{E}^\Phi[w_2 - 1]$ is some other linear operator on the same neighbourhood. The point is that for a tangential operator P we have

$$P(f + \sigma h) = Pf + \sigma \widetilde{P} h.$$

Thus along Σ the operator P is insensitive to how f is extended off Σ. It is easily seen that if P is a tangential differential operator then there is a formula for P, along Σ, involving only derivatives tangential to Σ, and the converse also holds.

Using Corollary 3.2 we see at once that certain powers of $I{\cdot}D$ act tangentially on appropriately weighted tractor bundles. We state this precisely. Suppose that Σ is a (non-degenerate) hypersurface in a conformal manifold $(M^{n+1}, \boldsymbol{c})$, and σ a defining density for Σ. Then recall $\Sigma = \mathcal{Z}(\sigma)$ and I^2 is nowhere zero in a neighbourhood of Σ, where $I_A := \frac{1}{n+1} D_A \sigma$ is the scale tractor. The following holds.

Theorem 3.12 *Let $\mathcal{E}^\Phi$ be any tractor bundle and $k \in \mathbb{Z}_{\geq 1}$. Then, for each $k \in \mathbb{Z}_{\geq 1}$, along Σ*

$$P_k : \mathcal{E}^\Phi\left[\frac{k-n}{2}\right] \to \mathcal{E}^\Phi\left[\frac{-k-n}{2}\right] \quad \textit{given by} \quad P_k := \left(-\frac{1}{I^2} I{\cdot}D\right)^k \tag{3.59}$$

is a tangential differential operator, and so determines a canonical differential operator $P_k : \mathcal{E}^\Phi[\frac{k-n}{2}]|_\Sigma \to \mathcal{E}^\Phi[\frac{-k-n}{2}]|_\Sigma$.

Proof The P_k are differential by construction. Thus the result is immediate from Corollary 3.2 with $\widetilde{P} = \left(-\frac{1}{I^2} I{\cdot}D\right)^k$. □

3.9.4. The Extension Problems and their Asymptotics

Henceforth we consider an almost pseudo-Riemannian structure (M, c, σ) with σ a defining density for a hypersurface Σ and I^2 nowhere zero. We consider the problem of solving, off Σ asymptotically,

$$I{\cdot}Df = 0\,,$$

for $f \in \Gamma(\mathcal{E}^\Phi[w_0])$ and some given weight w_0. For simplicity we henceforth calculate on the side of Σ where σ is non-negative, so effectively this amounts to working locally along the boundary of a conformally compact manifold.

3.9.4.1. Solutions of the First Kind

Here we treat an obvious Dirichlet-like problem where we view $f|_\Sigma$ as the initial data. Suppose that f_0 is an arbitrary smooth extension of $f|_\Sigma$ to a section of $\mathcal{E}^\Phi[w_0]$ over M. We seek to solve the following problem:

Problem 3.1 Given $f|_\Sigma$, and an arbitrary extension f_0 of this to $\mathcal{E}^\Phi[w_0]$ over M, find $f_i \in \mathcal{E}^\Phi[w_0 - i]$ (over M), $i = 1, 2, \cdots$, so that

$$f^{(\ell)} := f_0 + \sigma f_1 + \sigma^2 f_2 + \cdots + O(\sigma^{\ell+1})$$

solves $I{\cdot}Df = O(\sigma^\ell)$, off Σ, for $\ell \in \mathbb{N} \cup \infty$ as high as possible.

Remark 3.19 For $i \geq 1$ we do not assume that the f_i are necessarily non-vanishing along Σ. Also, note that the stipulation 'given $f|_\Sigma$, and an arbitrary extension f_0 of this' can be rephrased as 'given f_0 in the space of sections with a fixed restriction to Σ (denoted $f|_\Sigma$)'. ∎

We write $h_0 = d + 2w_0$ so that $hf_0 = h_0 f_0$, for example. The existence or not of a solution at generic weights is governed by the following result.

Lemma 3.5 *Let $f^{(\ell)}$ be a solution of Problem 3.1 to order $\ell \in \mathbb{Z}_{\geq 0}$. Then provided $\ell \neq h_0 - 2 = n + 2w_0 - 1$ there is an extension*

$$f^{(\ell+1)} = f^{(\ell)} + \sigma^{\ell+1} f_{\ell+1},$$

unique modulo $\sigma^{\ell+2}$, which solves

$$I{\cdot}Df^{(\ell+1)} = 0 \quad \textit{modulo} \quad O(\sigma^{\ell+1}).$$

If $\ell = h_0 - 2$ then the extension is obstructed by $P_{\ell+1} f_0|_\Sigma$, where $P_{\ell+1} = (-\frac{1}{I^2} I{\cdot}Df)^{\ell+1}$ is the tangential operator on densities of weight w_0 given by Theorem 3.12.

Proof Note that $I{\cdot}Df = 0$ is equivalent to $-\frac{1}{I^2} I{\cdot}Df = 0$ and so we can recast this as a formal problem using the Lie algebra $\mathfrak{g} = \langle x, y, h\rangle$ from Proposition 3.20. Using the notation from there

$$yf^{(\ell+1)} = yf^{(\ell)} - x^\ell(\ell+1)(h+\ell) f_{\ell+1} + O(x^{\ell+1}).$$

Now $hf_{\ell+1} = \big(h_0 - 2(\ell+1)\big) f_{\ell+1}$, thus

$$yf^{(\ell+1)} = yf^{(\ell)} - x^\ell(\ell+1)(h_0 - \ell - 2) f_{\ell+1} + O(x^{\ell+1}). \tag{3.60}$$

By assumption $yf^{(\ell)} = O(x^\ell)$, thus if $\ell \neq h_0 - 2$ we can solve $yf^{(\ell+1)} = O(x^{\ell+1})$ and this uniquely determines $f_{\ell+1}|_\Sigma$.

On the other hand if $\ell = h_0 - 2$ then (3.60) shows that, modulo $O(x^{\ell+1})$,

$$yf^{(\ell)} = y\big(f^{(\ell)} + x^{\ell+1} f_{\ell+1}\big),$$

regardless of $f_{\ell+1}$. It follows that the map $f_0 \mapsto x^{-\ell} yf^{(\ell)}$ is tangential and $x^{-\ell} yf^{(\ell)}|_\Sigma$ is the obstruction to solving $yf^{(\ell+1)} = O(x^{\ell+1})$. By a simple induction this is seen to be a non-zero multiple of $y^{\ell+1} f_0|_\Sigma$. □

Thus by induction we conclude the following.

Proposition 3.21 *For $h_0 \notin \mathbb{Z}_{\geq 2}$ Problem* (3.1) *can be solved to order $\ell = \infty$. For $h_0 \in \mathbb{Z}_{\geq 2}$ the solution is obstructed by $[P_{h_0-1} f]|_\Sigma$; if, for a particular f, $[P_{h_0-1} f]|_\Sigma = 0$ then there is a family of solutions to order $\ell = \infty$ parametrised by sections $f_{h_0-1} \in \Gamma\mathcal{E}^\Phi[-d - w_0 + 1]|_\Sigma$.*

Appendix: Conformal Killing Vector Fields and Adjoint Tractors

Here we discuss conformal Killing vector fields and the prolongation of the conformal Killing equation which is given by a connection on a conformal tractor bundle called the adjoint tractor bundle. Although the adjoint bundle is not a new tractor bundle *per se* (being the bundle of skew endomorphisms of $\mathcal{E}^A$, naturally isomorphic to $\mathcal{E}_{[AB]}$), it is a very important tractor bundle and is worthy of our further attention; we discuss briefly the connection of the conformal Cartan bundle with the adjoint tractor bundle and its calculus. We finish by applying the adjoint tractor calculus and the prolongation of the conformal Killing equation to Lie derivatives of tractors with respect to conformal Killing vector fields.

A.1. The Conformal Cartan Bundle and the Adjoint Tractor Bundle

In the lectures we have been discussing tractor calculus on the so-called standard tractor bundle on a conformal manifold $(M, \boldsymbol{c})$, which can be seen as the associated bundle $\mathcal{G} \times_P \mathbb{R}^{p+1,q+1}$ to the P-principal conformal Cartan bundle $\mathcal{G} \to M$. We have not discussed the construction of the conformal Cartan bundle in the lectures because it is easier to construct the standard tractor bundle directly from the underlying conformal structure. Indeed, having understood the conformal standard tractor bundle one could simply define the conformal Cartan bundle as an adapted frame bundle for the standard tractor bundle $\mathcal{T}$; from this point of view the conformal Cartan connection simply encodes the tractor parallel transport of the frames, cf. (3.40). Although for calculational purposes the standard tractor approach presented in the lectures is optimal, we have seen (e.g. in Section 3.6.3.1) that the Cartan geometric point of view can afford additional insights. This is also the case when it comes to discussing the adjoint tractor bundle.

If $P \subset G = \mathrm{O}(p,q)$ is the stabiliser of a null ray in $\mathbb{R}^{p+1,q+1}$ then P contains a subgroup which may be identified with $\mathrm{CO}(p,q)$. The conformal Cartan bundle of (M, c) is then easily defined as the extension of the conformal orthogonal frame bundle $\mathcal{G}_0 \to M$ to a principal P-bundle $\mathcal{G} \to M$ corresponding to the inclusion $\mathrm{CO}(p,q) \subset P$. The total space of this bundle is then $\mathcal{G}_0 \times_{CO(p,q)} P$.

The conformal Cartan connection ω is a 1-form on $\mathcal{G}$ taking values in the Lie algebra of G, which can be written as a (vector space) direct sum

$$\mathbb{R}^d \oplus \mathfrak{co}(p,q) \oplus \left(\mathbb{R}^d\right)^*$$

where $d = p + q$. The $\mathbb{R}^d$ component of ω is simply the trivial extension of the *soldering form* of $\mathcal{G}_0$ (which is tautologically defined on any reduction of the frame bundle of a manifold) to $\mathcal{G}$; this component corresponds to the terms $-\mu_a$ and $+\boldsymbol{g}_{ab}\rho$ in the formula (3.22) for the standard tractor connection. The next component arises from noting that sections of the bundle $\mathcal{G} \to \mathcal{G}_0$ are in one-to-one correspondence with affine connections on M which preserve the conformal metric $\boldsymbol{g}$ (*Weyl connections*); the $\mathfrak{co}(p,q)$ component of ω is the 1-form $\boldsymbol{\gamma}$ on $\mathcal{G}$ whose pull-back to $\mathcal{G}_0$ by any section is the connection 1-form for the corresponding Weyl connection. In order to obtain the formula for the tractor connection from that of the Cartan connection one must first pull everything down from $\mathcal{G}$ to $\mathcal{G}_0$ using a Weyl connection, allowing one to break tractors (in associated bundles to $\mathcal{G}$) up into weighted tensors (in associated bundles to $\mathcal{G}_0$). If one does this using the Levi-Civita connection ∇ of a metric g then one can see that the component $\boldsymbol{\gamma}$ of ω gives rise to the three terms involving ∇ in (3.22). The final component of ω gives rise to the two terms involving the Schouten tensor in (3.22) and is determined by the other two components of ω and an algebraic condition on the curvature

$$\mathrm{d}\omega + \omega \wedge \omega$$

of ω (which can be identified with the tractor curvature). For more details see the first chapter of [12].

The conformal Cartan connection has the following three properties:

- $\omega_u : T_u\mathcal{G} \to \mathcal{G} \times \mathfrak{g}$ is a linear isomorphism for each $u \in \mathcal{G}$;
- ω is P-equivariant, that is $(r^p)^*\omega = Ad(p^{-1}) \circ \omega$ for all $p \in P$ (where r^p denotes the right action of p on $\mathcal{G}$);
- ω returns the generators of fundamental vector fields, that is
$$\omega\left(\left.\frac{d}{dt}\right|_0 u \cdot exp(tX)\right) = X$$
for all $u \in \mathcal{G}$ and $X \in \mathfrak{p} = \mathrm{Lie}(P)$.

These properties more generally define what is called a *Cartan connection* of type (G,P) on a P-principal bundle $\mathcal{G}$. In the model case the Lie group G can be seen as a P-principal bundle over the model space G/P and the Maurer-Cartan 1-form ω_{MC}, which evaluates left invariant vector fields at the identity, is a Cartan connection which has vanishing curvature by the Maurer–Cartan structure equations

$$\mathrm{d}\omega_{MC} + \omega_{MC} \wedge \omega_{MC} = 0.$$

The *adjoint tractor bundle* $\mathcal{A} \to M$ of a conformal manifold $(M, \boldsymbol{c})$ is the associated bundle to the conformal Cartan bundle $\mathcal{G}$ corresponding to the adjoint representation of G (restricted to P), that is $\mathcal{A} = \mathcal{G} \times_P \mathfrak{g}$. Since $\mathfrak{g} = \mathfrak{so}(p+1,q+1)$ can be identified with the skew-symmetric endomorphisms of $\mathbb{R}^{p+1,q+1}$ the adjoint tractor bundle can similarly be identified with the bundle of skew-symmetric endomorphisms of the standard tractor bundle $\mathcal{T}$ (with respect to the tractor metric). Clearly then we may identify $\mathcal{A}$ with $\mathcal{E}_{[AB]}$ by lowering a tractor index. An adjoint tractor $\mathbb{L}^A{}_B$ can be written in terms of the direct sum decomposition of $\mathcal{T}$ (and hence $\mathcal{A}$) as

$$\begin{pmatrix} -\nu & -l_b & 0 \\ -\rho^a & \mu^a{}_b & l^a \\ 0 & \rho_b & \nu \end{pmatrix} \tag{3.61}$$

where $\mu_{ab} = \mu_{[ab]}$ and the matrix acts on standard tractors from the left, as the tractor curvature does in (3.32) (the tractor curvature is in fact an adjoint tractor valued 2-form). It is easy to see that the two appearances of l^a make up the 'top slot' of $\mathbb{L}^A{}_B$, so that there is an invariant projection Π from $\mathcal{A}$ to TM that takes $\mathbb{L}^A{}_B$ to l^a.

By writing an adjoint tractor $\mathbb{L}^A{}_B$ in terms of the splitting tractors (X^A, Z^A_a, Y^A) corresponding to a choice of metric g one can easily obtain the formula for the tractor connection acting on $\mathbb{L}^A{}_B$ using (3.40). If $\mathbb{L}^A{}_B$ is given in matrix form (w.r.t. g) by (3.61) we then have

$$\nabla_a \mathbb{L}^B{}_C \stackrel{g}{=} \begin{pmatrix} * & * & 0 \\ * & \boldsymbol{g}^{bb'}(\nabla_a \mu_{b'c} - 2\mathrm{P}_{a[b'}l_{c]} + 2\boldsymbol{g}_{a[b'}\rho_{c]}) & \nabla_a l^b + \mu_a{}^b + \nu\delta^b_a \\ 0 & \nabla_a \rho_c - \mathrm{P}^b_a \mu_{bc} + \nu \mathrm{P}_{ac} & \nabla_a \nu - \mathrm{P}^b_a l_b - \rho_a \end{pmatrix} \tag{3.62}$$

where the entries marked with a $*$ are determined by skew-symmetry.

A.2. Prolonging the Conformal Killing Equation

Note that if $\mathbb{L}$ is a parallel adjoint tractor, and $l = \Pi(\mathbb{L})$, then from the above display we must have

$$\nabla_a l_b + \mu_{ab} + \nu \boldsymbol{g}_{ab} = 0$$

where μ_{ab} is skew. This implies that $\nabla_{(a}l_{b)_o} = 0$, in other words l^a is a *conformal Killing vector field.* A conformal Killing vector field k^a is a non-zero solution of the conformally invariant equation

$$\nabla_{(a}k_{b)_o} = 0,$$

where $k_b = \boldsymbol{g}_{bc}k^c$. Geometrically this equation says that the local flow of k preserves any metric $g \in \boldsymbol{c}$ up to a conformal factor, equivalently, the Lie derivative of any metric $g \in \boldsymbol{c}$ with respect to k is proportional to g. It is natural to ask whether there is a one-to-one correspondence between conformal Killing vector fields and (non-zero) parallel adjoint tractor fields – the answer to this question turns out to be no, except for on the flat model (where one can use this correspondence to easily write the $(d+2)(d+1)/2$-dimensional space of Killing vector fields explicitly).

One can however construct a different conformally invariant connection on the adjoint tractor bundle which does prolong the conformal Killing equation, that is for which parallel sections are in one-to-one correspondence with solutions. One can obtain this system directly by (fixing g and $\nabla = \nabla^g$ and then) writing the equation $\nabla_{(a}k_{b)_o} = 0$ as

$$\nabla_a k_b = \mu_{ab} + \nu \boldsymbol{g}_{ab},$$

introducing the new variables $\mu_{ab} \in \Gamma(\mathcal{E}_{[ab]}[2])$ and $\nu \in \Gamma(\mathcal{E})$. Beyond this the key step is to introduce the fourth variable

$$\rho_a = \nabla_a \nu + \mathsf{P}_{ab} k^b \in \Gamma(\mathcal{E}_a)$$

(rather than $\rho_a = \nabla_a \nu$), cf. (3.62). The remainder of the prolongation process simply involves taking covariant derivatives of the above two displays and then skew-symmetrising over certain pairs of indices in them in order to bring out curvature terms (as well as using Bianchi identities to simplify expressions); from these 'differential consequences' of the above two displays one may derive expressions for $\nabla_a \mu_{bc}$ and $\nabla_a \rho_b$ which are linear in the other three variables (respectively). (The way to accomplish this process efficiently is to suppose you have a flat Levi-Civita connection first and go through the process to obtain both expressions, then go back through the same steps and take into account the non-vanishing curvature for the general case.) The result is the following system of differential equations

$$\begin{aligned}
\nabla_a k_b &= \nu \boldsymbol{g}_{ab} + \mu_{ab} \\
\nabla_a \mu_{bc} &= -2\mathsf{P}_{a[b}k_{c]} - 2\boldsymbol{g}_{a[b}\rho_{c]} + W_{dabc}k^d \\
\nabla_a \nu &= \rho_a - \mathsf{P}_{ab}k^b \\
\nabla_a \rho_b &= -\mathsf{P}_a^c \mu_{bc} - \mathsf{P}_{ab}\nu - C_{cab}k^c
\end{aligned}$$

where $C_{abc} = 2\nabla_{[a}\mathsf{P}_{bc]}$ is the Cotton tensor.

From the above system we can see that if we define the connection $\tilde{\nabla}$ on $\mathcal{A}$ by

$$\tilde{\nabla}\mathbb{L} = \nabla\mathbb{L} + i_{\Pi(\mathbb{L})}\kappa \quad \text{for all} \quad \mathbb{L} \in \Gamma(\mathcal{A}) \tag{3.63}$$

then there is a one-to-one correspondence between (non-zero) $\tilde{\nabla}$-parallel sections of $\mathcal{A}$ and conformal Killing vector fields on $(M, \boldsymbol{c})$. To check this we simply calculate (in a scale g) that

$$\tilde{\nabla}_a \begin{pmatrix} * & * & 0 \\ * & \mu^b{}_c & l^b \\ 0 & \rho_c & \nu \end{pmatrix} = \begin{pmatrix} * & * & 0 \\ * & \boldsymbol{g}^{bb'}(\nabla_a \mu_{b'c} - 2\mathsf{P}_{a[b'}l_{c]} + 2\boldsymbol{g}_{a[b'}\rho_{c]}) & \nabla_a l^b + \mu_a{}^b + \nu \delta^b_a \\ 0 & \nabla_a \rho_c - \mathsf{P}^b_a \mu_{bc} + \nu \mathsf{P}_{ac} & \nabla_a \nu - \mathsf{P}^b_a l_b - \rho_a \end{pmatrix}$$
$$+ \begin{pmatrix} * & * & 0 \\ * & W_{da}{}^b{}_c l^d & 0 \\ 0 & -C_{dac} l^d & 0 \end{pmatrix}$$

and observe that by setting the right hand side equal to zero (and substituting $k^a = -l^a$) we recover our prolonged system for the conformal Killing equation. Note that it is possible to take a much more abstract and theoretical approach to obtaining this prolonged system and the corresponding connection $\tilde{\nabla}$ on $\mathcal{A}$ (see, e.g. [6, 12]). From the general theory (or direct observation) we also have the invariant linear differential operator $L : TM \to \mathcal{A}$ which takes a vector field l^a on M to the adjoint tractor $\mathbb{L}^A{}_B$ given in a scale g by (3.61) with $\mu_{ab} = -\nabla_{[a} l_{b]}$, $\nu = -\frac{1}{d} \nabla_a l^a$ and $\rho_a = \nabla_a \nu - \mathsf{P}_{ab} l^b$. Clearly $\Pi \circ L = \mathrm{id}_{TM}$ and consequently L is referred to as a *differential splitting operator*.

A.3. The Fundamental Derivative and Lie Derivatives of Tractors

Let $\mathbb{V}$ be a representation of P and let $\mathcal{V} = \mathcal{G} \times_P \mathbb{V}$. If a function $\tilde{v} : \mathcal{G} \to \mathbb{V}$ satisfies

$$\tilde{v}(u \cdot p) = p^{-1} \cdot \tilde{v}(u) \quad \text{for all} \quad u \in \mathcal{G}, p \in P$$

then we say that $\tilde{v}$ is *P-equivariant*. Observe that if $\tilde{v} : \mathcal{G} \to \mathbb{V}$ is a smooth P-equivariant map then the map $v : M \to \mathcal{V}$ which takes x to $[(u, v(u))]$ for some $u \in \mathcal{G}_x$ defines a smooth section of $\mathcal{V} \to M$ (being independent of the choice of $u \in \mathcal{G}_x$ for each x). It is easy to see that sections of $\mathcal{V} \to M$ are in one-to-one correspondence with such functions. (We have used this already in discussing conformal densities where $\mathbb{R}_+$ replaces P and $\mathcal{Q}$ replaces $\mathcal{G}$, see Section 3.3.4.) Now observe that the conformal Cartan connection ω allows us to view smooth P-equivariant functions $\tilde{\mathbb{L}} : \mathcal{G} \to \mathfrak{g}$ as smooth vector fields $V_{\mathbb{L}}$ on $\mathcal{G}$ (since $\omega_u : T_u\mathcal{G} \to \mathfrak{g}$ is an isomorphism for each $u \in \mathcal{G}$ and ω is smooth). From the properties of the Cartan connection it is not hard to see that any such vector field $V_{\mathbb{L}}$ must be P-invariant, that is,

$$(r^p)^* V_{\mathbb{L}} = V_{\mathbb{L}} \quad \text{for all} \quad p \in P$$

where $r^p : \mathcal{G} \to \mathcal{G}$ denotes the right action of $p \in P$. Moreover, the Cartan connection gives a one-to-one correspondence between P-equivariant functions $\tilde{\mathbb{L}} : \mathcal{G} \to \mathfrak{g}$ and P-invariant vector fields on $\mathcal{G}$. Combining this with the previous observation we see that one may naturally identify $\Gamma(\mathcal{A})$ with the space $\mathfrak{X}(\mathcal{G})^P$ of P-invariant vector fields on $\mathcal{G}$.

The observations of the preceding paragraph allow us to define a new canonical differential operator acting on sections of any vector bundle $\mathcal{V}$ associated with the conformal Cartan bundle $\mathcal{G}$. Fix $\mathbb{L} \in \Gamma(\mathcal{A})$ and let $v \in \Gamma(\mathcal{V}) = \Gamma(\mathcal{G} \times_P \mathbb{V})$, then $V_{\mathbb{L}}\tilde{v} = d\tilde{v}(V_{\mathbb{L}})$ is again a P-equivariant function from $\mathcal{G}$ to $\mathbb{V}$ and thus defines a section $D_{\mathbb{L}}v$ of $\mathcal{V}$. Thus for each section $\mathbb{L} \in \Gamma(\mathcal{A})$ we have a first order differential operator $D_{\mathbb{L}} : \mathcal{V} \to \mathcal{V}$. It is easy to see that $D_{f\mathbb{L}} = fD_{\mathbb{L}}$ for any $f \in C^\infty(M)$ so that we really have a (first order) differential operator taking sections of $\mathcal{V}$ to sections of $\mathcal{A}^* \otimes \mathcal{V}$. The operator

$$D : \mathcal{V} \to \mathcal{A}^* \otimes \mathcal{V}$$

defined in this way is referred to as the *fundamental derivative* (or *fundamental D operator*). This operator was introduced by Čap and Gover in [9].

Now if k is a conformal Killing vector field on $(M, \boldsymbol{c})$ and v is a section of $\mathcal{V} = \mathcal{G} \times_P \mathbb{V}$ then we may talk about the Lie derivative of v with respect to k; since $\mathcal{V} \to M$ is a natural bundle in the category of conformal manifolds (with diffeomorphisms as maps) one may pull the section v of $\mathcal{V}$ back by the (local) flow of k and define the Lie derivative by

$$\mathcal{L}_k v = \left.\frac{d}{dt}\right|_{t=0} (Fl^k_t)^*(v).$$

Notice that if $\mathbb{L}$ is a section of the adjoint tractor bundle then $V_{\mathbb{L}}\tilde{v}$ (which gives the P-equivariant function corresponding to $D_{\mathbb{L}}v$) may be written as the Lie derivative $\mathcal{L}_{V_{\mathbb{L}}}\tilde{v}$, it should not come as a total surprise then that the Lie derivative and the fundamental derivative are connected. In fact, one has that

$$\mathcal{L}_k v = D_{L(k)} v$$

for any conformal Killing vector field k on $(M, \boldsymbol{c})$ and any section v of a natural bundle $\mathcal{V} = \mathcal{G} \times_P \mathbb{V}$ (this is proven in [14]).

Using the results of [9] (and carefully comparing sign conventions) we have that

$$D_{\mathbb{L}}\tau \stackrel{g}{=} \nabla_l \tau + w\nu\tau \tag{3.64}$$

for sections τ of the density bundle $\mathcal{E}[w]$ (which can be thought of as an associated bundle to $\mathcal{G}_0$ and hence to $\mathcal{G}$), where $\mathbb{L}$ is given in the scale g by (3.61). We also have that

$$D_{\mathbb{L}} V = \nabla^{\mathcal{T}}_{\Pi(\mathbb{L})} V - \mathbb{L}(V)$$

for all $\mathbb{L} \in \Gamma(\mathcal{A})$ and $V \in \Gamma(\mathcal{T})$. The operator $D_{\mathbb{L}}$ is easily seen to satisfy the Leibniz property and hence for any tractor field $T^{A\cdots B}{}_{C\cdots D}$ we have

$$D_{\mathbb{L}}T^{A\cdots B}{}_{C\cdots D} = \nabla^{\mathcal{T}}_{\Pi(\mathbb{L})}T^{A\cdots B}{}_{C\cdots D} - \mathbb{L}^{A}{}_{A'}T^{A'\cdots B}{}_{C\cdots D} - \cdots - \mathbb{L}^{B}{}_{B'}T^{A\cdots B'}{}_{C\cdots D}$$
$$+ \mathbb{L}^{C'}{}_{C}T^{A\cdots B}{}_{C'\cdots D} + \cdots + \mathbb{L}^{D'}{}_{D}T^{A\cdots B}{}_{C\cdots D'}.$$

From all of this we can finally write down an explicit formula for the Lie derivative of a standard tractor field V^A in terms of 'slots': if k is a conformal Killing vector field on $(M,\boldsymbol{c})$ and $V^A \stackrel{g}{=} (\sigma,\mu^a,\rho)$ then

$$\mathcal{L}_k V^A = k^b\nabla_b V^B + \mathbb{K}^A{}_B V^B$$
$$\stackrel{g}{=} k^b \begin{pmatrix} \nabla_b\sigma - \mu_b \\ \nabla_b\mu^a + \rho\delta^a_b + \sigma \mathsf{P}^a_b \\ \nabla_b\rho - \mathsf{P}_{ab}\mu^a \end{pmatrix} + \begin{pmatrix} -\nu & k_b & 0 \\ -\rho^a & \mu^a{}_b & -k^a \\ 0 & \rho_b & \nu \end{pmatrix} \begin{pmatrix} \sigma \\ \mu^b \\ \rho \end{pmatrix}$$
$$\stackrel{g}{=} \begin{pmatrix} k^b\nabla_b\sigma - \nu\sigma \\ k^b\nabla_b\mu^a + \mu^a{}_b\mu^b - \sigma\nabla^a\nu \\ k^b\nabla_b\rho + \nu\rho + \mu^a\nabla_a\nu \end{pmatrix} = \begin{pmatrix} \mathcal{L}_k\sigma \\ \mathcal{L}_k\mu^a - \sigma\nabla^a\nu \\ \mathcal{L}_k\rho + \mu^a\nabla_a\nu \end{pmatrix}$$

where $\mathbb{K} = L(-k)$ and we have used that $\rho_a = \nabla_a\nu + \mathsf{P}_{ab}k^b$ and that $\mathcal{L}_k = D_{-\mathbb{K}}$ on densities so that by (3.64) we have $\mathcal{L}_k\sigma = \nabla_k\sigma - \nu\sigma$, $\mathcal{L}_k\rho = \nabla_k\rho + \nu\rho$, and

$$\begin{aligned} \mathcal{L}_k\mu^a &= (k^b\nabla_b\mu^a - \mu^b\nabla_b k^a) + \nu\mu^a \\ &= k^b\nabla_b\mu^a - \mu^b(\mu_b{}^a + \nu\delta^a_b) + \nu\mu^a \\ &= k^b\nabla_b\mu^a + \mu^a{}_b\mu^b \end{aligned}$$

since μ^a has conformal weight -1.

Note that one can also, of course, calculate the expression for $\mathcal{L}_k$ on densities and on standard tractor fields directly from the definition (which was done in [14]) by looking at what you get when you pull back densities and standard tractors by the local flow of k. As a check of our above formula for $\mathcal{L}_k V^A$ we observe that in the case where k is a Killing vector field for g then the flow of k preserves g and hence also preserves the splitting tractors X^A, Z^A_a and Y^A so that $\mathcal{L}_k X^A = 0$, $\mathcal{L}_k Z^A_a = 0$ and $\mathcal{L}_k Y^A = 0$; thus by the Leibniz property and linearity one immediately has that if $V^A = \sigma X^A + \mu^a Z^A_a + \rho Y^A$ then $\mathcal{L}_k V^A = (\mathcal{L}_k\sigma)X^A + (\mathcal{L}_k\mu^a)Z^A_a + (\mathcal{L}_k\rho)Y^A$ which is consistent with our above formula for $\mathcal{L}_k V^A$ since $\mathcal{L}_k g = 0$ forces $\nu = \frac{1}{d}\nabla^g_a k^a$ to be zero.

Remark 3.20 On vector fields the fundamental derivative acts according to

$$D_{\mathbb{L}}v^b = l^a\nabla_a v^b + (\mu_a{}^b + \nu\delta^b_a)v^a$$

where $\mathbb{L}$ is given in the scale g by (3.61) and $\nabla = \nabla^g$. Thus if k is a conformal Killing vector field then applying $\mathcal{L}_k = D_{L(k)}$ on vector fields simply returns the usual formula for the Lie derivative in terms of a torsion free (in this case Levi-Civita) connection:

$$\mathcal{L}_k v^b = k^a\nabla_a v^b - (\nabla_a k^b)v^a.$$

Similarly, using the Leibniz property of the fundamental derivative, we have that

$$
\begin{aligned}
&D_{\mathbb{L}} T^{b\cdots c}{}_{d\cdots e} \\
&= l^a \nabla_a T^{b\cdots c}{}_{d\cdots e} - (\mu_a{}^b + \nu\delta_a^b) T^{a\cdots c}{}_{d\cdots e} - \cdots - (\mu_a{}^c + \nu\delta_a^c) T^{b\cdots a}{}_{d\cdots e} \\
&\qquad + (\mu_d{}^a + \nu\delta_d^a) T^{b\cdots c}{}_{a\cdots e} + \cdots + (\mu_e{}^a + \nu\delta_e^a) T^{b\cdots c}{}_{d\cdots a},
\end{aligned}
$$

and again applying $\mathcal{L}_k = D_{L(k)}$ for a conformal killing vector field k simply yields the standard formula for the Lie derivative. These observations further demonstrate the consistency of our claims. What's more, we may now calculate the fundamental derivative of any weighted tensor-tractor field $T^{b\cdots c}{}_{d\cdots e}{}^{B\cdots C}{}_{D\cdots E}$. For instance, by writing $\boldsymbol{g}_{ab}$ as $\sigma^2 g_{ab}$ and using the Leibniz property one may easily show that

$$D_{\mathbb{L}} \boldsymbol{g}_{ab} = 0$$

for all $\mathbb{L} \in \Gamma(\mathcal{A})$. ∎

A.3.1. Static and Stationary Spacetimes

As an application of the above, we observe that if $g \in \boldsymbol{c}$ is an Einstein metric with corresponding scale tractor I and k is a Killing vector field for g with $\mathbb{K} = L(-k)$ then $\mathbb{K}(I) = 0$. This follows from the fact $\mathcal{L}_k I = 0$ since the flow of k preserves g (and hence also I), and from the above formula for the Lie derivative of I,

$$\mathcal{L}_k I^A = \nabla_k I^A + \mathbb{K}^A{}_B I^B = \mathbb{K}^A{}_B I^B,$$

since I is parallel. Moreover, if k is *hypersurface orthogonal* (i.e. its orthogonal distribution is integrable) then it is easy to see that $\mathbb{K}_{[AB}\mathbb{K}_{C]D}X^D = 0$ so that K_{AB} is simple and can be written as $2\nu_{[A}K_{B]}$ where $K_B = X^A \mathbb{K}_{AB}$ and $\nu_A K^A = 0$; on top of this, from $\mathbb{K}_{AB} I^B = 0$ we obtain $K_A I^A = 0$ and hence also $\nu_A I^A = 0$. These observations form the starting point for the development of conformal tractor calculus adapted to static and stationary spacetimes. In particular, we note that in the case where (M, g, k) is a static spacetime then $K_A = uN_A$ where N_A is the normal tractor to each of the spacelike hypersurfaces in the foliation given by $k^\perp \subset TM$ and u is the so called *static potential* (if we trivialise the density bundles using g); thus $K_A I^A = 0$ and $K_A \nu^A = 0$ imply that both I^A and ν^A lie in the intrinsic tractor bundle of the foliating spacelike hypersurfaces and one can carry out dimensional reduction using tractors. One can in fact still carry out dimensional reduction in the stationary case by identifying tractors which are Lie dragged by k and orthogonal to $K_B = X^A \mathbb{K}_{AB}$ with conformal tractors on the manifold of integral curves of k (see [14] for further details in both cases).

References

[1] M. Atiyah, R. Bott, and V.K. Patodi, *On the heat equation and the index theorem*, Inventiones Mathematicae, **19** (1973), 279–330.

[2] T.N. Bailey, M.G. Eastwood, and A.R. Gover, *Thomas's structure bundle for conformal, projective and related structures*, Rocky Mountain J. Math., **24** (1994), 1191–1217.

[3] T.N. Bailey, M.G. Eastwood, and C.R. Graham, *Invariant theory for conformal and CR geometry*, Annals of Mathematics, **139** (1994), 491–552.

[4] T. Branson, *Conformal structure and spin geometry*, in Dirac Operators: Yesterday and Today, (J.-P. Bourguignon, T. Branson, A. Chamseddine, O. Hijazi and R. Stanton, Eds.), International Press, (2005), 163–191.

[5] T. Branson and A.R. Gover, *Conformally invariant non-local operators*, Pacific Journal of Mathematics, **201** (2001), 19–60.

[6] A. Čap, *Infinitesimal automorphisms and deformations of parabolic geometries*, J. Eur. Math. Soc., **10** (2008), 415–437.

[7] A. Čap and A.R. Gover, *Standard tractors and the conformal ambient metric construction*, Annals Global Anal.Geom., **24** (2003), 231–259.

[8] A. Čap and A.R. Gover, *Tractor bundles for irreducible parabolic geometries*, in S.M.F. Colloques, Seminaires & Congres, **4** (2000), 129–154.

[9] A. Čap and A.R. Gover, *Tractor calculi for parabolic geometries*, Trans. Amer. Math. Soc., **354** (2002), 1511–1548.

[10] A. Čap, A.R. Gover, and M. Hammerl, *Projective BGG equations, algebraic sets, and compactifications of Einstein geometries*, J. London Math. Soc., **86** (2012), 433–454.

[11] A. Čap, A.R. Gover, and M. Hammerl, *Holonomy reductions of Cartan geometries and curved orbit decompositions*, Duke Math. J., **163** (2014), 5, 1035–1070.

[12] A. Čap, J. Slovák, Parabolic Geometries I: Background and General Theory, Mathematical Surveys and Monographs. American Mathematical Society, Providence, RI, 2009.

[13] P. Cherrier, *Problèmes de Neumann non linéaires sur les variètès riemanniennes*, J. Funct. Anal., **57** (1984), 154–206.

[14] S. Curry, Conformal Tractor Calculus for Stationary and Static Spacetimes, MSc thesis, University of Auckland, 2012.

[15] A.J. Di Scala and G. Manno, *On the extendability of parallel sections of linear connections*, Ann. Mat. Pura Appl. (4), **195** (2016), 1237–1253.

[16] P.A.M. Dirac, *The electron wave equation in de-Sitter space*, Ann. Math., **36** (1935), 657–669.

[17] C. Fefferman and C.R. Graham, The Ambient Metric, Annals of Mathematics Studies, 178. Princeton University Press, 2012. Available online: http://arxiv.org/pdf/0710.0919v2.pdf

[18] A. Fialkow, *Conformal geometry of a subspace,* Trans. Amer. Math. Soc., **56** (1944), 309–433.

[19] J. Frauendiener, *Conformal infinity*, Living Rev. Relativity, **7** (2004), 1. [Online Aricle]: cited 21/04/2014, http://www.livingreviews.org/lrr-2004-1

[20] J. Frauendiener and G.A.J. Sparling, *Local twistors and the conformal field equations*, J. Math. Phys., **41** (2000), 437–443.

[21] H. Friedrich, *Conformal Einstein evolution*, in Proceedings of the Tubingen Workshop on the Conformal Structure of Space-times, (H. Friedrich and J. Frauendiener, Eds.), Springer Lecture Notes in Physics, **604** (2002), 1–50.

[22] R. Geroch, *Local characterization of singularities in general relativity*, J. Math. Phys., **9** (1968), 450–465.

[23] R. Geroch, *Asymptotic structure of space-time*, in Asymptotic Structure of Space-Time, (F.P. Esposito and L. Witten, Eds.), Springer, (1977), 1–105.

[24] A.R. Gover, *Almost Einstein and Poincare–Einstein manifolds in Riemannian signature*, J. Geometry and Physics, **60** (2010), 182–204.

[25] A.R. Gover, *Aspects of parabolic invariant theory*, Supplemento ai Rendiconti del Circolo Matematico di Palermo, Serie II, **59** (1999), 25–47.

[26] A.R. Gover, *Conformal de Rham Hodge theory and operators generalising the Q-curvature*, Supplemento ai Rendiconti del Circolo Matematico di Palermo, Serie II, **75** (2005), 109–137.

[27] A.R. Gover, *Invariant theory and calculus for conformal geometries*, Advances in Mathematics, **163** (2001), 206–257.

[28] A.R. Gover, E. Latini, and A. Waldron, *Poincare–Einstein holography for forms via conformal geometry in the bulk*, Mem. Amer. Math. Soc., **235** (2014), 1–101.

[29] A.R. Gover and P. Nurowski, *Obstructions to conformally Einstein metrics in n dimensions*, J. Geom. Phys., **56** (2006), 450–484.

[30] A.R. Gover and L.J. Peterson, *Conformally invariant powers of the Laplacian, Q-curvature, and tractor calculus*, Communications in Mathematical Physics, **235** (2003), 339–378.

[31] A.R. Gover and L.J. Peterson, *The ambient obstruction tensor and the conformal deformation complex*, Pacific Journal of Mathematics, **226** (2006), 309–351.

[32] A.R. Gover and Y. Vyatkin, forthcoming.

[33] A.R. Gover and A. Waldron, *Boundary calculus for conformally compact manifolds*, Indiana Univ. Math. J., **63** (2014), 119–163.

[34] A.R. Gover and A. Waldron, *Generalising the Willmore equation: submanifold conformal invariants from a boundary Yamabe problem*, preprint, arXiv:1407.6742v1 [hep-th], (2014).

[35] A.R. Gover and A. Waldron, *Conformal hypersurface geometry via a boundary Loewner-Nirenberg-Yamabe problem*, arXiv:1506.02723 [math.DG] (2015).

[36] C.R. Graham, R. Jenne, L.J. Mason, and G.A.J. Sparling, *Conformally invariant powers of the Laplacian. I. Existence*, J. London Math. Soc., **46** (1992), 557–565.

[37] C.R. Graham and T. Willse, *Subtleties concerning conformal tractor bundles*, Cent. Eur. J. Math., **10** (2012), 1721–1732.

[38] D.H. Grant, A Conformally Invariant Third Order Neumann-Type Operator for Hypersurfaces, MSc thesis, University of Auckland, 2003.

[39] C. Kozameh, E.T. Newman, K.P. Tod, *Conformal Einstein spaces*, Gen. Relativity Gravitation, **17** (1985), 343–352.

[40] C. LeBrun, *Ambitwistors and Einstein's equations*, Class. Quant. Grav., **2** (1985), 555–563.

[41] F. Leitner, *Conformal Killing forms with normalisation condition*, Rend. Circ. Mat. Palermo Suppl. No. **75** (2005), 279–292.

[42] C. Lübbe, *A conformal extension theorem based on null conformal geodesics*, J. Math. Phys., **50**, 112502, (2009).

[43] C. Lübbe, *The conformal field equations and the tractor formalism in general relativity*, conference presentation, BritGrav 10, Dublin, 7th April 2010. Slides available online: http://www.dcu.ie/~nolanb/Luebbe.pdf

[44] C. Lübbe and K.P. Tod, *An extension theorem for conformal gauge singularities*, J. Math. Phys., **50**, 112501, (2009).

[45] R.S. Palais, Seminar on the Atiyah-Singer index theorem. With contributions by M. F. Atiyah, A. Borel, E. E. Floyd, R. T. Seeley, W. Shih and R. Solovay. Annals of Mathematics Studies, No. 57, Princeton University Press, 1965.

[46] S.M. Paneitz, *A quartic conformally covariant differential operator for arbitrary pseudo-Riemannian manifolds (Summary)*, SIGMA, **4** (2008), Paper 036, 3 pp.

[47] R. Penrose, *Asymptotic properties of fields and space-times*, Phys. Rev. Lett., **10** (1963), 66–68.

[48] R. Penrose, *The light cone at infinity*, in Relativistic Theories of Gravitation, (L. Infeld, Ed.), Pergamon Press, Oxford, UK, (1964), 369–373.

[49] R. Penrose, *Structure of space-time*, in Battelle Rencontres, (DeWitt, C.M., and Wheeler, J.A., Eds.), W.A. Benjamin, Inc., New York, NY, USA, (1968), 121–235.

[50] R. Penrose, *Zero rest-mass fields including gravitation: asymptotic behaviour*, Proc. R. Soc. Lond., **A284** (1965), 159–203.

[51] R. Penrose, M. MacCallum, *Twistor theory: an approach to the quantisation of fields and space-time*, Physics Reports (Section C of Physics Letters), **6** (1972), 241–316.

[52] R. Stafford, Tractor calculus and invariants for conformal submanifolds, MSc thesis, University of Auckland, 2006.

[53] P. Stredder, *Natural differential operators on Riemannian manifolds and representations of the orthogonal and special orthogonal groups*, J. Differential Geom., **10** (1975), 647–660.

[54] O. Veblen, *A conformal wave equation*, Proc. Nat. Acad. Sci. USA, **21** (1935), 484–487.

[55] Y. Vyatkin, Manufacturing Conformal Invariants of Hypersurfaces, PhD thesis, University of Auckland, 2013.

[56] R.M. Wald, General Relativity, University of Chicago Press, 1984.

SNC & ARG: Department of Mathematics, The University of Auckland, Private Bag 92019, Auckland 1142, New Zealand
E-mail address: `r.gover@auckland.ac.nz`
E-mail address: `sean.curry@auckland.ac.nz`

4

An Introduction to Quantum Field Theory on Curved Spacetimes

Christian Gérard

4.1. Introduction

The purpose of these notes is to provide an introduction to some recent aspects of quantum field theory on curved spacetimes, emphasizing its relations with partial differential equations and microlocal analysis.

4.1.1. Quantum Field Theory

Quantum field theory arose from the need to unify quantum mechanics with special relativity. However, trying to treat the two basic relativistic field equations, the Klein–Gordon equation

$$\partial_t^2\phi(t,x) - \Delta_x\phi(t,x) + m^2\phi(t,x) = 0,$$

and the Dirac equation

$$\gamma^0\partial_t\psi(t,x) + \gamma^i\partial_{x^i}\psi(t,x) - m\psi(t,x) = 0,$$

(where the γ^i are the Dirac matrices) in a way parallel to the non-relativistic Schroedinger equation

$$\partial_t\psi(t,x) - \frac{\mathrm{i}}{2m}\Delta_x\psi(t,x) + \mathrm{i}V(x)\psi(t,x) = 0$$

leads to difficulties (see e.g. [3]). For the Klein–Gordon equation, there exists a conserved scalar product

$$\langle\phi_1|\phi_2\rangle = \mathrm{i}\int_{\mathbb{R}^3}\partial_t\overline{\phi}_1(t,x)\phi_2(t,x) - \overline{\phi}_1(t,x)\phi_2(t,x)dx$$

which is however not positive definite, hence cannot lead to a probabilistic interpretation. However, we do have

$$\langle\phi|\mathrm{i}\partial_t\phi\rangle \geq 0, \text{ (positivity of the energy).}$$

For the Dirac equation the situation is the opposite: the conserved scalar product

$$\langle \psi_1 | \psi_2 \rangle = \int_{\mathbb{R}^3} \overline{\psi}_1(t,x) \cdot \psi_2(t,x) dx$$

is positive, but

$$\langle \psi | \mathrm{i}\partial_t \psi \rangle \text{ is indefinite.}$$

The reason behind these difficulties is that, although all these equations are partial differential equations, their nature is very different: the Klein–Gordon and Dirac equations are *classical* equations, while the Schroedinger equation is a *quantum* equation, obtained by quantizing the classical Newton equation

$$\ddot{x}(t) = -\nabla_x V(x(t)), \ x \in \mathbb{R}^n$$

or equivalently the Hamilton equations

$$\begin{cases} \dot{x}(t) = \partial_\xi h(x(t), \xi(t)), \\ \dot{\xi}(t) = -\partial_x h(x(t), \xi(t)) \end{cases}$$

for the classical Hamiltonian

$$h(x,\xi) = \frac{1}{2}\xi^2 + V(x).$$

We denote by $X = (x,\xi)$ the points in $T^*\mathbb{R}^n$ and introduce the coordinate functions

$$q : X \mapsto x, \ p : X \mapsto \xi.$$

If $\Phi(t) : T^*\mathbb{R}^n \to T^*\mathbb{R}^n$ is the flow of H_h and $q(t) := q \circ \Phi(t), p(t) := p \circ \Phi(t)$ then

$$\partial_t q(t) = p(t),$$

$$\partial_t p(t) = -\nabla V(q(t)),$$

where $\{\cdot,\cdot\}$ is the Poisson bracket. Note that

$$\{p_j(t), q_k(t)\} = \delta_{jk}, \ \{p_j(t), p_k(t)\} = \{q_j(t), q_k(t)\} = 0.$$

To quantize the Liouville equation means to find a Hilbert space $\mathcal{H}$ and functions $\mathbb{R} \ni t \mapsto p(t), q(t)$ with values in self-adjoint operators on $\mathcal{H}$ such that

$$[p_j(t), \mathrm{i}q_k(t)] = \delta_{jk}\mathbb{1}, \ [p_j(t), \mathrm{i}p_k(t)] = [q_j(t), \mathrm{i}q_k(t)] = 0, \partial_t q(t) = p(t),$$

$$\partial_t p(t) = -\nabla V(q(t)).$$

The last two equations are called Heisenberg equations. The solution is as follows.

(1) Find operators p, q satisfying

$$[p_j, \mathrm{i}q_k] = \delta_{jk}\mathbb{1}, \ [p_j, \mathrm{i}p_k] = [q_j, \mathrm{i}q_k] = 0.$$

(2) Construct the self-adjoint operator on $\mathcal{H}$

$$H = \frac{1}{2}p^2 + V(q).$$

(3) Then

$$q(t) := \mathrm{e}^{\mathrm{i}tH} q \mathrm{e}^{-\mathrm{i}tH}, \ p(t) := \mathrm{e}^{\mathrm{i}tH} p \mathrm{e}^{-\mathrm{i}tH}$$

solve Heisenberg equations.

The Stone–von Neumann theorem says that there is no choice in Step (1): modulo some technical conditions and multiplicity one has only one choice, up to unitary equivalence:

$$\mathcal{H} = L^2(\mathbb{R}^n), \ q = x, \ p = \mathrm{i}^{-1}\nabla_x.$$

Then $H = -\frac{1}{2}\Delta + V(x)$ is the Schroedinger operator. Step (2) is then a standard problem in the theory of self-adjoint operators, and Step (3) is straightforward.

The Klein–Gordon equation is also a Hamiltonian equation, though with an *infinite dimensional* phase space, which can be taken for example as $C_0^\infty(\mathbb{R}^d) \oplus C_0^\infty(\mathbb{R}^d)$. The classical Hamiltonian is then

$$h(\varphi, \pi) := \frac{1}{2}\int_{\mathbb{R}^d} \pi^2(x) + |\nabla_x\varphi(x)|^2 + m^2\varphi^2(x)dx,$$

for the linear case, or

$$h(\varphi, \pi) := \frac{1}{2}\int_{\mathbb{R}^d} \pi^2(x) + |\nabla_x\varphi(x)|^2 + m^2\varphi^2(x) + \varphi^n(x)dx,$$

for some nonlinear version. Here the symbols $\varphi(x)$, $\pi(x)$ are (coordinate) functions, parametrized by a point $x \in \mathbb{R}^d$, on the space of smooth solutions of the Klein–Gordon equation, with compactly supported Cauchy data. If ϕ is such a solution then

$$\varphi(x)(\phi) := \phi(0, x), \ \pi(x)(\phi) := \partial_t\phi(0, x).$$

It is well-known that these are *symplectic coordinates*, that is

$$\{\varphi(x), \varphi(x')\} = \{\pi(x), \pi(x')\} = 0, \ \{\pi(x), \phi(x')\} = \delta(x, x'), \ \forall\, x, x' \in \mathbb{R}^d.$$

One would like to follow the same path and consider families of operators on a Hilbert space $\mathcal{H}$, $\varphi(x), \pi(x), x \in \mathbb{R}^d$ such that

$$[\pi(x), \mathrm{i}\varphi(x')] = \delta(x - x')\mathbb{1}, \ [\varphi(x), \mathrm{i}\varphi(x')] = [\pi(x), \mathrm{i}\pi(x')] = 0, \ \forall\, x, x' \in \mathbb{R}^d.$$

The fundamental difference with non-relativistic quantum mechanics is that, since the phase space is infinite dimensional, the Stone–von Neumann theorem cannot be applied anymore: there exists an infinite number of inequivalent representations of commutation relations.

In other words, when one tries to quantize a classical field equation, the Hilbert space has to be constructed *together* with the quantum Hamiltonian: one cannot work on our familiar Hilbert space and then use tools from operator theory to construct the quantum Hamiltonian.

This is the reason why the rigorous construction of quantum field theory models is so difficult, except for *non-interacting* theories. For interacting theories this has been achieved only in two and three spacetime dimensions; see the construction of the $P(\varphi)_2$ and φ_3^4 models, which were the landmark successes of constructive field theory. In four spacetime dimensions one has to rely instead on perturbative methods.

Another lesson learned from quantum field theory (and also from quantum statistical mechanics) is that Hilbert spaces do not play such a central role anymore. Instead one focuses on *algebras* and *states*.

Let us finish this discussion by recalling a well-known anecdote. At the Solvay conference in 1927, Dirac told Bohr that he was trying to find a relativistic quantum theory of the electron (i.e. the Dirac equation). Bohr replied that this problem had already been solved by Klein, who had found the Klein–Gordon equation. We know now that these two equations are of a different nature, the first describing fermionic fields, the second bosonic ones, and that they can be interpreted as quantum equations only via quantum field theory.

4.1.2. Quantum Field Theory on Curved Spacetimes

Given the difficulties with the construction of interacting field theories on Minkowski spacetime, one may wonder why one should consider quantum field theories on *curved spacetimes*, which have no reason to be simpler.

One reason comes from attempts to quantize gravitation, where one starts by linearizing Einstein equations around a curved background metric g. Another argument is that there are several interesting quantum effects appearing in the presence of strong gravitational fields. The most famous one is the Hawking effect, which predicts that a black hole can emit quantum particles. There are several new challenges one has to face when moving from flat Minkowski spacetime to an arbitrary curved spacetime.

On the computational side, one cannot rely anymore on the Fourier transform and related analyticity arguments, which are natural and useful on Minkowski space, since the Klein–Gordon equation has then constant coefficients.

From a more conceptual angle, a curved spacetime does not have the large group of isometries (the Poincar group) of the Minkowski space. It follows that on a curved spacetime there seems to be no natural notion of a *vacuum state*, which is defined on Minkowski space as the unique state which is invariant under spacetime translations, and which has an additional *positive energy condition*.

In the 1980s, physicists managed to define a class of states, the so-called "Hadamard states," which were characterized by properties of their 2-point functions, which had to have a specific asymptotic expansion near the diagonal, connected with the well-known Hadamard parametrix construction for the Klein–Gordon equation on a curved spacetime. Later in 1995, in a seminal paper, Radzikowski reformulated the old Hadamard condition in terms of the *wave front set* of the 2-point function. The wave front set of a distribution, introduced in 1970 by Hörmander, is one of the important notions of *microlocal analysis*, a theory which was precisely developed to extend Fourier analysis, in the study of general partial differential equations.

The introduction of tools from microlocal analysis had a great influence on the field, leading for example to the proof of renormalizability of scalar interacting field theories by Brunetti and Fredenhagen.

The goal of these notes is to provide an introduction to the modern notion of Hadamard states for a mathematically oriented audience.

4.2. A Quick Introduction to Quantum Mechanics

This section provides a very quick introduction to the mathematical formalism of quantum mechanics, which is (or is expected to be) still relevant to quantum field theory.

4.2.1. Hilbert Space Approach

In ordinary quantum mechanics, the description of a physical system starts with a Hilbert space $\mathcal{H}$, whose scalar product is denoted by $(u|v)$. The *states* of the system are described by unit vectors $\psi \in \mathcal{H}$ with $\|\psi\| = 1$.

The various physical quantities which can be measured (like position, momentum, energy, spin) are represented by *self-adjoint operators* on $\mathcal{H}$, that is (forgetting about important issues with unbounded operators) linear operators A on $\mathcal{H}$, assumed to be bounded for simplicity, such that $A = A^*$ (where A^* is the *adjoint* of A), and which are called *observables*.

If $\psi \in \mathcal{H}$, $\|\psi\| = 1$ is a state vector, then the map

$$A \mapsto \omega_\psi(A) = (\psi|A\psi)$$

computes the *expectation value* of A in the state ψ which represents the average value of actual measurements of the physical quantity represented by A.

Rather quickly people were also led to consider *mixed states*, where the state of the system is only incompletely known. For example if ψ_i, $i \in \mathbb{N}$ is an orthonormal family and $0 \leq \rho_i \leq 1$ are real numbers with $\sum_{i=0}^{\infty} \rho_i = 1$, then we can consider the trace-class operator

$$\rho = \sum_{i=0}^{\infty} \rho_i |\psi_i)(\psi_i|, \ \mathrm{Tr}\rho = 1,$$

called a *density matrix*, and the map

$$A \mapsto \omega_\rho(A) := \mathrm{Tr}(\rho A)$$

which is called a *mixed state*. Vector states are also called *pure states*.

4.2.2. Algebraic Approach

The framework above is sufficient to cover all of non-relativistic quantum mechanics, that is in practice quantum systems consisting of a *finite* number of non-relativistic particles. However, when one considers systems with an *infinite* number of particles, as in statistical mechanics or quantum field theory, where the notion of particles is dubious, an algebraic framework is more relevant. It starts with the following observation about the space $B(\mathcal{H})$ of bounded operators on $\mathcal{H}$.

If we equip the space with the operator norm, it is a Banach space, and a Banach algebra, that is an algebra with the property such that $\|AB\| \leq \|A\|\,\|B\|$. It is also an involutive Banach algebra, that is the adjoint operation $A \mapsto A^*$ has the properties such that

$$(AB)^* = B^*A^*, \ \|A^*\| = \|A\|.$$

Finally one can easily check that

$$\|A^*A\| = \|A\|^2, \ A \in B(\mathcal{H}).$$

This last property has very important consequences; for example one can deduce from it the functional calculus and spectral theorem for self-adjoint operators.

An abstract algebra $\mathfrak{A}$ equipped with a norm and an involution with these properties, which is moreover complete, is called a C^**algebra*. The typical example of a C^* algebra is of course the algebra $B(\mathcal{H})$ of bounded operators on a Hilbert space.

If $\mathcal{H}$ is a Hilbert space, a $*-$homomorphism

$$\mathfrak{A} \ni A \mapsto \pi(A) \in B(\mathcal{H})$$

is called a *representation* of $\mathfrak{A}$ in $\mathcal{H}$. An injective representation is called *faithful*.

The need for such a change of point of view comes from the fact that a physical system, like a gas of electrons, can exist in many different physical realizations, for example at different temperatures. In other words it does not come equipped with a canonical Hilbert space.

Observables, like for example the electron density, have a meaning irrelevant of the realizations, and are described by self-adjoint elements in some C^* algebra $\mathfrak{A}$. However, the various Hilbert spaces and the representations of the observables on them are very different from one temperature to another.

One can also describe the possible physical realizations of a system with the language of *states*. A *state* ω on $\mathfrak{A}$ is a linear map

$$\omega : \mathfrak{A} \mapsto \mathbb{C}$$

such that

$$\omega(A^*A) \geq 0, \ A \in \mathfrak{A}.$$

Assuming that $\mathfrak{A}$ has a unit (which can always be assumed by adjoining one), one also requires that

$$\omega(\mathbb{1}) = 1.$$

The set of states on a C^* algebra is a convex set, its extremal points being called *pure states*. If $\mathfrak{A} \subset B(\mathcal{H})$ and ψ is a unit vector, or if ρ is a density matrix, then

$$\omega_\psi(A) := (\psi|A\psi), \ \omega_\rho(A) := \mathrm{Tr}(\rho A)$$

are states on $\mathfrak{A}$. If $\mathfrak{A} = B(\mathcal{H})$, then ω_ψ is a pure state. It is important to be aware of the fact that if $\mathfrak{A}$ is only a C^* subalgebra of $B(\mathcal{H})$, then ω_ψ may *not* be a pure state on $\mathfrak{A}$.

4.2.3. The Gelfand–Naimark–Segal Construction

After being told that one should use C^* algebras and states, one may wonder where the Hilbert spaces have gone. Given a C^* algebra $\mathfrak{A}$ and a state ω on it, it is quite easy to construct a canonical Hilbert space and a representation of $\mathfrak{A}$ on it, as proved by Gelfand, Naimark, and Segal. There exists a triple $(\mathcal{H}_\omega, \pi_\omega, \Omega_\omega)$ where $\mathcal{H}_\omega$ is a Hilbert space, $\pi_\omega : \mathfrak{A} \mapsto B(\mathcal{H}_\omega)$ is a faithful representation, and $\Omega_\omega \in \mathcal{H}_\omega$ is a unit vector such that

$$\omega(A) = (\Omega_\omega|\pi_\omega(A)\Omega_\omega),\ A \in \mathfrak{A}.$$

4.3. Notation

In this section we collect some notation that will be used in these notes. If $\mathcal{X}$ is a real or complex vector space we denote by $\mathcal{X}^\#$ its dual. Bilinear forms on $\mathcal{X}$ are identified with elements of $L(\mathcal{X}, \mathcal{X}^\#)$, which leads to the notation $x_1 \cdot bx_2$ for $b \in L(\mathcal{X}, \mathcal{X}^\#)$, $x_1, x_2 \in \mathcal{X}$. The space of symmetric (resp. anti-symmetric) bilinear forms on $\mathcal{X}$ is denoted by $L_{\rm s}(\mathcal{X}, \mathcal{X}^\#)$ (resp. $L_{\rm a}(\mathcal{X}, \mathcal{X}^\#)$).

If $\sigma \in L_{\rm s}(\mathcal{X}, \mathcal{X}^\#)$ is non-degenerate, we denote by $O(\mathcal{X}, \sigma)$ the linear (pseudo-)orthogonal group on $\mathcal{X}$. Similarly if $\sigma \in L_{\rm a}(\mathcal{X}, \mathcal{X}^\#)$ is non-degenerate, that is $(\mathcal{X}, \sigma)$ is a symplectic space, we denote by $Sp(\mathcal{X}, \sigma)$ the linear symplectic group on $\mathcal{X}$.

If $\mathcal{Y}$ is a complex vector space, we denote by $\mathcal{Y}_\mathbb{R}$ its *realification*, that is $\mathcal{Y}$ considered as a real vector space. We denote by $\overline{\mathcal{Y}}$ a *conjugate vector space* to $\mathcal{Y}$, that is a complex vector space $\overline{\mathcal{Y}}$ with an anti-linear isomorphism $\mathcal{Y} \ni y \mapsto \overline{y} \in \overline{\mathcal{Y}}$. The *canonical conjugate vector space* to $\mathcal{Y}$ is simply the real vector space $\mathcal{Y}_\mathbb{R}$ equipped with the complex structure $-{\rm i}$, if ${\rm i}$ is the complex structure of $\mathcal{Y}$. In this case the map $y \to \overline{y}$ is chosen as the identity. If $a \in L(\mathcal{Y}_1, \mathcal{Y}_2)$, we denote by $\overline{a} \in L(\overline{\mathcal{Y}}_1, \overline{\mathcal{Y}}_2)$ the linear map defined by

$$\overline{a}\overline{y}_1 := \overline{ay_1},\ \overline{y}_1 \in \overline{\mathcal{Y}}_1. \tag{4.1}$$

We denote by $\mathcal{Y}^*$ the *anti-dual* of $\mathcal{Y}$, that is the space of anti-linear forms on $\mathcal{Y}$. Clearly $\mathcal{Y}^*$ can be identified with $\overline{\mathcal{Y}^\#} \sim \overline{\mathcal{Y}}^\#$.

Sesquilinear forms on $\mathcal{Y}$ are identified with elements of $L(\mathcal{Y}, \mathcal{Y}^*)$, and we use the notation $(y_1|by_2)$ or $\overline{y_1} \cdot by_2$ for $b \in L(\mathcal{Y}, \mathcal{Y}^*)$, $y_1, y_2 \in \mathcal{Y}$.

The space of hermitian (resp. anti-hermitian) sesquilinear forms on $\mathcal{Y}$ is denoted by $L_{\rm s}(\mathcal{Y}, \mathcal{Y}^*)$ (resp. $L_{\rm a}(\mathcal{Y}, \mathcal{Y}^*)$).

If $q \in L_{\rm h}(\mathcal{Y}, \mathcal{Y}^*)$ is non-degenerate, that is $(\mathcal{Y}, q)$ is a pseudo-unitary space, we denote by $U(\mathcal{Y}, q)$ the linear pseudo-unitary group on $\mathcal{Y}$.

If b is a bilinear form on the real vector space $\mathcal{X}$, its canonical sesquilinear extension to $\mathbb{C}\mathcal{X}$ is by definition the sesquilinear form $b_\mathbb{C}$ on $\mathbb{C}\mathcal{X}$ given by

$$(w_1|b_\mathbb{C}w_2) := x_1 \cdot bx_2 + y_1 \cdot by_2 + {\rm i}x_1 \cdot by_2 - {\rm i}y_1 \cdot bx_2, \quad w_i = x_i + {\rm i}y_i$$

for $x_i, y_i \in \mathcal{X}$, $i = 1, 2$. This extension maps (anti-)symmetric forms on $\mathcal{X}$ onto (anti-)hermitian forms on $\mathbb{C}\mathcal{X}$.

Conversely if $\mathcal{Y}$ is a complex vector space and $\mathcal{Y}_{\mathbb{R}}$ is its realification, that is $\mathcal{Y}$ considered as a real vector space, then for $b \in L_{\mathrm{h/a}}(\mathcal{Y}, \mathcal{Y}^*)$ the form $\mathrm{Re}b$ belongs to $L_{\mathrm{s/a}}(\mathcal{Y}_{\mathbb{R}}, \mathcal{Y}_{\mathbb{R}}^{\#})$.

4.4. CCR and CAR Algebras

4.4.1. Introduction

It is useful to discuss the canonical commutation relations (CCR) and canonical anticommutation relations (CAR) without making reference to a Fock space. There are some mathematical subtleties with CCR algebras, coming from the fact that the field operators are "unbounded." These subtleties can mostly be ignored for our purposes.

4.4.2. Algebras Generated by Symbols and Relations

In physics many algebras are defined by specifying a set of generators and the relations they satisfy. This is completely sufficient to perform computations, but mathematicians may feel uncomfortable with such an approach. However, it is easy (and actually rather useless) to give a rigorous definition.

Assume that $\mathcal{A}$ is a set. We denote by $c_{\mathrm{c}}(\mathcal{A}, \mathbb{K})$ the vector space of functions $\mathcal{A} \to \mathbb{K}$ with finite support (usually $\mathbb{K} = \mathbb{C}$). If for $A \in \mathcal{A}$, we denote the indicator function $1\!\!1_{\{A\}}$ simply by A, we see that any element of $c_{\mathrm{c}}(\mathcal{A}, \mathbb{K})$ can be written as $\sum_{A \in \mathcal{B}} \lambda_A A$, $\mathcal{B} \subset \mathcal{A}$ finite, $\lambda_A \in \mathbb{K}$.

Then $c_{\mathrm{c}}(\mathcal{A}, \mathbb{K})$ can be seen as the vector space of finite linear combinations of elements of $\mathcal{A}$. We set

$$\mathfrak{A}(\mathcal{A}, 1\!\!1) := \overset{\mathrm{al}}{\otimes} c_{\mathrm{c}}(\mathcal{A}, \mathbb{K}),$$

called the *universal unital algebra over* $\mathbb{K}$ *with generators* $\mathcal{A}$. Usually one writes $A_1 \cdots A_n$ instead of $A_1 \otimes \cdots \otimes A_n$ for $A_i \in \mathcal{A}$.

Let us denote by $\overline{\mathcal{A}}$ another copy of $\mathcal{A}$. We denote by $\overline{a}$ the element $a \in \overline{\mathcal{A}}$. We then set $*a := \overline{a}$, $*\overline{a} := a$ and extend $*$ to $\mathfrak{A}(\mathcal{A} \sqcup \overline{\mathcal{A}}, 1\!\!1)$ by setting

$$(b_1 b_2 \cdots b_n)^* = b_n^* \cdots b_2^* b_1^*, \quad b_i \in \mathcal{A} \sqcup \overline{\mathcal{A}}, \ 1\!\!1 = 1\!\!1^*.$$

The algebra $\mathfrak{A}(\mathcal{A} \sqcup \overline{\mathcal{A}}, 1\!\!1)$ equipped with the involution $*$ is called the *universal unital* $*-$*algebra over* $\mathbb{K}$ *with generators* $\mathcal{A}$.

Now let $\mathfrak{R} \subset \mathfrak{A}(\mathcal{A}, \mathbb{1})$ (the set of "relations"). We denote by $\mathfrak{I}(\mathfrak{R})$ the ideal of $\mathfrak{A}(\mathcal{A}, \mathbb{K})$ generated by $\mathfrak{R}$. Then the quotient

$$\mathfrak{A}(\mathcal{A}, \mathbb{1})/\mathfrak{I}(\mathfrak{R})$$

is called the *unital algebra with generators $\mathcal{A}$ and relations $R = 0, \ R \in \mathfrak{R}$.*

Similarly if $\mathfrak{R} \subset \mathfrak{A}(\mathcal{A} \cup \overline{\mathcal{A}}, \mathbb{1})$ is $*$-invariant, then $\mathfrak{A}(\mathcal{A} \sqcup \overline{\mathcal{A}}, \mathbb{1})/\mathfrak{I}(\mathfrak{R})$ is called the *unital $*$-algebra with generators $\mathcal{A} \sqcup \overline{\mathcal{A}}$ and relations $R = 0, \ R \in \mathfrak{R}$.*

4.4.3. Polynomial CCR Algebra

We fix a (real) presymplectic space $(\mathcal{X}, \sigma)$, that is $\sigma \in L_{\mathrm{a}}(\mathcal{X}, \mathcal{X}^{\#})$ is not supposed to be injective.

Definition 4.1 The *polynomial CCR $*$-algebra over $\mathcal{X}$*, denoted by $\mathrm{CCR}^{\mathrm{pol}}(\mathcal{X}, \sigma)$, is defined to be the unital complex $*$-algebra generated by elements $\phi(x), x \in \mathcal{X}$, with relations

$$\phi(\lambda x) = \lambda\phi(x), \ \lambda \in \mathbb{R}, \ \ \phi(x_1 + x_2) = \phi(x_1) + \phi(x_2),$$

$$\phi^*(x) = \phi(x), \ \ \phi(x_1)\phi(x_2) - \phi(x_2)\phi(x_1) = \mathrm{i}x_1{\cdot}\sigma x_2 \mathbb{1}.$$

4.4.4. Weyl CCR Algebra

One problem with $\mathrm{CCR}^{\mathrm{pol}}(\mathcal{X}, \sigma)$ is that (unless $\sigma = 0$) its elements cannot be faithfully represented as bounded operators on a Hilbert space. To cure this problem one has to work with Weyl operators, which lead to the Weyl CCR $*-$algebra.

Definition 4.2 The *algebraic Weyl CCR algebra over $\mathcal{X}$* denoted by $\mathrm{CCR}^{\mathrm{Weyl}}(\mathcal{X}, \sigma)$ is the $*$-algebra generated by the elements $W(x)$, $x \in \mathcal{X}$, with relations

$$W(0) = \mathbb{1}, \ W(x)^* = W(-x),$$

$$W(x_1)W(x_2) = \mathrm{e}^{-\frac{\mathrm{i}}{2}x_1\cdot\sigma x_2} W(x_1 + x_2), \ x, x_1, x_2 \in \mathcal{X}.$$

It is possible to equip $\mathrm{CCR}^{\mathrm{Weyl}}(\mathcal{X}, \sigma)$ with a unique C^*-norm. Its completion for this norm is called the *Weyl CCR algebra over $\mathcal{X}$* and still denoted by $\mathrm{CCR}^{\mathrm{Weyl}}(\mathcal{X}, \sigma)$. We will mostly work with $\mathrm{CCR}^{\mathrm{pol}}(\mathcal{X}, \sigma)$, which will simply be denoted by $\mathrm{CCR}(\mathcal{X}, \sigma)$. Of course the formal relation between the two approaches is

$$W(x) = \mathrm{e}^{\mathrm{i}\phi(x)}, \ x \in \mathcal{X},$$

which does not make sense a priori, but from which mathematically correct statements can be deduced.

4.4.5. Charged Symplectic Spaces

Definition 4.3 A complex vector space $\mathcal{Y}$ equipped with a non-degenerate anti-hermitian sesquilinear form σ is called a *charged symplectic space*. We set

$$q := \mathrm{i}\sigma \in L_{\mathrm{h}}(\mathcal{Y}, \mathcal{Y}^*),$$

which is called the *charge*.

4.4.6. Kähler Spaces

Let $(\mathcal{Y}, \sigma)$ be a charged symplectic space. Its complex structure will be denoted by $\mathrm{j} \in L(\mathcal{Y}_{\mathbb{R}})$ (to distinguish it from the complex number $\mathrm{i} \in \mathbb{C}$). Note that $(\mathcal{Y}_{\mathbb{R}}, \mathrm{Re}\sigma)$ is a real symplectic space with $\mathrm{j} \in Sp(\mathcal{Y}_{\mathbb{R}}, \mathrm{Re}\sigma)$ and $\mathrm{j}^2 = -\mathbb{1}$. We have

$$\overline{y}_1 q y_2 = y_1 \cdot \mathrm{Re}\sigma \mathrm{j} y_2 + \mathrm{i} y_1 \cdot \mathrm{Re}\sigma y_2, \; y_1, y_2 \in \mathcal{Y}.$$

The converse construction is as follows. A real (pre-)symplectic space $(\mathcal{X}, \sigma)$ with a map $\mathrm{j} \in L(\mathcal{X})$ such that

$$\mathrm{j}^2 = -\mathbb{1}, \; \mathrm{j} \in Sp(\mathcal{X}, \sigma)$$

is called a *pseudo-Kähler space*. If in addition $\nu := \sigma \mathrm{j}$ is positive definite, it is called a *Kähler space*. We now set

$$\mathcal{Y} = (\mathcal{X}, \mathrm{j}),$$

which is a complex vector space, whose elements are logically denoted by y. If $(\mathcal{X}, \sigma, \mathrm{j})$ is a pseudo-Kähler space we can set

$$\overline{y}_1 q y_2 := y_1 \cdot \sigma \mathrm{j} y_2 + \mathrm{i} y_1 \cdot \sigma y_2, \; y_1, y_2 \in \mathcal{Y}$$

and check that q is sesquilinear hermitian on $\mathcal{Y}$ equipped with the complex structure j. One may consider the CCR algebra $\mathrm{CCR}^{\mathrm{pol}}(\mathcal{Y}_{\mathbb{R}}, \mathrm{Re}\sigma)$, with selfadjoint generators $\phi(y)$ and relations

$$[\phi(y_1), \phi(y_2)] = \mathrm{i} y_1 \cdot \mathrm{Re}\sigma y_2 \mathbb{1}.$$

But one can instead generate $\mathrm{CCR}^{\mathrm{pol}}(\mathcal{Y}_{\mathbb{R}}, \mathrm{Re}\sigma)$ by the *charged fields*

$$\psi(y) := \frac{1}{\sqrt{2}}(\phi(y) + \mathrm{i}\phi(\mathrm{j}y)), \; \psi^*(y) := \frac{1}{\sqrt{2}}(\phi(y) - \mathrm{i}\phi(\mathrm{j}y)), \; y \in \mathcal{Y}.$$

The map $\mathcal{Y} \ni y \mapsto \psi^*(y)$ (resp. $\mathcal{Y} \ni y \mapsto \psi(y)$) is $\mathbb{C}$−linear (resp. $\mathbb{C}$−anti-linear). The commutation relations take the form

$$[\psi(y_1), \psi(y_2)] = [\psi^*(y_1), \psi^*(y_2)] = 0,$$
$$[\psi(y_1), \psi^*(y_2)] = \overline{y}_1 \cdot q y_2 \mathbb{1},\ y_1, y_2 \in \mathcal{Y}.$$

Note the similarity with the CCR expressed in terms of creation/annihilation operators, the difference being the fact that q is not necessarily positive. In this context, it is natural to denote $\mathrm{CCR}(\mathcal{Y}_{\mathbb{R}}, \mathrm{Re}\sigma)$ by $\mathrm{CCR}(\mathcal{Y}, \sigma)$.

4.4.7. CAR Algebra

Here we fix a Euclidean space $(\mathcal{X}, \nu)$ (possibly infinite dimensional).

Definition 4.4 The *algebraic CAR algebra over* $\mathcal{X}$, denoted $\mathrm{CAR}^{\mathrm{alg}}(\mathcal{X}, \nu)$, is the complex unital $*$-algebra generated by elements $\phi(x)$, $x \in \mathcal{X}$, with relations

$$\phi(\lambda x) = \lambda\phi(x),\ \lambda \in \mathbb{R}, \quad \phi(x_1 + x_2) = \phi(x_1) + \phi(x_2),$$
$$\phi^*(x) = \phi(x), \quad \phi(x_1)\phi(x_2) + \phi(x_2)\phi(x_1) = 2x_1 \cdot \nu x_2 \mathbb{1}.$$

Again $\mathrm{CAR}^{\mathrm{alg}}(\mathcal{X}, \nu)$ has a unique C^*−norm, and its completion is denoted by $\mathrm{CAR}(\mathcal{X}, \nu)$.

4.4.8. Kähler Spaces

Now let $(\mathcal{Y}, \nu)$ be a hermitian space, denoting again its complex structure by j. Then $(\mathcal{Y}_{\mathbb{R}}, \mathrm{Re}\nu)$ is a Euclidean space, with $\mathrm{j} \in U(\mathcal{Y}_{\mathbb{R}}, \mathrm{Re}\nu)$. We have

$$\overline{y}_1 \cdot \nu y_2 = y_1 \cdot \mathrm{Re}\nu y_2 - \mathrm{i} y_1 \cdot \mathrm{Re}\mathrm{j} y_2,\ y_1, y_2 \in \mathcal{Y}.$$

Denoting by $\phi(y)$ the self-adjoint fields which generate $\mathrm{CAR}(\mathcal{Y}_{\mathbb{R}}, \mathrm{Re}\nu)$, we can introduce the charged fields

$$\psi(y) := \phi(y) + \mathrm{i}\phi(\mathrm{j}y),\ \psi^*(y) := \phi(y) - \mathrm{i}\phi(\mathrm{j}y),\ y \in \mathcal{Y}.$$

Again the map $\mathcal{Y} \ni y \mapsto \psi^*(y)$ (resp. $\mathcal{Y} \ni y \mapsto \psi(y)$) is $\mathbb{C}$−linear (resp. $\mathbb{C}$−anti-linear). The anti-commutation relations take the form

$$[\psi(y_1), \psi(y_2)]_+ = [\psi^*(y_1), \psi^*(y_2)]_+ = 0,$$
$$[\psi(y_1), \psi^*(y_2)]_+ = 2\overline{y}_1 \cdot \nu y_2 \mathbb{1},\ y_1, y_2 \in \mathcal{Y}.$$

4.5. States on CCR/CAR Algebras

4.5.1. Introduction

Let $\mathfrak{A}$ be a unital $*-$algebra. A *state* on $\mathfrak{A}$ is a linear map $\omega : \mathfrak{A} \to \mathbb{K}$ such that

$$\omega(A^*A) \geq 0, \ \forall A \in \mathfrak{A}, \ \omega(\mathbb{1}) = 1.$$

Elements of the form A^*A are called *positive*.

Let $\mathcal{X}$ be either a presymplectic or Euclidean space and let ω be a state on $\mathrm{CCR}(\mathcal{X}, \sigma)$ or $\mathrm{CAR}(\mathcal{X}, \nu)$. One can associate to ω a bilinear form on $\mathcal{X}$ called the *2-point function*:

$$\mathcal{X} \times \mathcal{X} \ni (x_1, x_2) \mapsto \omega(\phi(x_1)\phi(x_2)).$$

Of course to specify completely the state ω one also needs to know the *n-point functions*

$$\mathcal{X}^n \ni (x_1, \dots x_n) \mapsto \omega(\phi(x_1)\dots\phi(x_n)).$$

A particularly useful class of states are the *quasi-free* states, which are defined by the fact that all n-point functions are determined by the 2-point function.

4.5.2. Bosonic Quasi-Free States

Let $(\mathcal{X}, \sigma)$ be a presymplectic space and ω a state on $\mathrm{CCR}^{\mathrm{Weyl}}(\mathcal{X}, \sigma)$. The function

$$\mathcal{X} \ni x \mapsto \omega(W(x)) =: G(x)$$

is called the *characteristic function* of the state ω, and is a non-commutative version of the Fourier transform of a probability measure.

There is also a non-commutative version of Bochner's theorem (the theorem which characterizes these Fourier transforms).

Proposition 4.1 *A map $G : \mathcal{X} \to \mathbb{C}$ is the characteristic function of a state on $\mathrm{CCR}^{\mathrm{Weyl}}(\mathcal{X}, \sigma)$ iff for any $n \in \mathbb{N}$, $x_i \in \mathcal{X}$ the $n \times n$ matrix*

$$\left[G(x_j - x_i)\mathrm{e}^{\frac{\mathrm{i}}{2}x_i \cdot \sigma x_j}\right]_{1 \leq i,j \leq n}$$

is positive.

Proof $\Rightarrow$: for $x_1, \dots, x_n \in \mathcal{X}$, $\lambda_1, \dots, \lambda_n \in \mathbb{C}$ set

$$A := \sum_{j=1}^{n} \lambda_j W(x_j) \in \mathrm{CCR}^{\mathrm{Weyl}}(\mathcal{X}, \sigma).$$

Such A are dense in $\mathrm{CCR}^{\mathrm{Weyl}}(\mathcal{X},\sigma)$. One computes A^*A using the CCR and obtains

$$A^*A=\sum_{j,k=1}^{n}\overline{\lambda}_j\lambda_k W(x_j-x_k)\mathrm{e}^{\frac{\mathrm{i}}{2}x_j\cdot\sigma x_k},$$

from which $\Rightarrow$ follows.

$\Leftarrow$: one uses exactly the same argument, defining ω using G, and the above formula shows that ω is positive. □

Definition 4.5 (1) A state ω on $\mathrm{CCR}^{\mathrm{Weyl}}(\mathcal{X},\sigma)$ is a *quasi-free state* if there exists $\eta\in L_{\mathrm{s}}(\mathcal{X},\mathcal{X}^{\#})$ (a symmetric form on $\mathcal{X}$) such that

$$\omega\big(W(x)\big)=\mathrm{e}^{-\frac{1}{2}x\cdot\eta x},\quad x\in\mathcal{X}. \tag{4.2}$$

(2) The form η is called the *covariance of the quasi-free state* ω.

Quasi-free states should be considered as non-commutative versions of *Gaussian measures*. To explain this remark, consider the Gaussian measure on $\mathbb{R}^d$ with covariance η:

$$\mathrm{d}\mu_\eta:=(2\pi)^{d/2}\det\eta^{-\frac{1}{2}}\mathrm{e}^{-\frac{1}{2}y\cdot\eta^{-1}y}dy.$$

We have

$$\int\mathrm{e}^{\mathrm{i}x\cdot y}d\mu_\eta(y)=\mathrm{e}^{-\frac{1}{2}x\cdot\eta x}.$$

Note also that if $x_i\in\mathbb{R}^d$, then

$$\int\prod_{1}^{2n+1}x_i\cdot y\mathrm{d}\mu_\eta(y)=0,$$

$$\int\prod_{1}^{2n}x_i\cdot y\mathrm{d}\mu_\eta(y)=\sum_{\sigma\in\mathrm{Pair}_{2n}}\prod_{j=1}^{n}x_{\sigma(2j-1)}\cdot\eta x_{\sigma(2j)},$$

which should be compared with Definition 4.6.

Proposition 4.2 *Let $\eta\in L_{\mathrm{s}}(\mathcal{X},\mathcal{X}^{\#})$. Then the following are equivalent.*

(1) $\mathcal{X}\ni x\mapsto\mathrm{e}^{-\frac{1}{2}x\cdot\eta x}$ *is a characteristic function and hence there exists a quasi-free state satisfying (4.2).*
(2) $\eta_{\mathbb{C}}+\frac{\mathrm{i}}{2}\sigma_{\mathbb{C}}\geq 0$ *on $\mathbb{C}\mathcal{X}$, where $\eta_{\mathbb{C}},\sigma_{\mathbb{C}}\in L\big(\mathbb{C}\mathcal{X},(\mathbb{C}\mathcal{X})^*\big)$ are the canonical sesquilinear extensions of η,σ.*
(3) $|x_1\cdot\sigma x_2|\leq 2(x_1\cdot\eta x_1)^{\frac{1}{2}}(x_2\cdot\eta x_2)^{\frac{1}{2}},\ x_1,x_2\in\mathcal{X}.$

Proof The proof of (1) $\Rightarrow$ (2) is easy, by considering complex fields $\phi(w) = \phi(x_1) + \mathrm{i}\phi(x_2)$, $w = x_1 + \mathrm{i}x_2$ and noting that the positivity of ω implies that $\omega(\phi^*(w)\phi(w)) \geq 0$, for any $w \in \mathbb{C}\mathcal{X}$.

The proof of (2) $\Rightarrow$ (1) is more involved: let us fix $x_1, \ldots, x_n \in \mathcal{X}$ and set

$$b_{jk} = x_j \cdot \eta x_k + \frac{\mathrm{i}}{2} x_j \cdot \sigma x_k.$$

Then, for $\lambda_1, \ldots, \lambda_n \in \mathbb{C}$,

$$\sum_{1 \leq j,k \leq n} \overline{\lambda_j} b_{jk} \lambda_k = \overline{w} \cdot \eta_{\mathbb{C}} w + \frac{\mathrm{i}}{2} \overline{w} \cdot \omega_{\mathbb{C}} w, \quad w = \sum_{j=1}^{n} \lambda_j x_j \in \mathbb{C}\mathcal{X}.$$

By (2), the matrix $[b_{jk}]$ is positive. One has then to use an easy lemma, stating that the pointwise product of two positive matrices is positive. From this it follows also that $[\mathrm{e}^{b_{jk}}]$ is positive, and hence the matrix $[\mathrm{e}^{-\frac{1}{2}x_j \cdot \eta x_j} \mathrm{e}^{b_{jk}} \mathrm{e}^{-\frac{1}{2}x_k \cdot \eta x_k}]$ is positive. Hence:

$$\begin{aligned}
&\sum_{j,k=1}^{n} \mathrm{e}^{-\frac{1}{2}(x_k - x_j) \cdot \eta (x_k - x_j)} \mathrm{e}^{\frac{\mathrm{i}}{2} x_j \cdot \omega x_k} \overline{\lambda}_j \lambda_k \\
&= \sum_{j,k=1}^{n} \mathrm{e}^{-\frac{1}{2}x_j \cdot \eta x_j} \mathrm{e}^{b_{jk}} \mathrm{e}^{-\frac{1}{2}x_j \cdot \eta x_j} \overline{\lambda}_j \lambda_k \\
&= \sum_{j,k=1}^{n} G(x_j - x_k) \mathrm{e}^{\frac{\mathrm{i}}{2} x_j \cdot \sigma x_k} \geq 0,
\end{aligned}$$

for $G(x) = \mathrm{e}^{-\frac{1}{2}x \cdot \eta x}$. By Proposition 4.1, this means that G is a characteristic function. The proof of (2) $\Leftrightarrow$ (3) is an exercise in linear algebra. □

It is easy to deduce from ω the corresponding state acting on $\mathrm{CCR}^{\mathrm{pol}}(\mathcal{X}, \sigma)$, by setting

$$\begin{aligned}
&\omega(\phi(x_1) \cdots \phi(x_n)) \\
&:= \frac{\mathrm{d}}{dt_1} \cdots \frac{\mathrm{d}}{\mathrm{d}t_n} \omega(W(t_1 x_1 + \cdots + t_n x_n))_{|t_1 = \cdots t_n = 0}.
\end{aligned}$$

In particular

$$\omega(\phi(x_1)\phi(x_2)) = x_1 \cdot \eta x_2 + \frac{\mathrm{i}}{2} x_1 \cdot \sigma x_2.$$

The corresponding definition of a quasi-free state on $\mathrm{CCR}^{\mathrm{pol}}(\mathcal{X}, \sigma)$ is as follows.

Definition 4.6 A state ω on $\mathrm{CCR}^{\mathrm{pol}}(\mathcal{X},\sigma)$ is *quasi-free* if

$$\omega\big(\phi(x_1)\cdots\phi(x_{2m-1})\big)=0,$$
$$\omega\big(\phi(x_1)\cdots\phi(x_{2m})\big)=\sum_{\sigma\in\mathrm{Pair}_{2m}}\prod_{j=1}^{m}\omega\big(\phi(x_{\sigma(2j-1)})\phi(x_{\sigma(2j)})\big).$$

We recall that Pair_{2m} is the set of *pairings*, that is the set of partitions of $\{1,\ldots,2m\}$ into pairs. Any pairing can be written as

$$\{i_1,j_1\},\cdots,\{i_m,j_m\}$$

for $i_k<j_k$ and $i_k<i_{k+1}$, hence can be uniquely identified with a permutation $\sigma\in S_{2m}$ such that $\sigma(2k-1)=i_k$, $\sigma(2k)=j_k$.

4.5.3. Gauge-Invariant Quasi-Free States

Let us now assume that $(\mathcal{X},\sigma,\mathrm{j})$ is a pseudo-Kähler space, that is that there exists an anti-involution $\mathrm{j}\in Sp(\mathcal{X},\sigma)$. Note that we have

$$\mathrm{e}^{\mathrm{j}\theta}=\cos\theta+\mathrm{j}\sin\theta,\ \theta\in\mathbb{R},$$

and that the map

$$[0,2\pi]\ni\theta\mapsto\mathrm{e}^{\mathrm{j}\theta}\in Sp(\mathcal{X},\sigma)$$

is a 1-parameter group called the group of *(global) gauge transformations*.

A quasi-free state ω on $\mathrm{CCR}^{\mathrm{Weyl}}(\mathcal{X},\sigma)$ is called *gauge invariant* if

$$\omega(W(x))=\omega(W(\mathrm{e}^{\mathrm{j}\theta}x)),\ x\in\mathcal{X},\theta\in\mathbb{R}.$$

We can of course let ω act on $\mathrm{CCR}^{\mathrm{pol}}(\mathcal{X},\sigma)$. It is much more convenient then to use the *charged fields* $\psi^{(*)}(x)$ as generators of $\mathrm{CCR}^{\mathrm{pol}}(\mathcal{X},\sigma)$. We have then:

Proposition 4.3 *A state ω on $\mathrm{CCR}^{\mathrm{pol}}(\mathcal{X},\sigma)$ is gauge invariant quasi-free iff*

$$\omega\left(\Pi_1^n\psi^*(y_i)\Pi_1^p\psi(x_i)\right)=0,\ \textit{if}\ n\neq p$$
$$\omega\left(\Pi_1^n\psi^*(y_i)\Pi_1^n\psi(x_i)\right)=\sum_{\sigma\in S_n}\prod_{i=1}^{n}\omega(\psi^*(y_i)\psi(x_{\sigma(i)})).$$

It follows that a gauge invariant quasi-free state ω is uniquely determined by the sesquilinear form

$$\omega(\psi(y_1)\psi^*(y_2))=:\overline{y}_1\cdot\lambda_+y_2.$$

Clearly $\lambda_+\in L_{\mathrm{h}}(\mathcal{Y},\mathcal{Y}^*)$. Let

$$\omega(\psi^*(y_2)\psi(y_1))=:\overline{y}_1\cdot\lambda_-y_2,$$

with $\lambda_- \in L_{\rm h}(\mathcal{Y}, \mathcal{Y}^*)$. From the commutation relations we have of course

$$\lambda_+ - \lambda_- = q,$$

so λ_- is determined by λ_+, but nevertheless it is convenient to work with the pair $(\lambda_\pm)$ and to call $\lambda_\pm$ the *complex covariances* of ω.

The link between the real and complex covariances is as follows.

Lemma 4.1 *We have*

$$\eta = \mathrm{Re}\left(\lambda_\pm \mp \frac{1}{2}q\right), \ \hat{\eta} = \lambda_\pm \mp \frac{1}{2}q,$$

where $\hat{\eta} \in L_{\rm h}(\mathcal{Y}, \mathcal{Y}^*)$ *is given by*

$$\overline{y}_1 \hat{\eta} y_2 := y_1 \cdot \eta y_2 - \mathrm{i} y_1 \cdot \eta \mathrm{j} y_2.$$

Note that the fact that $\hat{\eta}$ is sesquilinear follows from the gauge invariance of ω. From the above lemma one easily gets the following characterization of complex covariances.

Proposition 4.4 *Let* $\lambda_\pm \in L_{\rm h}(\mathcal{Y}, \mathcal{Y}^*)$. *Then the following are equivalent.*

(1) $\lambda_\pm$ *are the covariances of a gauge-invariant quasi-free state on* $\mathrm{CCR}^{\rm pol}(\mathcal{Y}, q)$,
(2) $\lambda_\pm \geq 0$ *and* $\lambda_+ - \lambda_- = q$.

Proof Since ω is gauge invariant we have

$$\mathrm{j} \in O(\mathcal{Y}_{\mathbb{R}}, \eta) \cap Sp(\mathcal{Y}_{\mathbb{R}}, \mathrm{Re}\sigma) = O(\mathcal{Y}_{\mathbb{R}}, \eta) \cap O(\mathcal{Y}_{\mathbb{R}}, \mathrm{Re}q).$$

From this fact and Lemma 4.1 we deduce that $\eta \geq 0 \Leftrightarrow \lambda_+ \geq \frac{1}{2}q$, and that the second condition in Proposition 4.2 (with σ replaced by $\mathrm{Re}\sigma$) is equivalent to

$$\pm q \leq 2\lambda_+ - q \ \Leftrightarrow \ \lambda_\pm \geq 0.$$

This completes the proof of the proposition. □

4.5.4. Complexifying Bosonic Quasi-Free States

If $(\mathcal{X}, \sigma)$ is real symplectic, we can form $(\mathbb{C}\mathcal{X}, \sigma_{\mathbb{C}})$ which is charged symplectic. As real symplectic space it equals $(\mathcal{X}, \sigma) \oplus (\mathcal{X}, \sigma)$. If ω is a quasi-free state on $(\mathcal{X}, \sigma)$ with real covariance η, we form a state $\omega_{\mathbb{C}}$ on $((\mathbb{C}\mathcal{X})_{\mathbb{R}}, \mathrm{Re}\sigma_{\mathbb{C}})$, with covariance $\mathrm{Re}\eta_{\mathbb{C}}$, which is by definition gauge invariant. Hence, possibly after complexification, we can always reduce ourselves to gauge invariant quasi-free states.

4.5.5. Pure Quasi-Free States

In this subsection we discuss *pure* quasi-free states, which turn out to be precisely *vacuum states*. To start the discussion, let us recall that from Proposition 4.2 (3) one has

$$|x_1 \cdot \sigma x_2| \le 2(x_1 \cdot \eta x_1)^{\frac{1}{2}} (x_2 \cdot \eta x_2)^{\frac{1}{2}}, \ x_1, x_2 \in \mathcal{X}. \tag{4.3}$$

We can complete the real vector space $\mathcal{X}$ w.r.t. the (semi-definite) symmetric form η, after taking the quotient by the vectors of zero norm as usual. Note that from (4.3) σ passes to quotient and to completion.

Denoting once again $(\mathcal{X}/\mathrm{Ker}\eta)^{\mathrm{cpl}}$ by $\mathcal{X}$, we end up with the following situation: $(\mathcal{X}, \eta)$ is a real Hilbert space and σ is a bounded, anti-symmetric form on $\mathcal{X}$. However, σ may very well not be non-degenerate anymore, that is $(\mathcal{X}, \sigma)$ may just be presymplectic. One can show that if σ is degenerate, then the state on $\mathrm{CCR}(\mathcal{X}, \sigma)$ with covariance η is not pure. In the sequel we hence assume that σ is non-degenerate on $\mathcal{X}$. One can then prove the following theorem. The proof uses some more advanced tools, like the Araki-Woods representation and its properties.

Theorem 4.1 *The state ω of covariance η is pure iff the pair $(2\eta, \sigma)$ is Kähler, that is there exists $\mathrm{j} \in Sp(\mathcal{X}, \sigma)$ such that $\mathrm{j}^2 = \mathbb{1}$ and $2\eta = \sigma\mathrm{j}$.*

The link with Fock spaces and vacuum states is now as follows. If one equips $\mathcal{X}$ with the complex structure j and the scalar product

$$(x_1 | x_2) := x_1 \cdot 2\eta x_2 + \mathrm{i} x_1 \cdot \eta \mathrm{j} x_2,$$

then $\mathcal{Z} := (\mathcal{X}, (\cdot|\cdot))$ is a complex Hilbert space. One can build the bosonic Fock space $\Gamma_{\mathrm{s}}(\mathcal{Z})$ and the Fock representation $\mathcal{X} \ni x \mapsto \mathrm{e}^{\mathrm{i}\phi(x)} \in U(\Gamma_{\mathrm{s}}(\mathcal{Z}))$. This representation is precisely the GNS representation of the state ω, with GNS vector Ω_ω equal to the Fock vacuum.

For reference let us state the version of Theorem 4.1 using charged fields.

Theorem 4.2 *Let $\lambda_\pm \in L_{\mathrm{h}}(\mathcal{Y}, \mathcal{Y}^*)$. Then the following are equivalent.*

(1) *$\lambda_\pm$ are the covariances of a* pure *gauge-invariant quasi-free state on* $\mathrm{CCR}^{\mathrm{pol}}(\mathcal{Y}, \sigma)$.
(2) *There exists an involution $\kappa \in U(\mathcal{Y}, q)$ such that*

$$q\kappa \ge 0, \quad \lambda_\pm = \frac{1}{2} q(\kappa \pm \mathbb{1}).$$

(3) $\lambda_\pm \ge \pm\frac{1}{2}q$, $\lambda_\pm q^{-1} \lambda_\pm = \pm\lambda_\pm$, $\lambda_+ - \lambda_- = q$.

Theorem 4.2 can be easily deduced from Theorem 4.1.

4.5.6. Fermionic Quasi-Free States

We consider now a Euclidean space $(\mathcal{X}, \nu)$. Without loss of generality we can assume that $\mathcal{X}$ is complete, that is $(\mathcal{X}, \nu)$ is a real Hilbert space. We consider the CAR algebra $\mathrm{CAR}(\mathcal{X}, \nu)$, with the self-adjoint fermionic fields $\phi(x)$ as generators.

Definition 4.7 (1) A state ω on $\mathrm{CAR}(\mathcal{X}, \nu)$ is called *quasi-free* if

$$\omega\big(\phi(x_1)\cdots\phi(x_{2m-1})\big) = 0,$$

$$\omega\big(\phi(x_1)\cdots\phi(x_{2m})\big) = \sum_{\sigma\in\mathrm{Pair}_{2m}} \mathrm{sgn}(\sigma)\prod_{j=1}^{m}\omega\big(\phi(x_{\sigma(2j-1)})\phi(x_{\sigma(2j)})\big),$$

for all $x_1, x_2, \cdots \in \mathcal{X}$, $m \in \mathbb{N}$.

(2) The anti-symmetric form $\beta \in L_{\mathrm{a}}(\mathcal{X}, \mathcal{X}^{\#})$ defined by

$$x_1 \cdot \beta x_2 := \mathrm{i}^{-1}\omega([\phi(x_1), \phi(x_2)])$$

is called the *covariance* of the quasi-free state ω.

From the CAR it follows that

$$\omega\big(\phi(x_1)\phi(x_2)\big) = x_1\cdot\nu x_2 + \frac{\mathrm{i}}{2}x_1\cdot\beta x_2, \quad x_1, x_2 \in \mathcal{X}. \tag{4.4}$$

Proposition 4.5 *Let $\beta \in L_{\mathrm{a}}(\mathcal{X}, \mathcal{X}^{\#})$. Then the following are equivalent.*

(1) *β is the covariance of a fermionic quasi-free state ω;*
(2) *$\nu_{\mathbb{C}} + \frac{\mathrm{i}}{2}\beta_{\mathbb{C}} \geq 0$ on $\mathbb{C}\mathcal{X}$;*
(3) *$|x_1\cdot\beta x_2| \leq 2(x_1\cdot\nu x_1)^{\frac{1}{2}}(x_2\cdot\nu x_2)^{\frac{1}{2}}$, $x_1, x_2 \in \mathcal{X}$.*

Proof As in the bosonic case (1) $\Rightarrow$ (2) and (2) $\Leftrightarrow$ (3) are easy to prove. The proof of (2) $\Rightarrow$ (1) is more difficult, since it relies on the Jordan–Wigner representation of $\mathrm{CAR}(\mathcal{X}, \nu)$ for $\mathcal{X}$ finite dimensional. □

4.5.7. Gauge-Invariant Quasi-Free States

We now assume that $(\mathcal{X}, \nu)$ is equipped with a Kähler anti-involution j. This implies that $\mathrm{e}^{\mathrm{j}\theta} \in O(\mathcal{X}, \nu)$ hence that there exists the group of global gauge transformations τ_θ, $\theta \in [0, 2\pi]$ defined by

$$\tau_\theta\phi(x) := \phi(\mathrm{e}^{\mathrm{j}\theta}x),\ x \in \mathcal{X}.$$

As in the bosonic case, a state ω on $\mathrm{CAR}(\mathcal{X}, \nu)$ is called *gauge invariant* if $\omega\circ\tau_\theta = \omega$, $\forall\theta \in [0, 2\pi]$. Again it is more convenient to use the charged fields

$$\psi(y) := \phi(y) + \mathrm{i}\phi(\mathrm{j}y),\ \psi^*(y) := \phi(y) - \mathrm{i}\phi(\mathrm{j}y),$$

as generators of CAR($\mathcal{Y}$). We recall that one sets $q := 2\nu_{\mathbb{C}} \in L_{\mathrm{h}}(\mathcal{Y}, \mathcal{Y}^*)$, which is moreover positive definite.

Proposition 4.6 *A state ω on $\mathrm{CAR}^{C^*}(\mathcal{Y})$ is gauge-invariant quasi-free iff:*

$$\omega\left(\Pi_1^n \psi^*(y_i)\Pi_1^p \psi(x_i)\right) = 0, \text{ if } n \neq p$$

$$\omega\left(\Pi_1^n \psi^*(y_i)\Pi_1^n \psi(x_i)\right) = \sum_{\sigma \in S_n} \mathrm{sgn}(\sigma) \prod_{i=1}^{n} \omega(\psi^*(y_i)\psi(x_{\sigma(i)})).$$

Again we can introduce the two complex covariances

$$\omega(\psi(y_1)\psi^*(y_2)) =: \overline{y}_1 \cdot \lambda_+ y_2, \ \omega(\psi^*(y_2)\psi(y_1)) =: \overline{y}_1 \cdot \lambda_- y_2.$$

Proposition 4.7 *Let $\lambda_\pm \in L_{\mathrm{h}}(\mathcal{Y}, \mathcal{Y}^*)$. Then the following are equivalent.*

(1) *$\lambda_\pm$ are the covariances of a gauge-invariant quasi-free state on* CAR($\mathcal{Y}, q$)*;*
(2) *$\lambda_\pm \geq 0$ and $\lambda_+ + \lambda_- = q$.*

4.5.8. Pure Quasi-Free States

We discuss pure quasi-free states in the fermionic case. We consider the general case, that is we do not assume the states to be gauge invariant.

Theorem 4.3 *Let $\beta \in L_{\mathrm{a}}(\mathcal{X}, \mathcal{X}^\#)$. Then β is the covariance of a pure quasi-free state on* CAR($\mathcal{X}, \nu$) *iff $(\nu, \frac{1}{2}\beta)$ is Kähler, that is there exists $\mathrm{j} \in O(\mathcal{X}, \nu)$ such that $\mathrm{j}^2 = -\mathbb{1}$ and $\nu = \frac{1}{2}\beta\mathrm{j}$.*

We leave the formulation of the gauge-invariant version of this theorem as an exercise.

4.6. Lorentzian Manifolds

4.6.1. Causality

Let M be a smooth manifold of dimension $n = d + 1$. We assume that M is equipped with a Lorentzian metric g, that is a smooth map

$$M \ni x \mapsto g_{\mu\nu}(x) \in L_{\mathrm{s}}(T_xM, T_x^*M),$$

with signature $(-1, d)$. The inverse $g^{-1}(x) \in L_{\mathrm{s}}(T_x^*M, T_xM)$ is traditionally denoted by $g^{\mu\nu}(x)$. We also set $|g|(x) := |\det g_{\mu\nu}(x)|$.

Using the metric g we can define time-like, causal, etc. tangent vectors and vector fields on M as on the Minkowski spacetime. The set $\{v \in T_xM : vg(x)v = 0\}$ is called the *lightcone* at x.

M is then called *time-orientable* if there exists a global continuous time-like vector field v. Once a time-orientation is chosen, one can define future/past directed time-like vector fields.

One can similarly define time-like, causal etc. piecewise C^1 curves by requiring the said property to hold for all its tangent vectors.

Definition 4.8 Let $x \in M$. The *causal*, resp. *time-like future*, resp. *past* of x is the set of all $y \in M$ that can be reached from x by a causal, resp. time-like future, resp. past-directed curve, and is denoted $J^{\pm}(x)$, resp. $I^{\pm}(x)$. For $\mathcal{U} \subset M$, its *causal*, resp. *time-like future*, resp. *past* is defined as

$$J^{\pm}(\mathcal{U}) = \bigcup_{x \in \mathcal{U}} J^{\pm}(x), \quad I^{\pm}(\mathcal{U}) = \bigcup_{x \in \mathcal{U}} I^{\pm}(x).$$

We define also the *causal*, resp. *time-like shadow*:

$$J(\mathcal{U}) = J^{+}(\mathcal{U}) \cup J^{-}(\mathcal{U}), \quad I(\mathcal{U}) = I^{+}(\mathcal{U}) \cup I^{-}(\mathcal{U}).$$

The classification of tangent vectors in T_xM can be naturally extended to linear subspaces of T_xM.

Definition 4.9 A linear subspace E of T_xM is *space-like* if it contains only space-like vectors, *time-like* if it contains both space-like and time-like vectors, and *null* (or *light-like*) if it is tangent to the lightcone at x.

If $E \subset T_xM$ we denote by $E^{\perp} \subset T_xM$ its orthogonal for $g(x)$.

Lemma 4.2 *A subspace $E \subset T_xM$ is space-like, resp. time-like, resp. null iff $E^{\perp}$ is time-like, resp. space-like, resp. null.*

We refer to [9, Lemma 3.1.1] for the proof.

4.6.2. Globally Hyperbolic Manifolds

Definition 4.10 A *Cauchy hypersurface* is a hypersurface $\Sigma \subset M$ such that each inextensible time-like curve intersects Σ at exactly one point.

The following deep result is originally due to Geroch, with a stronger condition instead of (1b) in this form due to Bernal-Sanchez.

Theorem 4.4 *Let M be a connected Lorentzian manifold. The following are equivalent.*

(1) *The following two conditions hold:*
  (*1a*) *for any $x, y \in M$, $J^{+}(x) \cap J^{-}(y)$ is compact,*
  (*1b*) (causality condition) *there are no closed causal curves.*

(2) *There exists a Cauchy hypersurface.*
(3) *M is isometric to $\mathbb{R} \times \Sigma$ with metric $-\beta dt^2 + g_t$, where β is a smooth positive function, g_t is a Riemannian metric on Σ depending smoothly on $t \in \mathbb{R}$, and each $\{t\} \times \Sigma$ is a smooth space-like Cauchy hypersurface in M.*

Definition 4.11 A connected Lorentzian manifold satisfying the equivalent conditions of the above theorem is called *globally hyperbolic*.

The adjective "hyperbolic" comes from the connection with hyperbolic partial differential equations: roughly speaking globally hyperbolic spacetimes are those on which the Cauchy problem for the Klein–Gordon equation can be formulated and uniquely solved.

Definition 4.12 A function $f : M \to \mathbb{C}$ is called *space-compact* if there exists $K \Subset M$ such that $\mathrm{supp} f \subset J(K)$. It is called *future/past space-compact* if there exists $K \Subset M$ such that $\mathrm{supp} f \subset J^{\pm}(K)$. The spaces of such smooth functions are denoted by $C^\infty_{\mathrm{sc}}(M)$, resp. $C^\infty_{\pm\mathrm{sc}}(M)$.

4.7. Klein–Gordon Fields on Lorentzian Manifolds

4.7.1. The Klein–Gordon Operator

Let (M, g) be a Lorentzian manifold. The *Klein–Gordon* operator on M is

$$P(x, \partial_x) = -|g|^{-\frac{1}{2}} \partial_\mu |g|^{\frac{1}{2}} g^{\mu\nu}(x) \partial_\nu + r(x),$$

acting on functions $u : M \to \mathbb{R}$. Here $r \in C^\infty(M, \mathbb{R})$ represent a (variable) mass. Using the metric connection, one can write

$$P(x, \partial_x) = -\nabla_a \nabla^a + r(x).$$

One can generalize the Klein–Gordon operator to sections of vector bundles, the most important example being the bundle of 1-forms on M, which appears in the quantization of Maxwell's equation. One would like to interpret some space of smooth solutions of the Klein–Gordon equation

$$P(x, \partial_x) u = 0 \qquad \text{(KG)}$$

as a symplectic space. Following the discussion in Section 4.5 we will consider *complex* solutions.

4.7.2. Conserved Current

Definition 4.13 We set for $u_1, u_2 \in C^\infty(M)$:

$$J^a(u_1, u_2) := \overline{u}_1 . \nabla^a u_2 - \overline{\nabla^a u_1} u_2 \in C^\infty(M).$$

$J^a(u_1, u_2)$ is a vector field called a *current* in the physics literature. Using the rules to compute with connections we obtain

$$\begin{aligned}\nabla_a J^a(u_1, u_2) &= \nabla_a \overline{u}_1 \nabla^a u_2 + \overline{u}_1 \nabla_a \nabla^a u_2 - \nabla_a \nabla^a \overline{u}_1 u_2 - \nabla^a \overline{u}_1 \nabla_a u_2 \\ &= \overline{u}_1 \nabla_a \nabla^a u_2 - \nabla_a \nabla^a \overline{u}_1 u_2 = P\overline{u}_1 u_2 - \overline{u}_1 P u_2.\end{aligned}$$

It follows that if $Pu_i = 0$ the vector field $J^a(u_1, u_2)$ is divergence free. Moreover from the Gauss formula we obtain:

Lemma 4.3 (Green's formula) *Let $U \subset M$ be an open set with ∂U non characteristic. Then*

$$\int_U \overline{u}_1 P u_2 - \overline{P u}_1 u_2 d\mu_g = -\int_{\partial U} \left(\overline{u}_1 \nabla_a u_2 - \overline{\nabla_a u_1} u_2\right) n^a d\sigma_g.$$

4.7.3. Advanced and Retarded Fundamental Solutions

The treatment of the Klein–Gordon operator is quite similar to its Riemannian analog, that is the Laplace operator, and starts with the construction of *fundamental solutions*, that is inverses. An important difference is the hyperbolic nature of the Klein–Gordon operator: to obtain a unique solution of the equation

$$Pu = v,$$

say for $v \in C_0^\infty(M)$, one has to impose extra *support conditions* on u, since there exists plenty of solutions of $Pu = 0$. The condition that (M, g) is globally hyperbolic is the natural condition so as to be able to construct fundamental solutions. In the sequel we will assume that (M, g) is globally hyperbolic.

The main result is the following theorem, due to Leray.

Theorem 4.5 *For any $v \in C_0^\infty(M)$, there exist unique functions $u^\pm \in C^\infty_{\pm\mathrm{sc}}(M)$ that solve*

$$P(x, \partial_x)u^\pm = v.$$

Moreover,

$$u^\pm(x) = (E^\pm v)(x) := \int E^\pm(x, y) v(y) \mathrm{d}\mu_g(y),$$

where $E^\pm \in \mathcal{D}'(M \times M, L(\mathcal{V}))$ satisfy

$$P \circ E^\pm = E^\pm \circ P = \mathbb{1}, \quad \operatorname{supp} E^\pm \subset \left\{(x, y) \; : \; x \in J^\pm(y)\right\}.$$

Note that P is self-adjoint for the scalar product

$$(u|v) = \int_M \overline{u} v d\mu_g,$$

which by uniqueness implies that $(E^{\pm})^* = E^{\mp}$. This also implies that by duality $E^{\pm}$ can be applied to distributions of compact support.

Definition 4.14 E^+, resp. E^-, is called the *retarded*, resp. *advanced Green's function.*

$$E := E^+ - E^-$$

is called the *Pauli–Jordan function.*

Note that $E = -E^*$, that is E is anti-hermitian.

4.7.4. Cauchy Problem

Once the existence of $E^{\pm}$ is established, it is easy to solve the Cauchy problem (one can also proceed the other way around). Let us denote by $\mathrm{Sol}_{\mathrm{sc}}(KG)$ the space of smooth, space compact solutions of (KG). Let Σ be a smooth Cauchy hypersurface. We denote by $\rho : C^\infty_{\mathrm{sc}}(M) \to C^\infty_0(\Sigma) \oplus C^\infty_0(\Sigma)$ the map

$$\rho u := (\rho_0 u, \rho_1 u) = (u_{|\Sigma}, n^\mu \partial_\mu u)_{|\Sigma}).$$

Theorem 4.6 *Let $f \in C^\infty_0(\Sigma) \otimes \mathbb{C}^2$. Then there exists a unique $u \in \mathrm{Sol}_{\mathrm{sc}}(KG)$ such that $\rho u = f$. This satisfies* $\mathrm{supp} u \subset J(\mathrm{supp} f_0 \cup \mathrm{supp} f_1)$ *and is given by*

$$u(x) = -\int_\Sigma n^\mu \nabla_{y^\mu} E(x,y) f_0(y) d\sigma(y) + \int_\Sigma E(x,y) f_1(y) d\sigma(y). \quad (4.5)$$

We will set $u =: Uf$, for u given by (4.5).

Remark 4.1 Let us denote by $\rho_i^* : \mathcal{D}'(\Sigma) \to \mathcal{D}'(M)$ the adjoints of ρ_i. Let us also set

$$q_\Sigma = \begin{pmatrix} 0 & \mathbb{1} \\ -\mathbb{1} & 0 \end{pmatrix} \in L(C^\infty_0(\Sigma) \oplus C^\infty_0(\Sigma)).$$

Then (4.5) can be rewritten as

$$\mathbb{1} = E\rho^* q_\Sigma \rho, \text{ on } \mathrm{Sol}_{\mathrm{sc}}(KG), \quad (4.6)$$

or equivalently as

$$U = E \circ \rho^* \circ q_\Sigma, \text{ on } C^\infty_0(\Sigma) \oplus C^\infty_0(\Sigma). \quad (4.7)$$

Proof We will just prove that the solution u of the Cauchy problem is given by the above formula. We fix $f \in C^\infty_0(M)$ and apply Green's formula to $u_0 = E^{\mp} f$, $u_1 = u$, and $U = J^{\pm}(\Sigma)$. We obtain

$$\int_{J^+(\Sigma)} \bar{f} u d\mu_g = \int_\Sigma \left(\overline{E^- f} \nabla_a u - \overline{\nabla_a E^- f} u \right) n^a d\sigma_g,$$
$$\int_{J^-(\Sigma)} \bar{f} u d\mu_g = \int_\Sigma - \left(\overline{E^+ f} \nabla_a u - \overline{\nabla_a E^+ f} u \right) n^a d\sigma_g.$$

Summing these two identities we obtain

$$\int_M \bar{f} u d\mu_g = \int_\Sigma \left(-\overline{Ef} \nabla_a u + \overline{\nabla_a Ef} u \right) d\sigma_g.$$

To complete the proof of the formula, it suffices to introduce the distribution kernel $E(x, y)$ to express Ef as an integral and to use $E = -E^*$. Details are left to the reader. □

4.7.5. Symplectic Structure of the Space of Solutions

By Theorem 4.6 we know that

$$\rho : \mathrm{Sol}_{\mathrm{sc}}(KG) \to C_0^\infty(\Sigma) \oplus C_0^\infty(\Sigma)$$

is bijective, with inverse U. For $u, v \in \mathrm{Sol}_{\mathrm{sc}}(KG)$ with $\rho u =: f$, $\rho v =: g$ we set

$$\begin{aligned} \overline{u_1} \cdot \sigma u_2 &:= \int_\Sigma \bar{f}_1 g_0 - \bar{f}_0 g_1 d\sigma_g = \int_\Sigma J^a(u, v) n_a d\sigma_g \\ \bar{f} \sigma_\Sigma g &:= \int_\Sigma \bar{f}_1 g_0 - \bar{f}_0 g_1 d\sigma_g. \end{aligned} \tag{4.8}$$

It is obvious from Theorem 4.6 that $(\mathrm{Sol}_{\mathrm{sc}}(KG), \sigma)$ is a (complex) symplectic space. From the Gauss formula we know that σ is independent of the choice of the Cauchy surface Σ.

4.7.6. Spacetime Fields

Theorem 4.7 (1) *Consider* $E : C_0^\infty(M) \to C^\infty(M)$. *Then*

$$\mathrm{Ran} E = \mathrm{Sol}_{\mathrm{sc}}(KG), \quad \mathrm{Ker} E = P C_0^\infty(M).$$

(2) *One has*

$$\overline{Ef}_1 \cdot \sigma Ef_2 = (f_1 | Ef_2),\ f_i \in C_0^\infty(M).$$

Part (1) of the theorem can be nicely rephrased by saying that the sequence

$$0 \longrightarrow C_0^\infty(M) \xrightarrow{P} C_0^\infty(M) \xrightarrow{E} \mathrm{Sol}_{\mathrm{sc}}(KG) \xrightarrow{P} 0$$

is exact.

Proof (1) Since $PE^{\pm} = \mathbb{1}$, we have $PE = 0$, and taking adjoints also $EP = 0$. This shows that $EC_0^\infty(M) \subset \mathrm{Sol}_{\mathrm{sc}}(KG)$ and $PC_0^\infty(M) \subset \mathrm{Ker}E$. It remains to prove the converse inclusions.

(1a) $\mathrm{Sol}_{\mathrm{sc}}(KG) \subset EC_0^\infty(M)$: let $u \in \mathrm{Sol}_{\mathrm{sc}}(KG)$. Since u is space-compact, we can find cutoff functions $\chi^{\pm} \in C^\infty_{\pm\mathrm{sc}}(M)$ such that $\chi^+ + \chi^- = 1$ on $\mathrm{supp}u$. We have some compact sets $K \subset K_1 \subset K_2$ such that $\mathrm{supp}u \subset J(K)$, $\mathrm{supp}\chi^{\pm} \subset J^{\pm}(K_2)$, and $\chi^{\pm} = 1$ on $J^{\pm}(K_1)$. It follows that $\mathrm{supp}\nabla\chi^{\pm} \subset J^{\pm}(K_2)\backslash J^{\pm}(K_1)$, hence $\mathrm{supp}\nabla\chi^{\pm} \cap \mathrm{supp}u \subset J^{\mp}(K_2) \cap J^{\pm}(K)$. This set is compact by the global hyperbolicity of M.

We set now $u^{\pm} = \chi^{\pm}u$, $f = Pu^+ = -Pu^-$. By the above discussion $f \in C_0^\infty(M)$, hence $u^{\pm} = \pm E^{\pm}f$ and $u = u^+ + u^- = Ef$.

(1b) $\mathrm{Ker}E = PC_0^\infty(M)$: let $u \in C_0^\infty(M)$ such that $Eu = 0$ and let $f = E^{\pm}u$. Then $\mathrm{supp}f \subset J^+(\mathrm{supp}u) \cap J^-(\mathrm{supp}u)$, hence again by global hyperbolicity, $f \in C_0^\infty(M)$.

(2) From (4.6) we obtain

$$E = -E\rho^* q_\Sigma \rho E = (\rho \circ E)^* \circ q_\Sigma \circ (\rho \circ E).$$

This implies (2). □

We now summarize the discussion as follows:

Theorem 4.8 (1) *The following spaces are symplectic spaces:*

$$(C_0^\infty(M)/PC_0^\infty(M), E), \ (\mathrm{Sol}_{\mathrm{sc}}(KG), \sigma), \ (C_0^\infty(\Sigma) \oplus C_0^\infty(\Sigma), \sigma_\Sigma).$$

(2) *The following maps are symplectomorphisms:*

$$(C_0^\infty(M))/PC_0^\infty(M), E) \xrightarrow{E} (\mathrm{Sol}_{\mathrm{sc}}(KG), \sigma) \xrightarrow{\rho} (C_0^\infty(\Sigma) \oplus C_0^\infty(\Sigma), \sigma_\Sigma).$$

4.7.7. Quasi-Free States for the Free Klein–Gordon Field

We can now consider quasi-free states on any of the symplectic spaces in Theorem 4.8. The most natural one is

$$(C_0^\infty(M))/PC_0^\infty(M), E)$$

which leads to *spacetime fields*. The associated CCR algebra will be denoted simply by $\mathrm{CCR}(C_0^\infty(M), E)$, ignoring the need to pass to quotient to get a true symplectic space.

Strictly speaking we would write symbols like

$$\phi([f]), \ [f] \in C_0^\infty(M)/PC_0^\infty(M).$$

We write this as

$$\phi(f)" = " \int_M \phi(x)f(x)\mathrm{d}\mu_g" = "\langle\phi,f\rangle$$

if $P\phi(x) = 0$, that is the quantum field ϕ satisfies the KG equation.

Causality: denoting by $\phi(f)$ the (self-adjoint) fields associated with $[f] \in C_0^\infty(M))/PC_0^\infty(M)$, we have

$$[\phi(f), \phi(g)] = \mathrm{Re}(f|Eg) = 0,$$

if supp$f$, supp$g$ are *causally disjoint*. This follows from the fact that

$$\mathrm{supp}E(x,y) \subset \{(x,y) \in M \times M \ : \ x \in J(y)\}.$$

Let us now consider a gauge-invariant quasi-free state ω, defined by the complex covariances $(\lambda_\pm)$. Recall that these complex covariances $\lambda_\pm$ are sesquilinear forms on $C_0^\infty(M))/PC_0^\infty(M)$. One may assume that they are obtained from sesquilinear forms $\Lambda_\pm$ on $C_0^\infty(M)$ which pass to quotient, that is such that

$$P \circ \Lambda_\pm = \Lambda_\pm \circ P = 0. \tag{4.9}$$

It is also natural to assume that $\Lambda_\pm$ are continuous sesquilinear forms, hence by the Schwartz kernel theorem we have distributional kernels $\Lambda_\pm(x,y) \in \mathcal{D}'(M \times M)$. Then (4.9) becomes, using $P = P^*$:

$$P(x, \partial_x)\Lambda_\pm(x,y) = P(y, \partial_y)\Lambda_\pm(x,y) = 0. \tag{4.10}$$

4.8. Free Dirac Fields on Lorentzian Manifolds

4.8.1. The Dirac Operator

Let (M, g) be a Lorentzian manifold. To define correctly a Dirac operator we need a *spin structure*. We keep the discussion simple and use a framework in [7].

Let $\mathcal{V}$ be a complex, finite dimensional vector space. We assume that there exists a map $M \ni x \mapsto \gamma^a(x) \in L(\mathcal{V})$ such that

$$[\gamma^a(x), \gamma^b(x)]_+ = 2g^{ab}(x).$$

Of course to make this definition clean one should use the language of bundles. One can think of γ^a as x dependent Dirac matrices.

Definition 4.15 Let $M \ni x \mapsto m(x) \in L(\mathcal{V})$. The operator on $C^\infty(M; \mathcal{V})$

$$\mathbb{D} := \gamma^a \partial_{x^a} + m(x)$$

is called a *Dirac operator* on M. The *Dirac equation* is

$$\mathbb{D}\zeta = 0. \tag{D}$$

We need some more structure to be able to quantize the Dirac equation. We assume that there exists a smooth map

$$M \ni x \mapsto \lambda(x) \in L_{\mathrm{h}}(\mathcal{V}, \mathcal{V}^*) \tag{4.11}$$

such that

$$\begin{aligned} &\gamma^a(x) \text{ is self-adjoint for } \lambda(x), \\ &m(x) - \frac{1}{2}\nabla_a\gamma^a(x) \text{ is anti-self-adjoint for } \lambda(x). \end{aligned} \tag{4.12}$$

We equip $C_0^\infty(M; \mathcal{V})$ of the sesquilinear form

$$(\zeta_1|\zeta_2) = \int_M \overline{\zeta}_1(x) \cdot \lambda(x)\zeta_2(x)d\mu_g,$$

and we obtain that if (4.12) holds then $\mathbb{D}^* = -\mathbb{D}$.

4.8.2. Conserved Current

We set for $\zeta_1, \zeta_2 \in C^\infty(M; \mathcal{V})$

$$J^a(\zeta_1, \zeta_2) := \overline{\zeta}_1(x) \cdot \lambda(x)\gamma^a(x)\zeta_2(x).$$

One proves that if ζ_i are solutions of the Dirac equation, then

$$\nabla_a J^a(\zeta_1, \zeta_2) = 0.$$

We also obtain Green's formula:

Lemma 4.4 *Let $U \subset M$ be an open set with ∂U non-characteristic. Then*

$$\int_U \overline{\zeta}_1 \cdot \lambda\mathbb{D}\zeta_2 - \overline{\mathbb{D}\zeta}_1 \cdot \lambda\zeta_2 d\mu_g = -\int_{\partial U} \overline{\zeta}_1 \cdot \lambda\gamma_a\zeta_2 n^a d\sigma_g,$$

for $\gamma_a = g_{ab}\gamma^b$.

4.8.3. Advanced and Retarded Fundamental Solutions

We assume now that (M, g) is globally hyperbolic. It is easy to construct fundamental solutions for the Dirac operator: in fact $\mathbb{D}\mathbb{D}$ is a Klein–Gordon operator, acting on vector valued functions, but with a scalar principal part equal to $-\nabla_a\nabla^a$. The existence of fundamental solutions $E^\pm$ for $\mathbb{D}^2$ yields the fundamental solutions $S^\pm = \mathbb{D}E^\pm$. This is summarized in the next theorem.

Theorem 4.9 *For any $f \in C_0^\infty(M;\mathcal{V})$, there exist unique functions $\zeta^\pm \in C^\infty_{\pm sc}(M;\mathcal{V})$ that solve*

$$\mathbb{D}\zeta^\pm = f.$$

Moreover,

$$\zeta^\pm(x) = (S^\pm f)(x) := \int S^\pm(x,y) f vy) \mathrm{d}\mu_g(y),$$

where $S^\pm \in \mathcal{D}'(M \times M, L(\mathcal{V}))$ satisfy

$$\mathbb{D} \circ S^\pm = S^\pm \circ \mathbb{D} = \mathbb{1}, \quad \mathrm{supp} S^\pm \subset \{(x,y) \;:\; x \in J^\pm(y)\}.$$

If (4.12) holds then

$$\lambda(x) S^\pm(x,y) = -S^\mp(x,y)^* \lambda(y),$$

that is $S^{\pm *} = -S^\mp$, if we equip $C_0^\infty(M;\mathcal{V})$ with the (non-positive) scalar product obtained from λ.

Definition 4.16 $S^\pm$ are called *retarded/advanced* Green's functions.

$$S := S^+ - S^-$$

is called the *Pauli–Jordan* function.

Note that $S = S^*$, that is S, is hermitian.

4.8.4. Cauchy Problem

We state without proof the existence and uniqueness result for the Cauchy problem. We denote by $\rho : C^\infty_{sc}(M;\mathcal{V}) \to C_0^\infty(\Sigma;\mathcal{V})$ the trace on Σ and by $\mathrm{Sol}_{sc}(\mathbb{D})$ the space of smooth space-compact solutions of the Dirac equation.

Theorem 4.10 *Let Σ be a Cauchy surface. Then for any $f \in C_0^\infty(\Sigma;\mathcal{V})$ there exists a unique $\zeta \in \mathrm{Sol}_{sc}(\mathbb{D})$ such that $\rho\zeta = f$. This satisfies* $\mathrm{supp}\zeta \subset J(\mathrm{supp} f)$ *and is given by*

$$\zeta(x) = -\int_\Sigma S(x,y)\gamma^a(y) n_a(y) f(y) d\mu_g(y).$$

4.8.5. Hermitian Structure on the Space of Solutions

We would like to equip $\mathrm{Sol}_{sc}(\mathbb{D})$ with a (positive) hermitian structure. To do this we need an additional positivity condition. We assume that there exists a global, time-like future directed vector field v such that

$$\lambda(x)\gamma^a(x) v_a(x) > 0, \; \forall x \in M. \tag{4.13}$$

It can be shown that if this is true for one such vector field, it is automatically true for all such vector fields, in particular for the normal vector to a given Cauchy surface. For $\zeta_1, \zeta_2 \in \mathrm{Sol}_{\mathrm{sc}}(\mathbb{D})$ with $\rho\zeta_i = f_i$ we set

$$\begin{aligned}\overline{\zeta}_1 \cdot \nu\zeta_2 &:= \int_\Sigma J^a(\zeta_1, \zeta_2)n_a d\sigma_g = \int_\Sigma \overline{f}_1 \cdot \lambda(y)\gamma^a(y)f_2(y)n_a(y)d\sigma_g(y),\\ \overline{f}_1 \cdot \nu_\Sigma f_2 &= \int_\Sigma \overline{f}_1 \cdot \lambda(y)\gamma^a(y)f_2(y)n_a(y)d\sigma_g(y).\end{aligned} \tag{4.14}$$

From Theorem 4.10 we see that $(\mathrm{Sol}_{\mathrm{sc}}(\mathbb{D}), \nu)$ is a hermitian space, and from (4.13) ν is positive definite. Moreover from the Gauss formula ν is independent of the choice of a Cauchy surface.

4.8.6. Spacetime Fields

Theorem 4.11 (1) *Consider* $S : \mathcal{D}(M; \mathcal{V}) \to C^\infty(M; \mathcal{V})$. *Then* $\mathrm{Ran}S = \mathrm{Sol}_{\mathrm{sc}}(\mathbb{D})$ *and* $\mathrm{Ker}S = \mathbb{D}C_0^\infty(M; \mathcal{V})$.

(2) *One has*

$$\overline{Sf}_1 \cdot \nu Sf_2 = (f_1|Sf_2),\ f_i \in C_0^\infty(M; \mathcal{V}).$$

Proof (1) is left to the reader.

(2) Theorem 4.10 can be rewritten as

$$\mathbb{1} = S\rho^*\gamma^{a*}n_a\rho, \text{ on } \mathrm{Sol}_{\mathrm{sc}}(\mathbb{D}),$$

hence by (1) as

$$S = S\rho^*\gamma^{a*}n_a\rho S,$$

which implies (2) since $S^* = S$. □

Theorem 4.12 (1) *The following spaces are pre-Hilbert spaces:*

$$(C_0^\infty(M; \mathcal{V})/\mathbb{D}C_0^\infty(M; \mathcal{V}), S),\ (\mathrm{Sol}_{\mathrm{sc}}(\mathbb{D}), \nu),\ (C_0^\infty(\Sigma; \mathcal{V}), \nu_\Sigma).$$

(2) *The following maps are unitary:*

$$(C_0^\infty(M; \mathcal{V})/\mathbb{D}C_0^\infty(M; \mathcal{V}), S) \xrightarrow{S} (\mathrm{Sol}_{\mathrm{sc}}(\mathbb{D}), \nu) \xrightarrow{\rho} (C_0^\infty(\Sigma; \mathcal{V}), \nu_\Sigma).$$

4.8.7. Quasi-Free States for the Free Dirac Field

As for the Klein–Gordon case we choose the pre-Hilbert space

$$(C_0^\infty(M; \mathcal{V})/\mathbb{D}C_0^\infty(M; \mathcal{V}), S).$$

The associated CAR algebra is denoted by $\mathrm{CAR}(C_0^\infty(M; V), S)$.

The causality is a bit different: we obtain

$$[\phi(f), \phi(g)]_{+} = \mathrm{Re}(f|Sg) = 0,$$

if supp$f$, supp$g$ are *causally disjoint*, that is fields supported in causally disjoint regions *anti-commute*. This puzzle is solved by considering only *even* elements of $\mathrm{CAR}(C_0^\infty(M;\mathcal{V})/\mathbb{D}C_0^\infty(M;\mathcal{V}), S)$ as true physical observables.

Let us consider a gauge-invariant quasi-free state ω, defined by the complex covariances $(\lambda_\pm)$ which are sesquilinear forms on $C_0^\infty(M;\mathcal{V})/\mathbb{D}C_0^\infty(M;\mathcal{V}))$. Again one assumes that they are obtained from sesquilinear forms $\Lambda_\pm$ on $C_0^\infty(M;\mathcal{V})$ which pass to quotient, that is such that

$$\mathbb{D} \circ \Lambda_\pm = \Lambda_\pm \circ \mathbb{D} = 0. \tag{4.15}$$

Introducing as before the distributional kernels $\Lambda_\pm(x,y) \in \mathcal{D}'(M \times M) \otimes \mathcal{V} \otimes \mathcal{V}^*$. Then (4.15) becomes

$$\mathbb{D}(x, \partial_x)\Lambda_\pm(x,y) = \mathbb{D}(y, \partial_y)\Lambda_\pm(x,y) = 0. \tag{4.16}$$

4.9. Microlocal Analysis of Klein–Gordon Quasi-Free States

4.9.1. The Need for Renormalization

The stress-energy tensor for a classical Klein–Gordon field is given by

$$T_{\mu\nu}(x) = \nabla_\mu\phi(x)\nabla_\nu\phi(x) - \frac{1}{2}g_{\mu\nu}(x)\left(g^{ab}(x)\nabla_a\phi(x)\nabla_b\phi(x) - m^2\phi^2(x)\right).$$

For a quantized Klein–Gordon field one would like to be able to define $T_{\mu\nu}(x)$ as an operator valued distribution. This means the following.

We choose a state ω (say a quasi-free state) and fix $f_1, \ldots, f_n, g_1, \ldots, g_p \in C_0^\infty(M)$. Then

$$x \mapsto \omega\left(\prod_1^n \phi(f_i)T_{\mu\nu}(x)\prod_1^p \phi(g_j)\right)$$

should be a distribution on M. This is never the case, even on Minkowski space: for example $\omega(\phi^2(x))$ should be the trace on $x = x'$ of $\omega(\phi(x)\phi(x')) = \omega_2(x,x')$, which makes no sense. We need to subtract the singular part of $\omega_2(x,x')$ near the diagonal, that is to perform a renormalization.

Another requirement is that the renormalization procedure should be "covariant": it should depend only on the metric in an arbitrarily small neighborhood of x. This implies that it will be covariant under isometric embeddings.

The procedure is as follows. One first performs the "point-splitting," that is consider

$$T^{\mu\nu}(x,y) = \nabla_\mu\phi(x)\nabla_\nu\phi(y) - \frac{1}{2}g_{\mu\nu}(x)\left(g_{ab}(x)\nabla_a\phi(x)\nabla_b\phi(y) - m^2\phi(x)\phi(y)\right).$$

One then removes the singular part by setting

$$:\phi(x)\phi(y): = \phi(x)\phi(y) - c_{\rm Had}(x,y)\,\mathbb{1},$$

where $c_{\rm Had}(x,y)$ is a well chosen distributional kernel. Then one has to check that the distributions

$$\omega\left(\prod_1^n\phi(x_i)T_{\mu\nu}(x,y):\prod_1^p\phi(y_j)\right)$$

have well-defined restrictions to $x = y$.

4.9.2. Old Form of Hadamard States

Choose a Cauchy surface Σ. A causal normal neighborhood N of Σ in M is an open neighborhood of Σ such that Σ is a Cauchy surface of (N, g) and for each $x, x' \in N$ such that $x \in J^+(x')$ there exists a convex normal open set containing $J^+(x') \cap J^-(x)$. We fix a *time function* $T : M \to \mathbb{R}$, that is a smooth function which increases towards the future, and an open neighborhood $\mathcal{O} \subset M \times M$ of the set of pairs of causally related points (x, x') such that $J^\pm(x) \cap J^\mp(x')$ are contained in a convex, normal open neighborhood.

One fixes also $\mathcal{O}'$ an open neighborhood in $N \times N$ of the set of pairs of causally related points such that $\mathcal{O}' \subset \overline{\mathcal{O}}$.

The squared geodesic distance $\sigma(x,x')$ is smooth on $\mathcal{O}$ and well defined on the subset of $N \times N$ consisting of causally related points. The *van Vleck–Morette determinant* is

$$\Delta(x,x') := -\det(-\nabla_\alpha\nabla_{\beta'}\sigma(x,x'))|g|^{-\frac{1}{2}}(x)|g|^{-\frac{1}{2}}(x').$$

One can then construct a sequence of functions $v^{(n)} \in C^\infty(\mathcal{O}\times\mathcal{O})$ of the form

$$v^{(n)}(x,x') = \sum_1^n \sigma(x,x')^i v_i(x,x'),$$

where the $v_i \in C^\infty(\mathcal{O}\times\mathcal{O})$ are uniquely determined by some transport equations (the so-called Hadamard recursion relations).

One then defines a sequence of distributions $c^{(n)}_{\text{Had}} \in \mathcal{D}'(\mathcal{O}\times\mathcal{O})$ for $n \geq 1$ by

$$c^{(n)}_{\text{Had}}(x,x') = \lim_{\epsilon\to 0^+} \frac{(2\pi)^2\Delta(x,x')^{\frac{1}{2}}}{\sigma(x,x') + 2\mathrm{i}\epsilon(T(x)-T(x')) + \epsilon^2}$$
$$+ \lim_{\epsilon\to 0^+} v^{(n)}(x,x') \ln\left(\sigma(x,x') + 2\mathrm{i}\epsilon(T(x)-T(x'))\right).$$

Let us now give the old definition of Hadamard states, restricting ourselves to real Klein–Gordon fields.

Definition 4.17 A quasi-free state ω on $\mathrm{CCR}(C_0^\infty(M), E)$ is a Hadamard state if for any $m \in \mathbb{N}$ there exists $n \in \mathbb{N}$ such that $\omega_2 - c^{(n)}_{\text{Had}}$ is of class C^m in $\mathcal{O}\times\mathcal{O}$.

4.9.3. The Wavefront Set on a Manifold

Let M be a manifold and T^*M be its cotangent bundle. The zero section T^*_0M will be denoted by Z.

We recall the spaces: $\mathcal{D}(M)$ (smooth compactly supported functions), $\mathcal{D}'(M)$ (distributions), $\mathcal{E}(M)$ (smooth functions with well-known topology), $\mathcal{E}'(M)$ (distributions with compact support).

4.9.4. Operations on Conic Sets

A set $\Gamma \subset T^*M\backslash Z$ is *conic* if $(x,\xi) \in \Gamma \Rightarrow (x,t\xi) \in \Gamma$ for all $t > 0$. Let $\Gamma_i \subset T^*M\backslash Z$, $i = 1,2$ be conic sets. We set

$$-\Gamma := \{(x,-\xi)\ :\ (x,\xi)\in\Gamma\},$$

$$\Gamma_1 \oplus \Gamma_2 := \{(x,\xi_1+\xi_2)\ :\ (x,\xi_i)\in\Gamma_i\}.$$

Let M_i, $i = 1,2$ be two manifolds and $\Gamma \subset T^*M_1 \times M_2\backslash Z$ be a conic set. The elements of $T^*M_1\times M_2\backslash Z$ will be denoted by (x_1,ξ_1,x_2,ξ_2) which allows us to consider Γ as a relation between T^*M_2 and T^*M_1, still denoted by Γ. Clearly Γ maps conic sets into conic sets. We set:

$$\Gamma' := \{(x_1,\xi_1,x_2,-\xi_2)\ :\ (x_1,\xi_1,x_2,\xi_2)\in\Gamma\},$$

$${}_{M_1}\Gamma := \{(x_1,\xi_1)\ :\ \exists x_2 \text{ such that } (x_1,\xi_1,x_2,0)\in\Gamma\} = \Gamma(Z_2),$$

$$\Gamma_{M_2} := \{(x_2,\xi_2)\ :\ \exists x_1 \text{ such that } (x_1,0,x_2,\xi_2)\in\Gamma\} = \Gamma^{-1}(Z_1).$$

4.9.5. Operations on Distributions

(1) **Complex conjugation**: if $u \in \mathcal{D}'(M)$ then $WF\overline{u} = -WF(u)$.
(2) **Tensor product**: if $u_i \in \mathcal{D}'(M_i)$, $i = 1,2$ then

$$\begin{aligned} WF(u_1 \otimes u_2) \subset (WF(u_1) \times WF(u_2)) &\cup (\text{supp} u_1 \times \{0\}) \\ &\times WF(u_2) \cup WF(u_1) \times (\text{supp} u_2 \times \{0\}) \\ &\subset (WF(u_1) \times WF(u_2)) \cup Z_1 \times WF(u_2) \cup WF(u_1) \times Z_2. \end{aligned}$$

(3) **Restriction to a submanifold**: let $S \subset M$ be a submanifold. The conormal bundle $T^*_S M$ is defined as:

$$T^*_S M := \{(x,\xi) \in T^*M \backslash Z \ : \ x \in S, \xi \cdot v = 0 \ \forall v \in T_x S\}.$$

If $u \in \mathcal{D}'(M)$ the restriction $u_{|S}$ of u to S is well defined if

$$WFu \cap T^*_S M = \emptyset.$$

One has

$$WFu_{|S} \subset \{(x, \xi_{|T_x S}) \ : \ x \in S, \ (x,\xi) \in WF(u)\}.$$

(4) **Product**: if $u_i \in \mathcal{D}'(M)$, $i = 1,2$ then $u_1 u_2$ is well defined if $WF(u_1) \oplus WF(u_2) \cap Z = \emptyset$ and then

$$WF(u_1 u_2) \subset WF(u_1) \cup WF(u_2) \cup WF(u_1) \oplus WF(u_2).$$

(5) **Kernels**: let $K : \mathcal{D}(M_2) \to \mathcal{D}'(M_1)$ be linear continuous and $K(x_1, x_2) \in \mathcal{D}'(M_1 \times M_2)$ its distributional kernel.

Then Ku is well defined for $u \in \mathcal{E}'(M_2)$ if $WF(u) \cap WF'_{M_2}(K) = \emptyset$ and then

$$WF(Ku) \subset {}_{M_1}WF(K) \cup WF'(K) \circ WF(u).$$

(6) **Composition**: let $K_1 \in \mathcal{D}'(M_1 \times M_2)$, $K_2 \in \mathcal{D}'(M_2 \times M_3)$, where K_2 is properly supported, that is the projection $\text{supp}K_2 \to M_2$ is proper. Then $K_1 \circ K_2$ is well defined if

$$WF'(K_1)_{M_2} \ \cap \ {}_{M_2}WF(K_2) = \emptyset,$$

and then

$$WF'(K_1 \circ K_2) \subset WF'(K_1) \circ WF'(K_2) \cup ({}_{M_1}WF(K_1) \times Z_3) \cup (Z_1 \times WF(K_2)_{M_3}).$$

4.9.6. Parametrices for the Klein–Gordon Operator

Once we choose an orientation, we can define for $x \in M$ the open future/past light cones $V_x^{\pm} \subset T_x M$. We denote by $V^*_{x\pm} \subset T^*_x M$ the dual cones

$$V^*_{x\pm} = \{\xi \in T^*_x M \ : \ \xi \cdot v > 0, \ \forall v \in V_{x\pm}, \ v \neq 0\}.$$

For simplicity we write

$$\xi \rhd 0 \text{ (resp. } \xi \lhd 0) \text{ if } \xi \in V^*_{x+} \text{ (resp. } V^*_{x-}),$$

$$\xi \unrhd 0 \text{ (resp. } \xi \unlhd 0) \text{ if } \xi \in (V^*_{x+})^{\text{cl}} \text{ (resp. } (V^*_{x-})^{\text{cl}}).$$

The principal symbol of the Klein–Gordon operator P is $p(x,\xi) = \xi_\mu g^{\mu\nu}(x)\xi_\nu$ and one sets

$$N := p^{-1}(\{0\}) \cap T^*M \backslash Z,$$

called the *characteristic manifold* of P. Note that N splits into its two connected components (positive/negative energy shells):

$$N = N^+ \cup N^-, \; N^\pm = N \cap \{\pm\xi \unrhd 0\}.$$

We denote by H_p the Hamilton field of p. We denote by $X = (x,\xi)$ the points in $T^*M\backslash Z$. The bicharacteristic (Hamilton curve for p) passing through X is denoted by $B(X)$. For $X, Y \in N$ we write $X \sim Y$ if $Y \in B(X)$. Clearly this is an equivalence relation.

For $X \sim Y$, we write $X \succ Y$ (resp. $X \prec Y$) if X comes strictly after (before) Y w.r.t. the natural parameter on the bicharacteristic curve through X and Y.

We recall Hörmander's propagation of singularities theorem:

Theorem 4.13 *Let $u \in \mathcal{D}'(M)$ such that $Pu \in C^\infty(M)$. Then*

$$WF(u) \subset N, \; X \in WF(u) \Rightarrow B(X) \subset WF(u).$$

The *bicharacteristic relation* of P is the set

$$C := \{(X,Y) \in N \times N \; : \; X \sim Y\}.$$

We set

$$\Delta_N := \{(X,X)\} \cap N \times N$$

the diagonal in $N \times N$.

Parametrices (i.e. inverses modulo smoothing operators) of operators of real principal type (of which Klein–Gordon operators are an example) were studied by Duistermaat and Hörmander in the famous paper [8]. They introduced the notion of *distinguished parametrices*, that is parametrices which are uniquely determined (modulo smoothing terms of course) by the wavefront set of their kernels. Distinguished parametrices are in one-to-one correspondence with *orientations* of C, defined below.

Definition 4.18 An orientation of C is a partition of $C\backslash\Delta_N$ as $C^1 \cup C^2$ where C^i are open sets in $C\backslash\Delta_N$ and inverse relations (i.e. $\text{Exch}(C^1) = C^2$).

Note that $C^i \neq \emptyset$, $C^i \neq C\backslash\Delta_N$ (because they are inverse relations) and C^i are open and closed in $C\backslash\Delta_N$. Therefore C^i are a union of connected components of $C\backslash\Delta_N$.

Theorem 4.14 (D-H) *Let $C\backslash\Delta_N = C^1 \cup C^2$ be an orientation of C. Then there exists parametrices E^i, $i = 1, 2$ of P such that*

$$WF'(E^i) \subset \Delta^* \cup C^i,$$

where Δ^* is the diagonal in $T^*M\backslash Z \times T^*M\backslash Z$. Any left or right parametrix with the same property is equal to E^1 or E^2 modulo C^∞.

Orientations of C are themselves in one-to-one correspondence with the partitions of $N^1 \cup N^2 = N$ into open and closed sets, that is into connected components of N. For the Klein–Gordon operator N has two connected components:

$$N_\pm := \{X \in N \ : \xi \in V^*_{x\pm}\},$$

which are invariant under the bicharacteristic flow, hence two orientations, hence four distinguished parametrices. The two connected components are N_+, N_-. The two orientations are

$$C\backslash\Delta_N = C_+ \cup C_- \text{ for}$$

$$C_+ := \{(X, Y) \in C \ : \ X \succ Y\},\ C_- := \{(X, Y) \in C \ : \ X \prec Y\}$$

and

$$C\backslash\Delta_N = C^+ \cup C^- \text{ for}$$

$$C^+ := \{(X, Y) \in C \ : \ x \in J^+(y)\},\ C^- := \{(X, Y) \in C \ : \ x \in J^-(y)\}.$$

The associated parametrices are well-known in physics, we have already encountered two of them, namely $E^\pm$.
Feynman: denoted E_F:

$$\mathrm{WF}(E_F)' = \Delta^* \cup C_+.$$

Anti-Feynman: denoted $E_{\overline{F}}$:

$$\mathrm{WF}(E_{\overline{F}})' = \Delta^* \cup C_-.$$

Retarded: denoted E^+:

$$\mathrm{WF}(E^+)' = \Delta^* \cup C^+.$$

Advanced: denoted E^-:

$$\mathrm{WF}(E^-)' = \Delta^* \cup C^-.$$

The parametrices $E^{\pm}$ are more fundamental since they are used to define the symplectic form E. The parametrices E_F, $E_{\overline{F}}$ appear in connection with the vacuum state on Minkowski space and with Hadamard states on general curved spacetimes.

4.9.7. Examples

Assume that $P = \partial_t^2 + \epsilon^2$ (whatever ϵ^2 is, e.g. a real number). Then:

$$
\begin{aligned}
E^+(t) &= \theta(t)\frac{\sin \epsilon t}{\epsilon}, \\
E^-(t) &= -\theta(-t)\frac{\sin \epsilon t}{\epsilon}, \\
E_F(t) &= \frac{1}{2\mathrm{i}\epsilon}\left(\mathrm{e}^{\mathrm{i}t\epsilon}\theta(t) + \mathrm{e}^{-\mathrm{i}t\epsilon}\theta(-t)\right), \\
E_{\overline{F}}(t) &= -\frac{1}{2\mathrm{i}\epsilon}\left(\mathrm{e}^{-\mathrm{i}t\epsilon}\theta(t) + \mathrm{e}^{\mathrm{i}t\epsilon}\theta(-t)\right),
\end{aligned}
$$

for $\theta(t) =$ a Heaviside function.

We prove some properties of the distinguished parametrices.

Lemma 4.5 *We have*

$$
\begin{aligned}
&(1) \quad WF'(E^+ - E^-) = C, \\
&(2) \quad WF'(E^+ - E_F) = C \cap N_- \times N_-, \\
&(3) \quad WF'(E^- - E_F) = C \cap N_+ \times N_+.
\end{aligned}
$$

Proof (1) Since E^+ and E^- have disjoint wavefront sets above $\{x_1 \neq x_2\}$, we see that above $\{x_1 \neq x_2\}$

$$WF'(E^+ - E^-) = WF'(E^+) \cup WF'(E^-) = C \backslash \Delta_N.$$

Since $P_1(E^+ - E^-) \in C^\infty$ by propagation of singularities we obtain that $\Delta_N \subset WF'(E^+ - E^-)$. This proves (1).

(2) Above $\{(x_1, x_2) \ : \ x_1 \in J^-(x_2)\}$ we have

$$WF'(E^+ - E_F) = WF'(E_F) = \{(X_1, X_2) \ : \ x_1 \in J^-(x_2), \xi_1 \lhd 0\}.$$

Using again $P_1(E^+ - E_F) \in C^\infty$ we obtain that $WF'(E^+ - E_F) = C \cap N_- \times N_-$. The proof of (3) is similar. □

4.9.8. The Theorem of Radzikowski

Definition 4.19 Let $\Lambda_\pm : \mathcal{D}(M) \to \mathcal{E}(M)$ be linear continuous. The pair $\Lambda_\pm$ satisfies the *Hadamard condition* if

$$(Had)\ \mathrm{WF}(\Lambda_\pm)' = \{(X_1, X_2) \in N^\pm \times N^\pm : X_1 \sim X_2\}.$$

The pair $\Lambda_\pm$ satisfies the *generalized Hadamard condition* if there exists conic sets $\Gamma_\pm$ with $(X_1, X_2) \in \Gamma_\pm \Rightarrow \pm\xi_1 \rhd 0, \pm\xi_2 \rhd 0$ such that

$$(genHad)\ \ WF(\Lambda_\pm)' \subset \Gamma_\pm.$$

We introduce the following conditions on a pair $\Lambda_\pm$:

$$(KG)\ P \circ \Lambda_\pm, \Lambda_\pm \circ P \text{ smoothing},$$

$$(CCR)\ \Lambda_+ - \Lambda_- - \mathrm{i}E \text{ smoothing}.$$

The following theorem is the theorem of Radzikowski (extended to the complex case).

Theorem 4.15 *The following three conditions are equivalent:*

(1) $\Lambda_\pm$ *satisfy (Had), (KG), and (CCR),*
(2) $\Lambda_\pm$ *satisfy (genHad), (KG), and (CCR),*
(3) *one has*

$$\Lambda_\pm = \mathrm{i}(E_F - E_\mp) \textit{ modulo } C^\infty(M \times M).$$

Proof (1) $\Rightarrow$ (2): obvious.

(2) $\Rightarrow$ (3): set $S_\pm = \mathrm{i}(E_F - E_\mp)$. By Lemma 4.5 we have $\mathrm{WF}(S_\pm)' \subset C \cap (N_\pm \times N_\pm)$. By (KG) and Theorem 4.13 we obtain that $\mathrm{WF}(\Lambda_\pm)' \subset N \times N$. Using then (genHad) we obtain that $\mathrm{WF}(\Lambda_\pm) \subset \Gamma_\pm \cap N \times N \subset N_+ \times N_+$. This implies that

$$\mathrm{WF}(\Lambda_\pm - S_\pm)' \subset N_\pm \times N_\pm,$$

which implies in particular that

$$\mathrm{WF}(\Lambda_+ - S_+)' \cap \mathrm{WF}(\Lambda_- - S_-)' = \emptyset. \tag{4.17}$$

By (CCR) we have also

$$(\Lambda_+ - S_+) - (\Lambda_- - S_-) = \mathrm{i}E - (S_+ - S_-) = \mathrm{i}E - \mathrm{i}E \bmod C^\infty(M \times M).$$

By (4.17) this implies that both $\Lambda_\pm - S_\pm$ are smooth.

(3) $\Rightarrow$ (1): (KG) and (CCR) are obvious, (Had) follows from Lemma 4.5. □

There is a remaining painful step, which I will only briefly explain. I will consider real fields for simplicity. The conclusion of the above theorem for the real covariance $\omega_2(x,x')$ is that $\omega_2 = \mathrm{i}(E_F - E^+) \bmod C^\infty$. Then one has to perform some painful computations to prove that $\mathrm{i}(E_F - E^+) = c_{\mathrm{Had}} \bmod C^\infty$.

4.9.9. Return to the Stress-Energy Tensor

Let us forget the derivatives and juste consider $:\phi^2(x):$. We put just one field on each side (extension to several fields is straightforward).

We compute the expectation value of

$$\phi(x_1)\phi(x)\phi(x')\phi(x_2) - \phi(x_1)\phi(x_2)c_{Had}(x,x'),$$

in the state ω. By the quasi-free property, we obtain a sum of two types of terms:

$$\omega_2(x_1,x)\omega_2(x',x_2),\ \omega_2(x,x_2)\omega_2(x_1,x')$$

and

$$\omega_2(x_1,x_2) \times (\omega_2(x,x') - c_{Had}(x,x')).$$

The second term has a restriction to the diagonal $x = x'$, since $\omega_2 - c_{Had}$ is smooth.

For the first term we consider the wavefront set of the distribution. By the tensor product rule we obtain

$$\mathrm{WF}(\omega_2) \times \mathrm{WF}(\omega_2) \cup (\mathrm{supp}\omega_2 \times \{0\}) \times \mathrm{WF}(\omega_2) \cup \mathrm{WF}(\omega_2) \times (\mathrm{supp}\omega_2 \times \{0\}).$$

We compute the conormal to $S = \{x = x'\}$:

$$T^*_S M = \{(x_1,\xi_1,x,\xi,y,\eta,x_2,\xi_2)\ :\ \xi_1 = \xi_2 = 0,\ x = y,\ \xi = -\eta\}.$$

We have to show the intersection is empty, which is obvious since $(X,Y) \in WF(\omega_2)$ implies $\xi,\eta \neq 0$.

4.10. Construction of Hadamard States

It is not a priori obvious that Hadamard states exist on an arbitrary globally hyperbolic spacetime. In this section we explain some constructions of Hadamard states.

4.10.1. Ultrastatic Spacetimes

Consider an ultrastatic spacetime $(\mathbb{R}\times\Sigma, g)$, $g = -dt^2 + h$. Let $a = -\Delta_h + m^2$ on Σ and $\epsilon = a^{\frac{1}{2}}$. One can then define the *vacuum state* ω_{vac}, whose two-point function is given by the kernel

$$\Lambda_{\pm}^{\text{vac}}(t) = \frac{1}{2\epsilon}\mathrm{e}^{\pm it\epsilon},$$

that is writing $u \in C^{\infty}(\mathbb{R} \times \Sigma)$ as $\mathbb{R} \ni t \mapsto u(t) \in C^{\infty}(\Sigma)$:

$$\Lambda_{\pm}^{\text{vac}}u(t) = \int_{\mathbb{R}} \frac{1}{2\epsilon}\mathrm{e}^{\pm \mathrm{i}(t-s)\epsilon}u(s)ds.$$

Sahlmann and Verch [18] have shown that ω_{vac} is a Hadamard state. Another proof can be given by using the arguments of Section 4.10.3.

Similarly one can define the *thermal state* at temperature β^{-1}, $\beta > 0$ with the kernel

$$\Lambda_{\pm}^{\beta}(t) := \frac{1}{2\epsilon(1-\mathrm{e}^{-\beta\epsilon})}(\mathrm{e}^{\pm it\epsilon} + \mathrm{e}^{\mp it\epsilon}\mathrm{e}^{-\beta\epsilon}).$$

This is again a Hadamard state since $\mathrm{e}^{-\beta\epsilon}$ is a smoothing operator.

The conclusion is that vacuum or thermal states on ultrastatic (or static) spacetimes are Hadamard states.

4.10.2. The FNW Deformation Argument

Let (M, g) be a globally hyperbolic spacetime. The deformation argument of Fulling, Narcowich, and Wald [11] is based on two facts.

The first fact is the so-called *time slice property*: if $U \subset M$ is a neighborhood of a Cauchy surface Σ, then for any $u \in C_0^{\infty}(M)$ there exists $v \in C_0^{\infty}(U)$ such that $u - v \in PC_0^{\infty}(M)$. In other words

$$C_0^{\infty}(M)/PC_0^{\infty}(M) = C_0^{\infty}(U)/PC_0^{\infty}(M).$$

This implies that a pair of distributions $\Lambda_{\pm} \in C_0^{\infty}(U \times U)$ satisfying:

$$P \circ \Lambda_{\pm} = \Lambda_{\pm} \circ P = 0,$$

$$\Lambda_{+} - \Lambda_{-} = -\mathrm{i}E \text{ on } U \times U,$$

$$\Lambda_{\pm} \geq 0 \text{ on } C_0^{\infty}(U)$$

generate a quasi-free state on M.

The second fact is *Hörmander's propagation of singularities theorem* (Theorem 4.13): If $\Lambda_{\pm}$ satisfy (*Had*) (or (*genHad*)) over $U \times U$, and $P \circ \Lambda_{\pm} = \Lambda_{\pm} \circ P = 0$, then $\Lambda_{\pm}$ satisfy (*Had*) or (*genHad*) globally.

The conclusion of these two facts is that if g_1, g_2 are two Lorentzian metrics such that (M, g_i) is globally hyperbolic, if they have a common Cauchy surface Σ, and if they coincide in a neighborhood of Σ, then a Hadamard state for the Klein–Gordon field on (M, g_1) generates a Hadamard state for the Klein–Gordon field on (M, g_2).

One argues then as follows. Let us fix a Cauchy surface Σ for (M,g) and identify (M,g) with $(\mathbb{R}\times\Sigma, -c^2(t,x)dt^2+h_{ij}(t,x)dx^idx^j)$. We set $\Sigma_t = \{t\}\times\Sigma$. We fix a real function $r \in C^\infty(M)$ and consider $P = -\nabla^a\nabla_a + r(x)$ – the associated Klein–Gordon operator. One chooses an ultra-static metric

$$g_{\rm us} = -dt^2 + h_{jk,{\rm us}}({\rm x})dx^jdx^k, \quad r_{\rm us}(x) = m^2 > 0,$$

and an interpolating metric $g_{\rm int} = -c^2_{\rm int}(t,x)dt^2 + h_{jk,{\rm int}}(x)dx^jdx^k$, and real function $r_{\rm int} \in C^\infty(M)$ such that $(g_{\rm int}, r_{\rm int}) = (g,r)$ near Σ_T, $(g_{\rm int}, r_{\rm int}) = (g_{\rm us}, m^2)$ Σ_{-T}.

The vacuum state $\omega^{\rm vac}$ for $(M, g_{\rm us})$ is Hadamard for $P_{\rm us}$, hence generates a Hadamard state for $P_{\rm int}$, which itself generates a Hadamard state ω for P.

Using the Cauchy evolution operator from Σ_{-T} to Σ_T, one sees that ω is pure, since $\omega^{\rm vac}$ is pure.

4.10.3. Construction of Hadamard States by Pseudodifferential Calculus

Let us now briefly describe another construction given in a joint work with Michal Wrochna [12]. It relies on the choice of a Cauchy surface Σ and on the global pseudodifferential calculus on Σ.

We identify M with $\mathbb{R}\times\Sigma$, with the split metric $g = -c^2(t,x)dt^2 + h_{ij}(t,x)dx^idx^j$, and we set $\Sigma_s := \{s\}\times\Sigma \subset M$. For simplicity we will assume that Σ is either equal to $\mathbb{R}^d$ or to a compact manifolds, but much more general situations can be treated as well.

We denote by $\Psi^m(\Sigma)$ the space of (uniform) pseudodifferential operators of order m on Σ, corresponding to quantization of symbols in $S^m_{1,0}(T^*\Sigma)$, and by $\Psi_{\rm ph}(\Sigma)$ the subspace of pseudodifferential operators with poly-homogeneous symbols. We set

$$\Psi^\infty(\Sigma) = \bigcup_{m\in\mathbb{R}} \Psi^m(\Sigma),\ \Psi^{-\infty}(\Sigma) = \bigcap_{m\in\mathbb{R}} \Psi^m(\Sigma).$$

We denote by $H^s(\Sigma)$ the Sobolev space of order s on Σ and

$$H^\infty(\Sigma) = \bigcap_{s\in\mathbb{R}} H^s(\Sigma),\ H^{-\infty}(\Sigma) = \bigcup_{s\in\mathbb{R}} H^s(\Sigma).$$

We use the third version of the phase space, namely $(C_0^\infty(\Sigma)\otimes\mathbb{C}^2, q)$, where we recall that $q = {\rm i}\sigma_\Sigma$. We embed $C_0^\infty(\Sigma)\otimes\mathbb{C}^2$ into $\mathcal{D}'(\Sigma)\otimes\mathbb{C}^2$ using the natural scalar product on $C_0^\infty(\Sigma)\otimes\mathbb{C}^2$, that is we identify sesquilinear forms with operators.

Quasi-free states are now defined by a pair of covariances $\lambda_\pm$ on $C_0^\infty(\Sigma) \otimes \mathbb{C}^2$, and the relationship with the spacetime covariances $\Lambda_\pm$ is

$$\Lambda_\pm = (\rho \circ E)^* \circ \lambda_\pm \circ (\rho \circ E).$$

It is natural to restrict attention to states with covariances $\lambda_\pm \in \Psi^\infty(\Sigma) \otimes M(\mathbb{C}^2)$. It can be shown that if a state has pseudodifferential covariances on Σ_s for some s it has pseudodifferential covariances on Σ_s for any s.

The most important part is to characterize the Hadamard condition in terms of $\lambda_\pm$, which relies on the construction of a parametrix for the Cauchy problem on Σ, see Proposition 4.8 below.

4.10.3.1. The Model Klein–Gordon Equation

By a change of coordinates and conjugation with a convenient weight, the equation reduces to the general form

$$\partial_t^2 u + a(t, x, \partial_x) u = 0,$$

where $a(t, x, \partial_x)$ is a second order, elliptic self-adjoint operator on $L^2(\Sigma)$.

If $\Sigma = \mathbb{R}^d$, a natural hypothesis (which if needed can be rephrased in terms of the original metric g) is that

$$a(t, x, \partial_x) = -\sum_{ij} a_{jk}(t, x) \partial_{x^j} \partial_{x^k} + \sum_{a_j}(t, x) \partial_{x^j} + r(t, x),$$

where

$$C^{-1}\xi^2 \le a_{jk}(t, x)\xi_j\xi_k \le C\xi^2,$$

$$|\partial_t^m \partial_x^\alpha a_{jk}|,\ |\partial_t^m \partial_x^\alpha a_j],\ \partial_t^m \partial_x^\alpha r \text{ bounded locally uniformly in } t,$$

an assumption of course related to the uniform pseudo-differential calculus on Σ. Let us set for $s \in \mathbb{R}$ $\rho_s u := (u\restriction_{\Sigma_s}, \mathrm{i}^{-1}\partial_t u\restriction_{\Sigma_s})$ and U_s the solution of the Cauchy problem on Σ_s, that is

$$(\partial_t^2 + a(t))U_s = 0, \rho_s \circ U_s = \mathbb{1}.$$

4.10.3.2. Parametrices for the Cauchy Problem

Let $\mathbb{R} \ni t \mapsto b(t)$ be a map with values in linear operators on $L^2(\Sigma)$ with $\mathrm{Dom}\, b(t) = H^1(\Sigma)$ such that $b(t) - b^*(t)$ is bounded, and $\mathbb{R} \ni t \mapsto b(t)$ is norm continuous with values in $B(H^1(\Sigma), L^2(\Sigma))$. We denote by $\mathrm{Texp}(\mathrm{i} \int_s^t b(\sigma) d\sigma)$ the strongly continuous group with generator $b(t)$. A routine computation shows that

$$(\partial_t^2 + a(t))\mathrm{Texp}\left(\mathrm{i} \int_s^t b(\sigma) d\sigma\right) = 0$$

if and only if $b(t)$ solves the following Riccati equation:

$$\mathrm{i}\partial_t b(t) - b^2(t) + a(t) = 0. \tag{4.18}$$

This equation can be solved modulo $C^\infty(\mathbb{R}, \Psi^{-\infty}(\Sigma))$ by symbolic calculus, which amounts to solving transport equations.

Theorem 4.16 *There exists $b(t) \in C^\infty(\mathbb{R}, \Psi^1(\Sigma))$ which is unique modulo $C^\infty(\mathbb{R}, \Psi^{-\infty}(\Sigma))$ such that*

(*i*) $b(t) = \epsilon(t) + C^\infty(\mathbb{R}, \Psi^0(\Sigma))$,

(*ii*) $(b(t) + b^*(t))^{-1} = \epsilon(t)^{-\frac{1}{2}}(\mathbb{1} + r_{-1}(t))\epsilon(t)^{-\frac{1}{2}}$, $r_{-1}(t) \in C^\infty(\mathbb{R}, \Psi^{-1}(\Sigma))$,

(*iii*) $(b(t) + b^*(t))^{-1} \geq c(t)\epsilon(t)^{-1}$, *for some* $c(t) > 0$,

(*iv*) $\mathrm{i}\partial_t b - b^2 + a = r_{-\infty} \in C^\infty(\mathbb{R}, \Psi^{-\infty}(\Sigma))$.

Note that if $b^+(t) := b(t)$ solves (4.18), so does $b^-()t := -b^*(t)$. Setting $u^\pm(t,s) = \mathrm{Texpi}\int_s^t b^\pm(\sigma)d\sigma$, we obtain the following result.

Theorem 4.17 *Set*

$$\begin{aligned} r_s^{0\pm} &:= \mp(b^+(s) - b^-(s))^{-1}b^\mp(s) \in \Psi^0(\Sigma), \\ r_s^{1\pm} &:= \pm(b^+(s) - b^-(s))^{-1} \in \Psi^{-1}(\Sigma), \\ r_s^\pm f &:= r^{0\pm}f^0 + r^{1\pm}f^1,\ f = (f^0, f^1) \in H^\infty(\Sigma) \otimes \mathbb{C}^2. \end{aligned}$$

Then

$$U_s = u^+(\cdot, s)r_s^+ + u^-(\cdot, s)r_s^- + C^\infty(\mathbb{R}, \Psi^{-\infty}(\Sigma)).$$

4.10.3.3. Pure Hadamard States

The following proposition summarizes conditions implying that a pair of operators $\lambda_s^\pm$ are the Cauchy surface covariances (at time s) of a (pure) Hadamard state.

Proposition 4.8 *Let $\lambda_s^\pm : H^\infty(\Sigma) \otimes \mathbb{C}^2 \to H^\infty(\Sigma) \otimes \mathbb{C}^2$ be continuous. Then $\lambda_s^\pm$ are the Cauchy surface covariances of a Hadamard state ω if:*

(*i*) $\lambda_s^{\pm *} = \lambda_s^\pm, \lambda^\pm \geq 0$,

(*ii*) $\lambda_s^+ - \lambda_s^- = q$,

(*iii*) $f \in H^{-\infty}(\Sigma) \otimes \mathbb{C}^2 \cap \mathrm{Ker}\lambda^\mp \Rightarrow \mathrm{WF}(U_s f) \subset N^\pm$.

If additionally $c_s^\pm := \pm\mathrm{i}q^{-1} \circ \lambda_s^\pm$ *are* projections, *then* ω *is* pure.

Note that if $f \in H^{-\infty}(\Sigma)\otimes\mathbb{C}^2$ and $r_s^{\mp}f = 0$, then $\mathrm{WF}U_s f \subset N^{\pm}$. This allows us easily to construct a pure Hadamard state associated with the asymptotic solution $b(t)$ of (4.18).

In fact if we set

$$T_s(b) = (b(s) + b^*(s))^{-\frac{1}{2}} \begin{pmatrix} b^*(s) & 1\!\!1 \\ b(s) & 1\!\!1 \end{pmatrix},$$

we note that

$$q = \begin{pmatrix} 0 & 1\!\!1 \\ 1\!\!1 & 0 \end{pmatrix} = T_s(b)^* \begin{pmatrix} 1\!\!1 & 0 \\ 0 & 1 \end{pmatrix} T_s(b).$$

It follows that

$$\lambda^{\pm}(s) := T_s(b)^* \pi^{\pm} T_s(b),\ \pi^+ = \begin{pmatrix} 1\!\!1 & 0 \\ 0 & 0 \end{pmatrix},\ \pi^- = \begin{pmatrix} 0 & 0 \\ 0 & 1 \end{pmatrix},$$

are the Cauchy surface covariances of a pure Hadamard state.

Remark 4.2 One can show that if the Cauchy surface covariances of a state ω are pseudodifferential at some time s, the same is true at any other time s'.

Moreover one can show that *any* pure Hadamard state with pseudodifferential Cauchy surface covariances is of the form given above, for some $t \mapsto b(t)$ as in Theorem 4.16.

The above construction of Hadamard states by pseudodifferential calculus can be generalized to more delicate situations, like *linearized Yang–Mills fields*, where the deformation argument cannot be applied anymore, see [13].

4.10.4. Construction of Hadamard States by the Characteristic Cauchy Problem

It is possible to construct Hadamard states by replacing the (space-like) Cauchy surface Σ by some *null* hypersurface C; typically C is chosen to be the *forward lightcone* from some point $p \in M$. Then the interior M_0 of C, when equipped with g, is a globally hyperbolic spacetime in its own right. This approach was introduced by Valter Moretti [16] and generalized afterwards to various similar situations [4, 5, 6, 17] in order to construct a distinguished Hadamard state on *asymptotically flat* spacetimes, for the conformal wave equation.

In [14] we construct a large family of pure Hadamard states on the cone C, using again pseudodifferential calculus.

4.10.4.1. The Geometric Framework

Let (M, g) be a globally hyperbolic spacetime, $p \in M$ a distinguished point of M. Let

$$C := \partial J^+(p) \backslash \{p\}$$

be the future lightcone from p and

$$M_0 = I^+(p) \text{ its interior.}$$

One can show that (M_0, g) is also a globally hyperbolic spacetime. Moreover one can show that

$$J^+(K; M_0) = J^+(K; M),\ J^-(K; M_0) = J^-(K; M) \cap M_0,\ \forall\, K \subset M_0. \quad (4.19)$$

Some global conditions are needed to avoid singularities of C. One assumes that there exists $f \in C^\infty(M)$ such that:

(1) $C \subset f^{-1}(\{0\}),\ \nabla_a f \neq 0 \text{ on } C,\ \nabla_a f(p) = 0,\ \nabla_a \nabla_b f(p) = -2g_{ab}(p),$

(2) the vector field $\nabla^a f$ is complete on C.

Note that since C is a null hypersurface the vector field $\nabla^a f$ is tangent to C.

From this hypothesis it is easy to construct coordinates (f, s, θ) near C, with $f, s \in \mathbb{R}$, $\theta \in \mathbb{S}^{d-1}$ such that $C \subset \{f = 0\}$ and

$$g\restriction_C = -2dfds + h(s, \theta)d\theta^2, \quad (4.20)$$

where $h(s, \theta)d\theta^2$ is a Riemannian metric on $\mathbb{S}^{d-1}$.

Such choice of coordinates allows one to identify C with $\tilde{C} := \mathbb{R} \times \mathbb{S}^{d-1}$. A natural space of smooth functions on $\tilde{C}$ is then provided by $H^\infty(\tilde{C})$ – the intersection of Sobolev spaces of all orders, defined using the round metric $m(\theta)d\theta^2$ on $\mathbb{S}^{d-1}$.

4.10.4.2. Bulk to Boundary Correspondence

Let us consider the restriction P_0 of the Klein–Gordon operator P to $C^\infty(M_0)$, and $E_{0\pm}$ its advanced/retarded Green's functions. From (4.19), one obtains $E_{0\pm} = E_\pm\restriction_{M_0 \times M_0}$, hence:

$$E_0 = E\restriction_{M_0 \times M_0}.$$

This implies that any solution $\phi_0 \in \mathrm{Sol}_{\mathrm{sc}}(P_0)$ uniquely extends to $\phi \in \mathrm{Sol}_{\mathrm{sc}}(P)$. In particular its trace $\phi_0\restriction_C$ is well defined.

It is convenient to introduce the coordinates (s, θ) on C, and to set:

$$\begin{aligned} \rho : \mathrm{Sol}_{\mathrm{sc}}(P_0) &\to C^\infty(\mathbb{R} \times \mathbb{S}^{d-1}) \\ \phi &\mapsto \beta^{-1}(s, \theta)\phi\restriction_C (s, \theta), \end{aligned}$$

for

$$\beta(s,\theta) := |m|^{\frac{1}{4}}(\theta)|h|^{-\frac{1}{4}}(s,\theta),$$

where $h(s,\theta)d\theta^2$ is defined in (4.20) and $m(\theta)d\theta^2$ is the round metric on $\mathbb{S}^{d-1}$.

We equip $H^\infty(\tilde{C})$ with the symplectic form:

$$\overline{g}_1\sigma_C g_2 := \int_{\mathbb{R}\times\mathbb{S}^{d-1}} (\partial_s\overline{g}_1 g_2 - \overline{g}_1\partial_s g_2)|m|^{\frac{1}{2}}(\theta)dsd\theta,\ g_1, g_2 \in \mathcal{H}(\tilde{C}). \quad (4.21)$$

Introducing the *charge* $q := \mathrm{i}\sigma_C$ we have:

$$\overline{g}_1 q g_2 = 2(g_1|D_s g_2)_{L^2(\tilde{C})},\ g_1, g_2 \in \mathcal{H}(\tilde{C}), \quad (4.22)$$

where $D_s = \mathrm{i}^{-1}\partial_s$ is self-adjoint on $L^2(\tilde{C})$ on its natural domain. Clearly $(H^\infty(\tilde{C}), \sigma_C)$ is a complex symplectic space.

One can show the following result (the second statement follows from Stokes theorem), which summarizes the *bulk to boundary correspondence*:

Proposition 4.9 (1) *ρ maps $\mathrm{Sol}_{\mathrm{sc}}(P_0)$ into $\mathcal{H}(\tilde{C})$;*
(2) *$\rho : (\mathrm{Sol}_{\mathrm{sc}}(P_0), \sigma) \to (\mathcal{H}(\tilde{C}), \sigma_C)$ is a monomorphism, that is:*

$$\overline{\rho\phi_1}\sigma_C\rho\phi_2 = \overline{\phi}_1\sigma\phi_2,\ \forall\phi_1, \phi_2 \in \mathrm{Sol}_{\mathrm{sc}}(P_0).$$

4.10.4.3. Hadamard States on the Cone

From Proposition 4.9, we see that any quasi-free state ω_C on $\mathrm{CCR}(H^\infty(\tilde{C}), \sigma_C)$ generates a quasi-free state ω_0 on $\mathrm{CCR}(C_0^\infty(M_0)/PC_0^\infty(M_0), E_0)$. The task is now to give conditions on ω_C implying that ω_0 is Hadamard.

Let us denote by $x = (r, s, y)$ the coordinates (f, s, θ) near C and by $\xi = (\rho, \sigma\eta)$ the dual coordinates. The complex covariances of ω_C are denoted by $\lambda^\pm \in \mathcal{D}'(\tilde{C}\times\tilde{C})$. The associated covariances $\Lambda^\pm$ of ω_0 are then:

$$\Lambda^\pm := (\rho\circ E_0)^* \circ \lambda^\pm \circ (\rho\circ E_0).$$

One can show the following result.

Theorem 4.18 *Assume that $\lambda^\pm : H^\infty(\tilde{C}) \to H^\infty(\tilde{C})$ and ${}_{\tilde{C}}\mathrm{WF}(\lambda^\pm)' = \mathrm{WF}(\lambda^\pm)'_{\tilde{C}} = \emptyset$. Then if:*

$$(i)\quad \mathrm{WF}(\lambda^\pm)' \cap \{(Y_1, Y_2) : \pm\sigma_1 < 0 \text{ or } \pm\sigma_2 < 0\} = \emptyset,$$

$$(ii)\quad \mathrm{WF}(\lambda^\pm)' \cap \{(Y_1, Y_2) : \pm\sigma_1 > 0 \text{ and } \pm\sigma_2 > 0\} \subset \Delta,$$

$\Lambda^\pm$ satisfy (Had).

A state on $\mathrm{CCR}(H^\infty(\tilde{C}), \sigma_C)$ satisfying thc assumptions of Theorem 4.18 will be called a *Hadamard state on the cone*.

4.10.4.4. Hadamard States on the Cone and Pseudodifferential Calculus

Recall that q defined in (4.22) equals $2D_s$, whose resolvent $(2D_s - z)^{-1}$ is not an elliptic pseudodifferential operator on $\tilde{C}$, for the usual calculus. However, it belongs to a larger *bi-homogeneous class* $\Psi^{p_1,p_2}(\tilde{C})$, which can be loosely defined as $\Psi^{p_1}(\mathbb{R}) \otimes \Psi^{p_2}(\mathbb{S}^{d-1})$.

It is then possible to construct Hadamard states on the cone by mimicking the arguments in Section 4.10.3. Note that no construction of a parametrix for the characteristic Cauchy problem on $\tilde{C}$ is necessary, since the Hadamard condition on the cone is rather explicit.

4.10.4.5. Purity Inside the Cone

A natural issue with this construction is wether a pure state on the cone generates a pure state inside. This is not a priori obvious, since the map ρ is not surjective. Nevertheless it is shown in [14] that purity is preserved using the results of Hörmander [15] on the characteristic Cauchy problem. This is the only place where the solvability of the characteristic Cauchy problem is important.

References

[1] C. Bär, K. Fredenhagen, editors. *Quantum Field Theory on Curved Spacetimes*, Springer Lecture Notes in Physics, Springer, 2009.

[2] C. Bär, N. Ginoux, F. Pfäffle: *Wave Equations on Lorentzian Manifolds and Quantization*, ESI Lectures in Mathematics and Physics, Springer, 2007.

[3] Bjorken, J.D., Drell, S.D. *Relativistic Quantum Mechanics*, McGraw-Hill, 1964.

[4] C. Dappiaggi, V. Moretti and N. Pinamonti, Distinguished Quantum States in a Class of Cosmological Spacetimes and their Hadamard Property, J. Math. Phys. 50, 2009, 062304.

[5] C. Dappiaggi, V. Moretti and N. Pinamonti, Rigorous Construction and Hadamard Property of the Unruh State in Schwarzschild Spacetime, Adv. Theor. Math. Phys. 15, 2011, 355.

[6] C. Dappiaggi, D. Siemssen, *Hadamard States for the Vector Potential on Asymptotically Flat Spacetimes*, Rev. Math. Phys. 25, 2013, 1350002.

[7] J. Derezinski, C. Gérard. *Mathematics of Quantization and Quantum Fields*, Cambridge Monographs on Mathematical Physics, Cambridge University Press, 2013.

[8] Duistermaat, J.J., Hörmander, L. Fourier Integral Operators. II. Acta Math. 128, 183–269, 1972.

[9] Friedlander, F.G. *The Wave Equation on a Curved Space-time* Cambridge University Press, 1975.

[10] S. Fulling. *Aspects of Quantum Field Theory in Curved Space-Time*, London Mathematical Society Student Texts, Cambridge University Press, 1989.

[11] Fulling, S.A., Narcowich, F.J., Wald, R.M. Singularity Structure of the Two-point Function in Quantum Field Theory in Curved Space-time, II, Annals of Physics, 136, 243–272, 1981.

[12] C. Gérard, M. Wrochna, Construction of Hadamard States by Pseudo-differential Calculus, Comm. Math. Phys. 325, 713–755, 2014.

[13] C. Gérard, M. Wrochna, Hadamard States for the Linearized Yang-Mills Equation on Curved Spacetime, arXiv:1403.7153, 2014.

[14] C. Gérard, M. Wrochna, Construction of Hadamard States by Characteristic Cauchy Problem, arXiv:1409.6691, 2014.

[15] L. Hörmander, A Remark on the Characteristic Cauchy Problem, J. Funct. Anal. 93, 270–277, 1990.

[16] V. Moretti, Uniqueness Theorem for BMS-invariant States of Scalar QFT on the Null Boundary of Asymptotically Flat Spacetimes and Bulk-boundary Observable Algebra Correspondence, Comm. Math. Phys. 268, 727–756, 2006.

[17] V. Moretti, Quantum Out-states Holographically Induced by Asymptotic Flatness: Invariance under Space-time Symmetries, Energy Positivity and Hadamard Property, Comm. Math. Phys. 279, 31–75, 2008.

[18] Sahlmann, H., Verch, R. Passivity and Microlocal Spectrum Condition, Comm. Math. Phys. 214, 705–731, 2000.

Département de Mathématiques, Université de Paris XI, 91405 Orsay Cedex France
E-mail address: christian.gerard@math.u-psud.fr

5

A Minicourse on Microlocal Analysis for Wave Propagation

András Vasy

5.1. Introduction

This minicourse describes a microlocal framework for the linear analysis that has been useful for the global understanding of wave propagation phenomena. While it is useful to have some background in microlocal analysis since relatively sophisticated frameworks and notions are discussed and it may be easier for the reader to start with simpler cases, I will in fact cover the subject from scratch.

Section 5.2 covers the differential operator aspects of totally characteristic, or b-, operators, in particular pointing out the differences from the local theory, or from the global theory on compact manifolds without boundary.

The basics of microlocal analysis in Section 5.3 are covered roughly following Melrose's lecture notes [31] (which introduced me to the subject!), but in a generalized version. These generalizations concern both introducing the scattering algebra and allowing variable order operators. The former was first described systematically from a geometric perspective in [30] from the beginning; see [41, 37] for earlier descriptions in $\mathbb{R}^n$; here it serves to have a pseudodifferential algebra with Fredholm properties on a non-compact space right from the outset to orient the reader. The latter was discussed by

The author was supported in part by National Science Foundation grants DMS-1068742 and DMS-1361432. He is very grateful to the students at Stanford University in the Math 256B course he taught in Winter 2014, for which these notes were originally written, for their patience and comments. He is also grateful to the Université de Grenoble for its hospitality during the summer school in 2014 on mathematical aspects of general relativity, as well as to the participants; these notes again provided the backbone of the material covered. In addition, he is grateful to the hospitality of the Université de Nice in summer 2015 where he gave a lecture series based on the more advanced parts of these notes and where these notes were put in their final form. Thanks are also due to Peter Hintz for a careful reading of the final version. But most of all the author is grateful to Richard Melrose, from whom he learnt microlocal analysis, and whose notes are followed, in a generalized form, in Section 5.3.

Duistermaat and Unterberger [43, 7], though not quite in the setting we are interested in. I would remark that, apart from solving elliptic partial differential equations (PDEs), the elliptic scattering algebra results shown in this section also allow one to deal with the local geodesic X-ray transform [42], and were the key ingredient in showing boundary rigidity in a fixed conformal class [40].

Section 5.4 gives a thorough description of propagation phenomena, from generalizations of Hörmander's propagation of singularities theorem [28] to the scattering algebra (where it is due to Melrose in [30]) variable order settings, and including complex absorption and radial points. These are then used to discuss the limiting absorption principle in scattering theory as well as the Klein–Gordon equation on Minkowski-like spaces. The approach of this minicourse, given the limited space, is that the proofs are given in the basic microlocal setting; extension to the b-operators (presented in the recent literature in detail) is then straightforward, and I do not give the detailed proofs in that setting here. Note that I also use large parameter, or semiclassical, estimates, which are discussed briefly in Section 5.5. While their incorporation in the earlier section would have been straightforward, they result in further notational overheads, so it seemed better to sketch only the basic properties and leave the details to the reader, who may also refer to Zworski's recent treatment of the topic [58].

Section 5.5 shows, following [48, 47], how the tools developed can be applied to understand the analytic continuation of the resolvent of the Laplacian on conformally compact, asymptotically hyperbolic, spaces. While I do not discuss it here, another application of these microlocal tools is the work of Dyatlov and Zworski [11] to dynamical zeta functions for Anosov flows; I refer to [49] for a concise overview of some recent applications and to detailed references.

Finally Section 5.6 introduces the b-pseudodifferential operator algebra of Melrose [34] by local reduction to cylindrical models in Euclidean space. This actually does not give quite as much of the full "small calculus" as the analogous localization for the standard algebra on compact manifolds without boundary and for the scattering algebra, though it is sufficient for us. This completes the tools necessary for the linear analysis of waves on Lorentzian scattering spaces, which are generalizations of asymptotic Minkowski spaces [48, 2, 26, 12], as well as Kerr–de Sitter space, which was described in [48, 26], apart from a treatment of the trapped set in the latter case (for which we provide references at the end of the section). Indeed, it was the Kerr–de Sitter project [48] that started the work on the non-elliptic microlocal framework presented here.

In fact, these techniques extend to non-linear problems. The extension to semilinear wave equations was done in [26], to quasilinear problems without trapping in [24], and to quasilinear problems with trapping in [22]. The quasilinear works introduce b-pseudodifferential operators with coefficients with Sobolev regularity (rather than smoothness) in the spirit of Beals and Reed [3] and, to deal with trapping, tame linear estimates; it turns out that these extensions of the "smooth" microlocal framework described in these notes is not hard. Given these linear estimates, the non-linear techniques are rather straightforward by now; an easy version of the Nash–Moser iteration due to Saint Raymond [38] is used in [22] to deal with the losses of derivatives due to the trapping. These quasilinear results were then used in the recent proof of the stability of slowly rotating Kerr–de Sitter black holes by Hintz and Vasy [23].

I hope that these detailed lecture notes will provide a more accessible path, available in a single place, to the microlocal material that has proved so useful in our understanding of so many problems.

5.2. The Overview

The ultimate goal of this minicourse is to understand the analysis of not necessarily elliptic totally characteristic, or b-(pseudo)differential, operators. These are operators on manifolds M with boundaries or corners, though they can also arise on complete manifolds without boundary after one appropriately compactifies or bordifies them (i.e. makes them into a manifold with boundary or corners, not necessarily compact), as we discuss below. We will study both elliptic and hyperbolic operators, with the Laplacian or the d'Alembertian of a b-metric (i.e. a metric corresponding to this structure), and provide main examples. An important aspect of b-analysis is, however, that the principal symbol, on which ellipticity etc. is based, does not completely capture the problem: there is also an analytic family of operators at the boundary ∂M, depending on the dual of the normal variable to the boundary, the poles of whose inverse determine both Fredholm properties and asymptotic expansions.

To be more concrete, consider the vector space of vector fields $\mathcal{V}_b(M)$ tangent to ∂M, M an n-dimensional manifold with boundary. Recall that $V \in \mathcal{V}_b(M)$ is equivalent to the statement that for all $f \in C^\infty(M)$ with $f|_{\partial M} = 0$ one has $Vf|_{\partial M} = 0$; this in turn is equivalent to the same statement for a single f with a non-degenerate differential at ∂M (such a non-degenerate function, when non-negative, is called a *boundary defininingfunction*). This space $\mathcal{V}_b(M)$ is a left $C^\infty(M)$ module and a Lie algebra: if f is a function that vanishes at ∂M, $V, W \in \mathcal{V}_b(M)$, then VWf, WVf vanish at ∂M, thus so does

$$[V, W]f = (VW - WV)f.$$

Further, in a local coordinate chart mapping an open set O in M to an open set U in $[0, \infty)_x \times \mathbb{R}^{n-1}_y$, with ∂M mapped to $x = 0$ (thus x is a local boundary defining function), any smooth vector field has the form

$$V = b_0 \partial_x + \sum_{j=1}^{n-1} b_j \partial_{y_j}, \; b_j \in \mathcal{C}^\infty(M),$$

so the tangency of V to ∂M means $Vx|_{x=0} = 0$, that is $b_0|_{x=0} = 0$, so $b_0 = xa_0$, $a_0 \in \mathcal{C}^\infty(M)$. Correspondingly,

$$V = a_0(x\partial_x) + \sum_j a_j \partial_{y_j}, \; a_j \in \mathcal{C}^\infty(M), \tag{5.1}$$

locally; conversely, any vector field with such local coordinate expressions is in $\mathcal{V}_b(M)$.

Our main interest is in differential operators generated by such $V \in \mathcal{V}_b(M)$. Namely we let $\mathrm{Diff}^m_b(M)$ consist of finite sums of up to m-fold products of elements of $\mathcal{V}_b(M)$. With the usual Fourier analysis convention that $D = \frac{1}{i}\partial$, typical examples, in local coordinates, are operators such as

$$(xD_x)^2 + \sum_j D^2_{y_j}$$

which is elliptic, corresponding to a Riemannian space with a cylindrical end, as well as, after appropriate conjugation and division by a factor, to conic spaces, and

$$\Big(xD_x - \sum_j y_j D_{y_j}\Big)^2 - \sum_j D^2_{y_j},$$

which is hyperbolic, and corresponds to a neighborhood of the static patch in a de Sitter space.

Typically the spaces of interest are not presented as manifolds with boundary; we need to bordify them for this purpose. Thus, an exact cylindrical end metric is traditionally written as

$$g = dr^2 + h(y, dy), \; r \to +\infty,$$

with Laplacian

$$\Delta_g = D_r^2 + \Delta_h.$$

Letting $x = e^{-r}$ (see Figure 5.1) means $r \to +\infty$ corresponds to $x \to 0$, and $dr = -\frac{dx}{x}$, $D_r = -xD_x$ shows that such a metric and Laplacian indeed has

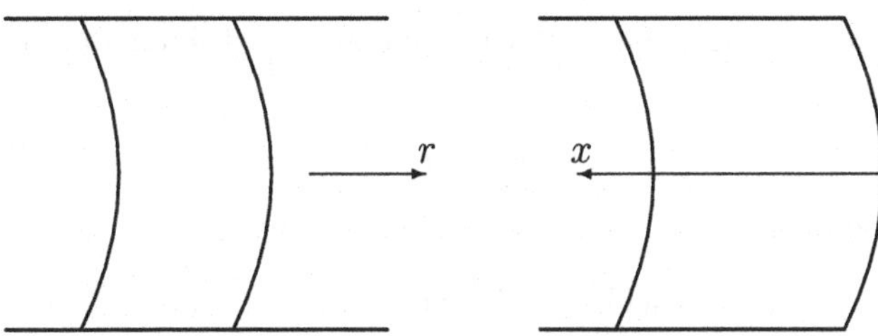

Figure 5.1. Left: a cylindrical end, with $r \to +\infty$. Right: the bordification, with $x = 0$ corresponding to $r = \infty$

the stated form if one bordifies the end by gluing $x = 0$ to it, that is replacing $(r_0, +\infty)_r \times Y$ by $[0, e^{-r_0})_x \times Y$, with the identification

$$(r, y) \mapsto (e^{-r}, y)$$

of

$$(r_0, +\infty)_r \times Y \to (0, e^{-r_0})_x \times Y;$$

combining this with the compact core of the original manifold one obtains the manifold with boundary M. (More precisely, if the original manifold is $\tilde{M}$, the manifold M is the disjoint union of $\tilde{M}$ and $[0, r_0^{-1}) \times Y$ with the identification of the cylindrical end in $\tilde{M}$ with $(0, r_0^{-1}) \times Y$ as above, with a base for the topology of M given by the images under the equivalence relation of open sets in either $\tilde{M}$ or $[0, r_0^{-1}) \times Y$, and with coordinate charts given by those in $\tilde{M}$ as well as $[0, r_0^{-1})$ times coordinate charts in Y.) Note that a smooth function $f \in C^\infty(M)$ has a Taylor series expansion at ∂M in terms of powers of x; this amounts to powers of e^{-r} in the original coordinates.

While we postpone detailed discussion, it turns out that wave, or more general Klein–Gordon, equations on asymptotic de Sitter spaces, or even Kerr–de Sitter spaces, describing rotating black holes in a background with a positive cosmological constant (reflecting our current understanding of the universe) also give rise to b-problems that the tools we develop can be used to analyze. A byproduct of this discussion is a new perspective to analyze asymptotically hyperbolic spaces, and indeed it sheds light on the role so-called even metrics play in these. Before continuing, we first comment on a different class of spaces, which do not seem to be b-spaces at the outset, where b-analysis in fact turns out to be useful.

On a manifold with boundary M, there is a conformally related class of vector fields, called scattering vector fields $\mathcal{V}_{sc}(M)$. These arise by letting ρ be a boundary defining function of M (i.e. $\rho \geq 0$, ρ vanishes exactly on ∂M, and

$d\rho$ is non-degenerate at ∂M; x was a local boundary defining function above), and letting

$$\mathcal{V}_{\mathrm{sc}}(M) = \rho\mathcal{V}_{\mathrm{b}}(M).$$

In local coordinates, such vector fields thus have the form

$$V = a_0(x^2\partial_x) + \sum_j a_j(x\partial_{y_j}),\ a_j \in \mathcal{C}^\infty(M),$$

locally; conversely, any vector field with such local coordinate expressions is in $\mathcal{V}_{\mathrm{sc}}(M)$. Again, we can define scattering differential operators by taking finite sums of finite products of these. A typical scattering differential operator is

$$(x^2D_x)^2 + (xD_y)^2 - \lambda,$$

where $\lambda \in \mathbb{C}$. Up to an inessential first order term, this is the spectral family of the Laplacian of a conic metric on the "large end" of a cone, such as Euclidean space near infinity. To see this, recall that an exact conic metric has the form

$$dr^2 + r^2\,h(y, dy),$$

and thus, up to first order terms, its Laplacian has the form $D_r^2 + r^{-2}\Delta_h$. If we let $x = r^{-1}$, $dr = -\frac{dx}{x^2}$, and $D_r = -x^2D_x$, this takes the form described above if we bordify by adding $x = 0$ to $(0, r_0^{-1})_x \times Y$. Note that the identification

$$(r_0, \infty) \times Y \to (0, r_0^{-1}) \times Y$$

is now via

$$(r, y) \to (r^{-1}, y),$$

so if the bordified space is denoted by M, elements of $\mathcal{C}^\infty(M)$ have an expansion in powers of x, that is of r^{-1}, unlike e^{-r} above. Thus, one has to pay attention to the bordification/compactification one uses. Now, if $\lambda = 0$, we can factor this operator as

$$x^2L,\ L = D_x x^2 D_x + \Delta_y \in \mathrm{Diff}_{\mathrm{b}}^2(M),$$

and thus b-analysis is applicable, but if $\lambda \neq 0$ this is no longer the case. Thus, for $\lambda \neq 0$, one needs to use the scattering framework to analyze the problem. (If $\lambda = 0$, the scattering framework can be used as a starting point, but the operator is degenerate then in an appropriate sense, which forces one to work with the b-framework at least implicitly.)

In the case of Euclidean space, a straightforward calculation shows that compactification of $\mathbb{R}^n$ by reciprocal polar coordinates, that is identifying $\mathbb{R}^n \setminus \{0\}$ with $(0, \infty)_r \times \mathbb{S}^{n-1}$, and compactifying to $M = \overline{\mathbb{R}^n} = \overline{\mathbb{B}^n}$ (a closed

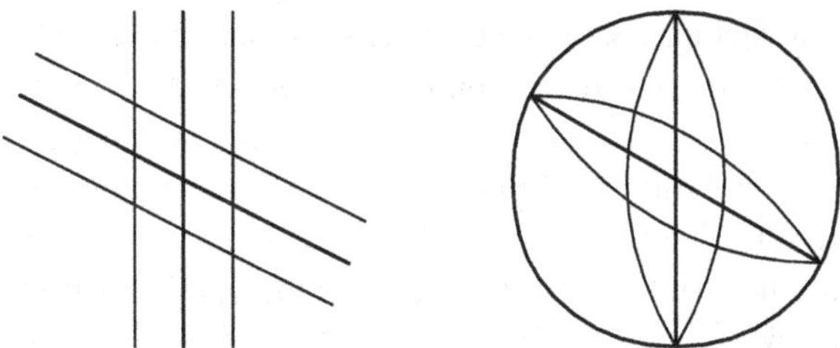

Figure 5.2. Left: two families of parallel lines in $\mathbb{R}^2$. Right: the compactification $\overline{\mathbb{R}^2}$ and the image of the two families of parallel lines. Note that parallel families end at the same two points at $\partial\overline{\mathbb{R}^2}$, while conic sectors, such as on the upper left between the two thick lines, map to wedge shaped domains

ball) using the map $x = r^{-1}$, see Figure 5.2, means not only that the spectral family of the Laplacian becomes an element of $\mathrm{Diff}^2_{\mathrm{sc}}(M)$, but more generally translation invariant differential operators become elements of $\mathrm{Diff}^m_{\mathrm{sc}}(M)$ and indeed become a basis, over $\mathcal{C}^\infty(M)$, of $\mathrm{Diff}^m_{\mathrm{sc}}(M)$. In particular, $\Box_g$, where g is the Minkowski metric, satisfies $\Box_g \in \mathrm{Diff}^2_{\mathrm{sc}}(M)$. Further, much like the Euclidean Laplacian, due to its homogeneity under dilations (which is what the b-structure relates to), $\Box_g = x^2 L$ with $L \in \mathrm{Diff}^2_{\mathrm{b}}(M)$. This means that the Minkowski wave operator, and indeed more general operators of similar form, called d'Alembertians of Lorentzian scattering metrics, are amenable to b-analysis. Note, however, that the Klein–Gordon operator $\Box_g - \lambda \in \mathrm{Diff}^2_{\mathrm{sc}}(M)$ does not factor in this way for $\lambda \neq 0$, and it is thus analyzable in the scattering framework.

The general approach I present to the analysis of differential operators P is via Fredholm estimates, such as

$$\|u\|_{\mathcal{X}} \leq C(\|Pu\|_{\mathcal{Y}} + \|u\|_{\tilde{\mathcal{X}}})$$

and

$$\|v\|_{\mathcal{Y}^*} \leq C(\|P^* v\|_{\mathcal{X}^*} + \|v\|_{\mathcal{Z}}),$$

where the inclusion of the spaces $\mathcal{X}$ into $\tilde{\mathcal{X}}$ and $\mathcal{Y}^*$ into $\mathcal{Z}$ is compact. Since our spaces correspond to geometrically complete non-compact manifolds, or compact manifolds but with differentiability encoded via complete vector fields such as $\mathcal{V}_{\mathrm{b}}(M)$, such a compact inclusion has two ingredients: gain in differentiability and gain in decay. (The simplest example is weighted Sobolev spaces $H^{s,r}(\mathbb{R}^n)$, with $H^s(\mathbb{R}^n)$ the standard Sobolev space, which is a weighted L^2-space on the Fourier transform side: $H^s(\mathbb{R}^n) = \mathcal{F}^{-1}\langle.\rangle^{-s}L^2(\mathbb{R}^n)$, and $H^{s,r}(\mathbb{R}^n) = \langle.\rangle^{-r}H^s(\mathbb{R}^n)$, with $\langle z\rangle = (1 + |z|^2)^{1/2}$; see (5.47). Then the inclusion $H^{s,r}(\mathbb{R}^n) \to H^{s',r'}(\mathbb{R}^n)$ being compact requires both $s > s'$ and $r > r'$.

These spaces in fact turn out to be the scattering Sobolev spaces $H^{s,r}_{\mathrm{sc}}(\overline{\mathbb{R}^n})$ and $H^{s',r'}_{\mathrm{sc}}(\overline{\mathbb{R}^n})$.) The gain in differentiability, much as in standard local (pseudo)-differential analysis, is based on properties of principal symbols, which capture the operators modulo lower (differential) order terms. However, the gain in decay is more subtle.

To illustrate this, notice a structural difference between the Lie algebras $\mathcal{V}_{\mathrm{b}}(M)$ and $\mathcal{V}_{\mathrm{sc}}(M)$:

$$[\mathcal{V}_{\mathrm{b}}(M), \mathcal{V}_{\mathrm{b}}(M)] \subset \mathcal{V}_{\mathrm{b}}(M),\ [\mathcal{V}_{\mathrm{sc}}(M), \mathcal{V}_{\mathrm{sc}}(M)] \subset \rho\mathcal{V}_{\mathrm{sc}}(M),$$

that is while $\mathcal{V}_{\mathrm{b}}(M)$ is merely a Lie algebra, $\mathcal{V}_{\mathrm{sc}}(M)$ is commutative to the leading (zeroth) order at ∂M. In order to analyze differential operators in $\mathrm{Diff}_{\mathrm{b}}(M)$ or $\mathrm{Diff}_{\mathrm{sc}}(M)$, we develop pseudodifferential algebras $\Psi_{\mathrm{b}}(M)$ and $\Psi_{\mathrm{sc}}(M)$. The leading order commutativity of $\mathcal{V}_{\mathrm{sc}}(M)$ means that $\Psi_{\mathrm{sc}}(M)$ has another symbol at ∂M which admits similar constructions and estimates as the standard principal symbol; the two symbols are indeed related via the Fourier transform on $\mathbb{R}^n$ as we discuss in the next sections covering standard and scattering pseudodifferential operators. On the other hand, to $\Psi_{\mathrm{b}}(M)$ corresponds an analytic family of operators on ∂M, called the indicial or normal family. For instance, on cylindrical ends, one conjugates the operator by the Mellin transform in x, that is the Fourier transform in $\log x = -r$:

$$(\mathcal{M}u)(\sigma, y) = \int x^{-i\sigma} u(x,y)\,\frac{dx}{x},$$

with

$$(\mathcal{M}^{-1}v)(x,y) = \frac{1}{2\pi}\int_{\operatorname{Im}\sigma=-\alpha} x^{i\sigma} v(\sigma,y)\,d\sigma,$$

where the choice of α corresponds to the weighted space we are working with. Here we are interested in functions supported near $x = 0$ (i.e. r near $+\infty$), so the behavior of u as $x \to +\infty$ is irrelevant, and the Mellin transform gives a result which is holomorphic in an upper half-plane: the more decay one has at $x = 0$, the larger the half-plane is. The Mellin transform changes xD_x to multiplication by the b-dual variable σ, so the indicial family of

$$P = (xD_x)^2 + \Delta_Y - \lambda$$

is

$$\hat{N}(P)(\sigma) = \sigma^2 + \Delta_Y - \lambda.$$

This family has a meromorphic inverse, with poles at $\sigma = \sigma_j$, $\sigma_j = \pm\sqrt{\lambda - \lambda_j}$, as λ_j runs over the eigenvalues of Δ_Y (here we are assuming Y is compact). Note that we are inverting operators (i.e. working with a non-commutative algebra), but on a manifold without boundary, that is on a simpler space. The Fredholm properties of P then will depend on the weighted b-Sobolev space we use, with the choice of weight corresponding to the choice of the imaginary part of σ: one needs to choose a weight x^α such that no poles $\hat{N}(P)(\sigma)^{-1}$ lie on the line $\operatorname{Im}\sigma = -\alpha$. (There are only finitely many poles in a strip $|\operatorname{Im}\sigma| < C$, so there are many good choices, but there are also a few bad choices.) Here the b-Sobolev spaces of non-negative integer differentiability order s and of real weight α are defined by

$$u \in H_{\mathrm{b}}^{s,\alpha}(M) \text{ iff } x^{-\alpha}u \in H_{\mathrm{b}}^{s,0}(M) = H_{\mathrm{b}}^{s}(M),$$
$$u \in H_{\mathrm{b}}^{s}(M) \text{ iff } \forall L \in \mathrm{Diff}_{\mathrm{b}}^{s}(M),\ Lu \in L_{\mathrm{b}}^{2}(M),$$

where $L_{\mathrm{b}}^2(M)$ is the L^2-space with respect to any non-degenerate b-density (see below), such as $|\frac{dx}{x}\,dy_1 \dots dy_{n-1}|$. Further, even if we have a Fredholm choice of space, the invertibility properties, even the Fredholm index, depend on this choice. This contrasts much with say formally self-adjoint elliptic problems on compact manifolds, for which any pair of Sobolev spaces, with orders differing by the order of the operator, give rise to a Fredholm pair of the same invertibility properties. One indication of this is that the kernel (nullspace) of elliptic operators on such manifolds is in $\mathcal{C}^\infty(M)$, that is in every Sobolev space, while in the b-setting, it is *not* in every weighted space, though the differentiability order does not matter. In fact, if $Pu \in \dot{\mathcal{C}}^\infty(M)$, say, that is $\mathcal{C}^\infty$ and vanishes with all derivatives at ∂M, then typically u has an asymptotic expansion at ∂M of the form

$$\sum_{j:\ \operatorname{Im}\sigma_j < -\alpha} x^{i\sigma_j} u_j,\ u_j \in \mathcal{C}^\infty(M),$$

(with additional logarithmic terms, such as $\sum_{\ell=0}^{k_j} (\log x)^\ell x^{i\sigma_j} u_{j,\ell}$ in the general case), where α is such that $u \in H_{\mathrm{b}}^{s,\alpha}(M)$ for some s, and there are no σ_j with $\operatorname{Im}\sigma_j = -\alpha$. The σ_j are also called resonances; they thus determine the asymptotic behavior/expansion of solutions of P.

While we have discussed linear problems, with a bit of work the analysis can be extended to nonlinear problems as well. In the non-elliptic context (our main interest here) small data problems are quite well behaved: for semilinear problems one can even use Picard iteration typically (much as for ordinary differential equations (ODEs)) to obtain global solutions and asymptotic expansions; for quasilinear problems a bit more care is needed.

We now return to $\mathcal{V}_{\mathrm{b}}(M)$; many of the following considerations apply, with simple modifications, to $\mathcal{V}_{\mathrm{sc}}(M)$. Much like the set of all vector fields $\mathcal{V}(M)$ is the set of all smooth sections of TM, and $\mathcal{V}_{\mathrm{b}}(M)$ is the set of all smooth sections of a vector bundle ${}^{\mathrm{b}}TM$, called the b-tangent bundle. Indeed, a local basis of ${}^{\mathrm{b}}TM$ is given by $x\partial_x, \partial_{y_1}, \ldots, \partial_{y_{n-1}}$, that is in terms of coordinates (x, y) as above, $(x, y_1, \ldots, y_{n-1}, a_0, a_1, \ldots, a_{n-1})$ are local coordinates on ${}^{\mathrm{b}}TM$, cf. (5.1). In M°, ${}^{\mathrm{b}}TM$ can be naturally identified with TM (corresponding to tangency to ∂M being vacuous there); globally $\mathcal{V}_{\mathrm{b}}(M) \subset \mathcal{V}(M)$ gives a fiber-preserving map

$$\iota : {}^{\mathrm{b}}TM \to TM;$$

with $(x, y_1, \ldots, y_{n-1}, b_0, b_1, \ldots, b_{n-1})$ coordinates on TM as in the above parameterization of vector fields, then the map is, in local coordinates,

$$\iota(x, y_1, \ldots, y_{n-1}, a_0, a_1, \ldots, a_{n-1}) = (x, y_1, \ldots, y_{n-1}, xa_0, a_1, \ldots, a_{n-1}),$$

corresponding to

$$b_0 = xa_0,\ b_j = a_j,\ j = 1, \ldots, n-1.$$

Note that ι is not injective at $p \in \partial M$; its kernel ${}^{\mathrm{b}}N_p\partial M$ is the span of $x\partial_x$, the space of normal b-vector fields to the boundary, while its range is $T_p\partial M$, the tangent space to ∂M (as a subspace of T_pM). Note that, unlike for T_pM, the natural subspace of ${}^{\mathrm{b}}T_pM$ is *not* vector fields tangent to ∂M, but vector fields b-normal to ∂M.

The dual bundle, ${}^{\mathrm{b}}T^*M$, called the b-cotangent bundle, of ${}^{\mathrm{b}}TM$, then has a local basis $\frac{dx}{x}, dy_1, \ldots, dy_{n-1}$, that is smooth sections locally have the form

$$\sigma\,\frac{dx}{x} + \sum \eta_j\,dy_j,$$

with $\sigma, \eta \in C^\infty(M)$. The adjoint of ι is the map $\pi : T^*M \to {}^{\mathrm{b}}T^*M$, which is sometimes (especially for boundary value problems) called the compressed cotangent bundle map, and ${}^{\mathrm{b}}T^*M$ the compressed cotangent bundle. In coordinates, with $(\xi, \zeta_1, \ldots, \zeta_{n-1})$ being dual coordinates to $(x, y_1, \ldots, y_{n-1})$,

$$\pi(x, y_1, \ldots, y_{n-1}, \xi, \zeta_1, \ldots, \zeta_{n-1}) = (x, y_1, \ldots, y_{n-1}, x\xi, \zeta_1, \ldots, \zeta_{n-1}),$$

corresponding to

$$\xi\,dx + \sum \zeta_j\,dy_j = (x\xi)\,\frac{dx}{x} + \sum_j \zeta_j\,dy_j.$$

The kernel of π on T_p^*M is now $N_p^*\partial M$, the conormal bundle fiber of ∂M, and the range is the span of the dy_j, which is the cotangent bundle of ∂M within ${}^{\mathrm{b}}T_p^*M$. Note that the natural subspace of ${}^{\mathrm{b}}T_p^*M$ is the cotangent, not the conormal, bundle of the boundary.

Intuitively, π projects out the normal momentum (momentum being dual to position), but preserves the tangential momentum, and thus can be used to encode the law of reflection for boundary problems. In fact, if we had more time in this minicourse, in its final part we would study exactly this problem (even on manifolds with corners): how do waves on, say, $M = Y \times \mathbb{R}_t$, (Y, h) Riemannian with boundary, $g = dt^2 - h$, behave at ∂M? Notice that $\Box_g \in \mathrm{Diff}^2(M)$, but is not in $\mathrm{Diff}^2_{\mathrm{b}}(M)$. Nonetheless, the tools we use to analyze $\Box_g$ are in the b-category. The reason for this is that "standard" pseudo-differential operators do not quite make sense on manifolds with boundary, and even in the cases when they do, they typically do not preserve boundary conditions. Thus, we use a two-algebra approach: the operator to analyze is in $\mathrm{Diff}(M)$, but the toolkit we use is in $\Psi_{\mathrm{b}}(M)$. I refer to [52] for details and further references. In fact, I would point out that b-pseudodifferential operators originated in Melrose's work on propagation problems on manifolds with boundary [33].

We now turn to tensorial considerations. A smooth non-degenerate symmetric bilinear form on ${}^{\mathrm{b}}TM$, that is a non-degenerate section of $\mathrm{Sym}^2\, {}^{\mathrm{b}}T^*M = {}^{\mathrm{b}}T^*M \odot {}^{\mathrm{b}}T^*M$ is, for each $p \in M$, a (symmetric, bilinear) map $g_p : {}^{\mathrm{b}}T_pM \times {}^{\mathrm{b}}T_pM \to \mathbb{R}$ such that if $V \in {}^{\mathrm{b}}T_pM$ and $g_p(V, W) = 0$ for all $W \in {}^{\mathrm{b}}T_pM$, then $V = 0$. Equivalently, any $g \in \mathrm{Sym}^2\, {}^{\mathrm{b}}T^*M$ gives a map ${}^{\mathrm{b}}T_pM \to {}^{\mathrm{b}}T^*_pM$, and then non-degeneracy is equivalent to the injectivity of this map. In view of the equality of dimensions of the domain and target spaces, this in turn is equivalent to its invertibility. We call such a non-degenerate symmetric bilinear form a metric. The signature of such a metric is $(k, n-k)$ if the maximal subspace on which it is positive definite is k; note that the signature is constant. Taking the negative of g reverses the signature, so for example $(n, 0)$ and $(0, n)$ are essentially equivalent signatures, corresponding to Riemannian metrics. Lorentzian metrics have signature $(1, n-1)$ or $(n-1, 1)$; we mostly use the former convention here (so time has a positive sign, space negative signs). One can also takes metrics of other signatures (if the dimension n is at least four); these can arise naturally.

In view of the invertibility, one obtains a dual bilinear form on ${}^{\mathrm{b}}T^*M$, as well as on tensorially related bundles, such as the form bundles or the symmetric bundles. One also has a metric density, which is a non-degenerate section of the density bundle ${}^{\mathrm{b}}\Omega M$. The latter has basis $|\frac{dx}{x}\, dy_1 \ldots d_{y_{n-1}}|$ in local coordinates; the metric density is

$$|dg| = |\det g| \left|\frac{dx}{x}\, dy_1 \ldots d_{y_{n-1}}\right|,$$

where one takes the determinant of the matrix of coefficients of g with respect to the b-frame $x\partial_x, \partial_{y_1}, \ldots, \partial_{y_{n-1}}$. In particular, this gives rise to a positive definite inner product on $\dot{C}^\infty(M)$, the space of C^∞ functions vanishing to infinite order at ∂M:

$$\langle u, v\rangle = \int u\overline{v}\,|dg|,$$

as well as a non-positive definite inner product on, say, differential forms

$$\langle u, v\rangle = \int G_p(u, \overline{v})\,|dg|,$$

where G is the dual metric on differential forms. While this inner product is not positive definite, this is only due to the fiber inner product G_p (on finite dimensional spaces), rather than to $|dg|$, so in particular formal adjoints of differential operators are well-defined differential operators. Moreover, the exterior derivative d maps $\dot{C}^\infty(M)$ to $\dot{C}^\infty(M, {}^{\mathrm{b}}T^*M)$, and more generally $\dot{C}^\infty(M, {}^{\mathrm{b}}\Lambda^k M)$ to $\dot{C}^\infty(M, {}^{\mathrm{b}}\Lambda^{k+1} M)$, and thus given a metric g we can define the Laplacian or d'Alembertian

$$\Box = d^*d + dd^*,$$

with adjoint $*$ taken relative to g.

5.3. The Basics of Microlocal Analysis

In this section we discuss basic properties of pseudodifferential and scattering pseudodifferential operators, introduced in this generality by Melrose [30], formerly discussed by Parenti and Shubin on $\mathbb{R}^n$ [41, 37], where it can be also considered an example of Hörmander's Weyl calculus [27]. These operators generalize differential operators of the form

$$A = \sum_{|\alpha|\leq m} a_\alpha D^\alpha, \text{ with } a_\alpha \in C^\infty(\overline{\mathbb{R}^n}), \tag{5.2}$$

as we show below in (5.31). Indeed, the conditions on the coefficients a_α are relaxed to be "symbolic," so that for instance $a_0(z) = \phi(z)|z|^{-\rho}$, $\phi \equiv 0$ near the origin, $\equiv 1$ near infinity is allowed. Thus, in particular operators such as $\Delta + V$, where V is the Coulomb potential, without its singularity at the origin, fit into this framework. (The singularity at the origin would make the problem into an elliptic b-problem, such as those discussed in Section 5.6, near 0, but we do not discuss this here.)

More generally, we can consider Riemannian metrics g with $g_{ij} \in C^\infty(\overline{\mathbb{R}^n})$ such that for all $z \in \overline{\mathbb{R}^n}$, $\sum_{ij} g_{ij}(z)\zeta_i\zeta_j = 0$ implies $\zeta = 0$, that is g is positive definite on the compact manifold $\overline{\mathbb{R}^n}$. Then, with V as above and with $\sigma \in \mathbb{C}$, $\Delta_g + V - \sigma$ is of the form (5.2) with $m = 2$.

The extension of this class to scattering pseudodifferential operators allows one to construct approximate inverses (parametrices), showing Fredholm properties, for operators that are elliptic *in this class*. Ellipticity here also encodes behavior at spatial infinity, so for instance $\Delta + V - \sigma$, where V may be Coulomb type with $\rho > 0$, is elliptic for $\sigma \in \mathbb{C} \setminus [0, \infty)$, but is not elliptic for $\sigma \in [0, \infty)$. It also allows one to develop tools to study non-elliptic operators. For instance, the limiting absorption principle, that is the existence of the limits

$$R(\sigma \pm i0) = \lim_{\epsilon \to 0+} (\Delta + V - (\sigma \pm i\epsilon))^{-1}$$

for V real valued and $\sigma > 0$ fits very nicely into this framework.

5.3.1. The Outline

Since there are technicalities along the way, we begin with an outline of this section. First, for $m, \ell, \ell' \in \mathbb{R}$, $\delta, \delta' \in [0, 1/2)$, we define two kinds of function spaces,

$$S^{m,\ell}_{\delta,\delta'}(\mathbb{R}^n; \mathbb{R}^n) \subset S^{m,\ell}_{\infty,\delta}(\mathbb{R}^n; \mathbb{R}^n) \subset C^\infty(\mathbb{R}^{2n}),$$

as well as analogues on $\mathbb{R}^{3n}$:

$$S^{m,\ell_1,\ell_2}_{\delta,\delta'}(\mathbb{R}^n; \mathbb{R}^n; \mathbb{R}^n) \subset S^{m,\ell_1,\ell_2}_{\infty,\delta}(\mathbb{R}^n; \mathbb{R}^n; \mathbb{R}^n) \subset C^\infty(\mathbb{R}^{3n}).$$

The elements of these spaces are called *symbols*; the important point is the behavior of these symbols at infinity. Here the spaces become larger with increasing m, ℓ, and ℓ_j, and $\delta = 0 = \delta'$ gives the standard classes also denoted by

$$S^{m,\ell}_{0,0}(\mathbb{R}^n; \mathbb{R}^n) = S^{m,\ell}(\mathbb{R}^n; \mathbb{R}^n),\ S^{m,\ell}_{\infty,0}(\mathbb{R}^n; \mathbb{R}^n) = S^{m,\ell}_{\infty}(\mathbb{R}^n; \mathbb{R}^n),$$

and similarly for the $\mathbb{R}^{3n}$ versions. *The cases $\delta = 0 = \delta'$ are by far the most important ones.* We have projections $\pi_L, \pi_R : \mathbb{R}^{3n} \to \mathbb{R}^{2n}$, with π_L dropping the second factor of $\mathbb{R}^{3n}$ and π_R dropping the first factor:

$$\pi_L(z, z', \zeta) = (z, \zeta),\ \pi_R(z, z', \zeta) = (z', \zeta);$$

the subscripts L and R refer to z, resp. z', being the left, resp. right, "base" or "position" variable. (The variable ζ will be the "dual" or "momentum" variable.) Then π_L^*, π_R^* pull back elements of the $\mathbb{R}^{2n}$ spaces to the corresponding $\mathbb{R}^{3n}$ spaces (with $\ell_1 = \ell$, $\ell_2 = 0$, resp, $\ell_2 = \ell$, $\ell_1 = 0$). With $\mathcal{S}$ denoting

Schwartz functions on $\mathbb{R}^n$, $\mathcal{S}'$ denoting tempered distributions on $\mathbb{R}^n$, and $\mathcal{L}$ denoting continuous linear operators, we define an oscillatory integral map:

$$I : S^{m,\ell_1,\ell_2}_{\infty,\delta}(\mathbb{R}^n;\mathbb{R}^n;\mathbb{R}^n) \to \mathcal{L}(\mathcal{S},\mathcal{S}),$$

and also show by duality that

$$I : S^{m,\ell_1,\ell_2}_{\infty,\delta}(\mathbb{R}^n;\mathbb{R}^n;\mathbb{R}^n) \to \mathcal{L}(\mathcal{S}',\mathcal{S}'),$$

and that the range of I is closed under Fréchet space or L^2-based adjoints. The compositions

$$q_L = I \circ \pi_L^*,\ q_R = I \circ \pi_R^*,$$

are called the left and right quantization maps. Now, it turns out that I is redundant, and its range on $S^{m,\ell_1,\ell_2}_{\delta,\delta'}(\mathbb{R}^n;\mathbb{R}^n;\mathbb{R}^n)$, resp. $S^{m,\ell_1,\ell_2}_{\infty,\delta}(\mathbb{R}^n;\mathbb{R}^n;\mathbb{R}^n)$, is that of q_L on $S^{m,\ell}_{\delta,\delta'}(\mathbb{R}^n;\mathbb{R}^n)$, resp. $S^{m,\ell}_{\infty,\delta}(\mathbb{R}^n;\mathbb{R}^n)$ with $\ell = \ell_1 + \ell_2$; the analogous statement also holds with q_L replaced by q_R. This is called left, resp. right, reduction; see Proposition 5.1. One defines pseudodifferential operators, $\Psi^{m,\ell}_{\delta,\delta'}$, resp. $\Psi^{m,\ell}_{\infty,\delta}$, to be the range of q_L (or equivalently q_R) on the spaces $S^{m,\ell}_{\delta,\delta'}(\mathbb{R}^n;\mathbb{R}^n)$, resp. $S^{m,\ell}_{\infty,\delta}(\mathbb{R}^n;\mathbb{R}^n)$, and writes

$$\Psi^{m,\ell} = \Psi^{m,\ell}_{0,0},\ \Psi^{m,\ell}_{\infty} = \Psi^{m,\ell}_{\infty,0}.$$

Once this reducibility is shown it is straightforward to see (using the general I, which is why it is introduced) that $A \in \Psi^{m,\ell}_{\delta,\delta'}$, $B \in \Psi^{m',\ell'}_{\delta,\delta'}$ implies $AB \in \Psi^{m+m',\ell+\ell'}_{\delta,\delta'}$, that is that $\Psi^{\infty,\infty}_{\delta,\delta'} = \cup_{m,\ell}\Psi^{m,\ell}_{\delta,\delta'}$ is an order-filtered algebra, with the analogous statements holding for $\Psi^{\infty,\infty}_{\infty,\delta}$ as well. One also shows that composition is commutative to the leading order, that is

$$A \in \Psi^{m,\ell}_{\delta,\delta'},\ B \in \Psi^{m',\ell'}_{\delta,\delta'} \Longrightarrow [A,B] = AB - BA \in \Psi^{m+m'-1+2\delta,\ell+\ell'-1+2\delta'}_{\delta,\delta'};$$

the analogous statement here is

$$A \in \Psi^{m,\ell}_{\infty,\delta},\ B \in \Psi^{m',\ell'}_{\infty,\delta} \Longrightarrow [A,B] = AB - BA \in \Psi^{m+m'-1+2\delta,\ell+\ell'}_{\infty,\delta},$$

that is the gain is only in the first order. This is conveniently encoded by the *principal symbol* maps

$$\sigma_{m,\ell} : \Psi^{m,\ell}_{\delta,\delta'} \to S^{m,\ell}_{\delta,\delta'}/S^{m-1+2\delta,\ell-1+2\delta'}_{\delta,\delta'},\ \sigma_{\infty,m,\ell} : \Psi^{m,\ell}_{\infty,\delta} \to S^{m,\ell}_{\infty,\delta}/S^{m-1+2\delta,\ell}_{\infty,\delta},$$

which are multiplicative (homomorphisms of filtered algebras); the leading order commutativity of pseudodifferential operators correspond to the commutativity of function spaces under multiplication. Here δ, δ' are suppressed in the principal symbol notation. An immediate consequence is the elliptic parametrix construction: for operators $A \in \Psi^{m,\ell}_{\delta,\delta'}$ with invertible principal symbol, which are called *elliptic*, one can construct an approximate inverse $B \in \Psi_{\delta,\delta'}^{-m,-\ell}$ such that $AB - \mathrm{Id}, BA - \mathrm{Id} : \mathcal{S}' \to \mathcal{S}$ are continuous, that is completely regularizing. In the case of $A \in \Psi_{\infty,\delta}^{m,\ell}$, we only have $AB - \mathrm{Id}$, $BA - \mathrm{Id} : \mathcal{S}' \to C^\infty(\mathbb{R}^n)$, that is are smoothing, but do not give decay at infinity. Since completely regularizing operators are compact from any weighted Sobolev space to any other weighted Sobolev space, and since we show that (recalling the weighted Sobolev spaces from Section 5.2)

$$A \in \Psi_{\infty,\delta}^{m,\ell} \Longrightarrow A \in \mathcal{L}(H^{r,s}, H^{r-m,s-\ell})$$

for all $r, s \in \mathbb{R}$ (so analogous statements hold for $\Psi_{\delta,\delta'}^{m,\ell} \subset \Psi_{\infty,\delta}^{m,\ell}$), we deduce that *elliptic* $A \in \Psi_{\delta,\delta'}^{m,\ell}$ are *Fredholm* on any weighted Sobolev space, with the nullspace of both A and A^* lying in $\mathcal{S}(\mathbb{R}^n)$, and being independent of the choice of the weighted Sobolev space. In particular, if $A \in \Psi_{\delta,\delta'}^{m,0}$, $m > 0$, elliptic, is symmetric with respect to the L^2 inner product, then one immediately concludes that $A \pm i\,\mathrm{Id}$ is invertible as a map $H^{m,0} \to L^2$, and thus A is self-adjoint with domain $H^{m,0}$.

Another important direction we explore is *microlocalization*, by introducing the notion of the operator wavefront set, $\mathrm{WF}'(A)$, or $\mathrm{WF}'_\infty(A)$, which measures where in phase space A is "trivial." Thus, while $\sigma_{m,\ell}, \sigma_{\infty,m,\ell}$ capture the leading order behavior of operators, that is their behavior modulo one order lower operators, $\mathrm{WF}'(A)$ and $\mathrm{WF}'_{\infty,\ell}(A)$ give the locations where A is not residual, that is in $\Psi^{-\infty,-\infty}$, resp. $\Psi_\infty^{-\infty,\ell}$, so for instance the emptiness of $\mathrm{WF}'(A)$ implies $A \in \Psi^{-\infty,-\infty}$. One should think of these as an analogue of the singular support of distributions, which measures where a distribution is not C^∞, except that its location will not be in the base space $\mathbb{R}^n$, but rather at infinity in *phase space*, $\mathbb{R}^n \times \mathbb{R}^n$. To make this concrete, it is useful to compactify $\mathbb{R}^n \times \mathbb{R}^n$ to $\overline{\mathbb{R}^n} \times \overline{\mathbb{R}^n}$, see Figure 5.3; then for $A \in \Psi^{m,\ell}$, $\mathrm{WF}'(A) \subset \partial(\overline{\mathbb{R}^n} \times \overline{\mathbb{R}^n})$ while for $A \in \Psi_\infty^{m,\ell}$, $\mathrm{WF}'_{\infty,\ell}(A) \subset \overline{\mathbb{R}^n} \times \partial\overline{\mathbb{R}^n}$. Then one can perform a microlocal version of the elliptic parametrix construction, that is one that is localized, in the sense of WF', near points at which the operator A is elliptic; this is a first step towards understanding non-elliptic operators.

It turns out that it is convenient to generalize the class of operators considered here to allow their orders m and ℓ to vary, namely $m = \mathsf{m}$ is a function on $\partial\overline{\mathbb{R}^n} \times \overline{\mathbb{R}^n}$ and $\ell = \mathsf{l}$ a function on $\overline{\mathbb{R}^n} \times \partial\overline{\mathbb{R}^n}$, so at different points microlocally one has an operator of different order. This is the reason we consider $\delta, \delta' > 0$ here; we naturally end up with the classes $S_{\delta,\delta'}^{\mathsf{m},\ell}$ and $S_{\infty,\delta}^{\mathsf{m},\ell}$ where δ, δ' can be taken to be arbitrarily small but positive. (There is also the possibility of taking logarithmic weight losses below, but we do not discuss it here.)

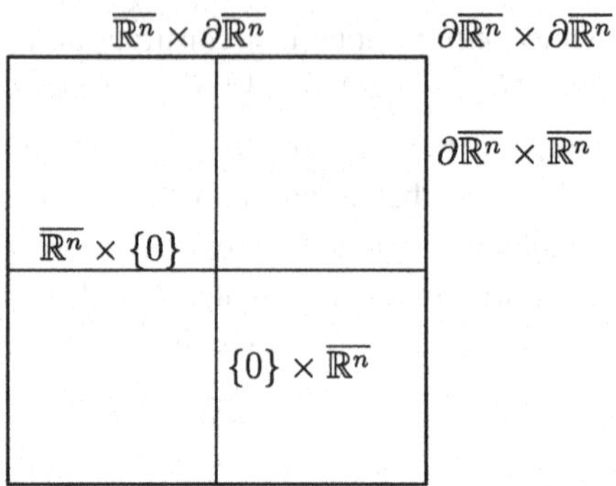

Figure 5.3. The product compactified phase space, $\overline{\mathbb{R}^n} \times \overline{\mathbb{R}^n}$. The whole boundary $\partial(\overline{\mathbb{R}^n} \times \overline{\mathbb{R}^n}) = (\overline{\mathbb{R}^n} \times \partial\overline{\mathbb{R}^n}) \cup (\partial\overline{\mathbb{R}^n} \times \overline{\mathbb{R}^n})$ carries $\mathrm{WF}'(A)$, while only $\overline{\mathbb{R}^n} \times \partial\overline{\mathbb{R}^n}$ carries $\mathrm{WF}'_{\infty,\ell}(A)$

5.3.2. The Definition of Pseudodifferential Operators and Oscillatory Integrals

We now go through the details. Thus, starting with $\mathbb{R}^n$, we consider operators of the form

$$Au(z) = (I(a)u)(z) = (2\pi)^{-n} \int_{\mathbb{R}^n \times \mathbb{R}^n} e^{i\zeta\cdot(z-z')} a(z, z', \zeta)\, u(z')\, dz',\ u \in \mathcal{S}(\mathbb{R}^n), \tag{5.3}$$

where a is a *product-type* symbol of class $S^{m,\ell_1,\ell_2}_{\delta,\delta'}$, $m, \ell_1, \ell_2 \in \mathbb{R}$, $\delta, \delta' \in [0, 1/2)$, that is differentiation in z, resp. z', resp. ζ, provides extra decay in the respective variables:

$$\begin{aligned} &a \in S^{m,\ell_1,\ell_2}_{\delta,\delta'}(\mathbb{R}^n_z; \mathbb{R}^n_{z'}; \mathbb{R}^n_\zeta) \\ &\quad \Longleftrightarrow a \in C^\infty(\mathbb{R}^n_z \times \mathbb{R}^n_{z'} \times \mathbb{R}^n_\zeta), \\ &|D^\alpha_z D^\beta_{z'} D^\gamma_\zeta a| \leq C_{\alpha\beta\gamma} \langle z\rangle^{\ell_1-|\alpha|} \langle z'\rangle^{\ell_2-|\beta|} \langle\zeta\rangle^{m-|\gamma|} (\langle z\rangle + \langle z'\rangle)^{\delta'|(\alpha,\beta,\gamma)|} \langle\zeta\rangle^{\delta|(\alpha,\beta,\gamma)|} \end{aligned}$$

with

$$|(\alpha, \beta, \gamma)| = |\alpha| + |\beta| + |\gamma|$$

and

$$\langle\cdot\rangle = (1 + |\cdot|^2)^{1/2}.$$

One writes

$$\begin{aligned} \|a\|_{S^{m,\ell_1,\ell_2}_{\delta,\delta'},N} = \sum_{|\alpha|+|\beta|+|\gamma|\leq N} &\sup \langle z\rangle^{-\ell_1+|\alpha|} \langle z'\rangle^{-\ell_2+|\beta|} (\langle z\rangle + \langle z'\rangle)^{-\delta'|(\alpha,\beta,\gamma)|} \\ &\times \langle\zeta\rangle^{-m+|\gamma|-\delta|(\alpha,\beta,\gamma)|} |D^\alpha_z D^\beta_{z'} D^\gamma_\zeta a|; \end{aligned}$$

as N runs over $\mathbb{N}$, these give a family of seminorms on S^{m,ℓ_1,ℓ_2}, giving it a Fréchet topology.

Note that the orders on S are reversed compared to the order of the factors, that is z, z', ζ; this is done in part to conform with the usual notation. Moreover, $(\langle z\rangle + \langle z'\rangle)^{\delta'|(\alpha,\beta,\gamma)|}$ can be replaced by $\langle (z,z')\rangle^{\delta'|(\alpha,\beta,\gamma)|}$. Also, z and z' play an equivalent role since, as mentioned before, and as we show below, one can even eliminate, say, the z' dependence. In fact, it turns out that the behavior of a is essentially irrelevant in the region where $\frac{\langle z\rangle}{\langle z'\rangle}$ is *not* bounded between M^{-1} and M, $M > 1$ is any fixed number, in that if one cuts a off to be supported outside such a set, one obtains an element of $\Psi^{-\infty,-\infty}_{\delta,\delta'}$, see (5.25), but since this is due to the oscillatory nature of the integral in ζ, this is not obvious at this point. However, we have already pointed out that fixing some $\chi \in C^\infty_c(\mathbb{R})$, $\chi \equiv 1$ on $[\frac{1}{2}, 2]$, supported in $[\frac{1}{4}, 4]$, for $a \in S^{m,\ell_1,\ell_2}_{\delta,\delta'}$ we have the decomposition as

$$a = a_1 + a_2,\ a_1 = \chi\left(\frac{\langle z\rangle}{\langle z'\rangle}\right) a,\ a_2 = \left(1 - \chi\left(\frac{\langle z\rangle}{\langle z'\rangle}\right)\right) a, \tag{5.4}$$

with a_j depending continuously on a in the $S^{m,\ell_1,\ell_2}_{\delta,\delta'}$ topology; (5.25) shows that the contribution of a_2 is essentially irrelevant in the sense stated above.

In fact, in the beginning it is better to start with a larger (at least if $\delta' = 0$) class of symbols, without extra decay in the z, z' variables upon differentiation: for $\delta \in [0, 1/2)$,

$$a \in S^{m,\ell_1,\ell_2}_{\infty,\delta}(\mathbb{R}^n_z; \mathbb{R}^n_{z'}; \mathbb{R}^n_\zeta) \Longleftrightarrow a \in C^\infty(\mathbb{R}^n_z \times \mathbb{R}^n_{z'} \times \mathbb{R}^n_\zeta),$$
$$|D^\alpha_z D^\beta_{z'} D^\gamma_\zeta a| \le C_{\alpha\beta\gamma} \langle z\rangle^{\ell_1} \langle z'\rangle^{\ell_2} \langle \zeta\rangle^{m-|\gamma|+\delta|(\alpha,\beta,\gamma)|}.$$

One writes

$$\|a\|_{S^{m,\ell_1,\ell_2}_{\infty,\delta},N} = \sum_{|\alpha|+|\beta|+|\gamma|\le N} \sup \langle z\rangle^{-\ell_1} \langle z'\rangle^{-\ell_2} \langle \zeta\rangle^{-m+|\gamma|-\delta|(\alpha,\beta,\gamma)|} |D^\alpha_z D^\beta_{z'} D^\gamma_\zeta a|.$$

For $\ell_1 = \ell_2 = 0$, this is Hörmander's uniform symbol class of type $1 - \delta, \delta$ (i.e. ρ, δ with $\rho = 1 - \delta$). Note that

$$S^{m,\ell_1,\ell_2}_{\delta,0} \subset S^{m,\ell_1,\ell_2}_{\infty,\delta},$$

and the inclusion map

$$\iota : S^{m,\ell_1,\ell_2}_{\delta,0} \hookrightarrow S^{m,\ell_1,\ell_2}_{\infty,\delta}$$

is continuous, with

$$\|a\|_{S^{m,\ell_1,\ell_2}_{\infty,\delta},N} \le \|a\|_{S^{m,\ell_1,\ell_2}_{\delta,0},N}$$

for all N.

Note that $\ell_j \leq \ell'_j$, $m \leq m'$ implies

$$S^{m,\ell_1,\ell_2}_{\delta,\delta'} \subset S^{m',\ell'_1,\ell'_2}_{\delta,\delta'},$$

and similarly with S_∞. Further, if $\delta \leq \tilde{\delta}$, $\delta' \leq \tilde{\delta}'$ then

$$S^{m,\ell_1,\ell_2}_{\delta,\delta'} \subset S^{m,\ell_1,\ell_2}_{\tilde{\delta},\tilde{\delta}'}.$$

One writes

$$S^{-\infty,\ell_1,\ell_2}_{\delta,\delta'} = \cap_{m\in\mathbb{R}} S^{m,\ell_1,\ell_2}_{\delta,\delta'},\ S^{-\infty,\ell_1,-\infty}_{\delta,\delta'} = \cap_{m\in\mathbb{R},\ell_2\in\mathbb{R}} S^{m,\ell_1,\ell_2}_{\delta,\delta'},$$

and similarly again with S_∞. Notice that for all $\delta, \delta' \in [0, 1/2)$,

$$S^{-\infty,-\infty,-\infty}_{\delta,\delta'} = \mathcal{S}(\mathbb{R}^{3n})$$

while $S^{-\infty,0,0}_{\infty,\delta}$ consists of C^∞ functions on $\mathbb{R}^{2n}_{z,z'}$ which are bounded with all derivatives, and take values in $\mathcal{S}(\mathbb{R}^n)$. Thus, these *residual* spaces are independent of δ, δ'. One also writes

$$S^{\infty,\infty,\infty}_{\delta,\delta'} = \cup_{m,\ell_1,\ell_2\in\mathbb{R}} S^{m,\ell_1,\ell_2}_{\delta,\delta'}.$$

Further, note that $S^{\infty,\infty,\infty}_{\delta,\delta'}$ forms a commutative filtered *-algebra in the sense that in addition to $S^{m,\ell_1,\ell_2}_{\delta,\delta'}$ being a vector space for each m, ℓ_1, ℓ_2, closed under complex conjugation, the (function-theoretic, i.e. pointwise) product (which is commutative) satisfies

$$a \in S^{m,\ell_1,\ell_2}_{\delta,\delta'},\ b \in S^{m',\ell'_1,\ell'_2}_{\delta,\delta'} \Rightarrow ab \in S^{m+m',\ell_1+\ell'_1,\ell_2+\ell'_2}_{\delta,\delta'},$$

as follows from Leibniz's rule. Similarly $S^{\infty,\infty,\infty}_{\infty,\delta}$ forms a commutative filtered *-algebra as well. Notice also that for $\delta' = 0$,

$$a \in S^{m,\ell_1,\ell_2}_{\delta,0} \Rightarrow D^\alpha_z D^\beta_{z'} D^\gamma_\zeta a \in S^{m-|\gamma|+\delta|(\alpha,\beta,\gamma)|,\ell_1-|\alpha|,\ell_2-|\beta|}_{\delta,0}, \tag{5.5}$$

while for general δ', the a_1 piece, as defined in (5.4), satisfies

$$a_1 \in S^{m,\ell_1,\ell_2}_{\delta,\delta'} \Rightarrow D^\alpha_z D^\beta_{z'} D^\gamma_\zeta a_1 \in S^{m-|\gamma|+\delta|(\alpha,\beta,\gamma)|,\ell_1-|\alpha|+\delta'|(\alpha,\beta,\gamma)|,\ell_2-|\beta|}_{\delta,\delta'}, \tag{5.6}$$

where by the support property of a_1, $\delta'|(\alpha, \beta, \gamma)|$ could also be shifted to the last order (and recall that a_2 will be shown to be essentially irrelevant). The analogue of (5.5) also holds for $S^{\infty,\infty,\infty}_{\infty,\delta}$, in which case ℓ_1 and ℓ_2 are unaffected by derivatives.

It is also useful to note the following lemma:

Lemma 5.1 *For $m' > m$, the residual spaces $S^{-\infty,\ell_1,\ell_2}_{\infty,\delta} = \cap_{\tilde{m}\in\mathbb{R}} S^{\tilde{m},\ell_1,\ell_2}_{\infty,\delta}$, resp. $S^{-\infty,\ell_1,\ell_2}_{\delta,\delta'} = \cap_{\tilde{m}\in\mathbb{R}} S^{\tilde{m},\ell_1,\ell_2}_{\delta,\delta'}$, are dense in $S^{m,\ell_1,\ell_2}_{\infty,\delta}$, resp. $S^{m,\ell_1,\ell_2}_{\delta,\delta'}$, in the topology of $S^{m',\ell_1,\ell_2}_{\infty,\delta}$, resp. $S^{m',\ell_1,\ell_2}_{\delta,\delta'}$.*

Proof Let $\chi \in C_c^\infty(\mathbb{R}^n)$ be such that $0 \leq \chi \leq 1$, $\chi(\zeta) = 1$ for $|\zeta| \leq 1$, $\chi(\zeta) = 0$ for $|\zeta| \geq 2$, and let $a_j(z, z', \zeta) = \chi(\zeta/j)a(z, z', \zeta)$, where $a \in S^{m,\ell_1,\ell_2}_{\infty,\delta}$. Then

$$D_z^\alpha D_{z}^{\prime\beta} D_\zeta^\gamma (a_j - a) = \sum_{\mu+\nu=\gamma} C_{\mu\nu} j^{-|\mu|}(D_\zeta^\mu(\chi - 1))(\zeta/j)(D_z^\alpha D_{z}^{\prime\beta} D_\zeta^\nu a)(z, z', \zeta),$$

with $C_{\mu\nu}$ combinatorial constants. The $\mu = 0$ term is supported in $|\zeta| \geq j$, the $\mu \neq 0$ terms are supported in $j \leq |\zeta| \leq 2j$. Correspondingly, for $\mu = 0$, the summand is bounded by

$$C_{0\gamma}\langle\zeta\rangle^{m-|\gamma|+\delta|(\alpha,\beta,\gamma)|}\langle z\rangle^{\ell_1}\langle z\rangle^{\ell_2}, \tag{5.7}$$

while for $\mu \neq 0$, $j \sim |\zeta|$ on the support, so the summand is bounded by a constant multiple of

$$\langle\zeta\rangle^{m-|\mu|-|\nu|+\delta|(\alpha,\beta,\nu)|}\langle z\rangle^{\ell_1}\langle z\rangle^{\ell_2}. \tag{5.8}$$

Multiplying by

$$\langle\zeta\rangle^{-m'+|\gamma|-\delta|(\alpha,\beta,\gamma)|}\langle z\rangle^{-\ell_1}\langle z\rangle^{-\ell_2},$$

in either case we obtain a quantity bounded by a constant multiple of $\langle\zeta\rangle^{-(m'-m)}$. Since the difference is supported in $|\zeta| \geq j$, and since $m' > m$, this goes to 0 as $j \to \infty$, proving the claim.

The proof for $a \in S^{m,\ell_1,\ell_2}_{\delta,\delta'}$ is similar, with (5.7) replaced by

$$C_{0\gamma}\langle\zeta\rangle^{m-|\gamma|+\delta|(\alpha,\beta,\gamma)|}\langle z\rangle^{\ell_1}\langle z\rangle^{\ell_2}\langle(z, z')\rangle^{\delta'|(\alpha,\beta,\gamma)|}, \tag{5.9}$$

and (5.8) replaced by

$$\langle\zeta\rangle^{m-|\mu|-|\nu|+\delta|(\alpha,\beta,\nu)|}\langle z\rangle^{\ell_1}\langle z\rangle^{\ell_2}\langle(z, z')\rangle^{\delta'|(\alpha,\beta,\nu)|}, \tag{5.10}$$

so multiplication by

$$\langle\zeta\rangle^{-m'+|\gamma|-\delta|(\alpha,\beta,\gamma)|}\langle z\rangle^{-\ell_1}\langle z\rangle^{-\ell_2}\langle(z, z')\rangle^{-\delta'|(\alpha,\beta,\gamma)|},$$

gives the desired result. □

As examples, recall that if a is a polynomial of order ℓ_1, ℓ_2 and m in the three variables, then certainly $a \in S^{m,\ell_1,\ell_2} = S^{m,\ell_1,\ell_2}_{0,0}$. More interestingly, if $a \in C^\infty(\overline{\mathbb{R}^n} \times \overline{\mathbb{R}^n} \times \overline{\mathbb{R}^n}) = C^\infty(\overline{\mathbb{R}^n}^3)$ then $a \in S^{0,0,0} = S^{0,0,0}_{0,0}$, so

$$a \in \langle z\rangle^{\ell_1}\langle z\rangle^{\ell_2}\langle\zeta\rangle^m C^\infty((\overline{\mathbb{R}^n})^3) \Rightarrow a \in S^{m,\ell_1,\ell_2} = S^{m,\ell_1,\ell_2}_{0,0}.$$

Such a are called *classical symbols*; one writes

$$S^{m,\ell_1,\ell_2}_{\mathrm{cl}} = \langle z\rangle^{\ell_1}\langle z\rangle^{\ell_2}\langle\zeta\rangle^m C^\infty((\overline{\mathbb{R}^n})^3).$$

Thus, $S^{m,\ell_1,\ell_2}_{\delta,\delta'}$ is a $C^\infty(\overline{\mathbb{R}^n}^3)$-module. A particular example is $a = |z|^{-\rho}\phi(z)$, where $\phi \equiv 0$ near 0, $\phi \equiv 1$ near ∞, then $a \in S^{-\rho,0,0}$, such an a can be thought of as a potential which may decay only slowly at infinity; $\rho = 1$ would give the Coulomb potential without its singularity at the origin.

On the flipside, we can rewrite the estimates for S^{m,ℓ_1,ℓ_2}:

$$|\alpha'| \le |\alpha|,\ |\beta'| \le |\beta|,\ |\gamma'| \le |\gamma| \Rightarrow |z^{\alpha'} D_z^\alpha (z')^{\beta'} D_{z'}^\beta \zeta^{\gamma'} D_\zeta^\gamma a| \\ \le C_{\alpha\beta\gamma} \langle z\rangle^{\ell_1} \langle z'\rangle^{\ell_2} \langle \zeta\rangle^m.$$

Since $z_i\partial_{z_j}$ and ∂_{z_j} generate all C^∞ vector fields over $C^\infty(\overline{\mathbb{R}^n})$ which are tangent to $\partial\overline{\mathbb{R}^n}$, whose set is denoted by $\mathcal{V}_b(\overline{\mathbb{R}^n})$, we can rewrite this equivalently as follows: let $V_{j,k} \in \mathcal{V}_b(\overline{\mathbb{R}^n})$, $j = 1, 2, 3$, $N_j \in \mathbb{N}$ (possibly 0) and $1 \le k \le N_j$ acting in the jth factor, then

$$\langle z\rangle^{-\ell_1} \langle z'\rangle^{-\ell_2} \langle \zeta\rangle^{-m} \prod_{j=1}^{3} \prod_{k=1}^{N_j} V_{j,k} a \in L^\infty.$$

This could be further rephrased, in terms of vector fields on $\overline{\mathbb{R}^n}^3$, tangent to all boundary faces: if V_j are such, $1 \le j \le N$ (possibly $N = 0$), then

$$\langle z\rangle^{-\ell_1} \langle z'\rangle^{-\ell_2} \langle \zeta\rangle^{-m} V_1 \dots V_N a \in L^\infty.$$

Since one can use any vector fields tangent to the various boundary faces, in any product decomposition $[0,1)_{r^{-1}} \times \mathbb{S}^{n-1}$ near the boundary of each factor $\overline{\mathbb{R}^n}$, one automatically has smoothness in the various angular variables; in the radial variables one has iterated regularity with respect to $r\partial_r$. We contrast this *conormal or symbolic* regularity with the *classical* regularity $a \in S^{m,\ell_1,\ell_2}_{\mathrm{cl}}$, which means

$$V_1 \dots V_N \langle z\rangle^{-\ell_1} \langle z'\rangle^{-\ell_2} \langle \zeta\rangle^{-m} a \in L^\infty$$

for *all* vector fields on $\overline{\mathbb{R}^n}^3$, without the tangency requirement. In particular, in terms of a product decomposition $[0,1)_{r^{-1}} \times \mathbb{S}^{n-1}$ near the boundary of each factor $\overline{\mathbb{R}^n}$, one has smoothness in the various angular variables *and* in the radial variables, that is one has iterated regularity with respect to ∂_r.

We are also interested in the generalization of this setting in which the orders m, ℓ_1, ℓ_2 are allowed to vary. Concretely, to set this up, suppose that $\mathsf{m}, \mathsf{l}_j \in S^{0,0,0}$ are real valued symbols. We write

$$\begin{aligned} &a \in S^{\mathsf{m},\mathsf{l}_1,\mathsf{l}_2}_{\delta,\delta'}(\mathbb{R}^n_z; \mathbb{R}^n_{z'}; \mathbb{R}^n_\zeta) \\ &\Longleftrightarrow a \in C^\infty(\mathbb{R}^n_z \times \mathbb{R}^n_{z'} \times \mathbb{R}^n_\zeta), \\ &|D_z^\alpha D_{z'}^\beta D_\zeta^\gamma a| \le C_{\alpha\beta\gamma} \langle z\rangle^{\mathsf{l}_1 - |\alpha|} \langle z'\rangle^{\mathsf{l}_2 - |\beta|} \langle \zeta\rangle^{\mathsf{m} - |\gamma|} (\langle z\rangle + \langle z'\rangle)^{\delta'|(\alpha,\beta,\gamma)|} \langle \zeta\rangle^{\delta|(\alpha,\beta,\gamma)|}. \end{aligned}$$

Notice that replacing m by m' where $\mathsf{m}-\mathsf{m}' \in S^{-\epsilon,0,0}$ for some $\epsilon > 0$ does not change the class since $\langle\zeta\rangle^{\mathsf{m}-\mathsf{m}'} = e^{(\mathsf{m}-\mathsf{m}')\log\langle\zeta\rangle}$, and $(\mathsf{m}-\mathsf{m}')\log\langle\zeta\rangle$ is a bounded function in this case. Since we are interested only in $\mathsf{m}, \mathsf{l}_j \in \mathcal{C}^\infty(\overline{\mathbb{R}^n} \times \overline{\mathbb{R}^n} \times \overline{\mathbb{R}^n})$, we regard m as a function on $\overline{\mathbb{R}^n} \times \overline{\mathbb{R}^n} \times \partial\overline{\mathbb{R}^n}$, and take an arbitrary (smooth) extension to $\overline{\mathbb{R}^n} \times \overline{\mathbb{R}^n} \times \overline{\mathbb{R}^n}$; we proceed similarly with the ℓ_j. Thus, with

$$m = \sup \mathsf{m}, \ \ell_j = \sup \mathsf{l}_j,$$

where the sup may be taken over the appropriate boundary of the compactification only, we have

$$a \in S^{\mathsf{m},\mathsf{l}_1,\mathsf{l}_2}_{\delta,\delta'} \Rightarrow a \in S^{m,\ell_1,\ell_2}_{\delta,\delta'}.$$

One can also define

$$a \in S^{\mathsf{m},\mathsf{l}_1,\mathsf{l}_2}_{\infty,\delta}(\mathbb{R}^n_z;\mathbb{R}^n_{z'};\mathbb{R}^n_\zeta) \Longleftrightarrow a \in \mathcal{C}^\infty(\mathbb{R}^n_z \times \mathbb{R}^n_{z'} \times \mathbb{R}^n_\zeta),$$
$$|D^\alpha_z D^\beta_{z'} D^\gamma_\zeta a| \le C_{\alpha\beta\gamma} \langle z\rangle^{\mathsf{l}_1} \langle z'\rangle^{\mathsf{l}_2} \langle\zeta\rangle^{\mathsf{m}-|\gamma|+\delta|(\alpha,\beta,\gamma)|},$$

so with m, ℓ_j as above

$$a \in S^{\mathsf{m},\mathsf{l}_1,\mathsf{l}_2}_{\infty,\delta} \Rightarrow a \in S^{m,\ell_1,\ell_2}_{\infty,\delta}.$$

However, these variable order spaces provide more precise information than simply taking $m = \sup \mathsf{m}$, etc., much like the S^{m,ℓ_1,ℓ_2} spaces provide more precise information that $S^{m,\ell_1,\ell_2}_\infty$. Further, we note that we have introduced the subscript δ and δ' (limiting the gains under differentiation) since the function $b = \langle\zeta\rangle^{\mathsf{m}} = e^{\mathsf{m}\log\langle\zeta\rangle}$ is in $S^{\mathsf{m},0,0}_{\delta,0}$ for all $\delta > 0$, but not for $\delta = 0$. Indeed, differentiating in, say, z_j, gives

$$D_{z_j} b = (D_{z_j}\mathsf{m})(\log\langle\zeta\rangle)\langle\zeta\rangle^{\mathsf{m}},$$

so there is a logarithmic loss (unless m is constant). On the other hand, we formally state the regularity result as a lemma:

Lemma 5.2 *Let $b(z,z',\zeta) = \langle\zeta\rangle^{\mathsf{m}(z,z',\zeta)}$. Then $b \in S^{\mathsf{m},0,0}_{\delta,0}$ for all $\delta > 0$.*

Proof Observe that $f = \mathsf{m}\log\langle\zeta\rangle \in S^{\epsilon,0,0}$ for all $\epsilon > 0$ since this holds for $\log\langle\zeta\rangle$, and as $\mathsf{m} \in S^{0,0,0}$. Further, if $f \in S^{\epsilon_0,\epsilon_1,\epsilon_2}$ with $0 \le \epsilon_0, \epsilon_1, \epsilon_2 < 1$ then

$$e^{-f} D^\alpha_z D^\beta_{z'} D^\gamma_\zeta e^f \in S^{-|\gamma|+\epsilon_0|(\alpha,\beta,\gamma)|,-|\alpha|+\epsilon_1|(\alpha,\beta,\gamma)|,-|\beta|+\epsilon_2|(\alpha,\beta,\gamma)|},$$

as follows by induction on $|\alpha|+|\beta|+|\gamma|$. Indeed, it holds when α, β, γ all vanish. Further,

$$e^{-f} D_{z_j}(D^\alpha_z D^\beta_{z'} D^\gamma_\zeta e^f) = D_{z_j}(e^{-f} D^\alpha_z D^\beta_{z'} D^\gamma_\zeta e^f) + (D_{z_j} f)(e^{-f} D^\alpha_z D^\beta_{z'} D^\gamma_\zeta e^f),$$

and $e^{-f}D_z^\alpha D_{z'}^\beta D_\zeta^\gamma e^f \in S^{-|\gamma|+\epsilon_0|(\alpha,\beta,\gamma)|,-|\alpha|+\epsilon_1|(\alpha,\beta,\gamma)|,-|\beta|+\epsilon_2|(\alpha,\beta,\gamma)|}$ by the inductive hypothesis, and then the first term on the right hand side improves the second order by 1 keeping all others unchanged, while $D_{z_j}f \in S^{\epsilon_0,\epsilon_1-1,\epsilon_2}$, so the second term on the right hand side adds $\epsilon_0, \epsilon_1 - 1, \epsilon_2$ to the orders, while $|\alpha|$ is increased by 1 in both cases. The argument is symmetric for all other derivatives, giving the conclusion. Applying this with $\epsilon_1 = \epsilon_2 = 0$, $\epsilon_0 = \epsilon$, $\epsilon > 0$ arbitrary, we deduce that for all $\delta > 0$ (namely, we take $\epsilon = \delta$), $\langle\zeta\rangle^{\mathsf{m}} \in S_{\delta,0}^{\mathsf{m},0,0}$ indeed. □

We still have, analogously to the constant order setting, that

$$a \in S_{\delta,\delta'}^{\mathsf{m},\mathsf{l}_1,\mathsf{l}_2},\ b \in S_{\delta,\delta'}^{\mathsf{m}',\mathsf{l}_1',\mathsf{l}_2'} \Rightarrow ab \in S_{\delta,\delta'}^{\mathsf{m}+\mathsf{m}',\mathsf{l}_1+\mathsf{l}_1',\mathsf{l}_2+\mathsf{l}_2'},$$

and for $\delta' = 0$

$$a \in S_{\delta,0}^{\mathsf{m},\mathsf{l}_1,\mathsf{l}_2} \Rightarrow D_z^\alpha D_{z'}^\beta D_\zeta^\gamma a \in S_{\delta,0}^{\mathsf{m}-|\gamma|+\delta|(\alpha,\beta,\gamma)|,\mathsf{l}_1-|\alpha|,\mathsf{l}_2-|\beta|}, \tag{5.11}$$

while for general δ', the a_1 piece, as defined in (5.4), satisfies

$$a_1 \in S_{\delta,\delta'}^{\mathsf{m},\mathsf{l}_1,\mathsf{l}_2} \Rightarrow D_z^\alpha D_{z'}^\beta D_\zeta^\gamma a_1 \in S_{\delta,\delta'}^{\mathsf{m}-|\gamma|+\delta|(\alpha,\beta,\gamma)|,\mathsf{l}_1-|\alpha|+\delta'|(\alpha,\beta,\gamma)|,\mathsf{l}_2-|\beta|}, \tag{5.12}$$

where by the support property of a_1, $\delta'|(\alpha,\beta,\gamma)|$ could also be shifted to the last order). The analogue of (5.11) also holds for $S_{\infty,\delta}^{\infty,\infty,\infty}$, in which case l_1 and l_2 are unaffected by derivatives.

Having discussed symbols in some detail, we now turn to operators, starting with the constant order $S_{\infty,\delta}$-type setting. Note that unless $m < -n$, the integral (5.3) with $a \in S_{\infty,\delta}^{m,\ell_1,\ell_2}$ is not absolutely convergent; if $m < -n$, it is, with the result $Au \in C(\mathbb{R}^n)$, and for $M > \ell_2 + n$,

$$\sup |\langle z\rangle^{-\ell_1} Au(z)| \leq C\|a\|_{S_{\infty,\delta}^{m,\ell_1,\ell_2},0}\|u\|_{\mathcal{S},0,M},$$

where C is a universal constant (independent of a and u) and

$$\|u\|_{\mathcal{S},k,M} = \sum_{|\alpha|\leq k}\sum_{|\beta|\leq M} \sup |z^\beta D_z^\alpha u|$$

are the Schwartz seminorms. However, if $m < -n$, one can also integrate by parts as usual in z', noting that $(1+\Delta_{z'})e^{i\zeta\cdot(z-z')} = \langle\zeta\rangle^2 e^{i\zeta\cdot(z-z')}$, so

$$\begin{aligned} Au(z) &= (2\pi)^{-n}\int_{\mathbb{R}^n\times\mathbb{R}^n} \langle\zeta\rangle^{-2N}(1+\Delta_{z'})^N e^{i\zeta\cdot(z-z')} a(z,z',\zeta)\,u(z')\,d\zeta\,dz' \\ &= (2\pi)^{-n}\int_{\mathbb{R}^n\times\mathbb{R}^n} e^{i\zeta\cdot(z-z')}\langle\zeta\rangle^{-2N}(1+\Delta_{z'})^N (a(z,z',\zeta)\,u(z'))\,d\zeta\,dz'. \end{aligned} \tag{5.13}$$

Expanding $(1+\Delta_{z'})^N(a(z,z',\zeta)\,u(z'))$, one deduces that

$$|(1+\Delta_{z'})^N(a(z,z',\zeta)\,u(z'))| \leq \langle z\rangle^{\ell_1}\langle z'\rangle^{\ell_2-M}\langle\zeta\rangle^{m+2N\delta}\|a\|_{S^{m,\ell_1,\ell_2}_{\infty,\delta},2N}\|u\|_{\mathcal{S},2N,M}, \tag{5.14}$$

so for just $m+2N\delta < -n+2N$, that is

$$2(1-\delta)N > m+n,$$

the right hand side of (5.13) is integrable, and defining $Au \in C(\mathbb{R}^n)$ to be the result,

$$\sup|\langle z\rangle^{-\ell_1}Au(z)| \leq C\|a\|_{S^{m,\ell_1,\ell_2}_{\infty,\delta},2N}\|u\|_{\mathcal{S},2N,M}. \tag{5.15}$$

This gives an extension of $A = I(a)$ to $S^{m,\ell_1,\ell_2}_{\infty,\delta}$. Since $S^{-\infty,\ell_1,\ell_2}_{\infty,\delta}$ is dense in $S^{m,\ell_1,\ell_2}_{\infty,\delta}$ in the topology of $S^{m,\ell_1,\ell_2'}_{\infty,\delta}$ for $m' > m$, and since for $m < -n$, the expressions (5.13) for various N are all equal, the continuity property (5.15) shows that A is independent of the choice of N provided $m < -n+2(1-\delta)N$ (since one can then take $m' \in (m, -n+2(1-\delta)N)$, and use the m'-continuity and density statements).

Now at least $Au \in C(\mathbb{R}^n)$, with a suitable bound, is defined, but in fact it is in $\mathcal{S}(\mathbb{R}^n)$. To see this, first note that $D_z^\alpha e^{i\zeta\cdot(z-z')} = \zeta^\alpha$, so for N sufficiently large, so that $m+|\alpha| < -n+2(1-\delta)N$, differentiating under the integral sign and using Leibniz's rule,

$$\begin{aligned}(D_z^\alpha Au)(z) &= \sum_{\gamma+\lambda\leq\alpha} C_{\gamma\lambda}(2\pi)^{-n}\int_{\mathbb{R}^n\times\mathbb{R}^n} D_z^\gamma(e^{i\zeta\cdot(z-z')})\langle\zeta\rangle^{-2N} \\ &\qquad (1+\Delta_{z'})^N(D_z^\lambda a(z,z',\zeta)\,u(z'))\,d\zeta\,dz' \\ &= \sum_{\gamma+\lambda\leq\alpha} C_{\gamma\lambda}(2\pi)^{-n}\int_{\mathbb{R}^n\times\mathbb{R}^n} e^{i\zeta\cdot(z-z')}\zeta^\gamma\langle\zeta\rangle^{-2N} \\ &\qquad (1+\Delta_{z'})^N(D_z^\lambda a(z,z',\zeta)\,u(z'))\,d\zeta\,dz',\end{aligned} \tag{5.16}$$

with $C_{\gamma\lambda}$ combinatorial constants, so by (5.14) with a replaced by $D_z^\lambda a$, with $M > n+\ell_2$ still,

$$\sup|\langle z\rangle^{-\ell_1}(D_z^\alpha Au)(z)| \leq C\|a\|_{S^{m,\ell_1,\ell_2}_{\infty,\delta},2N+|\alpha|}\|u\|_{\mathcal{S},2N,M}.$$

Further, $z_j e^{i\zeta\cdot(z-z')} = z_j' e^{i\zeta\cdot(z-z')} + D_{\zeta_j}e^{i\zeta\cdot(z-z')}$, so

$$z^\beta e^{i\zeta\cdot(z-z')} = (z'+D_\zeta)^\beta e^{i\zeta\cdot(z-z')} = \sum_{\mu+\nu\leq\beta} C_{\mu\nu}(z')^\mu D_\zeta^\nu e^{i\zeta\cdot(z-z')},$$

so integration by parts in ζ gives

$$(z^\beta D_z^\alpha Au)(z) = \sum_{\gamma+\lambda\leq\alpha}\sum_{\mu+\nu\leq\beta} C_{\gamma\lambda}C_{\mu\nu}(2\pi)^{-n}\int_{\mathbb{R}^n\times\mathbb{R}^n} e^{i\zeta\cdot(z-z')}$$
$$D_\zeta^\nu\big(\zeta^\gamma\langle\zeta\rangle^{-2N}(z')^\mu(1+\Delta_{z'})^N(D_z^\lambda a(z,z',\zeta)\,u(z'))\big)\,d\zeta\,dz'$$
$$= \sum_{\gamma+\lambda\leq\alpha}\sum_{\mu+\nu\leq\beta}\sum_{\nu'+\nu''\leq\nu} C_{\gamma\lambda}C_{\mu\nu}C_{\nu'\nu''}(2\pi)^{-n}\int_{\mathbb{R}^n\times\mathbb{R}^n} e^{i\zeta\cdot(z-z')}$$
$$D_\zeta^{\nu'}(\zeta^\gamma\langle\zeta\rangle^{-2N})(z')^\mu(1+\Delta_{z'})^N(D_\zeta^{\nu''}D_z^\lambda a(z,z',\zeta)\,u(z')))\,d\zeta\,dz'. \tag{5.17}$$

Thus with

$$M > n+\ell_2+|\beta| \text{ and } m+|\gamma|-|\nu'|-2N+(2N+|\nu''|+|\lambda|)\delta < -n,$$

the latter of which is implied by

$$m+|\alpha|+|\beta|\delta < -n+2(1-\delta)N,$$

we have

$$\sup|\langle z\rangle^{-\ell_1}z^\beta D_z^\alpha Au(z)| \leq C\|a\|_{S^{m,\ell_1,\ell_2}_{\infty,\delta},2N+|\alpha|}\|u\|_{\mathcal{S},2N,M},$$

with C independent of a, u. Now for $\ell_1 \leq 0$, $\langle z\rangle^{-\ell_1}$ can simply be dropped, while for $\ell_1 > 0$ the $\langle z\rangle^{-\ell_1}$ factor can be absorbed into a sum $z^{\beta'}$ terms with $|\beta'| \leq M'$ where $M' \geq \ell_1$, so we obtain that for

$$M' \geq \max(0,\ell_1),\ M > n+\ell_2+|\beta|+M',\ m+|\alpha|+|\beta|\delta < -n+2(1-\delta)N$$

we have

$$\sup|z^\beta D_z^\alpha Au(z)| \leq C\|a\|_{S^{m,\ell_1,\ell_2}_{\infty,\delta},2N+|\alpha|}\|u\|_{\mathcal{S},2N,M},$$

so $Au \in \mathcal{S}(\mathbb{R}^n)$, and the map $A : \mathcal{S} \to \mathcal{S}$ is continuous, and in fact the stronger continuity property, namely that

$$S^{m,\ell_1,\ell_2}_{\infty,\delta} \times \mathcal{S} \ni (a,u) \mapsto I(a)u \in \mathcal{S}$$

is continuous, holds. Thus, we have the first claim of the following lemma, as well as the second in case $\delta' = 0$:

Lemma 5.3 *The maps*

$$S^{m,\ell_1,\ell_2}_{\infty,\delta} \times \mathcal{S} \ni (a,u) \mapsto I(a)u \in \mathcal{S},$$
$$S^{m,\ell_1,\ell_2}_{\delta,\delta'} \times \mathcal{S} \ni (a,u) \mapsto I(a)u \in \mathcal{S},$$

are continuous.

Proof To deal with general (not necessarily vanishing) $\delta' \in [0, 1/2)$, proceed by using $\chi \in C_c^\infty(\mathbb{R})$, $\chi \equiv 1$ on $[\frac{1}{2}, 2]$, supported in $[\frac{1}{4}, 4]$. Then we can write $a \in S^{m,\ell_1,\ell_2}_{\delta,\delta'}$ as

$$a = a_1 + a_2,\ a_1 = \chi\left(\frac{\langle z\rangle}{\langle z'\rangle}\right) a,\ a_2 = \left(1 - \chi\left(\frac{\langle z\rangle}{\langle z'\rangle}\right)\right) a,$$

with a_j depending continuously on a in the $S^{m,\ell_1,\ell_2}_{\delta,\delta'}$ topology. Now, since

$$\langle z\rangle \sim \langle z'\rangle \sim \langle (z, z')\rangle$$

on $\operatorname{supp} a_1$, and since differentiation is local, a_1 satisfies estimates

$$|D_z^\alpha D_{z'}^\beta D_\zeta^\gamma a_1| \leq C_{\alpha\beta\gamma} \langle z\rangle^{\ell_1+\ell_2-|\alpha|-|\beta|+\delta'|(\alpha,\beta,\gamma)|} \langle \zeta\rangle^{m-|\gamma|+\delta|(\alpha,\beta,\gamma)|}.$$

Denoting the corresponding seminorms by $\|.\|_{\tilde S^{m,\ell_1+\ell_2}_{\delta,\delta'},N}$ temporarily, note that a_1 in $\tilde S^{m,\ell_1+\ell_2}_{\delta,\delta'}$ depends continuously on a. The right hand side of (5.14) becomes

$$\langle z\rangle^{\ell_1+\ell_2+2N\delta'} \langle z'\rangle^{-M} \langle \zeta\rangle^{m+2N\delta} \|a_1\|_{\tilde S^{m,\ell_1,\ell_2}_{\delta,\delta'},2N} \|u\|_{\mathcal{S},2N,M},$$

so for $M > n$ and $m + 2N\delta < -n + 2N$ the right hand side of (5.13) is integrable, and (5.15) becomes

$$\sup |\langle z\rangle^{-\ell_1-\ell_2-2N\delta'} A_1 u(z)| \leq C \|a_1\|_{\tilde S^{m,\ell_1,\ell_2}_{\delta,\delta'},2N} \|u\|_{\mathcal{S},2N,M}. \tag{5.18}$$

In fact, using $\langle z\rangle \sim \langle z'\rangle$ on $\operatorname{supp} a_1$, taking $M > n + \ell_1 + \ell_2 + 2N\delta + |\beta|$, $m + 2N\delta < -n + 2N$ (i.e. first choose N sufficiently large, then M sufficiently large), this even gives

$$\sup |z^\beta A_1 u(z)| \leq C \|a\|_{\tilde S^{m,\ell_1,\ell_2}_{\delta,\delta'},2N} \|u\|_{\mathcal{S},2N,M}.$$

To deal with derivatives, use (5.16) and note that the integrand is bounded by a constant multiple of

$$\sup_{|\gamma|+|\lambda|=|\alpha|} \Big(\langle z\rangle^{\ell_1+\ell_2+2N\delta'+|\lambda|\delta'} \langle z'\rangle^{-M} \langle \zeta\rangle^{m+2N\delta-2N+|\gamma|+|\lambda|\delta} \|a_1\|_{\tilde S^{m,\ell_1,\ell_2}_{\delta,\delta'},2N+|\alpha|} \|u\|_{\mathcal{S},2N,M} \Big),$$

which in turn is bounded by

$$\sup_{|\gamma|+|\lambda|=|\alpha|} \langle z\rangle^{\ell_1+\ell_2+2N\delta'+|\alpha|\delta'} \langle z'\rangle^{-M} \langle \zeta\rangle^{m+2N\delta-2N+|\alpha|} \|a_1\|_{\tilde S^{m,\ell_1,\ell_2}_{\delta,\delta'},2N+|\alpha|} \|u\|_{\mathcal{S},2N,M},$$

so in view of the support of a_1 first choosing N such that $m+2N\delta-2N+|\alpha|<-n$ and then M such that $M>n+\ell_1+\ell_2+2N\delta'+|\alpha|\delta'+|\beta|$, the estimate

$$\sup|z^\beta D_z^\alpha A_1 u(z)|\leq C\|a_1\|_{S^{m,\ell_1,\ell_2}_{\delta,\delta'},2N}\|u\|_{\mathcal{S},2N,M}$$

follows, with C independent of a_1, u. This shows that a_1 satisfies the conclusion of the lemma.

Now, to deal with a_2, integrate by parts in ζ, starting with (5.13) for $A_2=I(a_2)$ in place of $A=I(a)$, using

$$e^{i(z-z')\cdot\zeta}=\langle z-z'\rangle^{-2}(1+\Delta_\zeta)e^{i(z-z')\cdot\zeta},$$

so first for $m<-n$

$$\begin{aligned}A_2u(z)&=(2\pi)^{-n}\int_{\mathbb{R}^n\times\mathbb{R}^n}e^{i\zeta\cdot(z-z')}\langle z-z'\rangle^{-2K}\\&\qquad(1+\Delta_\zeta)^K\Big(\langle\zeta\rangle^{-2N}(1+\Delta_{z'})^N(a_2(z,z',\zeta)\,u(z'))\Big)\,d\zeta\,dz'\\&=\sum_{|\mu|+|\nu|\leq 2K}\tilde{C}_{\mu\nu}(2\pi)^{-n}\int_{\mathbb{R}^n\times\mathbb{R}^n}e^{i\zeta\cdot(z-z')}\langle z-z'\rangle^{-2K}(D_\zeta^\mu\langle\zeta\rangle^{-2N})\\&\qquad(1+\Delta_{z'})^N(D_\zeta^\nu a_2(z,z',\zeta)\,u(z'))\,d\zeta\,dz',\end{aligned}\tag{5.19}$$

where $\tilde{C}_{\mu\nu}$ are combinatorial constants. On the support of a_2,

$$\langle z-z'\rangle\geq C'(\langle z\rangle+\langle z'\rangle)$$

for some $C'>0$, and now the integrand on the right hand side is bounded by a constant multiple of

$$\langle z\rangle^{\ell_1}\langle z'\rangle^{\ell_2-M}\langle(z,z')\rangle^{-2K+(2N+2K)\delta'}\langle\zeta\rangle^{-2N+m+(2N+2K)\delta}\\\|a_2\|_{S^{m,\ell_1,\ell_2}_{\delta,\delta'},2N+2K}\|u\|_{\mathcal{S},2N,M}.$$

For a given β, we can now even take $M=0$, and take N, K so that

$$2N\delta'-(1-\delta')2K<-n-|\beta|-\ell_1-\ell_2$$

and

$$-(1-\delta)2N+2K\delta+m<-n;$$

to see that such a choice exists, take $K=N$, in which case sufficiently large N works as $1-2\delta, 1-2\delta'>0$. We then deduce

$$\sup|z^\beta A_2u(z)|\leq C\|a\|_{S^{m,\ell_1,\ell_2}_{\delta,\delta'},2N+2K}\|u\|_{\mathcal{S},2N,M}.$$

To deal with derivatives, we again use a calculation similar to (5.16) to obtain that

$$\begin{aligned} D_z^\alpha A_2 u(z) = \sum_{\gamma+\kappa+\lambda\le\alpha} C_{\gamma\kappa\lambda} \sum_{|\mu|+|\nu|\le 2K} \tilde{C}_{\mu\nu}(2\pi)^{-n} \\ \int_{\mathbb{R}^n\times\mathbb{R}^n} \zeta^\gamma e^{i\zeta\cdot(z-z')}(D_z^\kappa\langle z-z'\rangle^{-2K})(D_\zeta^\mu\langle\zeta\rangle^{-2N}) \\ (1+\Delta_{z'})^N(D_\zeta^\nu D_z^\lambda a_2(z,z',\zeta)\,u(z'))\,d\zeta\,dz'. \end{aligned} \tag{5.20}$$

Since

$$D_z^\kappa\langle z-z'\rangle^{-2K} \le C\langle z-z'\rangle^{-2K}$$

(indeed, one even has a bound $C\langle z-z'\rangle^{-2K-|\kappa|}$), so now the integrand on the right hand side is bounded by a constant multiple of

$$\begin{aligned} \langle z\rangle^{\ell_1}\langle z'\rangle^{\ell_2-M}\langle(z,z')\rangle^{-2K+(2N+2K+|\alpha|)\delta'}\langle\zeta\rangle^{-2N+m+(2N+2K+|\alpha|)\delta} \\ \|a_2\|_{S^{m,\ell_1,\ell_2}_{\delta,\delta'},2N+2K+|\alpha|}\|u\|_{\mathcal{S},2N,M}, \end{aligned}$$

which gives

$$\sup|z^\beta D_z^\alpha A_2 u(z)| \le C\|a_2\|_{S^{m,\ell_1,\ell_2}_{\delta,\delta'},2N+2K+|\alpha|}\|u\|_{\mathcal{S},2N,M}$$

when $M=0$, and take N,K so that

$$2N\delta' - (1-\delta')2K + \delta'|\alpha| < -n-|\beta|-\ell_1-\ell_2$$

and

$$-(1-\delta)2N + 2K\delta + m + |\alpha|\delta < -n,$$

which can be arranged exactly as in the $\alpha = 0$ case above. This completes the proof of the lemma. □

Note that for such an A with $m < -n$ to start, $u\in\mathcal{S}$, $\phi\in\mathcal{S}$,

$$\begin{aligned} \int Au(z)\phi(z)\,dz &= \int u(z')\left(\int e^{i(-\zeta)\cdot(z'-z)}a(z,z',\zeta)\phi(z)\,dz\,d\zeta\right)dz' \\ &= \int u(z')\left(\int e^{i\zeta\cdot(z'-z)}a(z,z',-\zeta)\phi(z)\,dz\,d\zeta\right)dz' \\ &= \int u(z')(I(b)\phi)(z')\,dz', \end{aligned}$$

where $b(z,z',\zeta) = a(z',z,-\zeta)$, so $b\in S^{m,\ell_2,\ell_1}_{\infty,\delta}$. Let j to be the transposition map $j(z,z',\zeta) = (z',z,\zeta)$, ρ the reflection map $\rho(z,z',\zeta) = (z,z',-\zeta)$, so

$\rho^* : S^{m,\ell_1,\ell_2}_{\infty,\delta} \to S^{m,\ell_1\ell_2}_{\infty,\delta}, j^* : S^{m,\ell_1,\ell_2}_{\infty,\delta} \to S^{m,\ell_2,\ell_1}_{\infty,\delta}$ are continuous for all m, ℓ_1, ℓ_2. We then have at first for $m < -n$,

$$\int (I(a)u)\phi = \int u(I(\rho^* j^* a)\phi),$$

so both sides being continuous trilinear maps $S^{m,\ell_1,\ell_2}_{\infty,\delta} \times \mathcal{S} \times \mathcal{S} \to \mathbb{C}$ for all m, ℓ_1, ℓ_2, by the density of $S^{-\infty,\ell_1,\ell_2}_{\infty,\delta}$ in $S^{m,\ell_1,\ell_2}_{\infty,\delta}$ in the $S^{m',\ell_1,\ell_2}_{\infty,\delta}$ topology for $m' > m$, the identity extends to all m. Thus, the Fréchet space adjoint, $I(a)^\dagger : \mathcal{S}' \to \mathcal{S}'$, defined by

$$(I(a)^\dagger \phi)(u) = \phi(I(a)u),\ \phi \in \mathcal{S}',\ u \in \mathcal{S},$$

satisfies

$$I(a)^\dagger \phi = I(\rho^* j^* a)\phi,\ \phi \in \mathcal{S},$$

that is by the weak-* density of $\mathcal{S}$ in $\mathcal{S}'$, $I(a)^\dagger$ is the unique continuous extension of $I(\rho^* j^* a)$ from $\mathcal{S}$ to $\mathcal{S}'$; one simply writes $I(\rho^* j^* a) = I(a)^\dagger$ even as maps $\mathcal{S}' \to \mathcal{S}'$. Since $\rho^* j^* \rho^* j^* a = a$, we deduce that for any a, $I(a) = I(\rho^* j^* a)^\dagger : \mathcal{S}' \to \mathcal{S}'$ is continuous.

Here we have used the bilinear distributional pairing; if one uses the sesquilinear L^2-pairing, one has

$$\begin{aligned} \int Au(z)\overline{\phi(z)}\, dz &= \int u(z') \overline{\int e^{i\zeta\cdot(z'-z)} \overline{a(z,z',\zeta)} \phi(z)\, dz\, d\zeta}\, dz' \\ &= \int u(z')\, \overline{(I(\tilde{b})\phi)(z')}\, dz', \end{aligned}$$

$\tilde{b}(z,z',\zeta) = \overline{a(z',z,\zeta)}$, so using $*$ to denote the corresponding (Hilbert-space-type) adjoint

$$(I(a))^* = I(cj^* a), \tag{5.21}$$

where c is the complex conjugation map.

Note that if $a \in S^{m,\ell_1,\ell_2}_{\delta,\delta'}$ then $cj^* a \in S^{m,\ell_2,\ell_1}_{\delta,\delta'}$, thus the adjoint of operators given by our scattering symbols is still in the same class, with ℓ_2 and ℓ_1 reversed.

While we have two indices ℓ_1 and ℓ_2 for growth in the spatial variables, this is actually redundant, $\ell_1 + \ell_2$ is the relevant quantity, as we have already seen signs of in the proof of Lemma 5.3 in the case of $S^{m,\ell_1,\ell_2}_{\delta,\delta'}$: for the a_1 term the orders were interchangeable due to support properties, while the a_2 term was irrelevant.

Lemma 5.4 *Given $\ell \in \mathbb{R}$, the range of the map $a \mapsto I(a)$ is independent of the choice of ℓ_1 and ℓ_2 as long as $\ell_1 + \ell_2 = \ell$.*

Definition 5.1 We now define

$$\Psi^{m,\ell}_{\infty,\delta}(\mathbb{R}^n) = \{I(a) : a \in S^{m,\ell,0}_{\infty,\delta}\}$$

and

$$\Psi^{m,\ell}_{\delta,\delta'}(\mathbb{R}^n) = \{I(a) : a \in S^{m,\ell,0}_{\delta,\delta'}\};$$

we could have used $S^{m,\ell_1,\ell_2}_{\infty,\delta}$, resp. $S^{m,\ell_1,\ell_2}_{\delta,\delta'}$ instead for any ℓ_1, ℓ_2 with $\ell_1 + \ell_2 = \ell$.

Proof To see this lemma for $S^{m,\ell_1,\ell_2}_{\infty,\delta}$, we note as in the proof of Lemma 5.3 that $(1 + \Delta_\zeta)e^{i(z-z')\cdot\zeta} = \langle z - z'\rangle^2 e^{i(z-z')\cdot\zeta}$, so at first for $m < -n$, as usual, for $a \in S^{m,\ell_1,\ell_2}_{\infty,\delta}$,

$$\begin{aligned}(I(a)u)(z) &= (2\pi)^{-n}\int_{\mathbb{R}^n\times\mathbb{R}^n} \langle z - z'\rangle^{-2N}(1+\Delta_\zeta)^N(e^{i\zeta\cdot(z-z')})a(z,z',\zeta)\,u(z')\,dz',\\ &= (2\pi)^{-n}\int_{\mathbb{R}^n\times\mathbb{R}^n} e^{i\zeta\cdot(z-z')}(\langle z - z'\rangle^{-2N}(1+\Delta_\zeta)^N a(z,z',\zeta))\,u(z')\,dz'\\ &= (I(b)u)(z),\end{aligned} \tag{5.22}$$

where

$$b(z,z',\zeta) = \langle z - z'\rangle^{-2N}(1+\Delta_\zeta)^N a(z,z',\zeta). \tag{5.23}$$

Notice that

$$\langle z\rangle^2 = 1+|z|^2 \le 1+(|z-z'|+|z|)^2 \le 1+2|z'|^2+2|z-z'|^2 \le 2\langle z-z'\rangle^2\langle z'\rangle^2, \tag{5.24}$$

and the analogous inequality also holds with z and z' interchanged, and

$$D^\alpha_z D^\beta_{z'}\langle z - z'\rangle^{-2N} \le C_{\alpha\beta}\langle z - z'\rangle^{-2N},$$

so for any m, ℓ_1, ℓ_2, $a \in S^{m,\ell_1,\ell_2}_{\infty,\delta}$, with b defined by (5.23) satisfies $b \in S^{m,\ell_1+s,\ell_2-s}_{\infty,\delta}$ for $-2N \le s \le 2N$, and the map

$$S^{m,\ell_1,\ell_2}_{\infty,\delta} \ni a \mapsto b \in S^{m,\ell_1+s,\ell_2-s}_{\infty,\delta}$$

is continuous, hence $I(a) = I(b)$ holds for all m, ℓ_1, ℓ_2 (as it holds for $m < -n$). Given any s, choosing sufficiently large N, shows that the range of I on $S^{m,\ell_1,\ell_2}_{\infty,\delta}$ only depends on $\ell_1 + \ell_2$.

Now, if $a \in S^{m,\ell_1,\ell_2}_{\delta,\delta'}$ then b defined by (5.23) is usually not in $S^{m,\ell_1+s,\ell_2-s}_{\delta,\delta'}$, as derivatives in z and z' do not typically give extra decay when hitting $\langle z-z'\rangle^{-2N}$. However, for the decomposition $a = a_1 + a_2$ used in the proof of Lemma 5.3, on the support of the a_2 piece derivatives of $\langle z-z'\rangle^{-2N}$ have the required decay (indeed, one has decay in (z,z') jointly upon differentiation in either z or z'), so the corresponding b_2 satisfies $b_2 \in S^{m,\ell_1-s,\ell_2-s'}_{\delta,\delta'}$ if $s+s' \le 2N(1-\delta')$ (with δ' coming from the ζ derivatives), while for the a_1 piece the weights ℓ_1 and ℓ_2 are directly equivalent as $\langle z\rangle \sim \langle z'\rangle$ on $\operatorname{supp} a_1$. □

I use this opportunity to remark that for the a_2 piece $I(a_2)$ of $I(a)$ in fact one has

$$I(a_2) \in \cap_{m',\ell'\in\mathbb{R}} \Psi^{m',\ell'}_{\delta,\delta'} = \Psi^{-\infty,-\infty}_{\delta,\delta'}. \tag{5.25}$$

We have already seen above that the analogue of this holds with $m'=m$ fixed, $l'\in\mathbb{R}$. In order to see that m' can be taken arbitrarily as well, note that due to the support of a_2, we can use $\Delta_\zeta e^{i(z-z')\cdot\zeta} = |z-z'|^2 e^{i(z-z')\cdot\zeta}$ and integrate by parts in ζ (noting that the diagonal singularity of $|z-z'|^{-2}$ is irrelevant due to the support of a_2) to see that

$$\begin{aligned}(I(a_2)u)(z) &= (2\pi)^{-n}\int_{\mathbb{R}^n\times\mathbb{R}^n} |z-z'|^{-2N}\Delta_\zeta^N(e^{i\zeta\cdot(z-z')})a_2(z,z',\zeta)\,u(z')\,dz',\\ &= (2\pi)^{-n}\int_{\mathbb{R}^n\times\mathbb{R}^n} e^{i\zeta\cdot(z-z')}(|z-z'|^{-2N}\Delta_\zeta^N a_2(z,z',\zeta))\,u(z')\,dz'\\ &= (I(b_2)u)(z),\end{aligned} \tag{5.26}$$

where

$$b_2(z,z',\zeta) = |z-z'|^{-2N}\Delta_\zeta^N a_2(z,z',\zeta) \in S^{m-(1-\delta)2N,\ell_1-s,\ell_2-s'}_{\delta,\delta'} \tag{5.27}$$

if $s+s'\le 2N(1-\delta')$. This shows (5.25). The analogue also holds on $S^{m,\ell_1,\ell_2}_{\infty,\delta}$, namely in that case the similarly defined a_2 gives rise to $I(a_2)\in\Psi^{-\infty,-\infty}_{\infty,\delta}$.

5.3.3. Left and Right Reduction

One very useful property of $\Psi^{m,\ell}_{\infty,\delta}(\mathbb{R}^n)$ is that it is in fact exactly the range of I acting on symbols of a special form, namely those independent of z'. Thus, let

$$\begin{aligned}a\in S^{m,\ell}_{\infty,\delta}(\mathbb{R}^n_z;\mathbb{R}^n_\zeta) \Longleftrightarrow\; & a\in C^\infty(\mathbb{R}^n_z\times\mathbb{R}^n_\zeta),\\ & |D_z^\alpha D_\zeta^\gamma a| \le C_{\alpha\gamma}\langle z\rangle^\ell\langle\zeta\rangle^{m-|\gamma|+\delta|(\alpha,\gamma)|};\end{aligned}$$

so with

$$\pi_L:\mathbb{R}^n_z\times\mathbb{R}^n_{z'}\times\mathbb{R}^n_\zeta\to\mathbb{R}^n_z\times\mathbb{R}^n_\zeta$$

the projection map dropping z', $a\in S^{m,\ell}_{\infty,\delta}(\mathbb{R}^n_z;\mathbb{R}^n_\zeta)$ if and only if

$$\pi_L^* a\in S^{m,\ell,0}_{\infty,\delta}(\mathbb{R}^n_z;\mathbb{R}^n_{z'};\mathbb{R}^n_\zeta).$$

As usual, the seminorms

$$\|a\|_{S^{m,\ell}_{\infty,\delta},N} = \sum_{|\alpha|+|\gamma|\leq N} \sup \langle z\rangle^{-\ell}\langle\zeta\rangle^{-m+|\gamma|-\delta|(\alpha,\gamma)|}|D_z^\alpha D_\zeta^\gamma a|$$

give a Fréchet topology. With π_R the projection dropping the z variables, one also has $a \in S^{m,\ell}_{\infty,\delta}(\mathbb{R}^n;\mathbb{R}^n)$ if and only if $\pi_R^* a \in S^{m,0,\ell}_{\infty,\delta}(\mathbb{R}^n_z;\mathbb{R}^n_{z'};\mathbb{R}^n_\zeta)$.

Then:

Proposition 5.1 *For any $\ell = \ell_1+\ell_2$ and $a \in S^{m,\ell_1,\ell_2}_{\infty,\delta}(\mathbb{R}^n_z;\mathbb{R}^n_{z'};\mathbb{R}^n_\zeta)$ there exists a unique $a_L \in S^{m,\ell}_{\infty,\delta}(\mathbb{R}^n_z;\mathbb{R}^n_\zeta)$ such that $I(a) = I(\pi_L^* a_L)$; one writes $q_L = I\circ\pi_L^* : S^{m,\ell}_{\infty,\delta} \to \Psi^{m,\ell}_{\infty,\delta}$. Here a_L is called the* left reduced symbol *of $I(a)$, and q_L is the* left quantization *map.*

Similarly, for any $\ell = \ell_1+\ell_2$ and $a \in S^{m,\ell_1,\ell_2}_{\infty,\delta}(\mathbb{R}^n_z;\mathbb{R}^n_{z'};\mathbb{R}^n_\zeta)$ there exists a unique $a_R \in S^{m,\ell}_{\infty,\delta}(\mathbb{R}^n_z;\mathbb{R}^n_\zeta)$ such that $I(a) = I(\pi_R^* a_R)$; one writes $q_R = I\circ\pi_R^* : S^{m,\ell}_{\infty,\delta} \to \Psi^{m,\ell}_{\infty,\delta}$. Here a_R is called the right reduced symbol *of $I(a)$, and q_R is the* right quantization *map.*

Moreover, the maps $a \mapsto a_L, a \mapsto a_R$ are continuous.

Further, with $\iota : \mathbb{R}^n\times\mathbb{R}^n \to \mathbb{R}^n\times\mathbb{R}^n\times\mathbb{R}^n$ the inclusion map as the diagonal in the first two factors, that is $\iota(z,\zeta) = (z,z,\zeta)$,

$$a_L \sim \sum_\alpha \frac{i^{|\alpha|}}{\alpha!}\iota^* D_{z'}^\alpha D_\zeta^\alpha a, \tag{5.28}$$

and

$$a_R \sim \sum_\alpha \frac{(-i)^{|\alpha|}}{\alpha!}\iota^* D_z^\alpha D_\zeta^\alpha a,$$

with the summation asymptotic *in ζ, that is is modulo $S^{-\infty,\ell}_{\infty,\delta}$; see* (5.36).

If instead $a \in S^{m,\ell_1,\ell_2}_{\delta,\delta'}$, then the conclusions hold with $a_L, a_R \in S^{m,\ell}_{\delta,\delta'}$, with the asymptotic summation being asymptotic both in z and in ζ, that is is modulo $S^{-\infty,-\infty}$.

In the case of variable orders, stated for $S^{\mathsf{m},\mathsf{l}_1,\mathsf{l}_2}_{\delta,\delta'}$ only:

Corollary 5.1 *If $a \in S^{\mathsf{m},\mathsf{l}_1,\mathsf{l}_2}_{\delta,\delta'}$ then $a_L, a_R \in S^{\mathsf{m},\mathsf{l}}_{\delta,\delta'}$, where*

$$\mathsf{l}(z,\zeta) = \mathsf{l}_1(z,z,\zeta) + \mathsf{l}_2(z,z,\zeta).$$

This corollary is an immediate consequence of the asymptotic expansion in Proposition 5.1, for the αth term there is in $S^{\mathsf{m}-(1-2\delta)|\alpha|,\mathsf{l}-(1-2\delta')|\alpha|}_{\delta,\delta'}$.

Notice that for $a \in S^{m,\ell}_{\infty,\delta}$,

$$q_L(a)u(z) = (2\pi)^{-n} \int_{\mathbb{R}^n} e^{i\zeta\cdot z} a(z,\zeta)\,(\mathcal{F}u)(\zeta)\,d\zeta \tag{5.29}$$

for $m < -n$, but now, for $u \in \mathcal{S}$, the right hand side extends continuously to $S^{m,\ell}_{\infty,\delta}$ for all m, so one could have directly defined $q_L(a)$ for all m. Similarly,

$$q_R(a)u = \mathcal{F}^{-1}\Big(\zeta \mapsto \int_{\mathbb{R}^n} e^{-iz'\cdot\zeta} a(z',\zeta)\,u(z')\,dz'\Big), \tag{5.30}$$

where now the right hand side makes sense directly as a tempered distribution for all m. However, relating q_L and q_R, as well as performing other important calculations, would be rather hard without having defined the map I in general, via a continuity/regularization argument! Note that for $a \in S^{-\infty,-\infty}_{\infty}$, in either case, one deduces that directly that $q_R(a)u$ and $q_L(a)u$ are in $\mathcal{S}$.

We remark that if $a \in S^{m,\ell}_{\infty}$ is a polynomial in ζ, that is $a(z,\zeta) = \sum_{|\alpha|\le m} a_\alpha(z)\zeta^\alpha$, then one can pull the factors $a_\alpha(z)$ out of the integral (5.29), and thus $\zeta^\alpha \mathcal{F} = \mathcal{F}D^\alpha$ and the Fourier inversion formula yields

$$q_L(a)u(z) = \sum_{|\alpha|\le m} a_\alpha(z)(D^\alpha u)(z),$$

that is with a_α acting as multiplication operators,

$$q_L(a) = \sum_{|\alpha|\le m} a_\alpha D^\alpha. \tag{5.31}$$

Similarly,

$$q_R(a)u(z) = \sum_{|\alpha|\le m} (D^\alpha(a_\alpha u))(z),$$

that is

$$q_R(a) = \sum_{|\alpha|\le m} D^\alpha a_\alpha.$$

So differential operators of order m on $\mathbb{R}^n$ with coefficients in $S^\ell(\mathbb{R}^n)$ lie in $\Psi^{m,\ell}$. In particular, differential operators with coefficients in $\mathcal{C}^\infty(\overline{\mathbb{R}^n})$ lie in $\Psi^{m,0}(\mathbb{R}^n)$.

We now prove Proposition 5.1. We only consider the left reduction, that is the L subscript case, as the R case is completely analogous. First, we note that the uniqueness is straightforward. Any operator $A = I(a)$, $a \in S^{m,\ell_1,\ell_2}_{\infty,\delta}$, has a Schwartz kernel, $K_A \in \mathcal{S}'$ (as it is a continuous linear map $\mathcal{S} \to \mathcal{S}$, thus $\mathcal{S} \to \mathcal{S}'$). When $m < -n$, the Schwartz kernel satisfies

$$K_A(\phi \otimes u) = \int (Au)(z)\phi(z)\,dz$$
$$= (2\pi)^{-n}\int e^{i\zeta\cdot(z-z')}a(z,z',\zeta)\,u(z')\,\phi(z)\,d\zeta\,dz'\,dz \qquad (5.32)$$
$$= \int (\mathcal{F}_\zeta^{-1}a)(z,z',z-z')u(z')\phi(z)\,dz'\,dz,$$

where $\mathcal{F}_\zeta^{-1}$ is the inverse Fourier transform in the third variable, ζ. ($\mathcal{F}_3^{-1}$ is a logically better, but less self-explanatory, notation.) Thus, for such a, K_A is the polynomially bounded function (hence tempered distribution) given by

$$F_a(z,z') = (\mathcal{F}_\zeta^{-1}a)(z,z',z-z') = (\mathcal{F}_3^{-1}a)(z,z',z-z'). \qquad (5.33)$$

If $a \in S^{m,\ell}_{\infty,\delta}$, then, with 2 denoting that the inverse Fourier transform is in the second slot, we have

$$F_{\pi_L^* a}(z,z') = (\mathcal{F}_2^{-1}a)(z,z-z') = (G^*\mathcal{F}_2^{-1}a)(z,z')$$

where $G : \mathbb{R}^{2n} \to \mathbb{R}^{2n}$ is the invertible linear map $G(z,z') = (z,z-z')$, thus one can pull back tempered distributions by it. Thus,

$$K_{I(\pi_L^* a)} = G^*\mathcal{F}_2^{-1}a,$$

and correspondingly

$$a = \mathcal{F}_2(G^{-1})^* K_{I(\pi_L^* a)},$$

first for $m < -n$, but then as both sides are continuous maps $S^{m,\ell}_{\infty,\delta} \to \mathcal{S}'$, this identity holds in general. In particular, given $\tilde{a} \in S^{m,\ell_1,\ell_2}_{\infty,\delta}$ there exists at most one $a \in S^{m,\ell_1+\ell_2}_{\infty,\delta}$ such that $I(\pi_L^* a) = I(\tilde{a})$, for

$$a = \mathcal{F}_2(G^{-1})^* K_{I(\tilde{a})}. \qquad (5.34)$$

Now for existence. In principle (5.34) solves this problem, but then one needs to show that the a it provides, that is a_L in the notation of the proposition, is not merely a tempered distribution, but is in an appropriate symbol class. So we proceed differently.

For the following discussion it is useful to replace a by a_1; recall that $I(a_2) \in \Psi^{-\infty,-\infty}_{\infty,\delta}$ in this case, thus does not affect the argument below. Hence, to minimize subscripts, we simply write a below, but we actually apply the argument to a_1. With the notation of the proposition, one expands a in Taylor series in z' around the diagonal $z' = z$, with the integral remainder term:

$$a(z,z',\zeta)=\sum_{|\alpha|\le N-1}\frac{(z'-z)^\alpha}{\alpha!}((\partial_{z'})^\alpha a)(z,z,\zeta)+R_N(z,z',\zeta)$$

$$R_N(z,z',\zeta)=\sum_{|\alpha|=N}N\frac{(z'-z)^\alpha}{\alpha!}\int_0^1(1-t)^{N-1}((\partial_{z'})^\alpha a)(z,(1-t)z+tz',\zeta)\,dt. \tag{5.35}$$

Now, for $m<-n$, as $(z'_j-z_j)e^{i\zeta\cdot(z-z')}=-D_{\zeta_j}e^{i\zeta\cdot(z-z')}$,

$$\begin{aligned}(I((z'_j-z_j)a)u)(z)&=(2\pi)^{-n}\int(-D_{\zeta_j})e^{i\zeta\cdot(z-z')}a(z,z',\zeta)\,u(z')\,dz'\,d\zeta\\&=(2\pi)^{-n}\int e^{i\zeta\cdot(z-z')}(D_{\zeta_j}a)(z,z',\zeta)\,u(z')\,dz'\,d\zeta\\&=(I(D_{\zeta_j}a)u)(z),\end{aligned}$$

so as

$$S^{m,\ell_1,\ell_2}_{\infty,\delta}\times\mathcal{S}\ni(a,u)\mapsto I((z'_j-z_j)a)u\in\mathcal{S}$$

and

$$S^{m,\ell_1,\ell_2}_{\infty,\delta}\times\mathcal{S}\ni(a,u)\mapsto I(D_{\zeta_j}a)u\in\mathcal{S}$$

are both continuous bilinear maps, the density of $S^{m,\ell_1,\ell_2}_{\infty,\delta}$ in the topology of $S^{m',\ell_1,\ell_2}_{\infty,\delta}$ for $m'>m$ shows that

$$I((z'-z)^\alpha a)=I(D^\alpha_\zeta a)$$

for all m and $a\in S^{m,\ell_1,\ell_2}_{\infty,\delta}$.

Thus, for a as in (5.35),

$$I(a)=\sum_{|\alpha|\le N-1}\frac{1}{\alpha!}I((D_\zeta)^\alpha\iota^*\partial^\alpha_{z'}a)+I(R'_N),$$

$$R'_N(z,z',\zeta)=\sum_{|\alpha|=N}N\frac{1}{\alpha!}\int_0^1(1-t)^{N-1}(D^\alpha_\zeta(\partial_{z'})^\alpha a)(z,(1-t)z+tz',\zeta)\,dt.$$

But keeping in mind the support properties of a (recall that it stands for the a_1 piece!),

$$(D_\zeta)^\alpha\iota^*\partial^\alpha_{z'}a\in S^{m-(1-2\delta)|\alpha|,\ell_1,\ell_2}_{\infty,\delta},\ R'_N\in S^{m-(1-2\delta)N,\ell_1+\ell_2,0}_{\infty,\delta},$$

with the map

$$S^{m,\ell_1,\ell_2}_{\infty,\delta}\ni a\to(D_\zeta)^\alpha\iota^*\partial^\alpha_{z'}a\in S^{m-(1-2\delta)|\alpha|,\ell_1+\ell_2}_{\infty,\delta}$$

continuous, and similarly with R'_N. Since $(D_\zeta)^\alpha\iota^*\partial^\alpha_{z'}a$ is independent of z', and for this the original a and a_1 give exactly the same expression, this proves the following weaker version of Proposition 5.1: for all $a \in S^{m,\ell_1,\ell_2}_{\infty,\delta}$ and for all N there exists $a_N \in S^{m,\ell_1+\ell_2}_{\infty,\delta}$ such that

$$I(a) - I(a_N) = I(R'_N),\ R'_N \in S^{m-(1-2\delta)N,\ell_1+\ell_2,0}_{\infty,\delta}.$$

Notice that if $a \in S^{m,\ell_1,\ell_2}_{\delta,\delta'}$ then writing $a = a_1 + a_2$, we already know by (5.25) that for any m', ℓ'_1, ℓ'_2 we can write $I(a_2) = I(b_2)$, $b_2 \in S^{m',\ell'_1,\ell'_2}_{\delta,\delta'}$, while for a_1 the analogous conclusions to the $S^{m,\ell_1,\ell_2}_{\infty,\delta}$ setting hold but with

$$(D_\zeta)^\alpha \iota^* \partial^\alpha_{z'} a_1 \in S^{m-(1-2\delta)|\alpha|,\ell_1+\ell_2-(1-2\delta)|\alpha|}_{\delta,\delta'},$$

$$R'_{1,N} \in S^{m-(1-2\delta)N,\ell_1+\ell_2-(1-2\delta')N,0}_{\delta,\delta'}.$$

An *asymptotic summation* argument allows one to improve this. This notion means the following: suppose $a_j \in S^{m-(1-2\delta)j,\ell}_{\infty,\delta}$ for $j \in \mathbb{N}$. Then there exists $a \in S^{m,\ell}_{\infty,\delta}$ such that

$$a - \sum_{j=0}^{N-1} a_j \in S^{m-(1-2\delta)N,\ell}_{\infty,\delta}. \tag{5.36}$$

To see this, we take $\chi \in \mathcal{C}^\infty(\mathbb{R}^n)$ with $\chi(\zeta) = 1$ for $|\zeta| \geq 2$, $\chi(\zeta) = 0$ for $|\zeta| \leq 1$. For $0 < \epsilon_j < 1$ to be determined, but with $\epsilon_j \to 0$, consider

$$a(z,\zeta) = \sum_{j=0}^{\infty} \chi(\epsilon_j\zeta) a_j(z,\zeta);$$

the sum is finite for (z,ζ) with $|\zeta| \leq R$, with only the finitely many terms with $\epsilon_j \geq R^{-1}$ contributing. Thus, a is $\mathcal{C}^\infty$; the question is convergence in $S^{m,\ell}_{\infty,\delta}$, and the property (5.36). But by Leibniz's rule,

$$(D^\alpha_\zeta D^\beta_z a)(z,\zeta) = \sum_{j=0}^{\infty} \sum_{\gamma\leq\alpha} C_{\alpha\gamma} \epsilon_j^{|\gamma|} (D^\gamma\chi)(\epsilon_j\zeta)(D^{\alpha-\gamma}_\zeta D^\beta_z a_j)(z,\zeta).$$

To get convergence of the tail in $S^{m-(1-2\delta)N,\ell}_{\infty,\delta}$, we need to estimate the sup norm of

$$\langle\zeta\rangle^{-m+(1-2\delta)N-\delta(|\alpha|+|\beta|)+|\alpha|} \langle z\rangle^{-\ell} \left(D^\alpha_\zeta D^\beta_z \left(\sum_{j=N}^{\infty} \chi(\epsilon_j\zeta) a_j(z,\zeta) \right) \right)$$

$$= \sum_{j=N}^{\infty} \sum_{\gamma\leq\alpha} C_{\alpha\gamma} \langle\zeta\rangle^{-\delta|\gamma|} \epsilon_j^{(j-N)(1-2\delta)} \left(\langle\zeta\rangle^{|\gamma|+(N-j)(1-2\delta)} \epsilon_j^{(N-j)(1-2\delta)+|\gamma|} (D^\gamma\chi)(\epsilon_j\zeta) \right)$$
$$\left(\langle\zeta\rangle^{-m+(1-2\delta)j+(1-\delta)(|\alpha|-|\gamma|)-\delta|\beta|} \langle z\rangle^{-\ell} (D^{\alpha-\gamma}_\zeta D^\beta_z a_j)(z,\zeta) \right); \tag{5.37}$$

where we have used the above expansion. For $\gamma = 0$, we use $|\zeta| \geq \epsilon_j^{-1}$ on $\operatorname{supp} \chi(\epsilon_j .)$, so for $j \geq N$ (as $\delta \in [0, 1/2)$),

$$\epsilon_j^{(N-j)(1-2\delta)} \langle \zeta \rangle^{(N-j)(1-2\delta)} = (\epsilon_j^2 + \epsilon_j^2 |\zeta|^2)^{(1-2\delta)(N-j)/2} \leq 1,$$

while for $\gamma \neq 0$ we use $\epsilon_j^{-1} \leq |\zeta| \leq 2\epsilon_j^{-1}$ on $\operatorname{supp}(D^\gamma \chi)(\epsilon_j .)$, so

$$1 \leq \langle \zeta \rangle \epsilon_j = (\epsilon_j^2 + \epsilon_j^2 |\zeta|^2)^{1/2} \leq 5^{1/2}$$

on $\operatorname{supp}(D^\gamma \chi)(\epsilon_j .)$ for all $\gamma \neq 0$, and thus for $j \geq N$,

$$\langle \zeta \rangle^{|\gamma| + (N-j)(1-2\delta)} \epsilon_j^{(N-j)(1-2\delta) + |\gamma|} \leq 5^{|\gamma|/2}$$

there. Thus, adding up the terms with $|\alpha| + |\beta| = M$ as required by the symbolic seminorms, there are constants $C_M > 0$ (arising from finitely many combinatorial constants, from suprema of finitely many derivatives of χ and from finite powers of $5^{1/2}$) such that the series is absolutely summable, and hence convergent, if for all M

$$\sum_{j \geq N + (1-2\delta)^{-1}}^{\infty} C_M \epsilon_j \|a_j\|_{S_{\infty,\delta}^{m-(1-2\delta)j,\ell}, M}$$

converges; here ϵ_j is from $\epsilon_j^{(j-N)(1-2\delta)} \leq \epsilon_j$ on the right hand side of (5.37), taking advantage of $j \geq N + (1-2\delta)^{-1}$ in our sum. Now, if $\|a_j\|_{S_{\infty,\delta}^{m-(1-2\delta)j,\ell}, M} \leq R_{j,M}$, where $R_{j,M}$ are specified constants, then one can arrange the convergence by for instance requiring that for $j > M$, the corresponding summand is $\leq 2^{-j}$, that is for $j > M$,

$$\epsilon_j \leq 2^{-j} C_M^{-1} R_{j,M}^{-1}.$$

Note that for each j this is finitely many constraints (as only the values of M with $M < j$ matter), which can thus be satisfied. Correspondingly, the tail of the series converges for each N in $S_{\infty,\delta}^{m-(1-2\delta)N,\ell}$, and thus $a \in S_{\infty,\delta}^{m,\ell}$ and also (5.36) holds. This gives a continuous asymptotic summation map on arbitrary bounded subsets of the product of the symbol spaces. (One can make the map globally defined and continuous by letting ϵ_j be the minimum of, say,

$$2^{-j} C_M^{-1} (1 + \|a_j\|_{S_{\infty,\delta}^{m-(1-2\delta)j,\ell}, M})^{-1},$$

over $M = 0, 1, \ldots, j-1$, but this is actually not important below.)

Now, let

$$\tilde{a} \sim \sum_\alpha \frac{1}{\alpha!} (D_\zeta)^\alpha \iota^* \partial_{z'}^\alpha a \in S_{\infty,\delta}^{m,\ell_1+\ell_2};$$

asymptotic summation can be done so that the map $a \mapsto \tilde{a}$ is continuous. Then $\tilde{a} - a_N \in S^{m-(1-2\delta)N,\ell_1,\ell_2}_{\infty,\delta}$ for all N, and thus

$$I(a) - I(\tilde{a}) \in \cap_N I(S^{m-(1-2\delta)N,\ell_1,\ell_2}_{\infty,\delta}).$$

If $a \in S^{m,\ell_1,\ell_2}_{\delta,\delta'}$ then with

$$\tilde{a} \sim \sum_\alpha \frac{1}{\alpha!}(D_\zeta)^\alpha \iota^* \partial^\alpha_{z'} a \in S^{m,\ell_1+\ell_2}_{\delta,\delta'},$$

where we asymptotically sum both in the z and in the ζ variables (this can be done at the same time, adding a factor of $\chi(\epsilon_j z)$),

$$I(a) - I(\tilde{a}) \in \cap_N I(S^{m-(1-2\delta)N,\ell_1,\ell_2-(1-2\delta')N}_{\delta,\delta'}).$$

The following lemma then finishes the proof of Proposition 5.1:

Lemma 5.5 *Suppose* $b \in S^{m,\ell_1,\ell_2}_{\infty,\delta}$ *satisfies* $I(b) \in \cap_N I(S^{m-N,\ell_1,\ell_2}_{\infty,\delta})$, *that is for all* $N \in \mathbb{N}$ *there is* $b_N \in S^{m-N,\ell_1,\ell_2}_{\infty,\delta}$ *such that* $I(b) = I(b_N)$. *Then there exists* $c \in S^{-\infty,\ell_1+\ell_2}_{\infty,\delta}$ *such that* $I(c) = I(b)$. *Moreover, if there are continuous maps* $j_N : b \to b_N$, *then the map* $b \to c$ *is continuous.*

Suppose instead $b \in S^{m,\ell_1,\ell_2}_{\delta,\delta'}$ *satisfies* $I(b) \in \cap_N I(S^{m-N,\ell_1,\ell_2-N}_{\delta,\delta'})$, *that is for all* $N \in \mathbb{N}$ *there is* $b_N \in S^{m-N,\ell_1,\ell_2-N}_{\delta,\delta'}$ *such that* $I(b) = I(b_N)$. *Then there exists* $c \in S^{-\infty,-\infty}$ *such that* $I(c) = I(b)$. *Moreover, if there are continuous maps* $j_N : b \to b_N$, *then the map* $b \to c$ *is continuous.*

Proof The idea of the proof is to use (5.34), as in the present setting the Schwartz kernel can be shown to be well-behaved, so (5.34) immediately gives the appropriate symbolic properties of c. Thus, we note that for all N there is $b_N \in S^{m-N,\ell_1,\ell_2}_{\infty,\delta}$ such that $I(b) = I(b_N)$, so taking N such that $m - N < -n$, (5.32)–(5.33) give the Schwartz kernel (which is independent of N) as the continuous polynomially bounded function

$$K_{I(b_N)}(z, z') = (\mathcal{F}^{-1}_\zeta b_N)(z, z', z - z');$$

taking $m - N < -n - k$, this is in fact C^k with polynomial bounds up to the kth derivatives. Correspondingly, it satisfies, for $|\alpha| + |\beta| + \delta|\gamma| \leq k$, and writing D^α_j for the αth derivative in the jth slot, M^α_j for the multiplication by the αth coordinate in the jth slot,

$$\begin{aligned}
&\langle z\rangle^{-\ell_1}\langle z'\rangle^{-\ell_2}(z - z')^\gamma D^\alpha_z D^\beta_{z'} K_{I(b_N)}(z, z') \\
&\quad = \big(\langle . \rangle_1^{-\ell_1}\langle . \rangle_2^{-\ell_2} M^\gamma_3 (D_1 + D_3)^\alpha (D_2 - D_3)^\beta (\mathcal{F}^{-1}_3 b_N)\big)(z, z', z - z') \\
&\quad = \big(\mathcal{F}^{-1}_3 \langle . \rangle_1^{-\ell_1}\langle . \rangle_2^{-\ell_2} D^\gamma_3 (D_1 + M_3)^\alpha (D_2 - M_3)^\beta b_N\big)(z, z', z - z').
\end{aligned}$$

As

$$\langle . \rangle_1^{-\ell_1} \langle . \rangle_2^{-\ell_2} D_3^\gamma (D_1 + M_3)^\alpha (D_2 - M_3)^\beta b_N$$

is bounded in $C_\infty(\mathbb{R}^n \times \mathbb{R}^n; L^1(\mathbb{R}^n_\zeta))$ by a seminorm of b_N as $|\alpha|+|\beta|+\delta|\gamma| \leq k$, $m - N < -n - k$, where C_∞ stands for bounded continuous functions,

$$\mathcal{F}_3^{-1} \langle . \rangle_1^{-\ell_1} \langle . \rangle_2^{-\ell_2} D_3^\gamma (D_1 + M_3)^\alpha (D_2 - M_3)^\beta b_N$$

is bounded in $C_\infty(\mathbb{R}^n \times \mathbb{R}^n \times \mathbb{R}^n)$ by a seminorm of b_N, hence the same holds for the pull back by the map $(z, z') \mapsto (z, z', z - z')$. Since N is arbitrary, we can take arbitrary α, β, γ and deduce that

$$\sup |\langle z \rangle^{-\ell_1} \langle z' \rangle^{-\ell_2} (z - z')^\gamma (D_z^\alpha D_{z'}^\beta K_{I(b)})(z, z')| < \infty.$$

Using (5.24) and that γ is arbitrary, we deduce that

$$\sup |\langle z \rangle^{-\ell_1 - \ell_2} (z - z')^\gamma D_z^\alpha D_{z'}^\beta K_{I(b)}| < \infty. \tag{5.38}$$

Since we want $K_{I(c)} = K_{I(b)}$, we need

$$(\mathcal{F}_2^{-1} c)(z, z - z') = K_{I(b)}(z, z'),$$

that is with $w = z - z'$,

$$(\mathcal{F}_2^{-1} c)(z, w) = K_{I(b)}(z, z - w).$$

Now, a linear change of variables for $K_{I(b)}$ gives that

$$\sup |\langle z \rangle^{-\ell_1 - \ell_2} w^\gamma (D_z^\alpha D_w^\beta \mathcal{F}_2^{-1} c)(z, w)| < \infty,$$

so $\langle z \rangle^{-\ell_1 - \ell_2} D_z^\alpha \mathcal{F}_2^{-1} c$ is Schwartz in w, uniformly in z, and thus $\langle z \rangle^{-\ell_1 - \ell_2} D_z^\alpha c$ is Schwartz in the second variable, ζ, uniformly in z, that is $c \in S^{-\infty, \ell_1 + \ell_2}_{\infty, \delta}$. This also shows that any seminorm of c depends only on the seminorms of b_N for some N, and does so continuously, and thus depends on b continuously.

The argument in the case of $S^{m, \ell_1, \ell_2}_{\delta, \delta'}$ is completely analogous, but now even

$$\begin{aligned}
&\langle z \rangle^{-\ell_1} \langle z' \rangle^{-\ell_2} (z')^\mu (z - z')^\gamma D_z^\alpha D_{z'}^\beta K_{I(b_N)}(z, z') \\
&\quad = \big(\langle . \rangle_1^{-\ell_1} \langle . \rangle_2^{-\ell_2} M_2^\mu M_3^\gamma (D_1 + D_3)^\alpha (D_2 - D_3)^\beta (\mathcal{F}_3^{-1} b_N)\big)(z, z', z - z') \\
&\quad = \big(\mathcal{F}_3^{-1} \langle . \rangle_1^{-\ell_1} \langle . \rangle_2^{-\ell_2} M_2^\mu D_3^\gamma (D_1 + M_3)^\alpha (D_2 - M_3)^\beta b_N\big)(z, z', z - z'),
\end{aligned}$$

with the result that

$$\sup |\langle z \rangle^{-\ell_1} \langle z' \rangle^{-\ell_2} (z')^\mu (z - z')^\gamma (D_z^\alpha D_{z'}^\beta K_{I(b)})(z, z')| < \infty.$$

Using (5.24) and that γ, μ are arbitrary, we deduce that

$$\sup |(z')^\mu (z - z')^\gamma D_z^\alpha D_{z'}^\beta K_{I(b)}| < \infty.$$

This gives $K_{I(b)} \in \mathcal{S}(\mathbb{R}^{2n})$, and the argument is finished as before. This completes the proof of Lemma 5.5. □

As already mentioned, this completes the proof of Proposition 5.1.

As a corollary of the lemma, we note that elements of $\Psi^{-\infty,\ell}_{\infty,\delta}$ have a $\mathcal{C}^\infty$ Schwartz kernel, of the form $\mathcal{C}^\infty(\mathbb{R}^n_z;\mathcal{S}(\mathbb{R}^n_{z'}))$, and thus give continuous linear maps $\mathcal{S}' \to \mathcal{C}^\infty(\mathbb{R}^n)$, that is are *smoothing*. Note that this does not mean decay at infinity. On the other hand, elements of $\Psi^{-\infty,-\infty}$ are *completely regularizing*, as their Schwartz kernel is in $\mathcal{S}(\mathbb{R}^{2n})$, and thus they give maps $\mathcal{S}' \to \mathcal{S}$. Note that maps $\mathcal{S}' \to \mathcal{S}$ are actually compact on all polynomially weighted Sobolev spaces $H^{r,s}$.

The isomorphism $q_L : S^{m,\ell}_{\infty,\delta} \to \Psi^{m,\ell}_{\infty,\delta}$ can be used to topologize $\Psi^{m,\ell}_{\infty,\delta}$. Since $q_R^{-1} \circ q_L$, $q_L^{-1} \circ q_R$ are continuous, this is the same topology as that induced by q_R.

5.3.4. The Principal Symbol

Note that if $a \in S^{m,\ell_1,\ell_2}_{\infty,\delta}$ then $\iota^*a - a_L, \iota^*a - a_R \in S^{m-1+2\delta,\ell_1+\ell_2}_{\infty,\delta}$, while if $a \in S^{m,\ell_1,\ell_2}_{\delta,\delta'}$ then $\iota^*a - a_L, \iota^*a - a_R \in S^{m-1+2\delta,\ell_1+\ell_2-1+2\delta'}_{\delta,\delta'}$. We thus make the following definition:

Definition 5.2 The *principal symbol* $\sigma_{\infty,m,\ell}(q_L(a))$ in $\Psi^{m,\ell}_{\infty,\delta}$ of $q_L(a)$, $a \in S^{m,\ell}_{\infty,\delta}$, is the equivalence class $[a]_\infty$ of a in $S^{m,\ell}_{\infty,\delta}/S^{m-1+2\delta,\ell}_{\infty,\delta}$.

The *joint principal symbol* $\sigma_{m,\ell}(q_L(a))$ in $\Psi^{m,\ell}_{\delta,\delta'}$ of $q_L(a)$, $a \in S^{m,\ell}_{\delta,\delta'}$, is the equivalence class $[a]$ of a in $S^{m,\ell}_{\delta,\delta'}/S^{m-1+2\delta,\ell-1+2\delta'}_{\delta,\delta'}$.

In case the orders are variable, the principal symbols

$$\sigma_{\infty,\mathsf{m},\mathsf{l}}(q_L(a)),\ \text{resp.}\ \sigma_{\mathsf{m},\mathsf{l}}(q_L(a)),$$

are defined analogously in $S^{\mathsf{m},\mathsf{l}}_{\infty,\delta}/S^{\mathsf{m}-1+2\delta,\mathsf{l}}_{\infty,\delta}$, resp. $S^{\mathsf{m},\mathsf{l}}_{\delta,\delta'}/S^{\mathsf{m}-1+2\delta,\mathsf{l}-1+2\delta'}_{\delta,\delta'}$.

Thus, the principal symbol also satisfies

$$\sigma_{\infty,m,\ell}(q_R(a)) = [a]_\infty,\ \sigma_{m,\ell}(q_R(a)) = [a],$$

with analogues for variable orders.

For $a \in \mathcal{C}^\infty(\overline{\mathbb{R}^n} \times \overline{\mathbb{R}^n}) \subset S^{0,0}$, there is a natural identification of the equivalence class, namely the restriction of a to $\partial(\overline{\mathbb{R}^n} \times \overline{\mathbb{R}^n})$ can be identified with its equivalence class, namely changing a by any element of $\mathcal{C}^\infty(\overline{\mathbb{R}^n} \times \overline{\mathbb{R}^n})$ which vanishes on the boundary, and thus being in $S^{-1,-1}$ does not affect the equivalence class, so the map $a \mapsto [a]$ descends to $a|_{\partial(\overline{\mathbb{R}^n}\times\overline{\mathbb{R}^n})} \to [a]$, and the result is injective. Note that $\overline{\mathbb{R}^n} \times \overline{\mathbb{R}^n}$ is a manifold with corners with two boundary hypersurfaces, $\partial\overline{\mathbb{R}^n} \times \overline{\mathbb{R}^n}$ and $\overline{\mathbb{R}^n} \times \partial\overline{\mathbb{R}^n}$, so equivalently one can restrict to each of these separately, and keep in mind that the restrictions must agree at the corner, $\partial\overline{\mathbb{R}^n} \times \partial\overline{\mathbb{R}^n}$; see Figure 5.3. The restrictions to these two hypersurfaces are denoted by

$$\sigma_{\text{fiber},0,0}(q_L(a)) = a|_{\overline{\mathbb{R}^n} \times \partial\overline{\mathbb{R}^n}}$$

and

$$\sigma_{\text{base},0,0}(q_L(a)) = a|_{\partial\overline{\mathbb{R}^n} \times \overline{\mathbb{R}^n}},$$

with the subscript indicating whether we are considering the part of $\sigma_{0,0}$ at "fiber infinity," that is as $|\zeta| \to \infty$, or "base infinity," that is as $|z| \to \infty$.

In the case of σ_∞, a common way of understanding it is in terms of the $\mathbb{R}^+$-action by dilations on the second factor of $\overline{\mathbb{R}^n} \times (\mathbb{R}^n \setminus \{0\})$:

$$\mathbb{R}^+ \times \overline{\mathbb{R}^n} \times (\mathbb{R}^n \setminus \{0\}) \ni (t, z, \zeta) \mapsto (z, t\zeta) \in \overline{\mathbb{R}^n} \times (\mathbb{R}^n \setminus \{0\}).$$

The quotient of $\mathbb{R}^n \setminus \{0\}$ by the $\mathbb{R}^+$ action can be identified with the unit sphere $\mathbb{S}^{n-1}$: every orbit of the $\mathbb{R}^+$-action intersects the sphere at exactly one point. A different identification of this quotient (which is actually more relevant from the perspective of where our analysis actually takes place) is the sphere at infinity, $\partial\overline{\mathbb{R}^n}$. Thus, homogeneous degree zero C^∞ functions on $\overline{\mathbb{R}^n} \times (\mathbb{R}^n \setminus \{0\})$ are identified with either $C^\infty(\overline{\mathbb{R}^n} \times \mathbb{S}^{n-1})$ or $C^\infty(\overline{\mathbb{R}^n} \times \partial\overline{\mathbb{R}^n})$. So one can correspondingly identify the principal symbol of $A = q_L(a_L)$, $a_L \in C^\infty(\overline{\mathbb{R}^n} \times \overline{\mathbb{R}^n})$, as a function on $\overline{\mathbb{R}^n} \times \mathbb{S}^{n-1}$, or instead as a homogeneous degree zero function on $\overline{\mathbb{R}^n} \times (\mathbb{R}^n \setminus \{0\})$.

Returning to σ, for

$$a = \langle z\rangle^\ell \langle\zeta\rangle^m \tilde{a},\ \tilde{a} \in C^\infty(\overline{\mathbb{R}^n} \times \overline{\mathbb{R}^n}),$$

one cannot simply restrict a to the boundary, though as (given ℓ and m) a and $\tilde{a}$ are in a bijective correspondence, one could restrict $\tilde{a}$ and call it the principal symbol, that is the actual principal symbol, as we defined it, is given by any C^∞ extension of this restriction times $\langle z\rangle^\ell\langle\zeta\rangle^m$. In a more geometric context this is not quite natural (it depends on the differentials of choices of boundary defining functions, here $\langle z\rangle^{-1}$ and $\langle\zeta\rangle^{-1}$, at the boundary). Taking $\ell = 0$, as it is the most common case, in terms of the $\mathbb{R}^+$ action on the second factor, it is more common then to view the part of the principal symbol corresponding to $\overline{\mathbb{R}^n} \times \partial\overline{\mathbb{R}^n}$ as a *homogeneous degree m* function on $\overline{\mathbb{R}^n} \times (\mathbb{R}^n \setminus \{0\})$. In terms of $\tilde{a}$ and its identification with a homogeneous degree zero function on $\overline{\mathbb{R}^n} \times (\mathbb{R}^n \setminus \{0\})$, the part of the principal symbol corresponding to $\overline{\mathbb{R}^n} \times \partial\overline{\mathbb{R}^n}$ is

$$\sigma_{\text{fiber},m,0}(A) = |\zeta|^m \tilde{a}.$$

On the other hand, the part of the principal symbol corresponding to $\partial\overline{\mathbb{R}^n} \times \overline{\mathbb{R}^n}$ can be described by simply restricting it to $\partial\overline{\mathbb{R}^n} \times \mathbb{R}^n$, with the result being symbolic in the second variable:

$$\sigma_{\text{base},m,0}(A) = \langle\zeta\rangle^m \tilde{a}|_{\partial\overline{\mathbb{R}^n}\times\mathbb{R}^n}.$$

Concretely, if A is a differential operator, $A = \sum a_\alpha D^\alpha$, $a_\alpha \in \mathcal{C}^\infty(\overline{\mathbb{R}^n})$, then the two parts of the principal symbol under this identification are

$$\sigma_{\text{fiber},m,0}(A)(z,\zeta) = \sum_{|\alpha|=m} a_\alpha(z)\zeta^\alpha, \ (z,\zeta) \in \overline{\mathbb{R}^n} \times (\mathbb{R}^n \setminus \{0\}), \tag{5.39}$$

and

$$\sigma_{\text{base},m,0}(A)(z,\zeta) = \sum_{|\alpha|\le m} a_\alpha(z)\zeta^\alpha, \ (z,\zeta) \in \partial\overline{\mathbb{R}^n} \times \mathbb{R}^n. \tag{5.40}$$

As an example, if g is a Riemannian metric on $\mathbb{R}^n$ with $g_{ij} \in \mathcal{C}^\infty(\overline{\mathbb{R}^n})$, then for $V \in \langle z\rangle^{-1}\mathcal{C}^\infty(\overline{\mathbb{R}^n})$,

$$H = \Delta_g + V - \sigma \tag{5.41}$$

has principal symbol in these two senses given by

$$\sigma_{\text{fiber},2,0} = \sum g_{ij}\zeta_i\zeta_j, \ \sigma_{\text{base},2,0} = \sum g_{ij}\zeta_i\zeta_j - \sigma.$$

In the case of σ (as opposed to σ_∞), one could apply a similar construction for the restriction of the symbol of $A = q_L(a_L)$ to $\partial\overline{\mathbb{R}^n} \times \overline{\mathbb{R}^n}$; it is then either a homogeneous degree zero function on $(\mathbb{R}^n \setminus \{0\}) \times \overline{\mathbb{R}^n}$ where the action is in the first factor, or a function on $\mathbb{S}^{n-1} \times \overline{\mathbb{R}^n}$; the last version would be rarely considered. Thus, two different point of views would be needed for describing σ in terms of homogeneous functions, which is the reason for this being a less useful point of view in this case than in that of σ_∞.

That the principal symbol captures the leading order behavior of pseudo-differential operators is given in the following proposition.

Proposition 5.2 *The sequences*

$$0 \to \Psi_{\infty,\delta}^{m-1+2\delta,\ell} \to \Psi_{\infty,\delta}^{m,\ell} \to S_{\infty,\delta}^{m,\ell}/S_{\infty,\delta}^{m-1+2\delta,\ell} \to 0,$$

resp.

$$0 \to \Psi_{\delta,\delta'}^{m-1+2\delta,\ell-1+2\delta'} \to \Psi_{\delta,\delta'}^{m,\ell} \to S_{\delta,\delta'}^{m,\ell}/S_{\delta,\delta'}^{m-1+2\delta,\ell-1+2\delta'} \to 0,$$

are short exact sequences of topological vector spaces.

Here $\iota : \Psi^{m-1+2\delta,\ell-1+2\delta'}_{\delta,\delta'} \to \Psi^{m,\ell}_{\delta,\delta'}$ *is the inclusion map and*

$$\sigma_{m,\ell} : \Psi^{m,\ell}_{\delta,\delta'} \to S^{m,\ell}_{\delta,\delta'}/S^{m-1+2\delta,\ell-1+2\delta'}_{\delta,\delta'}$$

is the principal symbol map, with analogous definitions in the case of $\Psi_{\infty,\delta}$. *The analogous statements also hold if* $m = \mathsf{m}$, $\ell = \mathsf{l}$ *are variable.*

This is essentially tautological, given the short exact sequence

$$0 \to S^{m-1+2\delta,\ell-1+2\delta'}_{\delta,\delta'} \to S^{m,\ell}_{\delta,\delta'} \to S^{m,\ell}_{\delta,\delta'}/S^{m-1+2\delta,\ell-1+2\delta'}_{\delta,\delta'} \to 0,$$

and the isomorphisms $q_{L,m',\ell'} : S^{m',\ell'}_{\delta,\delta'} \to \Psi^{m',\ell'}_{\delta,\delta'}$ with $m' = m, m-1+2\delta$, $\ell' = \ell, \ell - 1 + 2\delta'$, and that these are consistent with the inclusion $\iota_S : S^{m-1+2\delta,\ell-1+2\delta'}_{\delta,\delta'} \to S^{m,\ell}_{\delta,\delta'}$, that is one has a commutative diagram $q_{L,m,\ell} \circ \iota_S = \iota \circ q_{L,m-1+2\delta,\ell-1+2\delta'}$.

5.3.5. The Operator Wavefront Set

We also define operator wavefront sets, for which variable orders are irrelevant. We first start with the microsupport of symbols:

Definition 5.3 Suppose $a \in S^{m,\ell}_{\delta,\delta'}(\mathbb{R}^n;\mathbb{R}^n)$. We say that $\alpha \in \partial(\overline{\mathbb{R}^n}\times\overline{\mathbb{R}^n})$ is not in $\mathrm{esssupp}(a)$ if there is a neighborhood U of α in $\overline{\mathbb{R}^n}\times\overline{\mathbb{R}^n}$ such that $a|_{U\cap(\mathbb{R}^n\times\mathbb{R}^n)}$ is $\mathcal{S} = S^{-\infty,-\infty}$ (i.e. satisfies Schwartz estimates in U).

Similarly, for $a \in S^{m,\ell}_{\infty,\delta}(\mathbb{R}^n;\mathbb{R}^n)$ we say that $\alpha \in \overline{\mathbb{R}^n}\times\partial\overline{\mathbb{R}^n}$ is not in $\mathrm{esssupp}_{\infty,\ell}(a)$ if there is a neighborhood U of α in $\overline{\mathbb{R}^n}\times\overline{\mathbb{R}^n}$ such that $a|_{U\cap(\mathbb{R}^n\times\mathbb{R}^n)}$ is $S^{-\infty,\ell}_{\infty,\delta}$ (i.e. satisfies the corresponding symbol estimates in U).

In either case, esssupp is called the *microsupport*, or *essential support*, of a.

Now for operators we define the wavefront set in terms of the microsupport of their left amplitudes a_L.

Definition 5.4 Suppose that $A \in \Psi^{m,\ell}_{\delta,\delta'}$, $A = q_L(a_L)$. We write

$$\mathrm{WF}'(A) = \mathrm{esssupp}(a),$$

that is we say that $\alpha \in \partial(\overline{\mathbb{R}^n}\times\overline{\mathbb{R}^n})$ is *not* in $\mathrm{WF}'(A)$, the *wavefront set* of A, if there is a neighborhood U of α in $\overline{\mathbb{R}^n}\times\overline{\mathbb{R}^n}$ such that $a_L|_{U\cap(\mathbb{R}^n\times\mathbb{R}^n)}$ is $\mathcal{S} = S^{-\infty,-\infty}$ (i.e. satisfies Schwartz estimates in U).

Similarly, for $A \in \Psi^{m,\ell}_{\infty,\delta}$, we write $\mathrm{WF}'_{\infty,\ell}(A) = \mathrm{esssupp}_{\infty,\ell}(A)$.

Note that directly from the definition, the complement of esssupp, and thus the wavefront set, is open, that is the wavefront set itself is closed. Further, even for $\mathrm{WF}'_{\infty,\ell}$, ℓ is only relevant for $\alpha \in \partial\overline{\mathbb{R}^n}\times\partial\overline{\mathbb{R}^n}$; one commonly simply writes WF'_∞, or indeed WF'. While the principal symbol captures the leading order behavior of a pseudodifferential operator, the (complement of the) wave front set captures where it is (not) "trivial."

As an example, if $a \in \mathcal{C}^\infty(\overline{\mathbb{R}^n} \times \overline{\mathbb{R}^n})$, $A = q_L(a)$, then $\mathrm{WF}'(A) \subset \operatorname{supp} a \cap \partial(\overline{\mathbb{R}^n} \times \overline{\mathbb{R}^n})$, since certainly in the complement of $\operatorname{supp} a$, a vanishes, and is thus a symbol of order $-\infty, -\infty$. However, notice that the containment is not an equality, as for example $a \in \mathcal{S}(\mathbb{R}^{2n})$ which never vanishes on $\mathbb{R}^{2n}$ (e.g. a Gaussian) has support everywhere, but $\mathrm{WF}'(q_L(a)) = \emptyset$. Thus, the more precise statement is that $\alpha \notin \mathrm{WF}'(A)$ for such a, A, if α has a neighborhood U in $\partial(\overline{\mathbb{R}^n} \times \overline{\mathbb{R}^n})$ on which the full Taylor series of a vanishes.

Again, as in the case of the principal symbol, one could consider $\mathrm{WF}'_{\infty,\ell}$ a subset of $\overline{\mathbb{R}^n} \times (\mathbb{R}^n \setminus \{0\})$ which is invariant under the $\mathbb{R}^+$-action (dilations in the second factor), that is which is *conic*; this is the standard point of view. The corresponding statement for WF' is, as in the case of the principal symbol, more awkward, and is thus less common.

In view of Proposition 5.1, one could also use a_R with $A = q_R(a_R)$ in place of a_L in the definition. Also, as $\partial(\overline{\mathbb{R}^n} \times \overline{\mathbb{R}^n})$ and $\overline{\mathbb{R}^n} \times \partial\overline{\mathbb{R}^n}$ are compact, so symbol estimates corresponding to an open cover imply symbol estimates everywhere; we have:

Lemma 5.6 *If* $A \in \Psi^{m,\ell}_{\delta,\delta'}$ *and* $\mathrm{WF}'(A) = \emptyset$, *then* $A \in \Psi^{-\infty,-\infty}$.

If $A \in \Psi^{m,\ell}_{\infty,\delta}$ *and* $\mathrm{WF}'_{\infty,\ell}(A) = \emptyset$, *then* $A \in \Psi^{-\infty,\ell}_{\infty}$.

The analogues also hold in variable order spaces.

We also have from (5.21) that

Proposition 5.3 *If* $A \in \Psi^{m,\ell}_{\infty,\delta}$ *then* $A^* \in \Psi^{m,\ell}_{\infty,\delta}$ *and*

$$\sigma_{\infty,m,\ell}(A^*) = \overline{\sigma_{\infty,m,\ell}(A)}, \ \mathrm{WF}'_{\infty}(A^*) = \mathrm{WF}'_{\infty}(A).$$

If $A \in \Psi^{m,\ell}_{\delta,\delta'}$ *then* $A^* \in \Psi^{m,\ell}_{\delta,\delta'}$ *and*

$$\sigma_{m,\ell}(A^*) = \overline{\sigma_{m,\ell}(A)}, \ \mathrm{WF}'(A^*) = \mathrm{WF}'(A).$$

The analogues also hold in variable order spaces.

We can also strengthen the surjectivity part of Proposition 5.2:

Proposition 5.4 *For* $a \in S^{m,\ell}_{\infty,\delta}$ *there exists* $A \in \Psi^{m,\ell}_{\infty,\delta}$ *with* $\sigma_{\infty,m,\ell}(A) = [a]$ *and* $\mathrm{WF}'_{\infty}(A) \subset \operatorname{esssupp}_{\infty} a$.

Similarly, for $a \in S^{m,\ell}_{\delta,\delta'}$ *there exists* $A \in \Psi^{m,\ell}_{\delta,\delta'}$ *with* $\sigma_{m,\ell}(A) = [a]$ *and* $\mathrm{WF}'(A) \subset \operatorname{esssupp} a$.

The analogues also hold in variable order spaces.

Indeed, taking $A = q_L(a)$ or $A = q_R(a)$ will do the job.

5.3.6. Composition and Commutators

The most important part of a treatment of pseudodifferential operators is their properties under composition and commutators:

Proposition 5.5 *If* $A \in \Psi^{m,\ell}_{\infty,\delta}$, $B \in \Psi^{m',\ell'}_{\infty,\delta}$, *then* $AB \in \Psi^{m+m',\ell+\ell'}_{\infty,\delta}$,

$$\sigma_{\infty,m+m',\ell+\ell'}(AB) = \sigma_{\infty,m,\ell}(A)\sigma_{\infty,m',\ell'}(B),$$

and

$$\mathrm{WF}'_\infty(AB) \subset \mathrm{WF}'_\infty(A) \cap \mathrm{WF}'_\infty(B).$$

If $A \in \Psi^{m,\ell}_{\delta,\delta'}$, $B \in \Psi^{m',\ell'}_{\delta,\delta'}$, *then* $AB \in \Psi^{m+m',\ell+\ell'}_{\delta,\delta'}$, *and*

$$\sigma_{m+m',\ell+\ell'}(AB) = \sigma_{m,\ell}(A)\sigma_{m',\ell'}(B),$$

and

$$\mathrm{WF}'(AB) \subset \mathrm{WF}'(A) \cap \mathrm{WF}'(B).$$

The analogues also hold in variable order spaces.

Thus, Ψ_∞ and Ψ are order-filtered $*$-algebras, and in case of Ψ_∞, composition is commutative to leading order in terms of the differential order, m, while in the case of Ψ, it is commutative to leading order in both the differential and the growth orders m and ℓ.

Proof This proposition is proved easily using Proposition 5.1, taking advantage of (5.29) and (5.30). To do so, first assume $a, b \in S^{-\infty,-\infty}_\infty$, then

$$\begin{aligned}
&(q_L(a)q_R(b)u)(z)\\
&\quad = (2\pi)^{-n}\int_{\mathbb{R}^n} e^{i\zeta\cdot z}a(z,\zeta)\Big(\mathcal{F}\mathcal{F}^{-1}\big(\zeta' \mapsto \int_{\mathbb{R}^n} e^{-iz'\cdot\zeta'}b(z',\zeta')\,u(z')\,dz'\big)\Big)\,d\zeta\\
&\quad = (2\pi)^{-n}\int_{\mathbb{R}^n}\int_{\mathbb{R}^n} e^{i\zeta\cdot(z-z')}a(z,\zeta)b(z',\zeta)\,u(z')\,dz'\,d\zeta = (I(c)u)(z),
\end{aligned}$$

with

$$c(z,z',\zeta) = a(z,\zeta)b(z',\zeta) \in S^{-\infty,-\infty,-\infty}_\infty.$$

However, with $c = c(a,b)$ so defined, the map

$$S^{m,\ell}_{\infty,\delta} \times S^{m',\ell'}_{\infty,\delta} \ni (a,b) \mapsto c \in S^{\ell,\ell',m+m'}_{\infty,\delta}$$

is continuous, so as both trilinear maps

$$(a,b,u) \mapsto q_L(a)q_R(b)u,\ (a,b,u) \mapsto I(c(a,b))u$$

are continuous

$$S^{m,\ell}_{\infty,\delta} \times S^{m',\ell'}_{\infty,\delta} \times \mathcal{S} \to \mathcal{S}$$

for all m, m', ℓ, ℓ', it follows that

$$q_L(a)q_R(b) = I(c(a,b)).$$

Since q_L, q_R are isomorphisms, the closedness of $\Psi^{m,\ell}_{\infty,\delta}$ under composition is immediate, as is the continuity of composition. As for the principal symbol, this statement follows since for $B \in \Psi^{m',\ell'}_{\infty,\delta}$, if $B = q_R(b)$, then $\sigma_{\infty,m',\ell'}(B) = b$, and then by (5.28), $I(c(a,b)) = q_L(c_L)$ with $c_L - ab \in S^{m+m'-1+2\delta,\ell+\ell'}_{\infty,\delta}$. The wavefront set statement is also immediate in view of (5.28).

In the case of Ψ, the same arguments go through, but corresponding to the improvement in (5.28), $c_L - ab \in S^{m+m'-1+2\delta,\ell+\ell'-1+2\delta'}_{\delta,\delta'}$. □

Going one order farther in the asymptotic expansion of compositions, one immediately obtains the principal symbol of the commutators. Here we recall the Poisson bracket on $\mathbb{R}^n_z \times \mathbb{R}^n_\zeta$, identified with $T^*\mathbb{R}^n$:

$$\{a,b\} = \sum_{j=1}^{n} \left((\partial_{\zeta_j} a)(\partial_{z_j} b) - (\partial_{z_j} a)(\partial_{\zeta_j} b)\right).$$

Proposition 5.6 *If $A \in \Psi^{m,\ell}_{\infty,\delta}$, $B \in \Psi^{m',\ell'}_{\infty,\delta}$, then $[A,B] \in \Psi^{m+m'-1+2\delta,\ell+\ell'}_{\infty,\delta}$, and*

$$\sigma_{\infty,m+m'-1+2\delta,\ell+\ell'}(AB) = \frac{1}{i}\{\sigma_{\infty,m,\ell}(A), \sigma_{\infty,m',\ell'}(B)\}.$$

If $A \in \Psi^{m,\ell}_{\delta,\delta'}$, $B \in \Psi^{m',\ell'}_{\delta,\delta'}$, then $[A,B] \in \Psi^{m+m'-1+2\delta,\ell+\ell'-1+2\delta'}$, and

$$\sigma_{m+m'-1+2\delta,\ell+\ell'-1+2\delta'}(AB) = \frac{1}{i}\{\sigma_{m,\ell}(A), \sigma_{m',\ell'}(B)\}.$$

The analogues also hold in variable order spaces.

5.3.7. Ellipticity

We now turn to the simplest consequences of the machinery we built up, such as the parametrix construction for elliptic operators.

Definition 5.5 We say that A is elliptic in $\Psi^{m,\ell}_{\infty,\delta}$, resp. $\Psi^{m,\ell}_{\delta,\delta'}$, if $[a]_\infty$, resp. $[a]$, is invertible, that is if there exists $[b]_\infty \in S^{-m,-\ell}_{\infty,\delta}/S^{-m-1+2\delta,-\ell}_{\infty,\delta}$, resp. $[b] \in S^{-m,-\ell}_{\delta,\delta'}/S^{-m-1+2\delta,-\ell-1+2\delta'}_{\delta,\delta'}$ with $[a]_\infty[b]_\infty = [1]$ in $S^{0,0}_{\infty,\delta}/S^{-1+2\delta,0}_{\infty,\delta}$, resp. $[a][b] = [1]$ in $S^{0,0}_{\delta,\delta'}/S^{-1+2\delta,-1+2\delta'}_{\delta,\delta'}$.

More generally, we make the analogous definition if $m = \mathsf{m}$, $l = \mathsf{l}$ are variable.

These definitions are equivalent to the statements that there exist $c > 0$, $R > 0$ such that

$$|a| \geq c\langle z\rangle^{\ell}\langle\zeta\rangle^{m},\ c > 0, |\zeta| > R, \tag{5.42}$$

resp.

$$|a| \geq c\langle z\rangle^{\ell}\langle\zeta\rangle^{m},\ c > 0, |\zeta| + |z| > R; \tag{5.43}$$

indeed, if a satisfies this, the reciprocal is easily seen to satisfy the appropriate conditions in $|\zeta| > R$, resp. $|z| + |\zeta| > R$, and the multiplying by a cutoff, identically 1 near infinity, in ζ, resp. (z, ζ), gives b. Conversely, if b exists, upper bounds for $|b|$ give the desired lower bounds for $|a|$.

Concretely, if $A = \sum_{|\alpha|\leq m} a_\alpha D^\alpha$ as in (5.2), then under the identification of the part of the principal symbol at $\overline{\mathbb{R}^n} \times \partial\overline{\mathbb{R}^n}$ with a homogeneous degree m function on $\overline{\mathbb{R}^n} \times (\mathbb{R}^n \setminus \{0\})$, while identifying the principal symbol at $\partial\overline{\mathbb{R}^n} \times \overline{\mathbb{R}^n}$ as an mth order symbol on $\partial\overline{\mathbb{R}^n} \times \overline{\mathbb{R}^n}$, ellipticity means:

$$z \in \overline{\mathbb{R}^n}, \zeta \neq 0 \Rightarrow \sum_{|\alpha|=m} a_\alpha \zeta^\alpha \neq 0,$$

and

$$z \in \partial\overline{\mathbb{R}^n}, \zeta \in \overline{\mathbb{R}^n} \Rightarrow \sum_{|\alpha|\leq m} a_\alpha \zeta^\alpha \neq 0.$$

For $H = \Delta_g + V - \sigma$ as in (5.41), ellipticity means

$$\begin{aligned} &(z, \zeta) \in \overline{\mathbb{R}^n} \times (\mathbb{R}^n \setminus \{0\}),\ \zeta \neq 0 \Rightarrow \sum g_{ij}(z)\zeta_i\zeta_j \neq 0, \\ &(z, \zeta) \in \partial\overline{\mathbb{R}^n} \times \mathbb{R}^n \Rightarrow \sum g_{ij}\zeta_i\zeta_j - \sigma \neq 0. \end{aligned} \tag{5.44}$$

Now the first is just the statement that g is a Riemannian metric on $\mathbb{R}^n$ in the uniform sense we discussed; the second holds if and only if $\sigma \notin [0, \infty)$. Note that if $V \in S^{-\rho}(\mathbb{R}^n)$ instead, $\rho \in (0, 1)$, then V does affect the principal symbol in the second sense, but it does *not* affect ellipticity.

If A is elliptic in $\Psi^{m,\ell}_{\delta,\delta'}$ (with the variable order case going through without changes), say, then one can construct a parametrix B with a residual, or completely regularizing, error, that is $B \in \Psi^{-m,-\ell}_{\delta,\delta'}$ such that

$$AB - \mathrm{Id}, BA - \mathrm{Id} \in \Psi^{-\infty,-\infty}.$$

Indeed, one takes any B_0 with $\sigma_{-m,-\ell}(B_0)$ being the inverse for $\sigma_{m,\ell}(A)$, so

$$\sigma_{0,0}(AB_0 - \mathrm{Id}) = \sigma_{m,\ell}(A)\sigma_{-m,-\ell}(B_0) - 1 = 0,$$

thus $E_0 = AB_0 - \mathrm{Id} \in \Psi_{\delta,\delta'}^{-1+2\delta,-1+2\delta'}$. Now, $AB_0 = \mathrm{Id} + E_0$, so one wants to invert $\mathrm{Id} + E_0$ approximately; this can be done by a finite Neumann series, $\mathrm{Id} + \sum_{j=1}^{N}(-1)^j E_0^j$, then

$$(\mathrm{Id} + E_0)\left(\mathrm{Id} + \sum_{j=1}^{N}(-1)^j E_0^j\right) - \mathrm{Id} \in \Psi_{\delta,\delta'}^{-(1-2\delta)(N+1),-(1-2\delta')(N+1)}.$$

This can be improved by writing $E_0^j = q_L(e_j)$, then computing the asymptotic sum

$$\tilde{e} \sim \sum_{j=1}^{\infty}(-1)^j e_j \in S_{\delta,\delta'}^{-1+2\delta,-1+2\delta'},$$

taking $\tilde{E} = q_L(\tilde{e})$, $(\mathrm{Id} + E_0)(\mathrm{Id} + \tilde{E}) - \mathrm{Id} \in \Psi^{-\infty,-\infty}$, so $B = B_0(\mathrm{Id} + \tilde{E})$ provides a right parametrix: $E = AB - \mathrm{Id} \in \Psi^{-\infty,-\infty}$. A left parametrix B' can be constructed similarly, and the standard identities showing the identity of left and right inverses in a semigroup, as applied to the quotient by completely regularizing operators, shows that $B - B' \in \Psi^{-\infty,-\infty}$, so one may simply replace B' by B. Indeed, if $B'A = \mathrm{Id} + E'$,

$$\begin{aligned} &B' = B'(AB - E) = (B'A)B - B'E = B - E'B - B'E, \\ &B'E, EB' \in \Psi^{-\infty,-\infty}. \end{aligned} \tag{5.45}$$

Notice that all of the constructions can be done uniformly as long as (5.43) is satisfied for a fixed c and R, that is one can construct the maps $A \mapsto B, E$ such that they are continuous from the set of elliptic operators to $\Psi_{\delta,\delta'}^{-m,-\ell}$ resp. $\Psi^{-\infty,-\infty}$.

If $A \in \Psi_{\infty,\delta}^{m,\ell}$ then the same argument only gains in the first order, m, so one obtains a parametrix $B \in \Psi_{\infty,\delta}^{-m,-\ell}$ with errors $E, E' \in \Psi_{\infty}^{-\infty,0}$.

We have thus proved:

Proposition 5.7 *If $A \in \Psi_{\delta,\delta'}^{m,\ell}$ is elliptic then there exists $B \in \Psi_{\delta,\delta'}^{-m,-\ell}$ such that $AB - \mathrm{Id}, BA - \mathrm{Id} \in \Psi^{-\infty,-\infty}$. Further, the maps $A \mapsto B \in \Psi_{\delta,\delta'}^{-m,-\ell}$ and $A \mapsto AB - \mathrm{Id}, BA - \mathrm{Id} \in \Psi^{-\infty,-\infty}$ can be taken to be continuous from the set of elliptic operators in $\Psi_{\delta,\delta'}^{m,\ell}$ (an open subset of $\Psi_{\delta,\delta'}^{m,\ell}$), equipped with the $\Psi_{\delta,\delta'}^{m,\ell}$ topology.*

If $A \in \Psi_{\infty,\delta}^{m,\ell}$ is elliptic then there exists $B \in \Psi_{\infty,\delta}^{-m,-\ell}$ such that $AB - \mathrm{Id}, BA - \mathrm{Id} \in \Psi_{\infty,\delta}^{-\infty,0}$. Again, the maps $A \mapsto B \in \Psi_{\infty,\delta}^{-m,-\ell}$ and $A \mapsto AB - \mathrm{Id}, BA - \mathrm{Id} \in \Psi_{\infty,\delta}^{-\infty,0}$ can be taken to be continuous from the set of elliptic operators in $\Psi_{\infty,\delta}^{m,\ell}$.

The analogous variable order statements also hold.

If $A \in \Psi^{m,\ell}_{\delta,\delta'}$ elliptic is invertible in the weak sense that there exist $G : \mathcal{S} \to \mathcal{S}'$ continuous such that $GA = \mathrm{Id} : \mathcal{S} \to \mathcal{S}$ and $AG = \mathrm{Id} : \mathcal{S} \to \mathcal{S}$ then (i.e. the left hand side, which a priori maps into $\mathcal{S}'$, actually maps into $\mathcal{S}$ with the claimed equality), with B a parametrix for A, $BA - \mathrm{Id} = E_L$, $AB - \mathrm{Id} = E_R$,

$$G = G(AB - E_R) = B - GE_R = B - (BA - E_L)GE_R = B - BE_R + E_LGE_R,$$

with the first two terms on the right in $\Psi^{-m,-\ell}_{\delta,\delta'}$, resp. $\Psi^{-\infty,-\infty}$, and the last term is residual as well since it is a continuous linear map $\mathcal{S}' \to \mathcal{S}$, and thus has Schwartz kernel in $\mathcal{S}(\mathbb{R}^{2n})$, thus is in $\Psi^{-\infty,-\infty}$. Hence $G \in \Psi^{-m,-\ell}$, and $G - B \in \Psi^{-\infty,-\infty}$. Thus, the inverses of actually invertible elliptic operators are pseudodifferential operators themselves.

As a corollary we have elliptic regularity:

Proposition 5.8 *If $A \in \Psi^{m,\ell}_{\delta,\delta'}$ (or more generally $A \in \Psi^{\mathsf{m},\mathsf{l}}_{\delta,\delta'}$) is elliptic and $Au \in \mathcal{S}$ for some $u \in \mathcal{S}'$ then $u \in \mathcal{S}$.*

Proof Let B be a parametrix for A with $BA - \mathrm{Id} = E \in \Psi^{-\infty,-\infty}$. Then

$$u = \mathrm{Id}\, u = (BA - E)u = B(Au) - Eu,$$

and $Eu \in \mathcal{S}$ as E is completely regularizing while $Au \in \mathcal{S}$ by assumption, hence $B(Au) \in \mathcal{S}$ as well. □

5.3.8. L^2 and Sobolev Boundedness

We can now discuss Hörmander's proof of L^2-boundedness of elements of $\Psi^{0,0}_{\delta,\delta'}$, or indeed $\Psi^{0,0}_{\infty,\delta}$, via a square root construction.

Lemma 5.7 *Suppose that $A \in \Psi^{0,0}_{\infty,\delta}$ is elliptic and symmetric ($A^* = A$) with principal symbol that has a positive (bounded below by a positive constant) representative a. Then there exists $B \in \Psi^{0,0}_{\infty,\delta}$ such that B is symmetric and $A = B^2 + E$ with $E \in \Psi^{-\infty,0}_{\infty}$. The maps $A \mapsto B \in \Psi^{0,0}_{\infty,\delta}$ and $A \mapsto E \in \Psi^{-\infty,0}_{\infty}$ can be taken continuously from the set of A satisfying these constraints (equipped with the $\Psi^{0,0}_{\infty,\delta}$ topology).*

The same result holds with the (∞,δ) subscript replaced by (δ,δ'), but with $E \in \Psi^{-\infty,-\infty}$.

Proof Let $b_0 = \sqrt{a}$; one easily checks that $b_0 \in S^{0,0}_{\infty}$. Let $\tilde{B}_0 \in \Psi^{0,0}_{\infty,\delta}$ have principal symbol b_0, and let $B_0 = \frac{1}{2}(\tilde{B}_0 + \tilde{B}_0^*)$, so B_0 still has principal symbol b_0 and is symmetric. Then $A - B_0^2$ has vanishing principal symbol, so $E_0 = A - B_0^2 \in \Psi^{-1+2\delta,0}_{\infty,\delta}$, providing the first step in the construction.

In general, for $j \in \mathbb{N}$, suppose one has found $B_j \in \Psi^{0,0}_{\infty,\delta}$ to be symmetric such that $E_j = A - B_j^2 \in \Psi^{-(1-2\delta)(j+1),0}_{\infty,\delta}$; we have shown this for $j = 0$. Let e_j be the principal symbol of E_j, and let $b_{j+1} = -\frac{1}{2b_0} e_j \in S^{-(1-2\delta)(j+1),0}_{\infty,\delta}$; this uses b_0 elliptic. Let $\tilde{B}_{j+1} \in \Psi^{-(1-2\delta)(j+1),0}_{\infty,\delta}$ have principal symbol b_{j+1}, $B'_{j+1} = \frac{1}{2}(\tilde{B}_{j+1} + \tilde{B}^*_{j+1})$, $B_{j+1} = B_j + B'_{j+1}$, so B_{j+1} is symmetric. Further, the principal symbol of

$$\begin{aligned} A - B_{j+1}^2 &= A - (B_j + B'_{j+1})^2 = A - B_j^2 - B_j B'_{j+1} - B'_{j+1} B_j - (B'_{j+1})^2 \\ &= E_j - B_j B'_{j+1} - B'_{j+1} B_j - (B'_{j+1})^2 \in \Psi^{-(1-2\delta)(j+1),0}_{\infty,\delta} \end{aligned}$$

is $e_j - 2b_0 b_{j+1} = 0$, so $E_{j+1} = A - B_{j+1}^2 \in \Psi^{-(1-2\delta)(j+2),0}_{\infty,\delta}$, providing the inductive steps. One can finish up by asymptotically summing, as in the elliptic case. □

Proposition 5.9 *Elements* $A \in \Psi^{0,0}_{\infty,\delta}$ *are bounded on* L^2.

Further, if a *is a representative of* $\sigma_{\infty,0,0}(A)$ *and* $C > \inf_{r \in S^{-1+2\delta,0}_{\infty,\delta}} \sup |a+r|$ *then there exists* $E \in \Psi^{-\infty,0}_{\infty}$ *such that*

$$\|Au\|_{L^2} \leq C\|u\|_{L^2} + |\langle Eu, u\rangle|.$$

Moreover, the map $A \mapsto E \in \Psi^{-\infty,0}_{\infty}$ *can be taken to be continuous, and thus the inclusion* $\Psi^{0,0}_{\infty,\delta} \to \mathcal{L}(L^2)$ *is continuous.*

Proof We reduce the proof to the boundedness of elements of $\Psi^{-\infty,0}_{\infty}$ on L^2, which is an easy consequence of Schur's lemma since by (5.38) the Schwartz kernel of elements of this space is a bounded continuous function in z with values in $\mathcal{S}(\mathbb{R}^n_z)$ (hence with values in $L^1(\mathbb{R}^n_{z'})$), and similarly with z and z' interchanged.

Now, suppose that $A \in \Psi^{0,0}_{\infty,\delta}$, so its principal symbol has a bounded representative a; let $M > \sup |a|$. Then $M^2 - |a|^2 \in S^{0,0}_{\infty,\delta}$ is bounded below by a positive constant and is thus elliptic. By Lemma 5.7, there exists $B \in \Psi^{0,0}_{\infty,\delta}$ symmetric such that $M^2 - A^*A = B^2 + E$, $E \in \Psi^{-\infty,0}_{\infty}$. Then, first for $u \in \mathcal{S}$, with inner products and norms the standard L^2 ones,

$$\langle M^2 u, u\rangle = \|Au\|^2 + \|Bu\|^2 + \langle Eu, u\rangle,$$

that is with $\|E\|_{\mathcal{L}(L^2)}$ the L^2 bound of E, which is finite as discussed above,

$$\|Au\|^2 \leq M^2\|u\|^2 + \|E\|_{\mathcal{L}(L^2)}\|u\|^2.$$

Since $\mathcal{S}$ is dense in L^2, this implies that A has a unique continuous extension to L^2; one still denotes it by A. Since $\mathcal{S}$ is also dense in $\mathcal{S}'$, and the inclusion

$L^2 \to \mathcal{S}'$ is continuous, this extension is the restriction of A acting on $\mathcal{S}'$. This proves the first part of the proposition.

For the second part we simply replace a by $a + r$, choosing $r \in S^{-1+2\delta,0}_{\infty,\delta}$ such that $C > \sup |a + r|$, then we can take $M = C$ in the argument above to complete the proof. □

While elements of $\Psi^{0,0}_{\delta,\delta'}$ are in $\Psi^{0,0}_{\infty,\delta}$ for $\delta' = 0$ and are thus L^2-bounded, it is useful to make the bound more explicit there as well, in addition to generalizing to $\delta' > 0$:

Proposition 5.10 *Elements $A \in \Psi^{0,0}_{\delta,\delta'}$ are bounded on L^2.*

Further, if a is a representative of $\sigma_{0,0}(A)$ and $C > \inf_{r \in S^{-1+2\delta,-1+2\delta'}_{\delta,\delta'}} \sup |a + r|$ then there exists $E \in \Psi^{-\infty,-\infty}$ such that

$$\|Au\|_{L^2} \le C\|u\|_{L^2} + |\langle Eu, u\rangle|. \tag{5.46}$$

Moreover, the map $A \mapsto E \in \Psi^{-\infty,-\infty}$ can be taken to be continuous.

Concretely, if $A = q_L(a)$ with $a \in C^\infty(\overline{\mathbb{R}^n} \times \overline{\mathbb{R}^n})$, then for any

$$C > \sup \left| a|_{\partial(\overline{\mathbb{R}^n} \times \overline{\mathbb{R}^n})} \right|,$$

(5.46) *holds.*

Proof This is the same argument as above, but constructing B in $\Psi^{0,0}_{\delta,\delta'}$. □

We now recall that the *weighted Sobolev spaces* are

$$H^{s,r} = \{u \in \mathcal{S}' : \langle z\rangle^r u \in H^s\}, \ \|u\|_{H^{s,r}} = \|\langle z\rangle^r u\|_{H^s}. \tag{5.47}$$

Further, with

$$\Lambda_s = \mathcal{F}^{-1}\langle\zeta\rangle^s \mathcal{F} \in \Psi^{s,0} \subset \Psi^{s,0}_\infty,$$

the standard *Sobolev spaces* are

$$H^s = \{u : \Lambda_s u \in L^2\} \text{ with } \|u\|_{H^s} = \|\Lambda^s u\|_{L^2}.$$

We note here that

$$\cup_{M,N\in\mathbb{R}} H^{M,N} = \mathcal{S}'.$$

Thus, $\Lambda_{s,r} = \Lambda_s\langle z\rangle^r : H^{s,r} \to L^2$ is an isometry, with inverse $\Lambda'_{-s,-r} = \langle z\rangle^{-r}\Lambda_{-s} : L^2 \to H^{s,r}$. Hence, the boundedness of some $A \in \Psi^{m,\ell}_{\infty,\delta}$ as a map $H^{s,r} \to H^{s',r'}$ is equivalent to the boundedness on L^2 of $\Lambda_{s',r'} A \Lambda'_{-s,-r}$ as

$$A = \Lambda'_{-s',-r'}(\Lambda_{s',r'} A \Lambda'_{-s,-r})\Lambda_{s,r}.$$

But $\Lambda_{s',r'} A \Lambda'_{-s,-r} \in \Psi^{m+s'-s,\ell+r'-r}_{\infty,\delta}$, so we conclude that

Proposition 5.11 *An operator $A \in \Psi^{m,\ell}_{\infty,\delta}$ is bounded $H^{s,r} \to H^{s',r'}$ if $m = s-s'$ and $\ell = r-r'$ (thus if $m \leq s-s'$ and $\ell \leq r-r'$).*

This gives a quantified version of elliptic regularity:

Proposition 5.12 *If $A \in \Psi^{m,\ell}_{\delta,\delta'}$ is elliptic and $Au \in H^{s,r}$ for some $u \in \mathcal{S}'$ then $u \in H^{s+m,r+\ell}$. In fact, for any M,N there is $C > 0$ such that*

$$\|u\|_{H^{s+m,r+\ell}} \leq C(\|Au\|_{H^{s,r}} + \|u\|_{H^{M,N}}).$$

If $A \in \Psi^{m,\ell}_{\infty,\delta}$ is elliptic and $Au \in H^{s,r}$ for some $u \in H^{k,r+\ell}$, $k \in \mathbb{R}$, then $u \in H^{s+m,r+\ell}$. In fact, for any k there is $C > 0$ such that

$$\|u\|_{H^{s+m,r+\ell}} \leq C(\|Au\|_{H^{s,r}} + \|u\|_{H^{k,r+\ell}}).$$

The point of the quantitative estimate is to allow M,N to be very negative, so for example $H^{s+m,r+\ell} \to H^{M,N}$ is compact. One thinks of $\|u\|_{H^{M,N}}$ as a "trivial" term correspondingly.

In the case of $\Psi^{m,\ell}_{\infty,\delta}$ ellipticity is too weak a notion to gain decay at infinity; one simply has a uniform gain of Sobolev regularity.

Proof Suppose $A \in \Psi^{m,\ell}_{\delta,\delta'}$. Let $B \in \Psi^{-m,-\ell}_{\delta,\delta'}$ be a parametrix for A with $BA - \mathrm{Id} = E \in \Psi^{-\infty,-\infty}$. Then

$$u = \mathrm{Id}\, u = (BA + E)u = B(Au) + Eu,$$

and $Eu \in \mathcal{S}$ while $Au \in H^{s,r}$ by assumption, hence $B(Au) \in H^{s+m,r+\ell}$, as claimed. The bound in the proposition follows from $E : H^{M,N} \to H^{s+m,r+\ell}$ being bounded.

If $A \in \Psi^{m,\ell}_{\infty,\delta}$, and $B \in \Psi^{-m,-\ell}_{\infty,\delta}$ is a parametrix, so $BA - \mathrm{Id} = E \in \Psi^{-\infty,0}_{\infty}$ then the same argument gives, using $E : H^{k,r+\ell} \to H^{s+m,r+\ell}$ bounded, the conclusion that $u \in H^{s+m,r+\ell}$, as well as the estimate. □

An immediate corollary is:

Proposition 5.13 *Any elliptic $A \in \Psi^{m,\ell}_{\delta,\delta'}$ is Fredholm as a map $H^{s,r} \to H^{s-m,r-\ell}$ for all $m,\ell,s,r \in \mathbb{R}$, that is has a closed range, finite dimensional nullspace and the range has finite codimension. Further, the nullspace is a subspace of $\mathcal{S}$, while the annihilator of the range in $H^{s-m,r-\ell}$ in the dual space $H^{-s+m,-r+\ell}$ is also in $\mathcal{S}$. Correspondingly, the nullspace of A as well as the annihilator of its range is independent of r,s; if A is invertible for one value of r,s, then it is invertible for all.*

Proof If B is a parametrix for A, then $B \in \mathcal{L}(H^{s-m,r-\ell}, H^{s,r})$ and $E_L = BA - \mathrm{Id}, E_R = AB - \mathrm{Id} \in \Psi^{-\infty,\infty}$. Thus E_L, E_R map $H^{s,r}$, resp. $H^{s-m,r-\ell}$, to $\mathcal{S}$

continuously, and are thus compact as maps in $\mathcal{L}(H^{s,r})$, resp. $\mathcal{L}(H^{s-m,r-\ell})$. Then standard arguments give the Fredholm property.

The property of the nullspace being in $\mathcal{S}$ is elliptic regularity. If v is in the annihilator as stated, that is $\langle v, Au\rangle = 0$ for all $u \in H^{s,r}$ then $\langle A^*v, u\rangle = 0$ for all $u \in H^{s,r}$, so $A^*v = 0$ in $H^{-s,-r}$. As A^* has principal symbol $\bar{a}$, elliptic regularity shows that $v \in \mathcal{S}$. □

Corollary 5.2 *Suppose $m, \ell > 0$, $A \in \Psi^{m,\ell}_{\delta,\delta'}$ is symmetric on L^2 and is elliptic. Then A is self-adjoint with domain $H^{m,\ell}$.*

Proof It suffices to show that $A - \sigma : H^{m,\ell} \to L^2$ are invertible for $\sigma \in \mathbb{C} \setminus \mathbb{R}$. As $m, \ell > 0$, these are elliptic regardless of σ, thus Fredholm as stated, with nullspace and cokernel being identified as the kernel of A^*, in $\mathcal{S}$. But the symmetry of A shows that for u in the kernel, $0 = \operatorname{Im}\langle (A-\sigma)u, u\rangle = -\operatorname{Im}\sigma\,\|u\|^2$, so $u = 0$, hence the kernel is trivial. Thus, the kernel of $A^* = A$ is also trivial, so A is surjective, thus the desired invertibility follows. □

Corollary 5.3 *Suppose $m \geq 0$, $\ell \geq 0$, $A \in \Psi^{m,\ell}_{\delta,\delta'}$ is symmetric on L^2 and $\sigma_{\mathrm{fiber},m.\ell}(A)$, resp. $\sigma_{\mathrm{base},m,\ell}(A)$, is elliptic if $m > 0$, resp. $\ell > 0$. Then A is self-adjoint with domain $H^{m,\ell}$.*

Proof We have already dealt with $m, \ell > 0$; $m, \ell = 0$ is standard, so it remains to deal with $m > 0$, $\ell = 0$ as $m = 0$, $\ell > 0$ is similar. Again, it suffices to show that $A - \sigma : H^{m,\ell} \to L^2$ are invertible for $\sigma \in \mathbb{C} \setminus \mathbb{R}$. The principal symbol has a real representative a (simply take the real part of any representative) and by ellipticity at fiber infinity there exist $c_0, R > 0$ such that $|a| \geq c_0|\zeta|^m$ if $|\zeta| > R$. We claim that

$$|a-\sigma|^2 = |a - \operatorname{Re}\sigma|^2 + |\operatorname{Im}\sigma|^2 \geq c\langle\zeta\rangle^{2m},\ c > 0.$$

Indeed, for $|a| \geq 2|\operatorname{Re}\sigma|$, $|a - \operatorname{Re}\sigma|^2 \geq (|a| - |\operatorname{Re}\sigma|)^2 \geq |a|^2/4$, so for $|\zeta| \geq R$ with $c_0|\zeta|^m > 2|\operatorname{Re}\sigma|$ the inequality follows. On the other hand, otherwise $|\zeta| \leq \max(R, (2c_0^{-1}|\operatorname{Re}\sigma|)^{1/m})$, so ζ is bounded, and then the $\operatorname{Im}\sigma$ term gives the desired inequality. So $A - \sigma$ is elliptic when $\operatorname{Im}\sigma \neq 0$, thus Fredholm as stated, with nullspace and cokernel in $\mathcal{S}$. Again, the symmetry of A shows that for u in the kernel, $0 = \operatorname{Im}\langle (A-\sigma)u, u\rangle = -\operatorname{Im}\sigma\,\|u\|^2$, so $u = 0$, hence the kernel of $A - \sigma$ is trivial. Thus, the kernel of $A^* = A$ is also trivial, so A is surjective, thus the desired invertibility follows. □

We summarize our results so far for the Schrödinger operators:

Proposition 5.14 *Let g be a Riemannian metric, $g_{ij} \in \mathcal{C}^\infty(\overline{\mathbb{R}^n})$, positive definite on $\overline{\mathbb{R}^n}$, $V \in S^{-\rho}(\mathbb{R}^n)$ with $\rho > 0$. Let $H = \Delta_g + V$. Then for*

$\sigma \in \mathbb{C} \setminus [0,\infty)$, $H - \sigma : H^{s,r} \to H^{s-2,r}$ *is Fredholm for all r, s, with nullspace in* $\mathcal{S}$. *If V is real-valued, then H is self-adjoint.*

5.3.9. Variable Order Sobolev Spaces

We can now define variable order Sobolev spaces.

Definition 5.6 Let $A \in \Psi^{\mathsf{m},\mathsf{l}}_{\delta,\delta'}$ be elliptic, $\mathsf{m} \geq m$, $\mathsf{l} \geq \ell$. Let $H^{\mathsf{m},\mathsf{l}}$ be a subspace of $H^{m,\ell}$ given by

$$H^{\mathsf{m},\mathsf{l}} = \{u \in H^{m,\ell} : Au \in L^2\},$$

with norm

$$\|u\|^2_{H^{\mathsf{m},\mathsf{l}}} = \|u\|^2_{H^{m,\ell}} + \|Au\|^2_{L^2}.$$

Then $H^{\mathsf{m},\mathsf{l}}$ is easily seen to be a complete space, thus a Hilbert space, which in the case of m, l being constant equal to m', ℓ', simply gives $H^{m',\ell'}$. Indeed, if $\{u_j\}_{j=1}^{\infty}$ is Cauchy in $H^{\mathsf{m},\mathsf{l}}$, then it is Cauchy in $H^{m,\ell}$, so it converges to some $u \in H^{m,\ell}$; in addition Au_j is Cauchy in L^2 so converges to some $v \in L^2$. But $A : \mathcal{S}' \to \mathcal{S}'$ is continuous, so $Au_j \to Au$ in $\mathcal{S}'$, so $v = Au \in L^2$, thus $u \in H^{\mathsf{m},\mathsf{l}}$. Further, as $Au_j \to Au$ in L^2, the completeness of $H^{\mathsf{m},\mathsf{l}}$ follows.

Moreover, different choices of both A and (m,ℓ) are equivalent in the sense that they give the same space with equivalent norms: if $\tilde{A} \in \Psi^{\mathsf{m},\mathsf{l}}_{\delta,\delta'}$ is elliptic as well, writing $B \in \Psi^{-\mathsf{m},-\mathsf{l}}_{\delta,\delta'}$ as a parametrix, with $E = BA - \mathrm{Id} \in \Psi^{-\infty,-\infty}_{\delta,\delta'}$,

$$\tilde{A}u = \tilde{A}(BA) - \tilde{A}Eu = (\tilde{A}B)Au - (\tilde{A}E)u$$

with $\tilde{A}B \in \Psi^{0,0}_{\delta,\delta'}$, $\tilde{A}E \in \Psi^{-\infty,-\infty}_{\delta,\delta'}$, we deduce that $\tilde{A}u \in L^2$, and $\|\tilde{A}u\|^2 \leq C(\|u\|^2_{H^{m,\ell}} + \|Au\|^2_{L^2})$, showing that the $\tilde{A}$-based norm is bounded by the A-based norm. A similar argument gives the converse estimate, thus the equivalence of norms.

We conclude:

Proposition 5.15 *An operator* $A \in \Psi^{\mathsf{m},\mathsf{l}}_{\delta,\delta'}$ *is bounded* $H^{\mathsf{s},\mathsf{r}} \to H^{\mathsf{s}',\mathsf{r}'}$ *if* $\mathsf{m} = \mathsf{s} - \mathsf{s}'$ *and* $\mathsf{l} = \mathsf{r} - \mathsf{r}'$ *(thus if* $\mathsf{m} \leq \mathsf{s} - \mathsf{s}'$ *and* $\mathsf{l} \leq \mathsf{r} - \mathsf{r}'$*).*

Proof Let s, r be such that $s \leq \mathsf{s}$, $r \leq \mathsf{r}$ and $m \geq \mathsf{m}$, $\ell \geq \mathsf{l}$. Such an $A \in \Psi^{\mathsf{m},\mathsf{l}}_{\delta,\delta'} \subset \Psi^{m,\ell}_{\delta,\delta'}$ maps $H^{s,r}$ to $H^{s-m,r-\ell}$ continuously. Further, if $\tilde{A} \in \Psi^{\mathsf{s},\mathsf{r}}_{\delta,\delta'}$, $\tilde{A}' \in \Psi^{\mathsf{s}',\mathsf{r}'}_{\delta,\delta'}$ are elliptic, then with $\tilde{B} \in \Psi^{-\mathsf{s},-\mathsf{r}}_{\delta,\delta'}$, $\tilde{B}\tilde{A} - \mathrm{Id} = \tilde{E} \in \Psi^{-\infty,-\infty}_{\delta,\delta'}$, then

$$\tilde{A}'Au = (\tilde{A}'A\tilde{B})\tilde{A}u - (\tilde{A}'A\tilde{E})u,$$

with $\tilde{A}'A\tilde{B} \in \Psi^{0,0}_{\delta,\delta'}$ and $\tilde{A}'A\tilde{E} \in \Psi^{-\infty,-\infty}_{\delta,\delta'}$, thus bounded on L^2, giving the conclusion. □

One then has a Fredholm and a self-adjointness statement as above for the variable order setting.

5.3.10. Microlocalization

The elliptic parametrix construction can be *microlocalized*, that is if the principal symbol of A is only assumed to be elliptic on (hence near) a closed subset K of $\partial(\overline{\mathbb{R}^n} \times \overline{\mathbb{R}^n})$, one still can construct a microlocal parametrix B, that is one whose errors $BA - \mathrm{Id}, AB - \mathrm{Id}$ as a parametrix have wavefront set disjoint from K. To make this precise, first we define microlocal ellipticity:

Definition 5.7 We say that $A \in \Psi^{m,\ell}_{\delta,\delta'}$, $\sigma_{m,\ell}(A) = [a]$, is elliptic at $\alpha \in \partial(\overline{\mathbb{R}^n} \times \overline{\mathbb{R}^n})$ if α has a neighborhood U in $\overline{\mathbb{R}^n} \times \overline{\mathbb{R}^n}$ such that $a|_{U\cap\mathbb{R}^n\times\mathbb{R}^n}$ is elliptic, that is satisfies (5.43) on U. We say that A is elliptic on a subset K of $\partial(\overline{\mathbb{R}^n} \times \overline{\mathbb{R}^n})$ if it is elliptic at each point of K. The elliptic set $\mathrm{Ell}(A)$ is the set of points at which A is elliptic; the characteristic set $\mathrm{Char}(A)$ is its complement.

We say that $A \in \Psi^{m,\ell}_{\infty,\delta}$, $\sigma_{\infty,m,\ell}(A) = [a]$, is elliptic at $\alpha \in \overline{\mathbb{R}^n} \times \partial\overline{\mathbb{R}^n}$ if α has a neighborhood U in $\overline{\mathbb{R}^n} \times \overline{\mathbb{R}^n}$ such that $a|_{U\cap\mathbb{R}^n\times\mathbb{R}^n}$ is elliptic, that is satisfies (5.42) on U. We say that A is elliptic on a subset K of $\overline{\mathbb{R}^n} \times \partial\overline{\mathbb{R}^n}$ if it is elliptic at each point of K. One defines $\mathrm{Ell}_\infty(A)$, $\mathrm{Char}_\infty(A)$ analogously to the above definition.

We also make the analogous definitions if $m = \mathsf{m}$, $\ell = \mathsf{l}$ are variable.

If $A \in \Psi^{m,\ell}_{\delta,\delta'}$ is elliptic on a closed (hence compact) K, then a covering argument shows that a satisfies (5.43) on a neighborhood of K. A similar statement holds for $A \in \Psi^{m,\ell}_{\infty,\delta}$.

Proposition 5.16 *If $A \in \Psi^{m,\ell}_{\delta,\delta'}$ (or $A \in \Psi^{\mathsf{m},\mathsf{l}}_{\delta,\delta'}$) is elliptic on a compact set K then there exists $B \in \Psi^{-m,-\ell}_{\delta,\delta'}$ (resp. $B \in \Psi^{-\mathsf{m},-\mathsf{l}}_{\delta,\delta'}$) such that $E_L = BA - \mathrm{Id}$, $E_R = AB - \mathrm{Id}$ satisfy $\mathrm{WF}'(E_L) \cap K = \emptyset$, $\mathrm{WF}'(E_R) \cap K = \emptyset$.*

Proof If A is elliptic on K, there is a neighborhood U of K in $\overline{\mathbb{R}^n} \times \overline{\mathbb{R}^n}$ such that $a|_{U\cap\mathbb{R}^n\times\mathbb{R}^n}$ is elliptic, that is satisfies (5.43) on U. We may shrink U so that $|z| + |\zeta| > R$ on U; thus $|a|_U|$ has a positive lower bound on all of U. Let $q \in C^\infty(\overline{\mathbb{R}^n} \times \overline{\mathbb{R}^n})$ be identically 1 near K, be supported in U, and let $Q \in \Psi^{0,0}$ be given by $Q = q_L(q)$. Thus, Q has principal symbol $\sigma_{0,0}(Q) = q|_{\partial(\overline{\mathbb{R}^n}\times\overline{\mathbb{R}^n})}$, and $\mathrm{WF}'(Q) \subset U$, $\mathrm{WF}'(\mathrm{Id} - Q) \cap K = \emptyset$. Now let $[a]$ be the principal symbol of A, let $b_0 = qa^{-1} \in S^{-m,-\ell}_{\delta,\delta'}$ since a is elliptic on U. Let $B_0 = q_L(b_0)$, so $\sigma_{-m,-\ell}(B_0) = b_0$ and $\mathrm{WF}'(B_0) \subset U$. Let $q_0 \in C^\infty(\overline{\mathbb{R}^n} \times \overline{\mathbb{R}^n})$ be identically 1 near K, have disjoint support from $1 - q$, so $q_0(1-q) = 0$, and let $Q_0 = q_L(q_0)$. Note that $\mathrm{WF}'(\mathrm{Id} - Q_0) \cap K = \emptyset$. Then

$E_{0,L} = Q_0(B_0A - \mathrm{Id}) \in \Psi^{0,0}_{\delta,\delta'}$, $E_{0,R} = (AB_0 - \mathrm{Id})Q_0 \in \Psi^{0,0}_{\delta,\delta'}$ have vanishing principal symbols, so $E_{0,L}, E_{0,R} \in \Psi^{-1+2\delta,-1+2\delta'}_{\delta,\delta'}$. As in the globally elliptic case, one may asymptotically sum the amplitudes $e_{L,j}$ of $(-1)^j E^j_{0,L}$ to obtain $\tilde{E}_L$ such that $F_N = \tilde{E}_L - \sum_{j=1}^N (-1)^j E^j_{0,L} \in \Psi^{-(1-2\delta)(N+1),-(1-2\delta')(N+1)}_{\delta,\delta'}$ for all N. Thus,

$$\begin{aligned}(\mathrm{Id} + \tilde{E}_L)Q_0B_0A &= (\mathrm{Id} + \tilde{E}_L)(E_{0,L} + \mathrm{Id}) + (\mathrm{Id} + \tilde{E}_L)(Q_0 - \mathrm{Id}) \\ &= \left(\mathrm{Id} + \sum_{j=1}^N (-1)^j E^j_{0,L} + F_N\right)(\mathrm{Id} + E_{0,L}) + (\mathrm{Id} + \tilde{E}_L)(Q_0 - \mathrm{Id}) \\ &= \mathrm{Id} + (-1)^{N+1}E^{N+1}_{0,L} + F_N(\mathrm{Id} + E_{0,L}) + (\mathrm{Id} + \tilde{E}_L)(Q_0 - \mathrm{Id}).\end{aligned}$$

Now,

$$(-1)^{N+1}E^{N+1}_{0,L} + F_N(\mathrm{Id} + E_{0,L}) \in \Psi^{-(1-2\delta)(N+1),-(1-2\delta')(N+1)}_{\delta,\delta'},$$

and is independent of N since when adding the identity we get

$$\left(\mathrm{Id} + \sum_{j=1}^N (-1)^j E^j_{0,L} + F_N\right)(\mathrm{Id} + E_{0,L}) = (\mathrm{Id} + \tilde{E}_L)(\mathrm{Id} + E_{0,L}),$$

so it is in $\Psi^{-\infty,-\infty}$, and $\mathrm{WF}'((\mathrm{Id} + \tilde{E}_L)(Q_0 - \mathrm{Id})) \subset \mathrm{WF}'(Q_0 - \mathrm{Id})$, which is disjoint from K. Thus, we may take

$$B_L = (\mathrm{Id} + \tilde{E}_L)Q_0B_0$$

as our microlocal left parametrix, and similarly obtain a microlocal right parametrix B_R. The parametrix identity (5.45) now shows that $\mathrm{WF}'(B_L - B_R) \cap K = \emptyset$, completing the proof.

The proof of the variable order case goes through without changes. □

One corollary is the following.

Corollary 5.4 *Suppose* $u \in \mathcal{S}'$, $A \in \Psi^{m,\ell}_{\delta,\delta'}$, *and* $Au \in H^{s,r}$ *then for* $Q \in \Psi^{0,0}_{\delta,\delta'}$ *with* $\mathrm{WF}'(Q) \cap \mathrm{Char}(A) = \emptyset$, $Qu \in H^{s+m,r+\ell}$. *Further, for all* M, N *there exists* $C > 0$ *such that*

$$\|Qu\|_{H^{s+m,r+\ell}} \le C(\|Au\|_{H^{s,r}} + \|u\|_{H^{M,N}}).$$

There is also an analogue with variable order spaces.

Proof Let B be a microlocal parametrix for A near $\mathrm{WF}'(Q)$. Then $BA - \mathrm{Id} = E$ with $\mathrm{WF}'(E) \cap \mathrm{WF}'(Q) = \emptyset$. Thus,

$$Qu = Q(BA - E)u = QB(Au) - (QE)u.$$

Now, $\mathrm{WF}'(QE) = \mathrm{WF}'(Q) \cap \mathrm{WF}'(E) = \emptyset$, so $QE \in \Psi^{-\infty,-\infty}$, and thus $QEu \in \mathcal{S}$, while $QB \in \Psi^{-m,-\ell}_{\delta,\delta'}$, so the proof is finished as for global elliptic regularity. □

Here the assumption $Au \in H^{s,r}$ is too strong; it only matters that Au is microlocally near $\mathrm{WF}'(Q)$. That is:

Corollary 5.5 *(Microlocal elliptic regularity; operator version.) Suppose $u \in \mathcal{S}'$, $A \in \Psi^{m,\ell}_{\delta,\delta'}$, and for some $Q' \in \Psi^{0,0}_{\delta,\delta'}$, $Q'(Au) \in H^{s,r}$. Then for $Q \in \Psi^{0,0}_{\delta,\delta'}$ with $\mathrm{WF}'(Q) \subset \mathrm{Ell}(A) \cap \mathrm{Ell}(Q')$, $Qu \in H^{s+m,r+\ell}$. Further, for all M, N there exists $C > 0$ such that*

$$\|Qu\|_{H^{s+m,r+\ell}} \leq C(\|Q'Au\|_{H^{s,r}} + \|u\|_{H^{M,N}}).$$

There is again an analogue with variable order spaces.

Proof We just note that $Q'A$ is elliptic on $\mathrm{Ell}(A) \cap \mathrm{Ell}(Q')$, so the previous corollary is applicable. □

One can restate the corollary in terms of microlocalizing the distributions instead of adding the microlocalizers explicitly as operators.

Definition 5.8 Suppose $\alpha \in \partial(\overline{\mathbb{R}^n} \times \overline{\mathbb{R}^n})$, $u \in \mathcal{S}'$. We say that $\alpha \notin \mathrm{WF}^{m,\ell}(u)$ if there exists $A \in \Psi^{0,0}_{\delta,\delta'}$ elliptic at α such that $Au \in H^{m,\ell}$. We say that $\alpha \notin \mathrm{WF}(u)$ if there exists $A \in \Psi^{0,0}_{\delta,\delta'}$ elliptic at α such that $Au \in \mathcal{S}$.

For $k, \ell, m \in \mathbb{R}$, $u \in H^{k,\ell}$, $\mathrm{WF}^{m,\ell}_{\infty}(u)$ is defined similarly: if $\alpha \in \overline{\mathbb{R}^n} \times \partial\overline{\mathbb{R}^n}$, we say $\alpha \notin \mathrm{WF}^{m,\ell}_{\infty}(u)$ if there exists $A \in \Psi^{0,0}_{\infty,\delta}$ elliptic at α such that $Au \in H^{m,\ell}$. We say that $\alpha \notin \mathrm{WF}_{\infty,\ell}(u)$ if there exists $A \in \Psi^{0,0}_{\infty,\delta}$ elliptic at α such that $Au \in H^{\infty,\ell}$.

We also make the analogous definition for variable order spaces.

Notice that a priori the notion of $\mathrm{WF}^{m,\ell}(u)$ depends on δ, δ', but in fact the arguments below show that it in fact has no such dependence, see Lemma 5.8.

The most important property of WF and pseudodifferential operators is microlocality:

Proposition 5.17 *If $A \in \Psi^{m,\ell}_{\delta,\delta'}$ and $u \in \mathcal{S}'$ then*

$$\mathrm{WF}^{s,r}(Au) \subset \mathrm{WF}'(A) \cap \mathrm{WF}^{s+m,r+\ell}(u)$$

and

$$\mathrm{WF}(Au) \subset \mathrm{WF}'(A) \cap \mathrm{WF}(u).$$

The variable order version of this statement also holds.

Proof We need to show that

$$\mathrm{WF}^{s,r}(Au) \subset \mathrm{WF}'(A) \text{ and } \mathrm{WF}^{s,r}(Au) \subset \mathrm{WF}^{s+m,r+\ell}(u).$$

We start with the former, which is straightforward. Suppose $\alpha \notin \mathrm{WF}'(A)$. Let $Q \in \Psi^{0,0}$ be elliptic at α but with $\mathrm{WF}'(Q) \cap \mathrm{WF}'(A) = \emptyset$; one can achieve this as $\mathrm{WF}'(A)$ is closed, so one simply needs to take $q \in C^\infty(\overline{\mathbb{R}^n} \times \overline{\mathbb{R}^n})$ equal to 1 near α and with essential support disjoint from $\mathrm{WF}'(A)$. Then $\mathrm{WF}'(QA) \subset \mathrm{WF}'(Q) \cap \mathrm{WF}'(A) = \emptyset$, so $QA \in \Psi^{-\infty,-\infty}$, thus $QAu \in \mathcal{S}$.

Now for the second inclusion. Suppose $\alpha \notin \mathrm{WF}^{s+m,r+\ell}(u)$. Then there exists $B \in \Psi^{0,0}_{\delta,\delta'}$ elliptic at α such that $Bu \in H^{s+m,r+\ell}$. Let $G \in \Psi^{0,0}_{\delta,\delta'}$ be a microlocal parametrix for B, so $GB = \mathrm{Id} + E$ with $\alpha \notin \mathrm{WF}'(E)$. Then $Au = AGBu - AEu$, and $AG \in \Psi^{m,\ell}_{\delta,\delta'}$, so $AGBu \in H^{s,r}$. On the other hand, $\alpha \notin \mathrm{WF}'(AE) \subset \mathrm{WF}'(E)$. So let $Q \in \Psi^{0,0}$ be elliptic at α but with $\mathrm{WF}'(Q) \cap \mathrm{WF}'(E) = \emptyset$. Then $QAE \in \Psi^{-\infty,-\infty}$, and thus

$$QAu = Q(AG)(Bu) - (QAE)u \in H^{s,r},$$

so $\alpha \notin \mathrm{WF}^{s,r}(u)$, completing the proof for $\mathrm{WF}^{s,r}(Au)$. The proof for $\mathrm{WF}(Au)$ is analogous. □

Note that the last part of the proof shows more:

Lemma 5.8 *If $\alpha \notin \mathrm{WF}^{s,r}(u)$ then there is a neighborhood U of α such that for all $Q \in \Psi^{0,0}_{\delta,\delta'}$ with $\mathrm{WF}'(Q) \subset U$, $Qu \in H^{s,r}$.*

Further, with the same U, for all $Q \in \Psi^{m,\ell}_{\delta,\delta'}$ with $\mathrm{WF}'(Q) \subset U$, $Qu \in H^{s-m,r-\ell}$.

The variable order version of this statement also holds.

Thus, while the wavefront set definition is a "there exists" statement, in fact it is equivalent to a "for all" statement, namely for all $Q \in \Psi^{0,0}_{\delta,\delta'}$ with $\mathrm{WF}'(Q)$ in a sufficiently neighborhood of α, $Qu \in H^{s,r}$. (The other direction is simply because these Q include those elliptic at α.)

Also, as immediate from the proof below, one can take U to be the elliptic set of the $B \in \Psi^{0,0}_{\delta,\delta'}$, elliptic at α, with $Bu \in H^{s,r}$, whose existence is guaranteed by $\alpha \notin \mathrm{WF}^{s,r}(u)$

Proof Suppose $\alpha \notin \mathrm{WF}^{s,r}(u)$. Then there exists $B \in \Psi^{0,0}_{\delta,\delta'}$ elliptic at α such that $Bu \in H^{s,r}$; let $G \in \Psi^{0,0}_{\delta,\delta'}$ be a microlocal parametrix for B, so $GB = \mathrm{Id} + E$ with $\alpha \notin \mathrm{WF}'(E)$. Let U be the complement of $\mathrm{WF}'(E)$; this is a neighborhood of α. Then for any $Q \in \Psi^{0,0}_{\delta,\delta'}$ with $\mathrm{WF}'(Q) \subset U$, $QE \in \Psi^{-\infty,-\infty}$, so $Qu = QGBu - QEu \in H^{s,r}$ as $QG \in \Psi^{0,0}_{\delta,\delta'}$.

The second statement is proved the same way, noticing that $QG \in \Psi^{m,\ell}_{\delta,\delta'}$ now. □

An immediate consequence is:

Lemma 5.9 *If* $u \in \mathcal{S}'$ *and* $\mathrm{WF}^{m,\ell}(u) = \emptyset$ *then* $u \in H^{m,\ell}$.

If $u \in H^{k,\ell}$ *and* $\mathrm{WF}^{m,\ell}_\infty(u) = \emptyset$ *then* $u \in H^{m,\ell}$.

The variable order version of this statement also holds.

Proof Suppose $u \in \mathcal{S}'$ and $\mathrm{WF}^{m,\ell}(u) = \emptyset$. Then for all $\alpha \in \partial(\overline{\mathbb{R}^n} \times \overline{\mathbb{R}^n})$ there exists U_α open such that for all $Q \in \Psi^{0,0}$ with $\mathrm{WF}'(Q) \subset U_\alpha$, $Qu \in H^{m,\ell}$. Now

$$\{U_\alpha : \alpha \in \partial(\overline{\mathbb{R}^n} \times \overline{\mathbb{R}^n})\}$$

is an open cover of the compact set $\partial(\overline{\mathbb{R}^n} \times \overline{\mathbb{R}^n})$, so there is a finite subcover, say $\{U_{\alpha_j} : j = 1, \ldots, N\}$. Let $\tilde{U}_{\alpha_j}$ be open in $\overline{\mathbb{R}^n} \times \overline{\mathbb{R}^n}$ with $\tilde{U}_{\alpha_j} \cap \partial(\overline{\mathbb{R}^n} \times \overline{\mathbb{R}^n}) = U_{\alpha_j}$. Then, with $\tilde{U}_{\alpha_0} = \mathbb{R}^n \times \mathbb{R}^n$,

$$\{\tilde{U}_{\alpha_j} : j = 0, 1, \ldots, N\}$$

is a finite open cover of $\overline{\mathbb{R}^n} \times \overline{\mathbb{R}^n}$. Let $\sum_{j=0}^N q_j = 1$ be a subordinate partition of unity, and let $Q_j = q_L(q_j)$. Then $\sum_{j=0}^N Q_j = \mathrm{Id}$, $Q_0 \in \Psi^{-\infty,-\infty}$ since q_0 has compact support, while for $j = 1, \ldots, N$, $\mathrm{WF}'(Q_j) \subset U_{\alpha_j}$ since $\operatorname{supp} q_j \subset \tilde{U}_{\alpha_j}$. Thus, $Q_j u \in H^{m,\ell}$ for all j, and thus $u = \sum Q_j u \in H^{m,\ell}$ as claimed.

The argument for WF_∞ is analogous. □

The distributional version of microlocal elliptic regularity then is:

Corollary 5.6 *(Microlocal elliptic regularity; distributional version.) Suppose* $u \in \mathcal{S}'$, $A \in \Psi^{m,\ell}_{\delta,\delta'}$. *Then*

$$\mathrm{WF}^{s+m,r+\ell}(u) \subset \mathrm{Char}(A) \cup \mathrm{WF}^{s,r}(Au).$$

The variable order version of this statement also holds.

Proof Suppose $\alpha \notin \mathrm{Char}(A) \cup \mathrm{WF}^{s,r}(Au)$, we need to show $\alpha \notin \mathrm{WF}^{s+m,r+\ell}(u)$. As $\alpha \notin \mathrm{WF}^{s,r}(Au)$ there exists $Q' \in \Psi^{0,0}_{\delta,\delta'}$ elliptic at α such that $Q'Au \in H^{s,r}$. Let $Q \in \Psi^{0,0}_{\delta,\delta'}$ be such that $\mathrm{WF}'(Q) \subset \mathrm{Ell}(A) \cap \mathrm{Ell}(Q')$, note that the set on the right is open and includes α. Then by Corollary 5.5, $Qu \in H^{s+m,r+\ell}$. Taking Q which is in addition elliptic at α completes the proof. □

The consequence of what we have proved so far for Schrödinger operators is:

Proposition 5.18 *Let g be a Riemannian metric,* $g_{ij} \in \mathcal{C}^\infty(\overline{\mathbb{R}^n})$, *positive definite on* $\overline{\mathbb{R}^n}$, $V \in S^{-\rho}(\mathbb{R}^n)$ *with* $\rho > 0$. *Let* $H = \Delta_g + V$. *Then for* $\sigma \in [0, \infty)$, $(H - \sigma)u \in H^{s,r}$ *implies*

$$\mathrm{WF}^{s+2,r}(u) \subset \left\{(z, \zeta) \in \partial\overline{\mathbb{R}^n} \times \mathbb{R}^n : \sum g_{ij}(z)\zeta_i\zeta_j = \sigma\right\}.$$

5.3.11. Diffeomorphism Invariance

Finally we note the diffeomorphism invariance of pseudodifferential operators.

Proposition 5.19 *Suppose $F : O \to U$ is a diffeomorphism between bounded open subsets O and U of $\mathbb{R}^n$. Suppose $A \in \Psi^{m,\ell}_{\infty,\delta}(\mathbb{R}^n)$, with Schwartz kernel supported in a compact subset of $U \times U$. Then $A_F = F^* A (F^{-1})^* \in \Psi^{m,\ell}_{\infty,\delta}$. Furthermore, with $DF(z)$ the Jacobian matrix of F, that is with kj entry $\partial_j F_k(z)$, and with $\dagger$ denoting $\mathbb{R}^n$-adjoint (i.e. j,k reversed),*

$$\mathrm{WF}'(A_F) = \{(z,\zeta) : (F(z), (DF)^\dagger(z)^{-1}\zeta) \in \mathrm{WF}'(A)\},$$

and

$$\sigma_{\infty,m,\ell}(A_F)(z,\zeta) = \sigma_{\infty,m,\ell}(A)(F(z), (DF)^\dagger(z)^{-1}\zeta).$$

Remark 5.1 The principal symbol here shows why we had a single parameter δ giving the losses in $\langle\zeta\rangle$ upon differentiation in either z or ζ: differentiation of the principal symbol of A_F in z gives rise to ζ derivatives as well as in that of A. Thus, to have the class diffeomorphism invariant, the losses under z derivatives have to be at least as large as those under ζ-derivatives. Thus, the ζ-derivatives (which are the derivatives tangent to the fibers of the cotangent bundle of $\mathbb{R}^n$, thus are invariantly defined) are necessarily better (in the sense of "no worse") behaved regarding these losses than the z-derivatives. If one also wants Fourier-invariance, one needs the opposite inequality as well, hence the equality.

Remark 5.2 Notice that if one writes a covector as $\sum_k \eta_k \, dw_k$, then its pull-back under the map F (with $F(z) = w$ for clarity) is $\sum_k \eta_k \, (\partial_j F_k)(z) \, dz_j$, that is

$$\zeta_j = \sum_k (\partial_j F_k)(z)\eta_k = ((DF)^\dagger(z)\eta)_j,$$

so $\zeta = (DF)^\dagger(z)\eta$. This means that $(DF)^\dagger(z)^{-1}\zeta \, dw$ is the pull-back of $\zeta \, dz$ by F^{-1}, that is the wavefront set and the principal symbol are well behaved (invariant) if we regard them as subsets of $T^*\mathbb{R}^n \setminus o$, resp. functions on $T^*\mathbb{R}^n \setminus o$: with $F^\sharp : T^*_U\mathbb{R}^n \to T^*_O\mathbb{R}^n$ the map induced by pull-back of covectors by F, and similarly for $(F^{-1})^\sharp : T^*_O\mathbb{R}^n \to T^*_U\mathbb{R}^n$, so $((F^{-1})^\sharp)^*$ maps functions on $T^*_U\mathbb{R}^n$ to those on $T^*_O\mathbb{R}^n$, then

$$\sigma_{\infty,m,\ell}(A_F) = ((F^{-1})^\sharp)^* \sigma_{\infty,m,\ell}(A),$$

and

$$\mathrm{WF}'(A_F) = ((F^{-1})^\sharp)^{-1}(\mathrm{WF}'(A)).$$

Proof Let $G = F^{-1}$ to simplify the notation.

First we consider the off-diagonal behavior. To do so, suppose more generally that $A : \mathcal{S} \to \mathcal{S}'$ continuous linear with Schwartz kernel supported in $U \times U$ (so A need not be a pseudo-differential operator). We claim that, with K_A the Schwartz kernel of A, the Schwartz kernel K_{A_F} of A_F is the (compactly supported) tempered distribution

$$K_{A_F} = ((F \times F)^* K_A)(\pi_R^* |\det(DF)|), \tag{5.48}$$

where $\pi_R : \mathbb{R}^n \times \mathbb{R}^n \to \mathbb{R}^n$ is the projection to the second factor. Indeed, if K_A is Schwartz (i.e. just $\mathcal{C}^\infty$, in view of the support) then, with $A_F u$ also considered as a distribution in the second expression,

$$\begin{aligned} K_{A_F}(u \otimes v) &= (A_F u)(v) = \int (A_F u)(z)\, v(z)\, dz \\ &= \int A(G^* u)(F(z)) v(z)\, dz = \int K_A(F(z), w') G^* u(w')\, v(z)\, dw'\, dz \\ &= \int K_A(F(z), w') u(G(w'))\, v(z)\, dw'\, dz \\ &= \int K_A(F(z), F(z')) u(z')\, v(z)\, |\det DF(z')|\, dz'\, dz, \end{aligned}$$

giving the above result for K_{A_F}. Since Schwartz functions with compact support in $O \times O$ are dense in tempered distributions supported in $O \times O$, and since the operations in (5.48) are continuous, the result follows for general tempered distributions K_A.

Applying this to the case of pseudodifferential operators A, which have $\mathcal{C}^\infty$ Schwartz kernel away from the diagonal, we conclude that A_F has $\mathcal{C}^\infty$ Schwartz kernel away from the diagonal. In particular, when considering the behavior near the diagonal, it suffices to work in a suitably small neighborhood of the diagonal.

We have from the definition of A,

$$\begin{aligned} A_F u(z) &= (A(G^* u))(F(z)) \\ &= (2\pi)^{-n} \int e^{i(F(z) - w')\cdot\eta} a(F(z), w', \eta) u(G(w'))\, dw'\, d\eta. \end{aligned}$$

Letting $z' = G(w')$, the change of variables formula for the integral gives

$$A_F u(z) = (2\pi)^{-n} \int e^{i(F(z) - F(z'))\cdot\eta} a(F(z), F(z'), \eta) u(z')\, |\det(DF)(z')|\, dz'\, d\eta.$$

This is almost of the desired form, except for the appearance of $F(z) - F(z')$ instead of $z - z'$ in the exponent. To deal with this, we use the easiest case of

Taylor's theorem (which really means the fundamental theorem of calculus in this context),

$$F_k(z) - F_k(z') = \sum_{j=1}^{n} (z_j - z_j') F_{kj}(z, z')$$

with

$$F_{kj}(z, z') = \int_0^1 (\partial_j F_k)(tz + (1-t)z')\, dt,$$

so

$$F_{kj}(z, z) = \partial_j F_k(z)$$

is the Jacobian matrix of F. More generally, let us write

$$\Phi(z, z') = (\partial_j F_k(z, z'))_{kj}$$

for this matrix. Thus, the exponent is

$$\sum_{k=1}^{n} \sum_{j=1}^{n} (z_j - z_j') F_{kj}(z, z') \eta_k = \sum_{j=1}^{n} (z_j - z_j') \zeta_j,$$

where

$$\zeta_j = \zeta_j(z, z', \eta) = \sum_{k=1}^{n} F_{kj}(z, z') \eta_k = (\Phi^\dagger(z, z')\eta)_j.$$

Note that the map

$$(z, z', \eta) \mapsto (z, z', \zeta(z, z', \eta))$$

is a diffeomorphism, linear in η, if (z, z') is close to the diagonal. Indeed, since F is a diffeomorphism, $\Phi(z, z)$ is invertible, and thus so is $\Phi(z, z')$ for (z, z') near the diagonal, so the inverse of the above map is simply

$$(z, z', \zeta) \mapsto (z, z', \Phi^\dagger(z, z')^{-1}\zeta).$$

Changing the variable of integration from η to ζ gives, as

$$|d\zeta| = |\det(\Phi(z, z'))^\dagger|\, |d\eta| = |\det \Phi(z, z')|\, |d\eta|,$$

$$\begin{aligned} A_F u(z) &= (2\pi)^{-n} \int e^{i(z-z')\cdot\zeta} a(F(z), F(z'), (\Phi^\dagger(z, z'))^{-1}\zeta) u(z') \\ &\qquad\qquad |\det \Phi(z, z')|^{-1} |\det(DF)(z')|\, dz'\, d\zeta \\ &= (2\pi)^{-n} \int e^{i(z-z')\cdot\zeta} a_F(z, z', \zeta) u(z')\, dz'\, d\zeta \end{aligned}$$

with

$$a_F(z,z',\zeta) = a(F(z),F(z'),(\Phi^\dagger(z,z'))^{-1}\zeta)|\det\Phi(z,z')|^{-1}|\det(DF)(z')|.$$

Thus, checking

$$a_F \in S^{m,\ell}_{\infty,\delta}$$

completes the proof. For this purpose the two determinant factors are irrelevant as they are $\mathcal{C}^\infty$. Thus, it remains to note that D_ζ applied to

$$a(F(z),F(z'),(\Phi^\dagger(z,z'))^{-1}\zeta)$$

again simply gives additional smooth factors, while D_z or $D_{z'}$ applied can either correspond to derivatives of a in the first or second slot, in which case they are harmless, or in the last slot when they give a factor in ζ, but also lower the symbolic order by 1, thus preserving the estimates.

The principal symbol statement follows from the cancelation of the determinant factors when one restricts to $z = z'$, and that $(\Phi^\dagger(z,z'))^{-1}$ is $(DF)^\dagger(z)^{-1}$ then; this also gives the wave front set statement. □

In fact, the same proof gives:

Proposition 5.20 *Suppose $F : O \to U$ is a diffeomorphism between open subsets O and U of $\overline{\mathbb{R}^n}$. Suppose $A \in \Psi^{m,\ell}_{\delta,\delta'}$, with Schwartz kernel supported in a compact subset of $U \times U$. Then $A_F = F^*A(F^{-1})^* \in \Psi^{m,\ell}_{\delta,\delta'}$. Furthermore, with $DF(z)$ the Jacobian matrix of F, that is with kj entry $\partial_j F_k(z)$, and with $\dagger$ denoting $\mathbb{R}^n$-adjoint (i.e. j,k reversed),*

$$\mathrm{WF}'(A_F) = \{(z,\zeta) : (F(z),(DF)^\dagger(z)^{-1}\zeta) \in \mathrm{WF}'(A)\},$$

and

$$\sigma_{m,\ell}(A_F)(z,\zeta) = \sigma_{\infty,m,\ell}(A)(F(z),(DF)^\dagger(z)^{-1}\zeta).$$

The point here is that for F as stated, DF is an elliptic symbol on O of order 0, and thus the near-diagonal argument goes through: in fact, one even gets the invertibility of $\Phi(z,z')$ for (z,z') in a conic neighborhood of the diagonal (as follows by working with valid coordinates on the compactification, and noting that a neighborhood in this compactified perspective gives a conic neighborhood without the compactification). The Schwartz kernel of pseudodifferential operators outside such a neighborhood is Schwartz, hence the off-diagonal piece pulls back correctly as well.

We can now use our results to analyze Fredholm problems in geometric settings. Note that the diffeomorphism invariance lets us define $\Psi^m_\delta(X)$ when X is a compact manifold:

Definition 5.9 For X a compact manifold (without boundary), $\delta \in [0, 1/2)$, $\Psi_\delta^m(X)$ consists of continuous linear maps $A : C^\infty(X) \to C^\infty(X)$, whose Schwartz kernel is C^∞ away from the diagonal in $X \times X$ and with the property that if U is a coordinate chart with $\Phi : U \to \tilde{U} \subset \mathbb{R}^n$ a diffeomorphism then for $\chi \in C_c^\infty(U)$, $(\Phi^{-1})^* \chi A \chi \Phi^* \in \Psi_{\delta,0}^{m,0}$.

Notice that we could have used $\Psi_{\delta,\delta'}^{m,\ell}$ in the definition for any $\delta' \in [0, 1/2)$ and $\ell \in \mathbb{R}$, or instead $\Psi_{\infty,\delta}^{m,\ell}$, without changing $\Psi_\delta^m(X)$ since the image of $\operatorname{supp} \chi$ under Φ is a compact subset of $\mathbb{R}^n$.

Notice also that if U, V are both coordinate charts with $\Phi : U \to \tilde{U}$, $\Xi : V \to \tilde{V}$ and if $\operatorname{supp} \chi \subset U \cap V$, then the statements that $(\Phi^{-1})^* \chi A \chi \Phi^* \in \Psi_\delta^{m,0}$ and $(\Xi^{-1})^* \chi A \chi \Xi^* \in \Psi_\delta^{m,0}$ are equivalent since if for instance $(\Phi^{-1})^* \chi A \chi \Phi^* \in \Psi_\delta^{m,0}$, then so is

$$(\Xi^{-1})^* \chi A \chi \Xi^* = (\Phi \circ \Xi^{-1})^* \big((\Phi^{-1})^* \chi A \chi \Phi^*\big) (\Xi \circ \Phi^{-1})^*$$

as $\Xi \circ \Phi^{-1} : \tilde{U} \to \tilde{V}$ is a diffeomorphism of subsets of $\mathbb{R}^n$ so Proposition 5.19 is applicable. Thus the "for all" statement (i.e. for all coordinate charts) in the definition can be replaced by an open cover and a subordinate partition of unity.

Finally, notice that the C^∞ off-diagonal statement is reasonable because if $B \in \Psi_{\delta,0}^{m,0}$ and $\psi \in C_c^\infty(\tilde{U})$ then $\Phi^* \psi B \psi (\Phi^{-1})^*$ has C^∞ Schwartz kernel away from the diagonal (since B has this property), so this is not an additional restriction, and we may simply regard pseudodifferential operators on $\mathbb{R}^n$ with support in a coordinate chart as pseudodifferential operators on X.

This also lets us define the principal symbol of A as a function on $T^*X \setminus o$: if $\Phi : U \to \tilde{U}$ is a coordinate chart, $K \subset U$ is compact and $\chi \in C_c^\infty(U)$ with $\chi \equiv 1$ on a neighborhood of K, then we let the principal symbol of A on T_K^*X be the pull-back of the principal symbol of $(\Phi^{-1})^* \chi A \chi \Phi^*$ on $T_{\tilde{U}}^* \mathbb{R}^n = \tilde{U} \times \mathbb{R}^n$ by $(\Phi^{-1})^\sharp$; as a consequence of Remark 5.2 this is well defined independently of the choices of Φ and χ. Here for $A \in \Psi_{\mathrm{cl}}^m(X)$ one can regard the principal symbol as a homogeneous degree m function on $T^*X \setminus o$, or if $m = 0$ then on $S^*X = (T^*X \setminus o)/\mathbb{R}^+$ (with the quotient corresponding to $\mathbb{R}^+$ acting on the fibers of T^*X via dilations); in general it is an element of $S_\delta^m(T^*X)/S_\delta^{m-1+2\delta}(T^*X)$, where the symbol space $S_\delta^m(T^*X) \subset C^\infty(T^*X)$ is locally the pull-back of $S_{\delta,0}^{m,0}(\mathbb{R}^n; \mathbb{R}^n)$ via $(\Phi^{-1})^\sharp$; again, different coordinate charts give the same space in the overlap. Similarly, one defines $\mathrm{WF}'(A)$ as the inverse image of $\mathrm{WF}'((\Phi^{-1})^* \chi A \chi \Phi^*)$ under $(\Phi^{-1})^\sharp$. In particular, the notion of the principal symbol allows us to talk about elliptic operators; an operator is elliptic if its principal symbol is invertible, or equivalently if the local coordinate version of the principal symbol is elliptic. One still has a short exact sequence

$$0 \to \Psi_\delta^{m-1+2\delta}(X) \to \Psi_\delta^m(X) \to S_\delta^m(T^*X)/S_\delta^{m-1+2\delta}(T^*X) \to 0,$$

with the key point being the surjectivity of the penultimate map. This follows by taking $a \in S_\delta^m(T^*X)$, using a partition of unity $\sum_k \chi_k = 1$ subordinate to a cover $\{U_k : k = 1, \dots, K\}$ by coordinate charts, $\Phi_k : U_k \to \tilde{U}_k$, and taking the quantization

$$q(a) = \sum_k \Phi_k^* \psi_k q_L\big((\Phi_k^{-1})^*(\chi_k a)\big)\psi_k(\Phi_k^{-1})^*,$$

where $\psi_k \in C_c^\infty(\tilde{U}_k)$ is identically 1 near the image of $\operatorname{supp}\chi_k$ under Φ_k. The statement $q(a) \in \Psi_\delta^m(X)$ follows from our remarks regarding the C^∞ off-diagonal behavior and that it suffices to check the pseudodifferential property by a single cover by coordinate charts; the principal symbol is then easily seen to be $\sum_k \Phi_k^*(\psi_k^2)\chi_k a = a$.

Thus, if X is a compact manifold, and $P \in \Psi_\delta^m(X)$ is an elliptic operator (i.e. its principal symbol is invertible everywhere), then we can construct a parametrix Q for P:

$$E_L = QP - \mathrm{Id}, E_R = PQ - \mathrm{Id} \in \Psi^{-\infty}(X).$$

Indeed, one simply repeats the construction on $\mathbb{R}^n$, by first inverting the principal symbol p of P to get $Q_0 = p^{-1}$, $E_0 = PQ_0 - \mathrm{Id} \in \Psi_\delta^{-1+2\delta}(X)$, then considering the Neumann series $\sum_{j=1}^\infty (-1)^j E_0^j$. In order to sum it, use a partition of unity $\sum_k \chi_k = 1$ corresponding to an open cover $\{U_k : k = 1, \dots, K\}$ of X by coordinate charts, let $\phi_k \in C_c^\infty(U_k)$ be identically 1 on $\operatorname{supp}\chi_k$, so $\phi_k E_0^j \chi_k$ is supported in $U_k \times U_k$ and $(\Phi_k^{-1})^* \phi_k E_0^j \chi_k \Phi_k^*$ is an element of $\Psi_\delta^{-j(1-2\delta)}$. Then we can use asymptotic summation on $\mathbb{R}^n$, that is write $(\Phi_k^{-1})^* \phi_k E_0^j \chi_k \Phi_k^* = q_L(e_{k,j})$ and for each k asymptotically sum in j to get $\tilde{e}_k \sim \sum_{j=1}^\infty (-1)^j e_{k,j}$, and let $\tilde{E}_k = q_L(\tilde{e}_k)$. Letting $\psi_k \in C_c^\infty(\tilde{U}_k)$ (with $\tilde{U}_k$ the image of U_k under Φ_k), ψ_k identically 1 near $\operatorname{supp}\phi_k$, $E_k = \Phi_k^* \psi_k \tilde{E}_k \psi_k (\Phi_k^{-1})^*$,

$$Q = Q_0 \left(\mathrm{Id} + \sum_{k=1}^K E_k\right)$$

provides a right parametrix. A left parametrix can be constructed similarly, and their equality modulo $\Psi^{-\infty}(X)$ can be shown as on $\mathbb{R}^n$.

Since $\Psi^{-\infty}(X)$ is bounded between any Sobolev spaces on X, we immediately obtain a Fredholm statement.

Proposition 5.21 *Any elliptic $A \in \Psi_\delta^m(X)$ is Fredholm as a map $H^s(X) \to H^{s-m}(X)$ for all $m, s \in \mathbb{R}$, that is has closed range, finite dimensional nullspace and the range has finite codimension. Further, the nullspace is a subspace of $C^\infty(X)$, while the annihilator of the range in $H^{s-m}(X)$ in the dual space $H^{-s+m}(X)$ is also in $C^\infty(X)$. Correspondingly, the nullspace of A as well as the annihilator of its range is independent of s; if A is invertible for one value of s, then it is invertible for all.*

There is an immediate analogue of all these results in the scattering algebra on manifolds with boundary.

Definition 5.10 For X a compact manifold with boundary, $\Psi^{m,\ell}_{\mathrm{sc},\delta,\delta'}(X)$ consists of continuous linear maps $A : \dot{C}^\infty(X) \to \dot{C}^\infty(X)$, whose Schwartz kernel is in $\dot{C}^\infty$ away from the diagonal in $X \times X$ and with the property that if U is a coordinate chart with $\Phi : U \to \tilde{U} \subset \overline{\mathbb{R}^n}$ a diffeomorphism then for $\chi \in C_c^\infty(U)$, $(\Phi^{-1})^*\chi A \chi \Phi^* \in \Psi^{m,\ell}_{\delta,\delta'}$. One writes $\Psi^{m,\ell}_{\mathrm{sc}}(X) = \Psi^{m,\ell}_{\mathrm{sc},0,0}(X)$.

Note that this definition states that the Schwartz kernels of elements vanish to infinite order, that is decay rapidly, away from the diagonal on $X \times X$, in particular near (y, y') if $y \neq y'$, $y, y' \in \partial X$. Again, this is a reasonable definition, for elements of $\Psi^{m,\ell}_{\delta,\delta'}$ have this property on $\overline{\mathbb{R}^n} \times \overline{\mathbb{R}^n}$, and thus for $B \in \Psi^{m,\ell}_{\delta,\delta'}$, $\chi \in C_c^\infty(\tilde{U})$, one has $\Phi^*\chi B \chi(\Phi^{-1})^* \in \Psi^{m,\ell}_{\delta,\delta'}(X)$ automatically. (This also uses the fact that, again in the overlap of coordinate charts, the pull-back pseudodifferential operator statements are equivalent due to the same argument as for the boundaryless case considered above.)

In this case the natural phase space is ${}^{\mathrm{sc}}T^*X$, which is locally, near a point on ∂X, spanned by $\frac{dx}{x^2}, \frac{dy_j}{x}$ if x is a local boundary defining function, and y_j are coordinates on ∂X. Alternatively, this is locally simply the pull-back of the bundle $\overline{\mathbb{R}^n_z} \times \mathbb{R}^n_\zeta \to \overline{\mathbb{R}^n_z}$ via Φ. Indeed, in local coordinates on $\overline{\mathbb{R}^n}$ near a point on $\partial\overline{\mathbb{R}^n}$, which can be taken as (x, y), $x = |z|^{-1} = r^{-1}$, y local coordinates on $\mathbb{S}^{n-1}$, $\zeta\, dz$ is a smooth non-degenerate linear combination of $\frac{dx}{x^2} = -dr$ and $\frac{dy_j}{x} = r\, dy_j$ as is well-known, showing that locally $\overline{\mathbb{R}^n_z} \times \mathbb{R}^n_\zeta$ is naturally identified with ${}^{\mathrm{sc}}T^*X$.

Then for $A \in \Psi^{m,\ell}_{\mathrm{sc},\delta,\delta'}(X)$, the principal symbol is naturally an element of

$$S^{m,\ell}_{\delta,\delta'}({}^{\mathrm{sc}}T^*X)/S^{m-1+2\delta,\ell-1+2\delta'}_{\delta,\delta'}({}^{\mathrm{sc}}T^*X).$$

One still has a short exact sequence.

One also has the scattering Sobolev spaces $H^{s,r}_{\mathrm{sc}}(X)$, defined naturally as Hilbert spaces up to equivalence of norms, by saying that a tempered distribution $u \in C^{-\infty}(X)$ is in $H^{s,r}_{\mathrm{sc}}(X)$ if for all coordinate charts $\Phi : U \to \tilde{U}$, and for all $\chi \in C_c^\infty(U)$, we have $(\Phi^{-1})^*(\chi u) \in H^{s,r}$. Equivalently, one may require that for some (and hence for all) elliptic $A \in \Psi_{\mathrm{sc}}^{s,r}(X)$, $Au \in L^2_{\mathrm{sc}}(X)$, where $L^2_{\mathrm{sc}}(X)$ is the scattering L^2-space, that is one given by a density $\overline{\mathbb{R}^n}$-locally equivalent to the standard L^2 density on $\mathbb{R}^n$, and which can thus be taken to be of the form $x^{-n-1}\nu$ where ν is a standard density on X, and x a boundary defining function. (Notice that locally $x^{-n-1}\,|dx\,dy_1\,\ldots\,dy_{n-1}| = r^{n-1}\,|dr\,dy_1,\ldots\,dy_{n-1}|$, showing the local equivalence to the Euclidean version.)

The elliptic parametrix construction also goes through, resulting in the Fredholm statement:

Proposition 5.22 *Any elliptic $A \in \Psi^{m,\ell}_{\mathrm{sc},\delta,\delta'}(X)$ is Fredholm as a map $H^{s,r}_{\mathrm{sc}}(X) \to H^{s-m,r-\ell}_{\mathrm{sc}}(X)$ for all $m, \ell, s, r \in \mathbb{R}$, that is has closed range, finite dimensional nullspace and the range has finite codimension. Further, the nullspace is a subspace of $\dot{C}^\infty(X)$, while the annihilator of the range in $H^{s-m,r-\ell}_{\mathrm{sc}}(X)$ in the dual space $H^{-s+m,-r+\ell}_{\mathrm{sc}}(X)$ is also in $\dot{C}^\infty(X)$. Correspondingly, the nullspace of A as well as the annihilator of its range is independent of r, s; if A is invertible for one value of r, s, then it is invertible for all.*

Further, tempered distributions $u \in C^{-\infty}(X)$ have wavefront sets $\mathrm{WF}_{\mathrm{sc}}(u)$, $\mathrm{WF}^{s,r}_{\mathrm{sc}}(u)$, which are subsets of $\partial\,\overline{^{\mathrm{sc}}T^*X}$, and can be defined either via local identification with $\overline{\mathbb{R}^n}$, or again directly by saying $\alpha \notin \mathrm{WF}^{s,r}_{\mathrm{sc}}(u)$ if there exists $A \in \Psi^{s,r}_{\mathrm{sc}}(X)$, elliptic at α, such that $Au \in L^2_{\mathrm{sc}}(X)$.

An immediate application is to the Laplacian of *Riemannian scattering metrics* (introduced by Melrose in [30]) which are Riemannian metrics g on X° which near ∂X have the form

$$g = \frac{dx^2}{x^4} + \frac{h}{x^2},$$

where h is a symmetric 2-cotensor on X such that, at ∂X, h restricts to be positive definite on $T\partial X$. These generalize the Euclidean metric on $\overline{\mathbb{R}^n}$ as taking $x = r^{-1}$ shows. Such g is a symmetric section on $\mathrm{Sym}^2\,{}^{\mathrm{sc}}T^*X$, and its dual gives a fiber metric on ${}^{\mathrm{sc}}T^*X$. Correspondingly, $\Delta_g = d_g^* d \in \mathrm{Diff}^2_{\mathrm{sc}}(X)$. For $V \in S^{-\rho}(X)$, $\rho > 0$, we then have $\Delta_g + V - \sigma$ elliptic if $\sigma \in \mathbb{C} \setminus [0, \infty)$, and we have the following analogue of Proposition 5.14 and Proposition 5.18:

Proposition 5.23 *Let g be a Riemannian scattering metric on X, $V \in S^{-\rho}(X)$ with $\rho > 0$. Let $H = \Delta_g + V$.*

Then for $\sigma \in \mathbb{C} \setminus [0, \infty)$, $H - \sigma : H^{s,r}_{\mathrm{sc}}(X) \to H^{s-2,r}_{\mathrm{sc}}(X)$ is Fredholm for all r, s, with nullspace in $\dot{C}^\infty(X)$. If V is real-valued, then H is self-adjoint.

Further, for $\sigma \in [0,\infty)$, $(H-\sigma)u \in H^{s,r}_{\mathrm{sc}}$ *implies*

$$\mathrm{WF}^{s+2,r}_{\mathrm{sc}}(u) \subset \{(z,\zeta) \in {}^{\mathrm{sc}}T^*_{\partial X}X : g_z^{-1}(\zeta,\zeta) = \sigma\}.$$

While we have not added vector bundles, this is straightforward using local trivializations in the spirit of Definitions 5.9–5.10, that is a pseudodifferential operator acting as a map between sections of two vector bundles is an operator with a C^∞, homomorphism valued, Schwartz kernel away from the diagonal which, in local coordinates that at the same time are trivializations of the bundles, is given by a matrix of pseudodifferential operators.

5.4. Propagation Phenomena

5.4.1. The Propagation of Singularities Theorem

We now understand elliptic operators in $\Psi^{m,\ell}$; the next challenge is to deal with non-elliptic operators. Let's start with classical operators, and indeed let's take $m=\ell=0$. Thus, $A=q_L(a)$, $a \in C^\infty(\overline{\mathbb{R}^n}\times\overline{\mathbb{R}^n})$, so $\sigma_{0,0}(A)$ is just the restriction of a to $\partial(\overline{\mathbb{R}^n}\times\overline{\mathbb{R}^n})$. Ellipticity is just the statement that $a_0 = a|_{\partial(\overline{\mathbb{R}^n}\times\overline{\mathbb{R}^n})}$ does not vanish. Thus, the simplest (or least degenerate/complicated) way an operator can be non-elliptic is if a_0 is real-valued and has a non-degenerate zero set. As $\partial(\overline{\mathbb{R}^n}\times\overline{\mathbb{R}^n})$ is *not* a smooth manifold at the corner, $\partial\overline{\mathbb{R}^n}\times\partial\overline{\mathbb{R}^n}$, one has to be a bit careful. Away from the corner non-degeneracy is the statement that $a_0(\alpha)=0$ implies $da_0(\alpha)\neq 0$; in this case the characteristic set, $\mathrm{Char}(A)=a_0^{-1}(\{0\})$, is a C^∞ codimension one embedded submanifold. At the corner, for $\alpha \in \partial\overline{\mathbb{R}^n}\times\partial\overline{\mathbb{R}^n}$, one can consider the two smooth manifolds with boundary $\overline{\mathbb{R}^n}\times\partial\overline{\mathbb{R}^n}$ and $\partial\overline{\mathbb{R}^n}\times\overline{\mathbb{R}^n}$, and denoting by $a_{0,\mathrm{fiber}}$ and $a_{0,\mathrm{base}}$ the corresponding restrictions, requiring that $a_0(\alpha)=0$ implies that $da_{0,\mathrm{fiber}}$ and $da_{0,\mathrm{base}}$ are *not* in the conormal bundle of the corner; in this case in both boundary hypersurfaces $\mathrm{Char}(A)$ is a smooth manifold transversal to the boundary. In many cases, such as stationary (elliptic) Schrödinger operators, the characteristic set is disjoint from the corner, $\mathrm{Char}(A)$ is disjoint from the corner, but this is not the case for wave propagation.

More generally, if A is classical, that is $a=\langle z\rangle^\ell\langle\zeta\rangle^m\tilde{a}$, $\tilde{a}\in C^\infty(\overline{\mathbb{R}^n}\times\overline{\mathbb{R}^n})$, we impose the analogous condition on $\tilde{a}$, that is that $\tilde{a}_0$ has a non-degenerate zero set. Note that if $b\in C^\infty(\overline{\mathbb{R}^n}\times\overline{\mathbb{R}^n})$ is elliptic, and $\tilde{a}_0$ has a non-degenerate zero set, then, with b_0 denoting the restriction of b to the boundary, the same holds for $b_0\tilde{a}_0$ as $d(b_0\tilde{a}_0)=b_0\,d\tilde{a}_0+\tilde{a}_0\,db_0$, so when $b_0\tilde{a}_0=0$, that is when $\tilde{a}_0=0$, $d(b_0\tilde{a}_0)$ is a non-vanishing multiple of $d\tilde{a}_0$.

Let's reinterpret this from the conic point of view, for instance corresponding to $\overline{\mathbb{R}^n}\times\partial\overline{\mathbb{R}^n}$, that is working on $\overline{\mathbb{R}^n}\times(\mathbb{R}^n\setminus\{0\})$. (Note that working with

$\partial\overline{\mathbb{R}^n} \times \mathbb{R}^n$, i.e. $(\mathbb{R}^n \setminus \{0\}) \times \overline{\mathbb{R}^n}$, is completely analogous.) Away from the corner, we may drop the compactification from the first factor, and thus we may assume $\ell = 0$. For A classical, then, $\sigma_{\text{fiber},m,0}(A)$ is homogeneous of degree m, given by $a_m = |\zeta|^m \tilde{a}_0$, where $\tilde{a}_0$ is considered as a homogeneous degree zero function. Now, $d(|\zeta|^m \tilde{a}_0) = |\zeta|^m \, d\tilde{a}_0 + \tilde{a}_0 \, d|\zeta|^m$, and $\text{Char}(A)$ is given by $\tilde{a}_0 = 0$, so it is now a conic (invariant under the $\mathbb{R}^+$-action) smooth codimension one embedded submanifold of $\mathbb{R}^n \times (\mathbb{R}^n \setminus \{0\})$.

The next relevant structure arises from the Hamilton vector field of A,

$$H_{a_m} = \sum_{j=1}^{n} \big((\partial_{\zeta_j} a_m)\partial_{z_j} - (\partial_{z_j} a_m)\partial_{\zeta_j} \big).$$

Note that $H_{a_m} a_m = 0$, thus this vector field is tangent to $\text{Char}(A)$, and thus defines a flow on $\text{Char}(A)$. Note that H_{a_m} is homogeneous of degree $m - 1$ in the sense that the push-forward of H_{a_m} under dilation in the fiber by $t > 0$, $M_t(z, \zeta) = (z, t\zeta)$, is $t^{-m+1} H_{a_m}$ since, using $\partial_{z_j} a_m$, resp. $\partial_{\zeta_j} a_m$, are homogeneous of degree m, resp. $m - 1$,

$$(H_{a_m} M_t^* f)(z,\zeta) = \sum_{j=1}^{n} \big(t^{1-m} (\partial_{\zeta_j} a_m)(z, t\zeta)(\partial_{z_j} f)(z, t\zeta) \\ - t^{-m} (\partial_{z_j} a_m)(z, t\zeta) t (\partial_{\zeta_j} f)(z, t\zeta) \big) = t^{1-m} (H_{a_m} f)(z, t\zeta).$$

In particular, integral curves of H_{a_m} through (z, ζ) and $(z, t\zeta)$ are the "same" up to reparameterization: denoting the first by γ, the second by $\tilde{\gamma}$, and denoting by $\tilde{M}_t$ dilations on $\mathbb{R}$: $\tilde{M}_t(s) = ts$,

$$\tilde{\gamma}(s) = (z(\gamma(t^{m-1}s)), t\zeta(\gamma(t^{m-1}s))) = (M_t \circ \gamma \circ \tilde{M}_{t^{m-1}})(s),$$

since $(\tilde{M}_t)_* \frac{d}{ds} = t \frac{d}{ds}$, so

$$(M_t \circ \gamma \circ \tilde{M}_{t^{m-1}})_* \frac{d}{ds} = t^{m-1} (M_t \circ \gamma)_* \frac{d}{ds} = t^{m-1} (M_t)_* H_{a_m} = H_{a_m},$$

as being an integral curve of H_{a_m} means exactly that the push-forward of $\frac{d}{ds}$ under the map is exactly H_{a_m}. In particular, up to reparameterization, the integral curves can be considered curves on $\mathbb{R}^n \times (\mathbb{R}^n \setminus \{0\})/\mathbb{R}^+$, that is on this quotient, the image of the curve is defined, but not the parameterization itself. The exception is if $m = 1$, in which case even the parameterization is well-defined on this quotient; in this case, H_{a_m} acts on homogeneous degree zero functions, and is thus a vector field on the quotient. In general, if m is arbitrary, one may consider instead the vector field $|\zeta|^{-m+1} H_{a_m}$, which is homogeneous of degree 0, and thus has well-defined integral curves on the quotient; this does depend on the choice of a positive homogeneous degree 1 function, such as $|\zeta|$, but a different choice only multiplies $|\zeta|^{-m+1}H_{a_m}$ by a smooth non-vanishing function, and thus simply reparameterizes the integral curves.

A different way of looking at this is that for $m = 1$, by the identification of $\mathbb{R}^n \times (\mathbb{R}^n \setminus \{0\})/\mathbb{R}^+$ with $\mathbb{R}^n \times \partial\overline{\mathbb{R}^n}$, one has a vector field on $\mathbb{R}^n \times \partial\overline{\mathbb{R}^n}$; if $m \neq 1$, then again this vector field is defined up to positive multiples.

Definition 5.11 The integral curves of H_{a_m} in $\mathrm{Char}(A) \subset \mathbb{R}^n \times (\mathbb{R}^n \setminus \{0\})$, as well as their projections to $\mathbb{R}^n \times (\mathbb{R}^n \setminus \{0\})/\mathbb{R}^+ = \mathbb{R}^n \times \mathbb{S}^{n-1} = \mathbb{R}^n \times \partial\overline{\mathbb{R}^n}$, are called null-bicharacteristics, or simply bicharacteristics.

The Hamiltonian version of classical mechanics corresponding to an energy function p is just the statement that particles move on integral curves of H_p. In physical terms, geometric optics for the wave equation is that at high frequencies waves follow paths given by classical mechanics. Concretely, including the time variable as part of our space, on $\mathbb{R}^n_z = \mathbb{R}^{n-1}_y \times \mathbb{R}_t$, one has a Lorentzian metric g, for instance a product-type metric, $g = dt^2 - h(y, dy)$, and then the dual metric function G on $\mathbb{R}^n_z \times \mathbb{R}^n_\zeta$, given by $G(z, \zeta) = G_z(\zeta, \zeta)$, with G_z the inverse of g_z. Note that G is homogeneous of degree 2. Then H_G is a vector field of homogeneity degree 1. The integral curves of H_G are the lifted geodesics; those inside $\{G = 0\}$ are the lifted null-geodesics. Now, the d'Alembertian

$$\Box_g = |\det g|^{-1/2} \sum D_i |\det g|^{1/2} G_{ij} D_j$$

has principal symbol G, while waves are solutions of $\Box_g u = 0$. Mathematically, the high frequency statement translates into singularities of solutions u. Concretely, we have the following theorem due to Hörmander and Duistermaat-Hörmander [28, 8]:

Theorem 5.1 *Suppose that $P \in \Psi^{m,0}$ and its principal symbol has a real homogeneous degree m representative p. Then in $\mathbb{R}^n \times \partial\overline{\mathbb{R}^n}$, $\mathrm{WF}^s(u) \setminus \mathrm{WF}^{s-m+1}(Pu)$ is a union of maximally extended (null)-bicharacteristics of p.*

Rather than giving a proof now, we prove a more general version.

We now turn to the general case, where the characteristic set $\mathrm{Char}(A)$ of $A \in \Psi^{m,\ell}_{\mathrm{cl}}$ possibly intersects the corner. So first we define the *rescaled Hamilton vector field*

$$\mathsf{H}_a = \mathsf{H}_{a,m,\ell} = \langle z\rangle^{-\ell+1} \langle\zeta\rangle^{-m+1} H_a,$$

and notice that as $\langle z\rangle\partial_{z_j}$ and $\langle\zeta\rangle\partial_{\zeta_j}$ are in $\mathcal{V}_b(\overline{\mathbb{R}^n_z})$, resp. $\mathcal{V}_b(\overline{\mathbb{R}^n_\zeta})$, and thus in $\mathcal{V}_b(\overline{\mathbb{R}^n_z} \times \overline{\mathbb{R}^n_\zeta})$,

$$\mathsf{H}_{a,m,\ell} \in \mathcal{V}_b(\overline{\mathbb{R}^n_z} \times \overline{\mathbb{R}^n_\zeta}).$$

As above, we note that if we replace $\langle z\rangle^{-\ell+1}\langle\zeta\rangle^{-m+1}$ by $b\langle z\rangle^{-\ell+1}\langle\zeta\rangle^{-m+1}$, where $b \in C^\infty(\overline{\mathbb{R}^n} \times \overline{\mathbb{R}^n})$ is positive on $\partial(\overline{\mathbb{R}^n} \times \overline{\mathbb{R}^n})$, then $\mathsf{H}_{a,m,\ell}$ is only multiplied by a positive factor, b, at $\mathrm{Char}(A) \subset \partial(\overline{\mathbb{R}^n} \times \overline{\mathbb{R}^n})$. Since $\mathsf{H}_{a,m,\ell}$ is tangent to the boundary, its integral curves are globally well-defined (i.e. they are well-defined on $\mathbb{R}$). We then make the definition:

Definition 5.12 The integral curves of $H_{a,m,\ell}$ in $\mathrm{Char}(A) \subset \partial(\overline{\mathbb{R}^n} \times \overline{\mathbb{R}^n})$ are called null-bicharacteristics, or simply bicharacteristics, and we consider them well-defined up to a direction-preserving reparameterization.

In general, we have the following theorem, which contains Theorem 5.1 as a special case:

Theorem 5.2 *Suppose that $P \in \Psi^{m,\ell}_{\mathrm{cl}}$, with real principal symbol p. Then in $\partial(\overline{\mathbb{R}^n} \times \overline{\mathbb{R}^n})$, $\mathrm{WF}^{s,r}(u) \setminus \mathrm{WF}^{s-m+1,r-\ell+1}(Pu)$ is a union of maximally extended (null)-bicharacteristics of p.*

A variant of this theorem in the variable order setting is the following:

Theorem 5.3 *Suppose that $P \in \Psi^{m,\ell}_{\mathrm{cl}}$, with real principal symbol p, and suppose $s \in C^\infty(\overline{\mathbb{R}^n} \times \partial\overline{\mathbb{R}^n})$, $r \in C^\infty(\partial\overline{\mathbb{R}^n} \times \overline{\mathbb{R}^n})$ are non-increasing along the rescaled Hamilton flow, that is $\mathsf{H}_{p,m,\ell}s \leq 0$, $\mathsf{H}_{p,m,\ell}r \leq 0$. Then in $\partial(\overline{\mathbb{R}^n} \times \overline{\mathbb{R}^n})$,*

$$\mathrm{WF}^{s,r}(u) \setminus \mathrm{WF}^{s-m+1,r-\ell+1}(Pu)$$

is a union of maximally forward extended (null)-bicharacteristics of p.

The analogous conclusion holds if s, r are non-decreasing and "forward" is replaced by "backward."

Note that all these theorems have empty statements at points at which $\mathsf{H}_{p,m,\ell}$ vanishes (or is radial in the conic setting, i.e. is a multiple of the generator of dilations in the conic variable), so there is nothing to prove at such points. Thus, the key point is to understand what happens near points at which $\mathsf{H}_{p,m,\ell}$ is non-vanishing.

The proof of these theorems relies on positive commutators, that is constructing a pseudodifferential operator A such that $i[P, A]$ is of the form B^*B, modulo terms that we can control by our assumptions. Such an estimate actually gives bounds for the microlocal $H^{s,r}$ norm of u. However, the bound can also be recovered from the regularity statement of the theorem via the closed graph theorem as we show now.

Theorem 5.4 *Suppose that $P \in \Psi^{m,\ell}_{\mathrm{cl}}$ with real principal symbol p. Suppose that $B, G, Q \in \Psi^{0,0}$, $\mathrm{WF}'(B) \subset \mathrm{Ell}(G)$ and for every $\alpha \in \mathrm{WF}'(B) \cap \mathrm{Char}(P)$, there is a point $\alpha' = \gamma(\sigma')$ on the bicharacteristic γ through α with $\gamma(0) = \alpha$*

such that $\alpha' \in \mathrm{Ell}(Q)$ *and such that for* $\sigma \in [0,\sigma']$ *(or* $\sigma \in [\sigma',0]$ *if* $\sigma' < 0$*),* $\gamma(\sigma) \in \mathrm{Ell}(G)$*. Then for any* M,N*, there is* $C > 0$ *such that if* $Qu \in H^{s,r}$*,* $GPu \in H^{s-m+1,r-\ell+1}$ *then* $Bu \in H^{s,r}$ *and*

$$\|Bu\|_{H^{s,r}} \leq C(\|Qu\|_{H^{s,r}} + \|GPu\|_{H^{s-m+1,r-\ell+1}} + \|u\|_{H^{M,N}}).$$

The analogous conclusion also holds in the variable order setting if either s,r *are non-increasing along the Hamilton flow and* $\sigma' < 0$ *or* s,r *are non-decreasing along the Hamilton flow and* $\sigma' > 0$*.*

Proof Under the assumptions, by Theorem 5.2, $Bu \in H^{s,r}$ since $\mathrm{WF}^{s,r}(Bu) = \emptyset$. Indeed, if $\alpha \notin \mathrm{WF}'(B)$, then $\alpha \notin \mathrm{WF}^{s,r}(Bu)$ automatically. If $\alpha \in \mathrm{WF}'(B)$, then if $\alpha \notin \mathrm{Char}(P)$ then $GPu \in H^{s-m+1,r-\ell+1}$ implies that $\mathrm{Ell}(G)$, thus $\mathrm{WF}'(B)$, are disjoint from $\mathrm{WF}^{s-m+1,r-\ell+1}(Pu)$, and thus by elliptic regularity, $\alpha \notin \mathrm{WF}^{s+1,r+1}(u)$. If $\alpha \in \mathrm{WF}'(B) \cap \mathrm{Char}(P)$, then Theorem 5.2 states $\alpha \notin \mathrm{WF}^{s,r}(u)$, finishing the proof of the claim.

Now,

$$\mathcal{X} = \{u \in H^{M,N} : Qu \in H^{s,r}, \ GPu \in H^{s-m+1,r-\ell+1}\}$$

is complete by the lemma below, as is

$$\mathcal{Y} = \{u \in H^{M,N} : Bu \in H^{s,r}\}.$$

From the previous paragraph, the identity map on $H^{M,N}$ restricts to a map $\iota : \mathcal{X} \to \mathcal{Y}$. Further, if $u_k \to u$ in $\mathcal{X}$ and $u_k = \iota(u_k) \to v$ in $\mathcal{Y}$ then in particular $u_k \to u$ in $H^{M,N}$ and $u_k \to v$ in $H^{M,N}$, so $\iota(u) = u = v$, that is the graph of ι is closed. The closed graph theorem thus implies that ι is continuous, which is exactly the estimate in the statement of the theorem. □

Lemma 5.10 *Suppose* $A_j \in \Psi^{m_j,\ell_j}$, $j = 1,\ldots,N$*. Let*

$$\mathcal{X} = \{u \in H^{r,s} : A_j u \in H^{r_j,s_j}, \ j = 1,\ldots,N\} \subset H^{r,s},$$

equipped with the norm

$$\|u\|_{\mathcal{X}}^2 = \|u\|_{H^{r,s}}^2 + \sum_{j=1}^{N} \|A_j u\|_{H^{r_j,s_j}}^2.$$

Then $\mathcal{X}$ *is complete.*

Here all spaces and operators may have variable orders.

Proof Suppose that $\{u_k\}_{k=1}^{\infty}$ is $\mathcal{X}$-Cauchy. Then u_k, resp. $A_j u_k$, are $H^{r,s}$, resp. H^{r_j,s_j}-Cauchy, and thus converge to some $v \in H^{r,s}$, resp. $v_j \in H^{r_j,s_j}$. But $A_j : H^{r,s} \to H^{r-m_j,s-\ell_j}$ is continuous, so $A_j u_k \to A_j v$ in $H^{r-m_j,s-\ell_j}$. Thus, $A_j u_k \to A_j v$ and $A_j u_k \to v_j$ in $\mathcal{S}'$, so $v_j = A_j v$. Thus, $A_j v \in H^{r_j,s_j}$, so $v \in \mathcal{X}$, and $u_k \to v$ in $\mathcal{X}$. □

5.4.2. Overview of the Proof of the Propagation Theorem

To motivate the proof of Theorem 5.2, we compute (for $A \in \Psi^{m',\ell'}$ with $A = A^*$, with non-variable order for now)

$$\langle Pu, Au\rangle - \langle Au, Pu\rangle = \langle (AP - P^*A)u, u\rangle = \langle ([A,P] + (P - P^*)A)u, u\rangle.$$

Note that if, for instance, $Pu = 0$, then the left hand side vanishes, and so if $[A,P] + (P - P^*)A$ has some definiteness properties, then we get an interesting conclusion. Note that $P - P^* \in \Psi^{m-1,\ell-1}$ since its principal symbol in $\Psi^{m,\ell}$ is $p - \overline{p} = 0$. The principal symbol of $i([A,P] + (P - P^*)A) \in \Psi^{m+m'-1,\ell+\ell'-1}$ is

$$\sigma_{m+m'-1,\ell+\ell'-1}(i([A,P] + (P - P^*)A)) = -H_p a - 2\tilde{p}a,$$

where $\tilde{p}$ is the principal symbol of $\frac{1}{2i}(P - P^*) \in \Psi^{m-1,\ell-1}$. Suppose we can arrange that

$$H_p a + 2\tilde{p}a = -b^2 + e,\ b \in S^{(m+m'-1)/2,(\ell+\ell'-1)/2},\ e \in S^{m+m'-1,\ell+\ell'-1}, \tag{5.49}$$

with e supported in the region where we have a priori estimates on u. Then letting $B \in \Psi^{(m+m'-1)/2,(\ell+\ell'-1)/2}$ be such that its principal symbol is b, $E \in \Psi^{m+m'-1,\ell+\ell'-1}$ such that its principal symbol is e,

$$i([A,P] + (P - P^*)A) = B^*B - E + F,$$

with $F \in \Psi^{m+m'-2,\ell+\ell'-2}$. Then we obtain

$$\langle iPu, Au\rangle - \langle iAu, Pu\rangle = \|Bu\|^2 - \langle Eu, u\rangle + \langle Fu, u\rangle, \tag{5.50}$$

that is we can control Bu in L^2 by controlling Pu, Eu, and Fu. Since F is lower order, it is dealt with inductively, gradually increasing m', ℓ', namely one assumes that u already has some a priori regularity, so that $\langle Fu, u\rangle$ is controlled (which is automatic for sufficiently low m', ℓ'), and then one obtains a bound for Bu in a stronger space than the a priori bound. Then one can increase m', ℓ' by 1 each to obtain new A', B', E', F'; the resulting F' will give $\langle F'u, u\rangle$ controlled provided $\mathrm{WF}'(F') \subset \mathrm{Ell}(B) \cup \mathrm{Ell}(E)$ (which follows if $\mathrm{WF}'(A') \subset \mathrm{Ell}(B) \cup \mathrm{Ell}(E)$) by the microlocal elliptic estimate. Thus, $B'u$ is controlled in L^2, which is a 1/2-order gain in terms of both orders over the control provided by Bu, and so on. Here one really needs to regularize the argument to make sense of its steps.

We remark that the sign in (5.49) is arbitrary;

$$H_p a + 2\tilde{p}a = b^2 - e,\ b \in S^{(m+m'-1)/2,(\ell+\ell'-1)/2},\ e \in S^{m+m'-1,\ell+\ell'-1}, \tag{5.51}$$

would also work, still giving rise to the control of $\|Bu\|^2$ in terms of the other quantities mentioned above.

5.4.3. The Commutant Construction

Before going through this in more detail, let's see whether we can arrange (5.49). Indeed, let's work with $m' = 0$, $\ell' = 0$ to start with, for the additional weights will not be an issue, and ignore $\tilde{p}$ as well. Thus, we essentially want to find $a \in \mathcal{C}^\infty(\overline{\mathbb{R}^n} \times \overline{\mathbb{R}^n})$ whose Hamilton derivative $\mathsf{H}_{p,m,\ell} a$ is negative (in the sense of non-positive, with some definiteness in the region of interest), apart from a region where e is supported: a is thus decreasing along the $\mathsf{H}_{p,m,\ell}$ orbits. But this is very simple to achieve locally when $\mathsf{H}_{p,m,\ell}$ is non-zero at α. Since the statement is slightly different (in terms of numerology) depending on whether α is at the corner or not, we consider these separately:

- If $\alpha \notin \partial\overline{\mathbb{R}^n} \times \partial\overline{\mathbb{R}^n}$, we can choose local coordinates
$$q_1, q_2, \ldots, q_{2n-1}, q_{2n}$$
near α on $\overline{\mathbb{R}^n} \times \overline{\mathbb{R}^n}$, with the chart O centered at α, such that
$$\mathsf{H}_{p,m,\ell} = \partial_{q_1},$$
and q_{2n} is a boundary defining function (i.e. vanishes non-degenerately at the unique boundary hypersurface, either $\partial\overline{\mathbb{R}^n} \times \overline{\mathbb{R}^n}$ or $\overline{\mathbb{R}^n} \times \partial\overline{\mathbb{R}^n}$, containing α; in the former case we can take $\rho_{\mathrm{base}} = q_{2n}$ in the latter $\rho_{\mathrm{fiber}} = q_{2n}$). We write $q' = (q_2, \ldots, q_{2n-1})$.
- If $\alpha \in \partial\overline{\mathbb{R}^n} \times \partial\overline{\mathbb{R}^n}$, we can choose local coordinates
$$q_1, q_2 \ldots, q_{2n-2}, q_{2n-1}, q_{2n}$$
near α on $\overline{\mathbb{R}^n} \times \overline{\mathbb{R}^n}$, with the chart O centered at α, such that
$$\mathsf{H}_{p,m,\ell} = \partial_{q_1},$$
and q_{2n-1}, q_{2n} are boundary defining functions (i.e. vanish non-degenerately at $\partial\overline{\mathbb{R}^n} \times \overline{\mathbb{R}^n}$, resp. $\overline{\mathbb{R}^n} \times \partial\overline{\mathbb{R}^n}$), so $\rho_{\mathrm{base}} = q_{2n-1}$, $\rho_{\mathrm{fiber}} = q_{2n}$ locally are acceptable choices. We write $q' = (q_2, \ldots, q_{2n-2})$.

Recall that in either case this is achieved by choosing a local hypersurface S transversal to $\mathsf{H}_{p,m,\ell}(\alpha)$ through α, and using the fact that the $\mathsf{H}_{p,m,\ell}$-exponential map from $S \times (-\epsilon, \epsilon)$ ($\epsilon > 0$ small) to a neighborhood of α is a diffeomorphism. Since the $\mathsf{H}_{p,m,\ell}$-flow is tangent to the boundary hypersurfaces, if we choose coordinates on S of which the last, resp. last two, are boundary defining functions, the pull-back under this diffeomorphism gives coordinates near α which are boundary defining functions, with q_{2n-1}, resp. q_{2n-2}, being the flow parameter in $(-\epsilon, \epsilon)$.

In fact, it does not really matter that $\mathsf{H}_{p,m,\ell} q_{2n} = 0$ (and $\mathsf{H}_{p,m,\ell} q_{2n-1} = 0$ in the second case); this vanishes at $q_{2n} = 0$ in any case if q_{2n} is *any* boundary

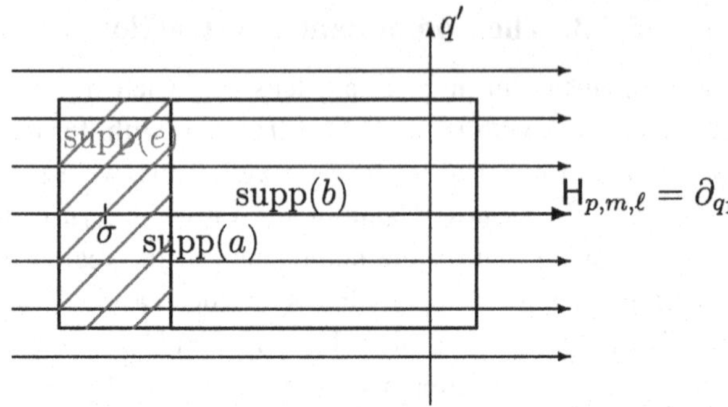

Figure 5.4. The supports of a, b, e in the first case. Here $\mathsf{H}_{p,m,\ell} = \partial_{q_1}$, q' are the vertical directions, while the q_{2n} axis points out of the page

defining function by the tangency of $\mathsf{H}_{p,m,\ell}$ to the boundary, so it is of the form $q_{2n}h$, with h smooth, and we can control such terms below just as we control the terms coming from $\tilde{p}$.

In the first case, with $q' = (q_2, \dots, q_{2n-1})$, see Figure 5.4, for

$$a = \chi_0(q_1)\chi_1(q')^2\chi_2(q_{2n})^2, \ \mathsf{H}_{p,m,\ell}a = \chi_0'(q_1)\chi_1(q')^2\chi_2(q_{2n})^2,$$

so we simply need to arrange that $\chi_0(q_1)\chi_1(q')^2\chi_2(q_{2n})^2$ is supported in the region of validity of the coordinate chart, which is arranged by making χ_0, χ_1, and χ_2 supported near 0, and such that $\chi_0' \leq 0$, or more precisely it is the negative of the square of a smooth function, away from the region where our error e is supported. Here χ_2 is not particularly important; being away from the corner, this can be taken constant 1 near the boundary, so any contribution to symbolic calculations is trivial.

Concretely, with γ the integral curve of $\mathsf{H}_{p,m,\ell}$ through α with $\gamma(0) = \alpha$, and $\gamma(\sigma) = \beta$ for some $\sigma < 0$ such that $\gamma([\sigma, 0]) \subset O$, and further if we are given a neighborhood U_2 of β and U_1 of $\gamma([\sigma, 0])$, which we may assume is a subset of O (otherwise replace these by their intersections with O), one can construct χ_0 and χ_1 as above with $\operatorname{supp} \chi_0\chi_1^2\chi_2^2 \subset U_1$, and with $\chi_0'\chi_1^2\chi_2^2$ the negative of the square of a smooth function outside a compact subset of U_2. Indeed, one simply chooses $\epsilon > 0$ such that in the coordinates q_j,

$$\begin{aligned} &\{q : \ |q'| < \epsilon, \ q_{2n} < \epsilon, \ q_1 \in (\sigma - \epsilon, \epsilon)\} \subset U_1, \\ &\{q : \ |q'| < \epsilon, \ q_{2n} < \epsilon, \ q_1 \in (\sigma - \epsilon, \sigma + \epsilon)\} \subset U_2, \end{aligned}$$

then one considers χ_1 supported in $\{q' : \ |q'| < \epsilon\}$, identically 1 near 0, χ_2 in $\{q_{2n} < \epsilon\}$, identically 1 near 0, and lets with $F > 0$ to be determined later as convenient,

$$\tilde{\chi}_0(t) = e^{F/(t-\epsilon/2)},\ t < \epsilon/2,\ \tilde{\chi}_0(t) = 0,\ t \geq \epsilon/2,$$

and

$$\psi_0(t) \in C^\infty(\mathbb{R}),\ \operatorname{supp}\psi_0 \subset (\sigma - \epsilon, \infty),\ \operatorname{supp}(1 - \psi_0) \subset (-\infty, \sigma + \epsilon),$$

and

$$\chi_0(t) = \tilde{\chi}_0(t)\psi_0(t)^2.$$

Thus, $\psi_0(q_1)$ is constant outside $\{q : q_1 \in (\sigma - \epsilon, \sigma + \epsilon)\}$, and $\tilde{\chi}_0' \leq 0$, indeed $\sqrt{-\tilde{\chi}_0'}$ is C^∞. Writing

$$b = \sqrt{-\tilde{\chi}_0'(q_1)}\psi_0(q_1)\chi_1(q')\chi_2(q_{2n}),$$
$$e = 2\tilde{\chi}_0(q_1)\psi_0(q_1)\psi_0'(q_1)\chi_1(q')^2\chi_2(q_{2n})^2,$$

we have

$$\mathsf{H}_{p,m,\ell}a = -b^2 + e$$

as desired. Now, choosing F large allows one to deal with $\tilde{p}$. The key point is that one can make $-\tilde{\chi}_0'$ dominate $\tilde{\chi}_0$ by choosing $F > 0$ large. Indeed,

$$\tilde{\chi}_0(t) = -F^{-1}(t - \epsilon/2)^2\tilde{\chi}_0'(t),$$

so as long as t is in a fixed compact set and $\tilde{\chi}_0$ is bounded by a small multiple of $-\tilde{\chi}_0'$, provided F is large. Since a strictly positive C^∞ function has a C^∞ square root, given $\tilde{\mathsf{p}} \in C^\infty$, such as $\tilde{\mathsf{p}} = \rho_{\text{fiber}}^{-m+1}(2\tilde{p})$, one can arrange that

$$\mathsf{H}_{p,m,\ell}a + \tilde{\mathsf{p}}a = -b^2 + e$$

by taking F large so that $1 - F^{-1}(q_1 - \epsilon/2)^2\tilde{\mathsf{p}} > 1/2$ on $\operatorname{supp} a$. While we assumed $m' = 0 = \ell'$, even if say $m' \neq 0$ and $\alpha \in \overline{\mathbb{R}^n} \times \partial\overline{\mathbb{R}^n}$, we simply add a factor of $q_{2n}^{-m'}$ as our weight, then $\mathsf{H}_{p,m,\ell}q_{2n}^{-m'} = 0$, so the weights do not contribute to the commutator, completing the construction. We remark that one only needs $\tilde{\mathsf{p}} \in S^{0,0}$, for then it is bounded, and for sufficiently large $F > 0$, $1 - F^{-1}(q_1 - \epsilon/2)^2\tilde{\mathsf{p}} > 1/2$ on $\operatorname{supp} a$, and thus

$$b = (1 - F^{-1}(q_1 - \epsilon/2)^2\tilde{\mathsf{p}})^{1/2}\sqrt{-\tilde{\chi}_0'(q_1)}\psi_0(q_1)\chi_1(q')\chi_2(q_{2n}) \in S^{0,0}. \quad (5.52)$$

Now, the variable order version is similar, except we use a weight q_{2n}^{-s} or q_{2n}^{-r}, with $s \in C^\infty(\overline{\mathbb{R}^n} \times \partial\overline{\mathbb{R}^n})$, $r \in C^\infty(\partial\overline{\mathbb{R}^n} \times \overline{\mathbb{R}^n})$, depending on where α lies. Here we need to extend s, r to functions defined at least locally in the region of interest on $\overline{\mathbb{R}^n} \times \overline{\mathbb{R}^n}$; we do so by making them independent of the boundary defining coordinate, q_{2n}. (In fact, for the argument below even $s, r \in S^{0,0}$ work.) Then notice that $\mathsf{H}_{p,m,\ell}q_{2n}^s = q_{2n}^s(-\log q_{2n})\mathsf{H}_{p,m,\ell}s$. Notice that $-\log q_{2n} \to +\infty$ as $q_{2n} \to 0$, so this term actually dominates all other terms where $\mathsf{H}_{p,m,\ell}s$ does not vanish, so it cannot be estimated away using $\tilde{\chi}_0'$. Correspondingly, we need to assume that $\mathsf{H}_{p,m,\ell}s$ has the correct sign, that is it is non-positive, for then

$$\mathsf{H}_{p,m,\ell}a = -b^2 - \tilde{b}^2 + e,\ \tilde{b}^2 = q_{2n}^s(-\log q_{2n})(-\mathsf{H}_{p,m,\ell}s),$$

allows for a similar argument as in the s constant case. (One either needs to arrange that $\sqrt{-\mathsf{H}_{p,m,\ell}s}$ is smooth, which is easily done since we have flexibility in choosing s, or instead use the sharp Gårding theorem.)

The construction in the second case, with α in the corner, is completely similar with $q' = (q_2, \ldots, q_{2n-2})$ now, taking

$$a = \chi_0(q_1)\chi_1(q')^2\chi_2(q_{2n-1})^2\chi_2(q_{2n})^2,$$
$$\mathsf{H}_{p,m,\ell}a = \chi_0'(q_1)\chi_1(q')^2\chi_2(q_{2n-1})^2\chi_2(q_{2n})^2.$$

Then with

$$b = \sqrt{-\tilde{\chi}_0'(q_1)\psi_0(q_1)}\chi_1(q')\chi_2(q_{2n-1})\chi_2(q_{2n}),$$

and

$$e = 2\tilde{\chi}_0(q_1)\psi_0(q_1)\psi_0'(q_1)\chi_1(q')^2\chi_2(q_{2n-1})^2\chi_2(q_{2n})^2,$$

we have

$$\mathsf{H}_{p,m,\ell}a = -b^2 + e.$$

More generally, with $\tilde{\mathsf{p}} \in S^{0,0}$,

$$b = (1-\digamma^{-1}(q_1-\epsilon/2)^2\tilde{\mathsf{p}})^{1/2}\sqrt{-\tilde{\chi}_0'(q_1)\psi_0(q_1)}\chi_1(q')\chi_2(q_{2n-1})\chi_2(q_{2n}) \in S^{0,0}, \tag{5.53}$$

gives

$$\mathsf{H}_{p,m,\ell}a + \tilde{\mathsf{p}}a = -b^2 + e.$$

Again, variable orders are fine, but these orders s, r need to satisfy $\mathsf{H}_{p,m,\ell}s \le 0$, $\mathsf{H}_{p,m,\ell}r \le 0$.

Note that if $\sigma > 0$, a similar construction works, but reversing the signs corresponding to (5.51), so in the variable order case we need s and r to have $\mathsf{H}_{p,m,\ell}s \ge 0$, $\mathsf{H}_{p,m,\ell}r \ge 0$ at the relevant boundary faces. Thus, with γ the integral curve of $\mathsf{H}_{p,m,\ell}$ through α with $\gamma(0) = \alpha$, and $\gamma(\sigma) = \beta$ for some $\sigma > 0$ such that $\gamma([0,\sigma]) \subset O$, and further if we are given a neighborhood U_2 of β and U_1 of $\gamma([0,\sigma])$, which we may again assume is a subset of O, one can construct χ_0, χ_1, and χ_2 as above with $\operatorname{supp} \chi_0\chi_1^2\chi_2^2 \subset U_1$, and with $\chi_0'\chi_1^2\chi_2^2$

the square of a smooth function outside a compact subset of U_2. Namely, χ_1, χ_2 are unchanged,

$$\tilde{\chi}_0(t) = e^{-F/(t+\epsilon/2)},\ t > -\epsilon/2,\ \tilde{\chi}_0(t) = 0,\ t \leq -\epsilon/2,$$

and

$$\psi_0(t) \in C^\infty(\mathbb{R}),\ \operatorname{supp} \psi_0 \subset (-\infty, \sigma + \epsilon),\ \operatorname{supp}(1 - \psi_0) \subset (\sigma - \epsilon, \infty),$$

and

$$\chi_0(t) = \tilde{\chi}_0(t)\psi_0(t)^2.$$

Thus, as above but with a sign switch, given a C^∞ function $\tilde{\mathsf{p}}$ on $\overline{\mathbb{R}^n} \times \overline{\mathbb{R}^n}$, or indeed simply $\tilde{\mathsf{p}} \in S^{0,0}$, one can arrange that

$$\mathsf{H}_{p,m,\ell}a + \tilde{\mathsf{p}}a = b^2 - e.$$

5.4.4. Regularization and Proof of the Propagation Theorem

Now, returning to $\sigma < 0$ for the sake of definiteness to deal with u that is not a priori nice, we regularize the argument. Thus, we replace A by a bounded family A_t, $t \in [0,1]$, $[0,1] \mapsto \Psi^{m',\ell'}$, continuous as a map $[0,1] \mapsto \Psi^{m'+\delta,\ell'+\delta}$ for $\delta > 0$, such that for $t > 0$, $A_t \in \Psi^{m'-K,\ell'-K}$ (for sufficiently large K), and such that $A_0 \in \Psi^{m',\ell'}$ is the operator denoted by A above. It will be convenient to shift orders, and thus we take $\Lambda_{s,r} \in \Psi^{s,r}$ to be elliptic and invertible, with principal symbol

$$\rho_{s,r} = \langle \zeta \rangle^s \langle z \rangle^r.$$

Further, for future purposes, we arrange that, as operators in $\Psi^{m',\ell'}$, the principal symbols a_t satisfy for some $M > 0$

$$\begin{gathered} H_p a_t + 2\tilde{p} a_t = -b_t^2 - M^2 \rho_{m-1,\ell-1} a_t + e_t, \\ b \in L^\infty([0,1]_t; S^{s,r}),\ e \in L^\infty([0,1]_t; S^{2s,2r}),\ \sqrt{a_t} \in L^\infty([0,1]_t; S^{m'/2,\ell'/2}); \end{gathered} \tag{5.54}$$

where, to simplify the notation, we let

$$s = (m + m' - 1)/2,\ r = (\ell + \ell' - 1)/2.$$

In applications we also need some microsupport and ellipticity conditions for the families, so we make the following definition:

Definition 5.13 For a bounded family $a = \{a_t : t \in [0,1]\}$ in $S^{m,\ell}$ and for $\alpha \in \partial(\overline{\mathbb{R}^n} \times \overline{\mathbb{R}^n})$ we say that $\alpha \notin \operatorname{esssupp}_{L^\infty}(a)$ if α has a neighborhood U in $\overline{\mathbb{R}^n} \times \overline{\mathbb{R}^n}$ such that $a_t|_{U \cap (\mathbb{R}^n \times \mathbb{R}^n)}$ is bounded in $S^{-\infty,-\infty}$ (i.e. each seminorm is bounded).

Further, we say that a bounded family $a = \{a_t : t \in [0,1]\}$ in $S^{m,\ell}$ is elliptic at $\alpha \in \partial(\overline{\mathbb{R}^n} \times \overline{\mathbb{R}^n})$ if α has a neighborhood U in $\overline{\mathbb{R}^n} \times \overline{\mathbb{R}^n}$ such that for some $c > 0$,

$$\left|a_t|_{U\cap(\mathbb{R}^n\times\mathbb{R}^n)}\right| \geq c\langle z\rangle^{m'}\langle\zeta\rangle^{\ell'}.$$

We then have the following lemma:

Lemma 5.11 *For $\alpha_0 \in \partial(\overline{\mathbb{R}^n} \times \overline{\mathbb{R}^n})$ with $\mathsf{H}_{p,m,\ell}$ non-vanishing at α_0. Then α_0 has a neighborhood O in $\partial(\overline{\mathbb{R}^n} \times \overline{\mathbb{R}^n})$ on which $\mathsf{H}_{p,m,\ell}$ is non-vanishing, and if $\alpha \in O$, γ the integral curve of $\mathsf{H}_{p,m,\ell}$ through α with $\gamma(0) = \alpha$, and $\gamma(\sigma) = \beta$ for some $\sigma < 0$ such that $\gamma([\sigma,0]) \subset O$, and further if we are given a neighborhood U_2 of β and U_1 of $\gamma([\sigma,0])$ contained in O, then there exists a_t, b_t, and e_t as in* (5.54) *such that*

$$\operatorname{esssupp}_{L^\infty} a, b \subset U_1, \ \operatorname{esssupp}_{L^\infty} e \subset U_2,$$

and b is elliptic on $\gamma([\sigma,0])$.

Proof We assume that α_0 is in the corner $\partial\overline{\mathbb{R}^n} \times \partial\overline{\mathbb{R}^n}$; otherwise one of the weights is irrelevant, and the corresponding coordinate can be taken to be one of the q' coordinates with no significant changes. We fix our coordinates so that $q_{2n-1} = \rho_{\text{base}}$, $q_{2n} = \rho_{\text{fiber}}$.

We first define χ_0, χ_1, χ_2 as above and let

$$\tilde{a} = q_{2n}^{-m'} q_{2n-1}^{-\ell'} \chi_0(q_1)\chi_1(q')^2\chi_2(q_{2n-1})^2\chi_2(q_{2n})^2.$$

For $t \in [0,1]$, let

$$\phi_t(\tau) = (1+t\tau)^{-K/2}, \ \tau \geq 0, \tag{5.55}$$

and note that

$$\frac{d\phi_t}{d\tau}(\tau) = f_t(\tau)\phi_t(\tau), \qquad f_t(\tau) = t(-K/2)(1+t\tau)^{-1}, \tag{5.56}$$

with $|f_t(\tau)| \leq K/2$, $|\tau f_t(\tau)| \leq K/2$. We then let

$$a_t = \phi_t(q_{2n-1}^{-2})\phi_t(q_{2n}^{-2})\tilde{a}.$$

Note that $a \in L^\infty([0,1]; S^{m',\ell'})$ satisfies $a_t \in S^{m'-K,\ell'-K}$ for $t > 0$, $\operatorname{esssupp}_{L^\infty}(a) \subset U_1$, and a is elliptic on $\gamma([\sigma,0])$. Further, $\sqrt{a} \in L^\infty([0,1]; S^{m'/2,\ell'/2})$. Now, $\mathsf{H}_{p,m,\ell}(\phi_t(q_{2n-1}^{-2})\phi_t(q_{2n}^{-2})) = 0$, and

$$\begin{aligned}
&q_{2n-1}^{\ell'+\ell-1} q_{2n}^{m'+m-1} H_p a_t \\
&= \phi_t(q_{2n}^{-2})\phi_t(q_{2n-1}^{-2})\big(\psi_0(q_1)\chi_1(q')\chi_2(q_{2n-1})\chi_2(q_{2n})\big)^2\tilde{\chi}_0'(q_1) \\
&\quad + 2\phi_t(q_{2n}^{-2})\phi_t(q_{2n-1}^{-2})\psi_0(q_1)\psi_0'(q_1)\big(\chi_1(q')\chi_2(q_{2n-1})\chi_2(q_{2n})\big)^2\tilde{\chi}_0(q_1).
\end{aligned}$$

Proceeding as above, we note that for $F > 0$ sufficiently large

$$
\begin{aligned}
&q_{2n-1}^{\ell'+\ell-1} q_{2n}^{m'+m-1}\Big((H_p a_t + 2\tilde{p} a_t) - M^2 \rho_{m-1,\ell-1} a_t\Big) \\
&= \phi_t(q_{2n}^{-2})\phi_t(q_{2n-1}^{-2})\big(\psi_0(q_1)\chi_1(q')\chi_2(q_{2n-1})\chi_2(q_{2n})\big)^2(\tilde{\chi}_0'(q_1) + r_t\tilde{\chi}_0(q_1)) \\
&\quad + 2\phi_t(q_{2n}^{-2})\phi_t(q_{2n-1}^{-2})\psi_0(q_1)\psi_0'(q_1)\chi_1(q')\chi_2(q_{2n-1})\chi_2(q_{2n}))^2\tilde{\chi}_0'(q_1) \\
&= -b_t^2 + e_t,
\end{aligned}
$$

with the desired properties. □

Let $\breve{a}_t = \sqrt{a_t}$, and let $\breve{A}_t = \frac{1}{2}(q_L(\breve{a}_t) + q_L(\breve{a}_t)^*)$, which is thus formally self-adjoint. Let $A_t = \breve{A}_t^2$. Then for $t > 0$

$$
\begin{aligned}
&\langle iPu, A_t u\rangle - \langle iA_t u, Pu\rangle \\
&\quad = \|B_t u\|^2 + M^2\|\Lambda_{(m-1)/2,(\ell-1)/2}\breve{A}_t u\|^2 - \langle E_t u, u\rangle + \langle F_t u, u\rangle,
\end{aligned}
\tag{5.57}
$$

where $F \in L^\infty([0,1];\Psi^{2s-1,2r-1})$. We write, with $\delta > 0$,

$$
\begin{aligned}
|\langle Pu, A_t u\rangle| &= |\langle(\Lambda_{(m-1)/2,(\ell-1)/2}^{-1})^*\breve{A}_t Pu, \Lambda_{(m-1)/2,(\ell-1)/2}\breve{A}_t u\rangle| \\
&\le \|(\Lambda_{(m-1)/2,(\ell-1)/2}^{-1})^*\breve{A}_t Pu\|\,\|\Lambda_{(m-1)/2,(\ell-1)/2}\breve{A}_t u\| \\
&\le \frac{1}{2\delta}\|(\Lambda_{(m-1)/2,(\ell-1)/2}^{-1})^*\breve{A}_t Pu\|^2 + \frac{\delta}{2}\|(\Lambda_{(m-1)/2,(\ell-1)/2})^*\breve{A}_t u\|^2.
\end{aligned}
$$

One has a similar bound for $\langle A_t u, Pu\rangle$. For $\delta > 0$ sufficiently small (namely $\delta < M^2$), $\delta\|\Lambda_{(m-1)/2,(\ell-1)/2}\breve{A}_t u\|^2$ can be absorbed into $M^2\|\Lambda_{(m-1)/2,(\ell-1)/2}\breve{A}_t u\|^2$, and thus we deduce from (5.57) that

$$
\|B_t u\|^2 \le |\langle E_t u, u\rangle| + |\langle F_t u, u\rangle| + \delta^{-1}\|(\Lambda_{(m-1)/2,(\ell-1)/2}^{-1})^*\breve{A}_t Pu\|^2. \tag{5.58}
$$

Now we let $t \to 0$; then assuming a priori control on the terms other than $\|B_t u\|^2$ we conclude that $B_t u$ is bounded in L^2. But $B_t u \to B_0 u$ in $\mathcal{S}'$, so using the weak compactness of the unit ball in L^2, and thus that $B_t u$ has a subsequence $B_{t_j} u$ converging L^2-weakly to some $v \in L^2$, and thus in $\mathcal{S}'$, we deduce that $B_0 u = v \in L^2$, yielding the desired regularity information.

In order to make the a priori control assumptions, we need a uniform version of the operator wavefront set:

Definition 5.14 For a bounded family $\mathcal{A} = \{A_t : t \in [0,1]\}$ in $\Psi^{m,\ell}$ and for $\alpha \in \partial(\overline{\mathbb{R}^n} \times \overline{\mathbb{R}^n})$ we say that $\alpha \notin \mathrm{WF}'_{L^\infty}(A_t)$ if α has a neighborhood U in $\overline{\mathbb{R}^n} \times \overline{\mathbb{R}^n}$ such that $A_t = q_L(a_t)$ and $a_t|_{U\cap(\mathbb{R}^n\times\mathbb{R}^n)}$ is bounded in $S^{-\infty,-\infty}$ (i.e. each seminorm is bounded).

Note that if $A_t = A_0$ for all t, then $\mathrm{WF}'_{L^\infty}(\mathcal{A}) = \mathrm{WF}'(A)$, that is this family wavefront set is an appropriate generalization of the standard operator wave front set.

Lemma 5.12 *If $\mathcal{A}$ and $\mathcal{B}$ are bounded families, then with $\mathcal{AB} = \{A_tB_t : t \in [0,1]\}$,*

$$\mathrm{WF}'_{L^\infty}(\mathcal{AB}) \subset \mathrm{WF}'_{L^\infty}(\mathcal{A}) \cap \mathrm{WF}'_{L^\infty}(\mathcal{B}).$$

In particular, if $Q \in \Psi^{m,\ell}$ then

$$\mathrm{WF}'_{L^\infty}(Q\mathcal{A}), \mathrm{WF}'_{L^\infty}(\mathcal{A}Q) \subset \mathrm{WF}'(Q) \cap \mathrm{WF}'_{L^\infty}(\mathcal{A}).$$

Further, $\alpha \notin \mathrm{WF}'_{L^\infty}(\mathcal{A})$ if and only if there exists $Q \in \Psi^{0,0}$ elliptic at α such that $Q\mathcal{A}$ is bounded in $\Psi^{-\infty,-\infty}$.

Proof The composition properties are automatic from the description of the product as an asymptotic sum. In particular, if $\alpha \notin \mathrm{WF}'_{L^\infty}(\mathcal{A})$ and Q is elliptic at α with $\mathrm{WF}'(Q) \cap \mathrm{WF}'_{L^\infty}(\mathcal{A}) = \emptyset$, then $Q\mathcal{A}$ is bounded in $\Psi^{-\infty,-\infty}$. Now, such a Q exists since $\mathrm{WF}'_{L^\infty}(\mathcal{A})$ is closed, and α is in its complement.

Conversely, if a Q as stated exists, let B be a microlocal parametrix for Q, so $BQ = \mathrm{Id} + E$ with $\alpha \notin \mathrm{WF}'(E)$. Then $A_t = BQA_t - EA_t$, with $\alpha \notin \mathrm{WF}'_{L^\infty}(EA_t)$ since $\alpha \notin \mathrm{WF}'(E)$, while $Q\mathcal{A}$ bounded in $\Psi^{-\infty,-\infty}$ implies that $BQ\mathcal{A}$ is also bounded there, thus $\alpha \notin \mathrm{WF}'_{L^\infty}(\mathcal{A})$. □

In summary, we have the following result, which is the basic microlocal propagation estimate, propagating control on $\mathrm{WF}'_{L^\infty}(E)$ to $\mathrm{WF}'(B_0)$:

Lemma 5.13 *Suppose that (5.54) is satisfied. Let $A_t = q_L(a_t)$, $B_t = q_L(b_t)$, $E_t = q_L(e_t)$, and let $Q_1, Q_2 \in \Psi^{0,0}$ such that Q_1 is elliptic on $\mathrm{WF}'_{L^\infty}(A)$, Q_2 is elliptic on $\mathrm{WF}'_{L^\infty}(E)$. If $Q_1u \in H^{s-1/2,r-1/2}$, $Q_1Pu \in H^{s-m+1,r-\ell+1}$, $Q_2u \in H^{s,r}$ then $B_0u \in L^2$ and for all M,N there is $C > 0$ such that*

$$\|B_0u\|_{L^2} \le C(\|Q_2u\|_{H^{s,r}} + \|Q_1Pu\|_{H^{s-m+1,r-\ell+1}} + \|Q_1u\|_{H^{s-1/2,r-1/2}} + \|u\|_{H^{M,N}}).$$

Proof We note that as Q_1 is elliptic on $\mathrm{WF}'_{L^\infty}(A)$, hence on $\mathrm{WF}'_{L^\infty}(F)$,

$$\|F_tu\|_{H^{-s+1/2,-r+1/2}} \le C(\|Q_1u\|_{H^{s-1/2,r-1/2}} + \|u\|_{H^{M,N}}),$$

and thus with $\tilde{Q}_1 \in \Psi^{0,0}$ with $\mathrm{WF}'(\mathrm{Id} - \tilde{Q}_1) \cap \mathrm{WF}'_{L^\infty}(A) = \emptyset$, Q_1 elliptic on $\mathrm{WF}'(\tilde{Q}_1)$, we have $(\mathrm{Id} - \tilde{Q}_1)F_t$ is bounded in $\Psi^{-\infty,\infty}$ and also

$$\|\tilde{Q}_1^*u\|_{H^{s-1/2,r-1/2}} \le C'(\|Q_1u\|_{H^{s-1/2,r-1/2}} + \|u\|_{H^{M,N}}),$$

so

$$|\langle F_tu, u\rangle| \le |\langle(\mathrm{Id} - \tilde{Q}_1)F_tu, u\rangle| + |\langle F_tu, \tilde{Q}_1^*u\rangle| \le C''(\|u\|^2_{H^{M,N}} + \|Q_1u\|^2_{H^{s-1/2,r-1/2}}).$$

Similarly,

$$|\langle E_t u, u\rangle| \leq C''(\|u\|^2_{H^{M,N}} + \|Q_2 u\|^2_{H^{s,r}}).$$

Next, using $m'/2 - (m-1)/2 = s - m + 1$, $\ell'/2 - (\ell-1)/2 = r - \ell + 1$,

$$\|(\Lambda^{-1}_{(m-1)/2,(\ell-1)/2})^* \check{A}_t P u\| \leq C(\|Q_1 P u\|_{H^{s-m+1,r-\ell+1}} + \|u\|_{H^{M,N}}).$$

Combining these estimates, we see that the right hand side of (5.58) remains uniformly bounded, and thus $B_0 u \in L^2$ as claimed. □

As a corollary:

Proposition 5.24 *For $\alpha_0 \in \partial(\overline{\mathbb{R}^n} \times \overline{\mathbb{R}^n})$ with $\mathsf{H}_{p,m,\ell}$ non-vanishing at α_0. Then α_0 has a neighborhood O in $\partial(\overline{\mathbb{R}^n} \times \overline{\mathbb{R}^n})$ on which $\mathsf{H}_{p,m,\ell}$ is non-vanishing, and if $\alpha \in O$, γ the integral curve of $\mathsf{H}_{p,m,\ell}$ through α with $\gamma(0) = \alpha$, and $\gamma(\sigma) = \beta$ for some $\sigma < 0$ such that $\gamma([\sigma,0]) \subset O$, and further if we are given a neighborhood U_2 of β and U_1 of $\gamma([\sigma,0])$ contained in O, $Q_1, Q_2 \in \Psi^{0,0}$ such that Q_1 is elliptic on U_1, Q_2 is elliptic on U_2, then there exists $Q_3 \in \Psi^{0,0}$ elliptic on $\gamma([\sigma,0])$ such that the following holds. If $Q_1 u \in H^{s-1/2,r-1/2}$, $Q_1 P u \in H^{s-m+1,r-\ell+1}$, $Q_2 u \in H^{s,r}$ then $Q_3 u \in H^{s,r}$ and for all M,N there is $C > 0$ such that*

$$\|Q_3 u\|_{H^{s,r}} \leq C(\|Q_2 u\|_{H^{s,r}} + \|Q_1 P u\|_{H^{s-m+1,r-\ell+1}} + \|Q_1 u\|_{H^{s-1/2,r-1/2}} + \|u\|_{H^{M,N}}).$$

The analogous result also holds for $\sigma > 0$.

Now propagation of singularities is an immediate consequence. Indeed, one can iterate, improving half an order at a time, taking $U_1' = \mathrm{Ell}(Q_3) \cap U_1$ and $U_2' = \mathrm{Ell}(Q_3) \cap U_2$ for the next iteration. This directly proves Theorem 5.4 when the bicharacteristic segment is contained in a single set O; in general a compactness argument proves it in a finite number of steps from this local version.

5.4.5. Complex Absorption

Although we required that $P \in \Psi^{m,\ell}_{\mathrm{cl}}$ have real principal symbol, this is not quite necessary. In particular, complex absorption has been a very useful technical tool, see for instance the work of Nonnenmacher and Zworski [35] (see also [5]): in this one "cuts off" a non-elliptic problem with real principal symbol by adding a complex term to the principal symbol which has support outside the region of interest.

So suppose instead that one is given a complex valued homogeneous function, $p - iq$, with p, q real-valued. Let $P \in \Psi^{m,\ell}_{\mathrm{cl}}$, resp. $Q \in \Psi^{m,\ell}_{\mathrm{cl}}$, have (real) principal symbols p, resp. q. Note that when $q \neq 0$, $p - iq$ is elliptic, while if q vanishes near a point α, then the previous microlocal estimate works. Thus, the key question is what happens at the characteristic set $\{p = 0,\ q = 0\}$ at points in supp q. The typical application here would involve a q that acts as microlocal "absorption" along the Hamilton flow; absorbing any singularity propagating along $\mathsf{H}_{p,m,\ell}$ in $\{p = 0\}$ away from supp u. Thus, one should consider q a bump function along the $\mathsf{H}_{p,m,\ell}$-bicharacteristics, whose sign will be very important; along a fixed $\mathsf{H}_{p,m,\ell}$-bicharacteristic γ in $\{p = 0\}$, where $q \neq 0$, the problem is elliptic, and thus if $(P - iQ)u$ is regular, so is u, where $q \equiv 0$, singularities propagate, so for a bump-function like q, the singularities are absorbed at the boundary of supp q. Since q is mostly a technical tool here, it is reasonable to make technically convenient assumptions on Q below.

Then for A formally self-adjoint,

$$\begin{aligned}&\langle i(P - iQ)u, Au\rangle - \langle iAu, (P - iQ)u\rangle \\ &\quad = \langle i([A, P] + (P - P^*)A)u, u\rangle + \langle (AQ + Q^*A)u, u\rangle, \qquad (5.59)\\ &i([A, P] + (P - P^*)A) \in \Psi^{m+m'-1,\ell+\ell'-1},\ AQ + Q^*A \in \Psi^{m+m',\ell+\ell'}.\end{aligned}$$

Now, if the principal symbol a of A is ≥ 0 as above, with the principal symbol of $i([A, P] + (P - P^*)A)$ being

$$-(H_p a + 2\tilde{p}a) = b^2 - e,$$

then for $q \geq 0$, the principal symbol of the second term on the right hand side of (5.59) has the same sign as that of the first. However, there is an issue with such an argument since the second term is of higher order than the first one. Thus, it is convenient to assume that $Q = T^2$, with $T \in \Psi^{m/2,\ell/2}$, $T = T^*$. Note that at the cost of changing P without changing its principal symbol, if $q = t^2$ for a non-negative symbol t, then this may always be arranged; simply take $T \in \Psi^{m/2,\ell/2}$ with principal symbol t and with $T = T^*$; then $Q - T^2 \in \Psi^{m-1,\ell-1}$, so replacing P by $P - i(Q - T^2)$, and Q by T^2, we have the desired form of Q. From now on we assume that

$$Q = T^2,\ T \in \Psi^{m/2,\ell/2},\ T = T^*.$$

When $a = \check{a}^2$ with $\check{a} \in S^{m'/2,\ell'/2}$, as arranged above, then with $\check{A} \in \Psi^{m'/2,\ell'/2}$ of principal symbol $\check{a}$ and $\check{A}^* = \check{A}$, let $A = \check{A}^2$ (so $A^* = A$) – we are simply a bit more careful in our specification of A. Then

$$AQ + Q^*A = \check{A}^2 Q + Q\check{A}^2 = 2\check{A}Q\check{A} + \check{A}[\check{A}, Q] + [Q, \check{A}]\check{A} = 2\check{A}T^2\check{A} + [\check{A}, [\check{A}, Q]],$$

and $[\check{A}, [\check{A}, Q]] \in \Psi^{m+m'-2,\ell+\ell'-2}$, which is thus the same order as F above, so is controlled by the a priori assumptions. On the other hand, $\langle \check{A}T^2\check{A}u, u\rangle = \|T\check{A}u\|^2$, so we obtain

$$\begin{aligned}&\langle i(P-iQ)u, Au\rangle - \langle iAu, (P-iQ)u\rangle\\ &\quad = \|Bu\|^2 - \langle Eu, u\rangle + \langle (F + [\check{A}, [\check{A}, Q]])u, u\rangle + 2\|T\check{A}u\|^2.\end{aligned}$$

Regularizing $\check{A}$ as $\check{A}_t$, all the previous arguments go through. Note also that if instead $\sigma > 0$, that is we have

$$-(H_p a + 2\tilde{p}a) = -b^2 + e,$$

then we need to change our requirements on Q, namely we need $q \leq 0$, and for the technically convenient setting we need $q = -t^2$, $t \geq 0$.

Proposition 5.25 *For $\alpha_0 \in \partial(\overline{\mathbb{R}^n} \times \overline{\mathbb{R}^n})$ with $\mathsf{H}_{p,m,\ell}$ non-vanishing at α_0. Suppose also that $Q = T^2$ as above. Then α_0 has a neighborhood O in $\partial(\overline{\mathbb{R}^n} \times \overline{\mathbb{R}^n})$ on which $\mathsf{H}_{p,m,\ell}$ is non-vanishing, and if $\alpha \in O$, γ the integral curve of $\mathsf{H}_{p,m,\ell}$ through α with $\gamma(0) = \alpha$, and $\gamma(\sigma) = \beta$ for some $\sigma < 0$ such that $\gamma([\sigma, 0]) \subset O$, and further if we are given a neighborhood U_2 of β and U_1 of $\gamma([\sigma, 0])$ contained in O, $Q_1, Q_2 \in \Psi^{0,0}$ such that Q_1 is elliptic on U_1, Q_2 is elliptic on U_2, then there exists $Q_3 \in \Psi^{0,0}$ elliptic on $\gamma([\sigma, 0])$ such that the following holds. If $Q_1 u \in H^{s-1/2,r-1/2}$, $Q_1(P - iQ)u \in H^{s-m+1,r-\ell+1}$, $Q_2 u \in H^{s,r}$ then $Q_3 u \in H^{s,r}$ and for all M, N there is $C > 0$ such that*

$$\begin{aligned}\|Q_3 u\|_{H^{s,r}} \leq C(&\|Q_2 u\|_{H^{s,r}} + \|Q_1(P - iQ)u\|_{H^{s-m+1,r-\ell+1}}\\ &+ \|Q_1 u\|_{H^{s-1/2,r-1/2}} + \|u\|_{H^{M,N}}).\end{aligned}$$

The analogous result also holds for $\sigma > 0$, provided $Q = -T^2$.

Thus, for $q \geq 0$, estimates propagate forward along the Hamilton flow; for $q \leq 0$, they propagate backwards. This means that singularities, as measured by the wavefront set, propagate in the *opposite direction*: if $q \geq 0$, and there is WF at α, then for $\sigma < 0$, there is also WF at $\gamma(\sigma)$, for otherwise our estimate (and the corresponding regularity statement) would give the absence of WF at α.

While we used $Q = T^2$ here, this was for a purely technical point. If $Q = Q^*$ (which may always be arranged at the cost of changing P without changing its principal symbol, namely replacing P by $P - i(Q - Q^*)/2$, and replacing Q by $(Q + Q^*)/2$) and $Q \geq 0$, then $\langle \check{A}Q\check{A}u, u\rangle = \langle Q\check{A}u, \check{A}u\rangle \geq 0$ still. In general, just because $q \geq 0$ and $Q = Q^*$, we do not have $Q \geq 0$, but by the sharp Gårding inequality, this holds modulo a one order lower error term, that is $\langle Qv, v\rangle \geq -C\|v\|^2_{H^{(m-1)/2,(\ell-1)/2}}$. Applying this with $v = \check{A}u$, we obtain a term from the right hand side that is controlled by $C'\|\Lambda_{(m-1)/2,(\ell-1)/2}\check{A}u\|^2$, which can be controlled as above by choosing $F > 0$ large to absorb this into $\|Bu\|^2$. For proofs of the sharp Gårding inequality we refer to [27, Theorem 18.1.14] in the currently considered setting, and [58, Theorem 4.32] in the semiclassical setting that is discussed in Section 5.5.

5.4.6. Fredholm Problems with Complex Absorption

This gives rise to the simplest non-elliptic Fredholm problem. So suppose that $P \in \Psi^{m,\ell}$ with real homogeneous principal symbol. Suppose also that $Q \in \Psi^{m,\ell}$ is such that its principal symbol q satisfies $q = t^2$ as above, and that for all $\alpha \in \partial(\overline{\mathbb{R}^n} \times \overline{\mathbb{R}^n})$ in $\Sigma(p) = \{p = 0\}$ the integral curve γ of H_p with $\gamma(0) = \alpha$ reaches $\{q > 0\}$ in finite time in both the forward and backward direction, that is there exist $T_\pm > 0$ such that $q(\gamma(T_\pm)) > 0$. Let

$$\mathcal{X}^{s,r} = \{u \in H^{s,r} : (P - iQ)u \in H^{s-m+1,r-\ell+1}\},\ \mathcal{Y}^{s,r} = H^{s,r}.$$

Then

$$P - iQ : \mathcal{X}^{s,r} \to \mathcal{Y}^{s-m+1,r-\ell+1},\ P^* + iQ^* : \mathcal{X}^{s,r} \to \mathcal{Y}^{s-m+1,r-\ell+1}$$

are Fredholm for all s, r. Note that this follows estimates of the kind

$$\|u\|_{H^{s,r}} \leq C(\|(P - iQ)u\|_{H^{s-m+1,r-\ell+1}} + \|u\|_{H^{M,N}}),$$
$$\|u\|_{H^{s,r}} \leq C(\|(P^* + iQ^*)u\|_{H^{s-m+1,r-\ell+1}} + \|u\|_{H^{M,N}}),$$

for sufficiently negative M, N, which in turn follow from the propagation of singularities result as discussed beforehand in the real principal symbol setting. Thus, we just need to check that if $u \in \mathcal{S}'$ and $(P - iQ)u \in H^{s-m+1,r-\ell+1}$ then $u \in H^{s,r}$, that is $\mathrm{WF}^{s,r}(u) = \emptyset$. But this is straightforward: if either p or q is elliptic at α, then $\alpha \notin \mathrm{WF}^{s+1,r+1}(u)$ and thus $\alpha \notin \mathrm{WF}^{s,r}(u)$ by microlocal elliptic regularity, and otherwise p vanishes at α, and thus the backward bicharacteristic γ from α reaches $q > 0$, where we know there is no $\mathrm{WF}^{s,r}(u)$, so propagation of singularities gives $\alpha \notin \mathrm{WF}^{s,r}(u)$. A similar argument works for the adjoint, using the forward bicharacteristic. This proves the claim.

5.4.7. Radial Points and Generalizations

The final ingredient for scattering theory is *radial points*. These are the points in the conic perspective where H_p is a multiple of the generator of dilations, and from the compactification point of view the points $\alpha \in \partial(\overline{\mathbb{R}^n} \times \overline{\mathbb{R}^n})$ where $\mathsf{H}_{p,m,\ell} = \langle\zeta\rangle^{-m+1}\langle z\rangle^{-\ell+1}H_p$, as a smooth vector field on $\overline{\mathbb{R}^n} \times \overline{\mathbb{R}^n}$, vanishes. At such points the argument given above breaks down, since there are no local coordinates on the boundary in which this rescaled Hamilton vector field would be a coordinate vector field. Also, at such points, one cannot use the derivative of a function on the boundary to dominate terms when H_p falls on weights or regularizers, with the result that the weights are required to produce the correct signs for the commutator argument.

Although here we are interested in rather large sets (higher dimensional submanifolds) of radial points, which were explicitly considered in a relatively general setting by Melrose in [30], as far as I am aware radial points made their first appearance in the work of Guillemin and Schaeffer [16] in a setting in which they are discrete. These discrete radial points also arise in scattering theory due to scattering by a symbolic potential of order 0 with non-degenerate critical points at infinity (i.e. at $\partial\overline{\mathbb{R}^n}$), where they were studied by Herbst and Skibsted [20, 21], and where microlocal techniques were applied by Hassell, Melrose, and Vasy [18, 19] obtaining even precise iterated regularity statements. In a less explicit way, such radial points have long played a role in scattering theory; indeed, the propagation set of Sigal and Soffer [39], which played a key role in their proof of the asymptotic completeness of short-range N-body Hamiltonians, is an N-particle generalization of radial sets, see [44, 45].

Following [48] it is convenient to proceed quite generally, using a generalization of this setting to a submanifold L of $\mathrm{Char}(P) \subset \partial(\overline{\mathbb{R}^n} \times \overline{\mathbb{R}^n})$ closed under the $\mathsf{H}_{p,m,\ell}$ flow, that is $\mathsf{H}_{p,m,\ell}$ is tangent to it, and which acts as a source or sink for the flow in a neighborhood of L within $\mathrm{Char}(P)$. Since we have corners here, we need to be more specific, and we require that L is a submanifold of one of the two boundary hypersurfaces which is transversal to the other boundary face (allowing of course that L does not intersect the other boundary face at all), see Figure 5.5. For the sake of definiteness, we assume that L is a subset of $\overline{\mathbb{R}^n} \times \partial\overline{\mathbb{R}^n}$, that is is at fiber infinity, and is transversal to $\partial\overline{\mathbb{R}^n} \times \partial\overline{\mathbb{R}^n}$, possibly via an empty intersection. Next, suppose that L is defined by $\{\rho_{1,j} = 0 : j = 1, \ldots, k\}$ in $\mathrm{Char}(P)$ within $\overline{\mathbb{R}^n} \times \partial\overline{\mathbb{R}^n}$, that is the $\rho_{1,j}$ have linearly independent differentials on their joint zero set, that is it is defined by $\{\hat{p} = 0,\ \rho_{1,j} = 0,\ j = 1, \ldots, k\}$ within $\overline{\mathbb{R}^n} \times \partial\overline{\mathbb{R}^n}$, where $\hat{p} = \rho_{\mathrm{fiber}}^m p$. Let

$$\rho_1 = \sum_{j=1}^{k} \rho_{1,j}^2,$$

which is thus a quadratic defining function of L, that is it vanishes there quadratically in a non-degenerate manner. Since $\mathsf{H}_{p,m,\ell}$ is tangent to L, $\mathsf{H}_{p,m,\ell}\rho_{1,j}$ vanishes at L, that is is a linear combination of the $\rho_{1,i}$ with C^∞ coefficients, so $\mathsf{H}_{p,m,\ell}\rho_1$ vanishes quadratically at L. The source(+)/sink(−) assumption is that there is a function $\beta_1 > 0$ such that

$$\mp\mathsf{H}_{p,m,\ell}\rho_1 = \beta_1\rho_1 + F_2 + F_3,$$

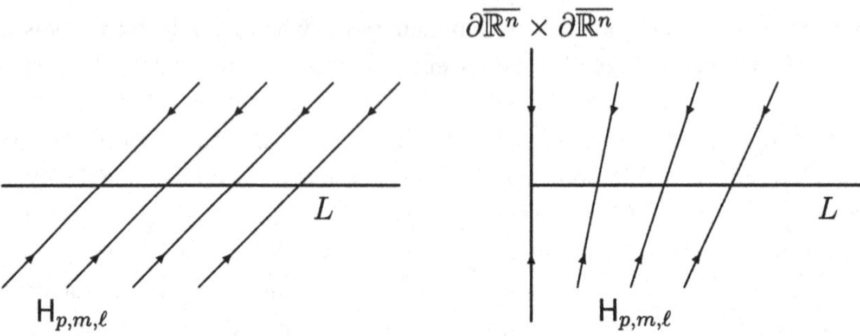

Figure 5.5. The Hamilton flow at radial sets L; in the examples both sinks. Left: $L \subset \mathbb{R}^n \times \partial\overline{\mathbb{R}^n}$, that is away from base infinity (the corner). Right: $L \subset \overline{\mathbb{R}^n} \times \partial\overline{\mathbb{R}^n}$ intersects $\partial\overline{\mathbb{R}^n} \times \partial\overline{\mathbb{R}^n}$ transversally

where $F_2 \geq 0$ and F_3 vanishes cubically at L. Under this assumption, L is a source or sink within Char(P), since ρ_1 is an increasing (+), resp. decreasing (−), function near L, as $|F_3| \leq C\rho_1^{3/2} \leq \beta_1\rho_1/2$ in a neighborhood of L. In particular, notice that if $\phi \in C_c^\infty([0,\infty))$ is such that ϕ is $\equiv 1$ near 0, $\phi \geq 0$ with $\sqrt{\phi}$ smooth and $\phi' \leq 0$ with $\sqrt{-\phi'}$ smooth, with support sufficiently close to 0 then

$$\pm\mathsf{H}_{p,m,\ell}\phi(\rho_1) = (-\phi'(\rho_1))(\beta_1\rho_1 + F_2 + F_3) \geq 0,$$

and further the second factor on the right hand side is > 0 on $\operatorname{supp}\phi'$, so indeed

$$\phi_1 = \sqrt{\pm\mathsf{H}_{p,m,\ell}\phi(\rho_1)}$$

is smooth, and vanishes near L.

To explain this source/sink condition note that if $\mathsf{H}_{p,m,\ell}$ actually vanishes at L, then one can consider its linearization at each point $\alpha \in L$, that is $\mathsf{H}_{p,m,\ell}$ then maps the ideal $\mathcal{I}$ of functions vanishing at α to itself, and thus $\mathcal{I}^2$ to itself, and thus acts on $\mathcal{I}/\mathcal{I}^2$, which can be identified with the cotangent space at α. Further, the conormal space of L at α is preserved by $\mathsf{H}_{p,m,\ell}$ as $\mathsf{H}_{p,m,\ell}\rho_{1,j}$ vanishes at L, that is $\mathsf{H}_{p,m,\ell}$ acts on this finite dimensional vector space. If the $\rho_{1,j}$ are such that $d\rho_{1,j}$ is an eigenvector with eigenvalue $\beta_{1,j}(\alpha)$ at each $\alpha \in L$, then $\mathsf{H}_{p,m,\ell}\rho_1$ is given by $\sum_{j=1}^k 2\beta_{1,j}\rho_{1,j}^2$ modulo cubically vanishing functions, and thus if all $\beta_{1,j}$ have the same non-zero sign, $\pm$, then letting β_1 to be $2\min_j|\beta_{1,j}|$, the above condition is satisfied (with $\pm$ corresponding to the sign $\pm$ here). In fact, in applications often $\mathsf{H}_{p,m,\ell}$ is a multiple of the identity operator on this finite dimensional vector space, and then all the $\beta_{1,j}$ are the same, and any defining functions $\rho_{1,j}$ can be used to construct ρ_1.

At the possible radial points in L, the only way one can have a positive commutator estimate is by taking into account the weights. Consider $\mathsf{H}_{p,m,\ell}$ as a vector field on $\overline{\mathbb{R}^n} \times \overline{\mathbb{R}^n}$ which is tangent to the boundary of this compact space, and recall that $\rho_{\text{base}} = \langle z \rangle^{-1}$ and $\rho_{\text{fiber}} = \langle \zeta \rangle^{-1}$ are defining functions of the boundary hypersurfaces. In particular, as already mentioned, $\mathsf{H}_{p,m,\ell}\rho_{\text{base}} = a_{\text{base}}\rho_{\text{base}}$ and $\mathsf{H}_{p,m,\ell}\rho_{\text{fiber}} = a_{\text{fiber}}\rho_{\text{fiber}}$ with $a_{\text{base}}, a_{\text{fiber}}$ well defined on $\partial(\overline{\mathbb{R}^n} \times \overline{\mathbb{R}^n})$ (though they depend on the choice of ρ_{base} and ρ_{fiber}). Moreover, if $\mathsf{H}_{p,m,\ell}$ actually vanishes at α, which is, say, at fiber infinity, then

$$\mathsf{H}_{p,m,\ell}(f\rho_{\text{fiber}}) = (\mathsf{H}_{p,m,\ell}f)\rho_{\text{fiber}} + f(\mathsf{H}_{p,m,\ell}\rho_{\text{fiber}}) = (a_{\text{fiber}}f + \mathsf{H}_{p,m,\ell}f)\rho_{\text{fiber}},$$

with $(a_{\text{fiber}}f + \mathsf{H}_{p,m,\ell}f)(\alpha) = a_{\text{fiber}}(\alpha)f(\alpha)$, shows that a_{fiber} is actually well-defined at such α, independent even of the choice of the boundary defining function. We then require that L is a source or sink even taking into account the infinitesimal "jet dynamics" at the boundary, that is a_{fiber} in our case (since we are working at fiber infinity) has the same sign as $\mathsf{H}_{p,m,\ell}\rho_1$: there is a defining function ρ_0 of $\overline{\mathbb{R}^n} \times \partial\overline{\mathbb{R}^n}$ such that

$$\mp \mathsf{H}_{p,m,\ell}\rho_0 = \beta_0\rho_0, \; \beta_0|_L > 0. \tag{5.60}$$

From the remarks above, if L consists of radial points, then one can simply take our preferred boundary defining function $\rho_0 = \rho_{\text{fiber}}$ in checking whether (5.60) holds, since all choices give the same result. Note that if (5.60) holds, then β_0 is thus bounded below by a positive constant in a neighborhood of L; we can always restrict work to such a neighborhood below.

We now consider the possibility that L intersects the corner (which the reader may ignore at first reading). We then also need to take care of weights in terms of ρ_{base}. Due to its importance in the analysis of Klein–Gordon type equations, we assume that we have a defining function (a positive smooth multiple of ρ_{base}) ρ_2 such that

$$\mathsf{H}_{p,m,\ell}\rho_2 = 2\beta_2\beta_0\rho_2, \qquad \beta_2|_L = 0;$$

in general β_2 can easily be handled as well.

Further, as before, $P - P^* \in \Psi^{m-1,\ell-1}$ also plays a role in these arguments, and at radial points it cannot be absorbed into other terms, just like the weight, that is powers of ρ_0, could not be thus absorbed. We normalize this and write

$$\tilde{p} = \sigma_{m-1,\ell-1}\left(\frac{1}{2i}(P - P^*)\right) = \pm\beta_0\tilde{\beta}\rho_0^{-m+1}.$$

Let $\phi_0 \in C_c^\infty(\mathbb{R})$ be identically 1 near 0, supported sufficiently close to 0; we use $\phi_0(\hat{p})$ to localize near $\mathrm{Char}(P)$. Note that

$$\mathsf{H}_{p,m,\ell}\phi_0(\hat{p}) = \phi_0'(\hat{p})\mathsf{H}_{p,m,\ell}\hat{p}$$

has support where $\hat{p}$ is in $\operatorname{supp}\phi_0'$, that is *away from the characteristic set*. Then we compute that with

$$a = \phi(\rho_1)^2 \phi_0(\hat{p})^2 \rho_0^{-m'} \rho_2^{-\ell'}$$

the principal symbol $-(H_p a + 2\tilde{p}a)$ of $i([A,P] + (P - P^*)A)$ is

$$\mp\rho_0^{-m'-m+1}\rho_2^{-\ell'-\ell+1}\big(\phi(\rho_1)\phi_1^2 \pm 2\phi^2\phi_0\phi_0' \mathsf{H}_{p,m,\ell}\hat{p} + \beta_0 m'\phi(\rho_1)^2 \\ + 2\beta_2\beta_0\ell'\phi(\rho_1)^2 + 2\beta_0\tilde{\beta}\phi(\rho_1)^2\big).$$

The second term in the parentheses is supported away from $\mathrm{Char}(P)$, and thus is controlled by microlocal elliptic estimates. The last three terms have essential support at L itself, and they have a definite sign if $m' + 2\beta_2\ell' + 2\tilde{\beta}$ has a definite sign. With $s = (m + m' - 1)/2$, $r = (\ell + \ell' - 1)/2$, this means that

$$s - (m-1)/2 + \beta_2(r - (\ell - 1)/2) + \tilde{\beta}$$

needs to have a definite sign when we prove $H^{s,r}$ regularity. Further, the first term in the parentheses has the same sign as these when the sign of $s-(m-1)/2+\beta_2+\tilde{\beta}$ is negative. Thus, when $s-(m-1)/2+\beta_2(r-(\ell-1)/2)+\tilde{\beta} > 0$, then all signs agree (apart from the off-the-characteristic-set term), and we have a result that has a different character from the standard propagation result in so far as not having to assume a priori H^s regularity anywhere to conclude H^s regularity near L. On the other hand, when $s - (m-1)/2 + \beta_2(r - (\ell - 1)/2) + \tilde{\beta} < 0$ we must assume regularity on $\operatorname{supp}\phi'$, that is in a punctured neighborhood of L, in order to conclude H^s regularity at L. In summary, using the fact that we can make ϕ have sufficiently small support so that the definite sign of this expression just at L implies the analogous conclusion on $\operatorname{supp}\phi$, modulo justifying the calculations via a regularization argument, we have

Proposition 5.26 *Suppose that L is as above, and $s-(m-1)/2+\tilde{\beta} > 0$ on L, and indeed suppose that $s' - (m-1)/2+\tilde{\beta} > 0$ on L for some $s' \in [s-1/2, s)$. Suppose U a neighborhood of L. Then there exists $Q \in \Psi^{0,0}$ elliptic on L such that for Q_1 elliptic on $\mathrm{WF}'(Q)$, with $\mathrm{WF}'(Q) \subset U$, and if $Q_1 u \in H^{s',r-1/2}$, $Q_1 Pu \in H^{s-m+1,r-\ell+1}$, then $Qu \in H^{s,r}$ and for all M,N there is $C > 0$ such that*

$$\|Qu\|_{H^{s,r}} \le C(\|Q_1 Pu\|_{H^{s-m+1,r-\ell+1}} + \|Q_1 u\|_{H^{s',r-1/2}} + \|u\|_{H^{M,N}}).$$

On the other hand, suppose now that $s - (m-1)/2 + \tilde{\beta} < 0$ on L. Suppose also that we are given neighborhoods U_1, U_2 of L with $\overline{U_1} \subset U_2$, and $Q_1, Q_2 \in \Psi^{0,0}$ with $\mathrm{WF}'(Q_2) \subset U_2 \setminus \overline{U_1}$, $\mathrm{WF}'(Q_1) \subset U_2$, Q_1 elliptic on U_1, and such that for every $\alpha \in (\mathrm{Char}(P) \cap U_1)\setminus L$, the bicharacteristic γ of p with $\gamma(0) = \alpha$ enters $\mathrm{Ell}(Q_2)$ while remaining in $\mathrm{Ell}(Q_1)$. Then there exists $Q \in \Psi^{0,0}$ elliptic *on L such that if $Q_2u \in H^{s,r}$, $Q_1u \in H^{s-1/2,r-1/2}$ and $Q_1Pu \in H^{s-m+1,r-\ell+1}$, then $Qu \in H^{s,r}$ and for all M, N there is $C > 0$ such that*

$$\|Qu\|_{H^{s,r}} \leq C(\|Q_2u\|_{H^{s,r}} + \|Q_1Pu\|_{H^{s-m+1,r-\ell+1}} + \|Q_1u\|_{H^{s-1/2,r-1/2}} + \|u\|_{H^{M,N}}).$$

Remark 5.3 In the second part of the above proposition, the 'there exists Q elliptic on L' statement can be replaced by 'for all Q with $\mathrm{WF}'(Q) \subset U_1$'. In addition, the first part of the proposition can also be rephrased, see the statement of Proposition 2.3 of [48].

Proof The first part of the proposition will follow immediately from the construction given below since the support of a can be arranged to be in the pre-specified set U, while the second part reduces to it taking into account that by the hypotheses u is microlocally in $H^{s,r}$ on $U_1 \setminus L$ by microlocal elliptic regularity and the (non-radial) propagation estimates, with an appropriate estimate holding in this space.

Modulo justifying the pairing argument, the standard calculation, (5.50) (with possibly a switch of the sign of the $\|Bu\|^2$ term) takes care of the $s-(m-1)/2+\tilde{\beta}|_L > 0$ case, with no E term in the sense of the real principal type result (since the main point there is that $\mathrm{WF}'(E) \cap \mathrm{Char}(P) \neq \emptyset$, i.e. one needs to make a priori assumptions on the characteristic set to propagate estimates), though with another potential error term $\tilde{E}$ with $\mathrm{WF}'(\tilde{E}) \cap \mathrm{Char}(P) = \emptyset$ arising from $\phi_0'(\hat{p})$ above, so

$$\tilde{e} = 2(\mp \mathsf{H}_{p,m,\ell}\hat{p})\phi(\rho_1)^2\phi_0(\hat{p})\phi_0'(\hat{p})\rho_0^{2s}\rho_2^{2r},$$

and B^*B replaced by $\sum_{j=1}^{2} B_j^*B_j$, with B_j having symbol b_j satisfying

$$b_1 = \sqrt{2\beta_0}\sqrt{s-(m-1)/2+\beta_2(r-(\ell-1)/2)+\tilde{\beta}}\,\phi(\rho_1)\phi_0(\hat{p})\rho_0^s\rho_2^r,$$
$$b_2 = \sqrt{\phi(\rho_1)\phi_1\phi_0(\hat{p})}\rho_0^s\rho_2^r.$$

Here we use the fact that one can make ϕ supported in a specified neighborhood of L; in particular in one in which $s-(m-1)/2+\beta_2(r-(\ell-1)/2)+\tilde{\beta}$ is bounded below by a positive constant since it is $s-(m-1)/2+\tilde{\beta}|_L > 0$ on L.

In a similar vein, but now regarding the ϕ_1 term as part of the error e, and still having an $\tilde{e}$ error term in the elliptic region, in the case $s-(m-1)/2+\tilde{\beta} < 0$, we can take

$$b = \sqrt{2\beta_0}\sqrt{s-(m-1)/2+\beta_2(r-(\ell-1)/2)+\tilde{\beta}}\,\phi(\rho_1)\phi_0(\hat{p})\rho_0^s\rho_2^r,$$
$$e = \phi(\rho_1)\phi_1^2\phi_0(\hat{p})^2\rho_0^{2s}\rho_2^{2r},$$
$$\tilde{e} = 2(\pm \mathsf{H}_{p,m,\ell}\hat{p})\phi(\rho_1)^2\phi_0(\hat{p})\phi_0'(\hat{p})\rho_0^{2s}\rho_2^{2r}.$$

In order to justify the argument, we use ϕ_t given in (5.55) with $\tau = \rho_0^{-1}$ and $K = 2(s - s')$. Then

$$\mp \mathsf{H}_{p,m,\ell}\phi_t(\rho_0^{-1}) = -\rho_0^{-1} f_t(\rho_0^{-1})\phi_t(\rho_0^{-1})\beta_0,$$

with

$$0 \le -\rho_0^{-1} f_t(\rho_0^{-1}) \le s - s'. \tag{5.61}$$

Thus, with

$$a_t = \phi(\rho_1)^2 \phi_0(\hat{p})^2 \rho_0^{-m'} \rho_2^{-\ell'} \phi_t(\rho_0^{-1})^2$$

the principal symbol $-(H_p a_t + 2\tilde{p} a_t)$ of $i([A_t, P] + (P - P^*)A_t)$ is

$$\mp \rho_0^{-m'-m+1} \rho_2^{-\ell'-\ell+1} \phi_t(\rho_0^{-1})^2 (\phi(\rho_1)\phi_1^2 + \beta_0 m' \phi(\rho_1)^2 + 2\beta_2\beta_0 \ell' \phi(\rho_1)^2 \\ + 2\beta_0\tilde{\beta}\phi(\rho_1)^2 - 2(-\rho_0^{-1} f_t(\rho_0^{-1}))\beta_0).$$

Then the above calculation of the commutator is unchanged provided either the new term $2(-\rho_0^{-1} f_t(\rho_0^{-1}))\beta_0$ has the same sign as the previous b_j or b terms, which indeed happens when $s - (m-1)/2 + \tilde{\beta}|_L < 0$, or if it can be absorbed in the b_j terms. For the latter we need

$$s - (m-1)/2 + \beta_2(r - (\ell - 1)/2) + \tilde{\beta} - (-\rho_0^{-1} f_t(\rho_0^{-1})) > 0,$$

which is satisfied if

$$s' - (m-1)/2 + \beta_2(r - (\ell - 1)/2) + \tilde{\beta} > 0$$

in view of (5.61). This proves the proposition, if the amount of regularization provided suffices for the proof to go through for $t > 0$. There are no issues with the argument provided we actually know that

$$\langle Pu, A_t u\rangle - \langle A_t u, Pu\rangle = \langle (A_t P - P^* A_t)u, u\rangle; \tag{5.62}$$

for the rest of the argument it is straightforward to check that all steps work, provided we take $s' = s - 1/2$ in the second case, and have s' as stated in the first case. (In fact, in the second case we may simply take $s' = s - 1$, and then the whole argument goes through directly, including the proof of (5.62); the same is true in the first case if $(s' - 1/2) - (m-1)/2 + \tilde{\beta} > 0$ on L with the notation of the statement of the proposition.)

So it remains to prove (5.62) for $t > 0$. The point is that while both sides make sense by the a priori assumptions, they are not necessarily equal in principle since for example $\langle P^* A_t u, u\rangle$ is not defined for u with just the property that $\mathrm{WF}^{s',r-1/2}(u)$ is disjoint from $\mathrm{WF}'_{L^\infty}(\{A_t\})$ (a bit stronger assumption is needed). This, however, is straightforward to overcome by using an additional regularizer family Λ_τ, for then

$$\begin{aligned}\langle Pu, A_t u\rangle - \langle A_t u, Pu\rangle &= \lim_{\tau\to 0}(\langle \Lambda_\tau Pu, A_t u\rangle - \langle \Lambda_\tau A_t u, Pu\rangle)\\ &= \lim_{\tau\to 0}(\langle A_t\Lambda_\tau Pu, u\rangle - \langle P^*\Lambda_\tau A_t u, u\rangle)\\ &= \lim_{\tau\to 0}\langle (A_t\Lambda_\tau P - P^*\Lambda_\tau A_t)u, u\rangle,\end{aligned}$$

so the argument is finished by showing that $(A_t\Lambda_\tau P - P^*\Lambda_\tau A_t) \to A_t P - P^* A_t$ strongly. But

$$(A_t\Lambda_\tau P - P^*\Lambda_\tau A_t) = \Lambda_\tau(A_t P - P^* A_t) + [A_t, \Lambda_\tau]P - [P^*, \Lambda_\tau]A_t,$$

and $A_t P - P^* A_t \in \Psi^{2s',2r-1}$, so $\Lambda_\tau(A_t P - P^* A_t) \to (A_t P - P^* A_t)$ in $\Psi^{2s'+\delta,2r-1+\delta}$ for $\delta > 0$ and is uniformly bounded in $\Psi^{2s',2r-1}$, while $[A_t, \Lambda_\tau]P$, $[P^*, \Lambda_\tau]A_t$ are uniformly bounded in $\Psi^{2s',2r-1}$ and converge to 0 in $\Psi^{2s'+\delta,2r-1+\delta}$ for $\delta > 0$, giving the desired strong convergence. This completes the proof of the proposition. □

As usual, one can iterate the argument and obtain:

Proposition 5.27 *Suppose that L is as above, and $s - (m-1)/2 + \tilde{\beta} > 0$ on L, and $s' - (m-1)/2 + \tilde{\beta} > 0$ on L and $r, r' \in \mathbb{R}$. Then there exist $Q \in \Psi^{0,0}$ elliptic on L such that for Q_1 elliptic on $\mathrm{WF}'(Q)$, if $Q_1 u \in H^{s',r'}$, $Q_1 Pu \in H^{s-m+1,r-\ell+1}$, then $Qu \in H^{s,r}$ and for all M, N there is $C > 0$ such that*

$$\|Qu\|_{H^{s,r}} \le C(\|Q_1 Pu\|_{H^{s-m+1,r-\ell+1}} + \|Q_1 u\|_{H^{s',r'}} + \|u\|_{H^{M,N}}).$$

On the other hand, suppose now that $s - (m-1)/2 + \tilde{\beta} < 0$ on L. Suppose also that we are given neighborhoods U_1, U_2 of L with $\overline{U_1} \subset U_2$, and $Q_1, Q_2 \in \Psi^{0,0}$ with $\mathrm{WF}'(Q_2) \subset U_2 \setminus \overline{U_1}$, $\mathrm{WF}'(Q_1) \subset U_2$, and such that for every $\alpha \in (\mathrm{Char}(p) \cap U_2) \setminus L$, the bicharacteristic γ of p with $\gamma(0) = \alpha$ enters $\mathrm{Ell}(Q_2)$ while remaining in $\mathrm{Ell}(Q_1)$. Then there exist $Q \in \Psi^{0,0}$ elliptic on L such that if $Q_2 u \in H^{s,r}$, and $Q_1 Pu \in H^{s-m+1,r-\ell+1}$, then $Qu \in H^{s,r}$ and for all M, N there is $C > 0$ such that

$$\|Qu\|_{H^{s,r}} \le C(\|Q_2 u\|_{H^{s,r}} + \|Q_1 Pu\|_{H^{s-m+1,r-\ell+1}} + \|u\|_{H^{M,N}}).$$

Notice that radial points are *not* structurally stable, unlike real principal type points in the characteristic set, that is where $\mathsf{H}_{p,m,\ell}$ is non-zero as a vector field on the boundary of the compactified phase space. Namely, if $P \in \Psi^{m,\ell}$ has real principal symbol p, $\mathsf{H}_{p,m,\ell}$ is non-radial at α then for any $\tilde{P} \in \Psi^{m,\ell}$ with real principal symbol $\tilde{p}$ and close to P in the pseudodifferential topology (i.e. $P, \tilde{P}$ are locally quantizations which are close in a suitable symbol seminorm), locally near α the characteristic set is near that of P and $\mathsf{H}_{\tilde{p},m,\ell}$ is also non-radial on it, and one has the real principal type propagation estimate. On the other hand, if α is a radial point for P, not only need α not be a radial point for $\tilde{P}$ (or be in the characteristic set): there may not even be radial points nearby! Even the generalized version we stated above with a submanifold L of normal sources/sinks need not apply. However, the *estimates* we proved are stable in the sense that if in the proof of Proposition 5.26 a certain choice of ϕ, ϕ_0 and powers of ρ_0, ρ_2 works, the same ϕ, ϕ_0 and powers of ρ_0, ρ_2 also give an estimate for $\tilde{P}$ when $\tilde{P}$ is close to P in the above sense, with real principal symbol. Proposition 5.27 is then similarly stable. See Remark 2.5 of [48].

5.4.8. Fredholm Problems with Radial Points

This gives rise to another non-elliptic Fredholm problem. So suppose that $P \in \Psi^{m,\ell}$ with real homogeneous principal symbol. Suppose also that $Q \in \Psi^{m,\ell}$ is such that its principal symbol q satisfies $q = t^2$ as above, $\mathrm{WF}'(Q) \cap L = \emptyset$, and that for all $\alpha \in \partial(\overline{\mathbb{R}^n} \times \overline{\mathbb{R}^n})$ in $\Sigma(p) = \{p = 0\}$ the integral curve γ of H_p with $\gamma(0) = \alpha$ reaches $\{q < 0\}$ in finite time in the backward direction, and tends to L in the forward direction. (Although we worked with $\overline{\mathbb{R}^n} \times \overline{\mathbb{R}^n}$ as our compactified phase space, all our arguments extend directly to the full manifold case, i.e. $\overline{{}^{\mathrm{sc}}T^*}X$, discussed in Section 5.3.11.) Again let

$$\mathcal{X}^{s,r} = \{u \in H^{s,r} : (P - iQ)u \in H^{s-m+1,r-\ell+1}\},\ \mathcal{Y}^{s,r} = H^{s,r}.$$

Then

$$P - iQ : \mathcal{X}^{s,r} \to \mathcal{Y}^{s-m+1,r-\ell+1},\ P^* + iQ^* : \mathcal{X}^{\tilde{s},\tilde{r}} \to \mathcal{Y}^{\tilde{s}-m+1,\tilde{r}-\ell+1} \tag{5.63}$$

are Fredholm for all r and for all s with

$$s - (m-1)/2 + \tilde{\beta}|_L > 0,\ \tilde{s} - (m-1)/2 - \tilde{\beta}|_L < 0. \tag{5.64}$$

Note that this is exactly the condition for "high regularity" $H^{s,r}$ estimates for P at the radial points, as well as for "low regularity" $H^{\tilde{s},\tilde{r}}$ estimates for P^* at these. Further, for

$$\tilde{s} = m - 1 - s,\ \tilde{r} = \ell - 1 - r, \tag{5.65}$$

the stated Sobolev spaces for $P^* + iQ^*$ are the duals of those stated for $P - iQ$, that is

$$H^{\tilde{s},\tilde{r}} = (H^{s-m+1,r-\ell+1})^*,\ H^{\tilde{s}-m+1,\tilde{r}-\ell+1} = (H^{s,r})^*.$$

Again, this Fredholm statement follows estimates of the kind

$$\|u\|_{H^{s,r}} \leq C(\|(P - iQ)u\|_{H^{s-m+1,r-\ell+1}} + \|u\|_{H^{M,N}}),$$
$$\|u\|_{H^{\tilde{s},\tilde{r}}} \leq C(\|(P^* + iQ^*)u\|_{H^{\tilde{s}-m+1,\tilde{r}-\ell+1}} + \|u\|_{H^{\tilde{M},\tilde{N}}}),$$

for $M < s$, $N < r$, $\tilde{M} < \tilde{s}$, $\tilde{N} < \tilde{r}$, which in turn follow from the propagation of singularities and the result at the radial points. Thus, now we just need to check that if $u \in H^{s',r'}$ for some r' and with s' as above, and $(P - iQ)u \in H^{s-m+1,r-\ell+1}$ then $u \in H^{s,r}$, that is $\mathrm{WF}^{s,r}(u) = \emptyset$. But if either p or q is elliptic at α, then $\alpha \notin \mathrm{WF}^{s+1,r+1}(u)$ and thus $\alpha \notin \mathrm{WF}^{s,r}(u)$ by microlocal elliptic regularity. Further, there is a neighborhood U of α such that $\alpha \in U$ implies $\alpha \notin \mathrm{WF}^{s,r}(u)$ by the high-regularity radial set result. Now propagating the regularity backwards (which is what forces the sign of q), if p vanishes at α, and thus the forward bicharacteristic γ from α reaches U, where we know there is no $\mathrm{WF}^{s,r}(u)$, so propagation of singularities gives $\alpha \notin \mathrm{WF}^{s,r}(u)$. For the adjoint, again we do not need to be concerned with elliptic points. If p vanishes at α but $\alpha \notin L$, then the backward bicharacteristic from α reaches $q < 0$, so the propagation of singularities result gives $\alpha \notin \mathrm{WF}^{\tilde{s},\tilde{r}}(u)$. Finally, at L, we use the low-regularity radial set result to complete the proof of the Fredholm claim.

This can be generalized when the characteristic set of p has two components $\Sigma_\pm$, and in Σ_+ the radial set L_+ is a source, in Σ_- it is a sink L_-. Namely, suppose that $P \in \Psi^{m,\ell}$ with real homogeneous principal symbol, $Q \in \Psi^{m,\ell}$ is such that its principal symbol q satisfies $q = t_+^2 - t_-^2$ with $\operatorname{supp} t_+ \cap \Sigma_- = \emptyset$, $\operatorname{supp} t_- \cap \Sigma_+ = \emptyset$, and that for all $\alpha \in \partial(\overline{\mathbb{R}^n} \times \overline{\mathbb{R}^n})$ in Σ_+, resp. Σ_-, the integral curve γ of H_p with $\gamma(0) = \alpha$ reaches $\{q > 0\}$, resp. $\{q < 0\}$, in finite time in the forward, resp. backward, direction, and tends to L_+, resp. L_-, in the backward, resp. forward, direction. Then $P - iQ$, $P^* + iQ^*$ are Fredholm as in (5.63) subject to (5.64).

This setup suffices for instance for the analysis of scattering theory on asymptotically hyperbolic spaces, which we will discuss in Section 5.5 as some geometric preliminaries are required.

But first, we discuss the geometrically simpler asymptotically Euclidean spaces in which no artificial complex absorption is needed, but for which instead we need to use variable order Sobolev spaces. For simplicity, let us take g to be a scattering metric actually asymptotic to the Euclidean metric, that is

$$g = \sum_{ij} g_{ij}(z)\, dz_i \otimes dz_j,$$

with $g_{ij} = \mathcal{C}^\infty(\overline{\mathbb{R}^n})$, with dual metric

$$G = \sum_{ij} G_{ij}(z)\, dz_i \otimes dz_j,$$

with $G_{ij} = \mathcal{C}^\infty(\overline{\mathbb{R}^n})$ still, and with

$$g_{ij} - \delta_{ij} \in \langle z\rangle^{-1}\mathcal{C}^\infty(\overline{\mathbb{R}^n}).$$

As discussed earlier, the principal symbol of

$$P = \Delta_g + V - \sigma,$$

$V \in \langle z\rangle^{-1}\mathcal{C}^\infty(\overline{\mathbb{R}^n})$ real-valued, is

$$\sum G_{ij}\zeta_i\zeta_j - \sigma,$$

modulo $S^{1,-1}$. Thus, regardless of σ, P is elliptic at fiber infinity; all possible issues are at the characteristic set, at finite ζ. If $\sigma \in \mathbb{C} \setminus [0,\infty)$, we have full ellipticity, as we have already discussed, so consider $\sigma > 0$ (with $\sigma = 0$ being a degenerate case). It is thus convenient to work in the homogeneous in the base framework; see Figure 5.6.

Since $G_{ij} = \delta_{ij}$ at base infinity, the Hamilton vector field there is then

$$H_p = 2\sum \zeta_j \partial_{z_j},$$

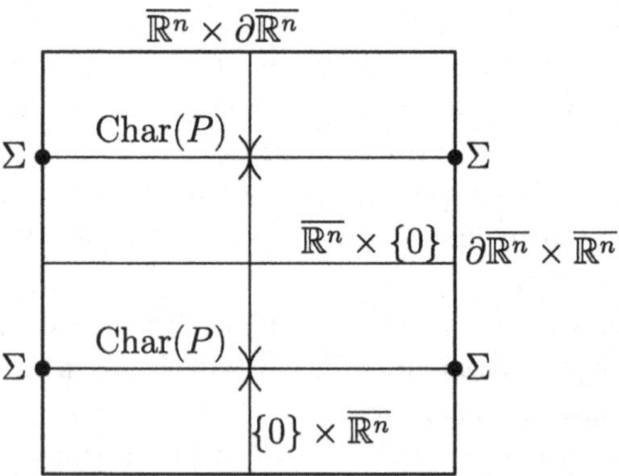

Figure 5.6. The characteristic set for $\sigma > 0$. From the compactified perspective, it is a subset of base infinity, $\partial\overline{\mathbb{R}^n} \times \overline{\mathbb{R}^n}$, denoted by Σ: it is an $(n-1)$-sphere bundle over the sphere at infinity, $\partial\overline{\mathbb{R}^n}$, thus a torus if $n = 2$; if $n = 1$ it is disconnected as shown, otherwise connected. From the homogeneous-in-the-base perspective, it is a subset of $(\mathbb{R}^n \setminus \{0\}) \times \mathbb{R}^n$, denoted by Char$(P)$ in the picture, with the parentheses at $\{0\} \times \mathbb{R}^n$ showing that this, that is the fiber of the cotangent bundle at 0, is *not* part of the homogeneous picture

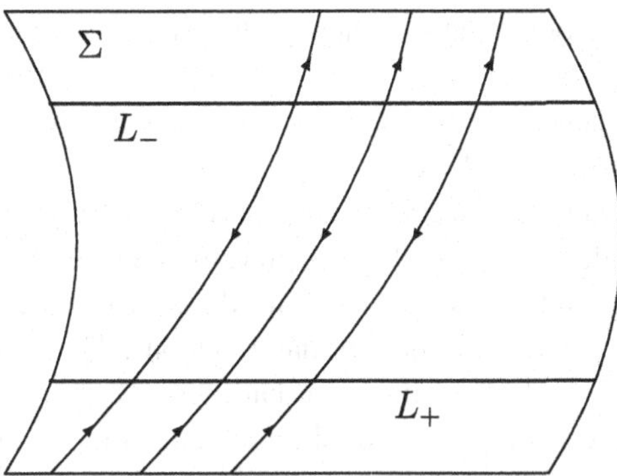

Figure 5.7. The characteristic set and the bicharacteristics for $\sigma > 0$ in the compactified picture; the figure is accurate for $n = 2$ when the characteristic set is a torus, on the picture shown cut along two circles, so the left and right side, resp. the top and bottom side, are identified. Here L_- is a source in the normal directions, while L_+ is a sink

and H_p is radial on

$$L_\pm = \{(z, \zeta) : z = c\zeta, \ |\zeta|^2 = \sigma, \ \pm c > 0\}.$$

Since the bicharacteristics are straight lines in $\mathbb{R}^n_z$, $t \mapsto (z_0 + 2t\zeta_0, \zeta_0)$, with ζ constant along them, it is easy to see that (in the compactified picture: recall that parallel lines tend to the same point at infinity, i.e. $\partial\overline{\mathbb{R}^n}$!) they approach $L_\pm$ as $t \to \pm\infty$, and indeed that all the assumptions at $L_\pm$ required above hold; see Figure 5.7.

Thus, if $n \geq 2$ the bicharacteristics connect the two components of the radial set $L_\pm$ (so $L_\pm$ lie inside the same component of the characteristic set); if $n = 1$ the characteristic set is disconnected. For the analysis, we note that, with $\ell = 0$, $\tilde{\beta} = 0$ (by the formal self-adjointness assumption due to the reality of V), if $r_- - (\ell - 1)/2 + \tilde{\beta}|_{L_-} > 0$, that is we are in the high regularity regime at L_-, then we can propagate H^{s,r_-} estimates out of L_- for elements of $H^{s',r'}$ with $r' - (\ell - 1)/2 + \tilde{\beta}|_{L_-} > 0$. These can be propagated to a punctured neighborhood of L_+. Now at L_+ we can propagate then to L_+ itself in $H^{s_+,r}$ provided $r_+ - (\ell - 1)/2 + \tilde{\beta}|_{L_+} < 0$, that is we are in the low regularity regime. Thus, we need spaces with differing regularity at $L_\pm$. Notice that the adjoint operator, on the dual spaces corresponding to these microlocal Sobolev spaces, will then proceed in the opposite direction, starting with high regularity estimates from L_+, propagating them to a punctured neighborhood at L_-, and then propagating these to L_- itself in the low regularity regime. (Of course, we could have reversed the choice of the function spaces for the operator, making the regularity high at L_+; then the high regularity would be at L_- for the adjoint.)

The geometric generalization is, as already discussed, Riemannian scattering metrics, see Proposition 5.23, or more generally $P = H - \sigma = \Delta_g + V - \sigma$, $V \in S^{-1}(X)$ real valued. One can then compute the bicharacteristics in terms of those of the boundary metric h; this was done by Melrose [30]. The conclusion is that there are again radial sets $L_\pm$, with the flow connecting them, that is L_+ is a sink, L_- a source, in the normal directions. The threshold regularity is $(\ell - 1)/2 = -1/2$. Choosing r so that $r > -1/2$ at L_-, $r < -1/2$ at L_+ or vice versa, gives rise to Fredholm problems. In this case one can show that in either case first $\mathrm{Ker}(H - \sigma) \subset \dot{C}^\infty(X)$ (with the kernel understood as the operator acting $H^{s,r}_{\mathrm{sc}}(X) \to H^{s-2,r}_{\mathrm{sc}}(X)$), and next that the kernel is actually trivial; without this one would in principle have a finite dimensional kernel. (The triviality of the kernel is called the "absence of embedded eigenvalues" in scattering theory, corresponding to $[0, +\infty)$ being the continuous spectrum of H, so the eigenvalues would be embedded in the continuous spectrum. We do not discuss this here, but refer to [30, Sections 10–13] for a discussion in the present setting. Conceptually, this absence is a combination of a positivity property, described in [50] more generally, obtained via a boundary pairing argument in [30], which allows one to conclude that the kernel is in $\dot{C}^\infty(X)$, and a unique continuation argument at infinity, obtained via a Carleman estimate, as in [27, Theorem 17.2.8].) Thus,

Proposition 5.28 *Let $r \in C^\infty({}^{\mathrm{sc}}T^*_{\partial X}X)$ be real valued, monotone in the characteristic set $\Sigma \subset {}^{\mathrm{sc}}T^*_{\partial X}X$ of P along the H_p flow and satisfying $r > -1/2$ resp. $r < -1/2$ at precisely one of L_+ and L_-. Then*

$$P : \{u \in H^{s,r}_{\mathrm{sc}}(X) : \; Pu \in H^{s-2,r+1}_{\mathrm{sc}}(X)\} \to H^{s-2,r+1}_{\mathrm{sc}}(X) \tag{5.66}$$

is invertible.

Further, if $r > -1/2$ at L_-, then

$$P^{-1} = (H - (\sigma + i0))^{-1} = \lim_{\epsilon \to 0} (H - (\sigma + i\epsilon))^{-1},$$

while if $r > -1/2$ at L_+ then $P^{-1} = (H - (\sigma - i0))^{-1}$.

All statements except the triviality of the kernel (which we do not discuss here) and the last claim have already been proved. As for the last claim, first note that

$$H - (\sigma + i\epsilon) : H^{s,r}_{\mathrm{sc}}(X) \to H^{s-2,r}_{\mathrm{sc}}(X)$$

is invertible for $\epsilon > 0$ by Proposition 5.23. Further, the propagation estimates giving rise to the Fredholm statement in (5.66) also hold for P replaced by

$H-(\sigma+i\epsilon)$, with $i\epsilon$ acting as complex absorption, *uniformly* in $\epsilon \in [0,1]$. This is how the limiting absorption principle corresponds to function spaces: $\epsilon > 0$ necessitates propagating the estimates *forward* along the Hamilton flow of P, that is r being larger than the threshold value at L_-. Then a simple argument shows that in fact $\lim_{\epsilon\to 0}(H-(\sigma+i\epsilon))^{-1}$ exists and is equal to P^{-1} on the corresponding space. The case of $(H-(\sigma-i0))^{-1}$ is analogous.

A very similar setup is applicable for the Klein–Gordon operator $\Box_g - m^2$, $m > 0$, on Minkowski space (with the signature of g being $(1, n-1)$), or on manifolds with Lorentzian scattering metrics, a notion introduced in [2]. (In [2] the $m = 0$ problem was considered, which we discuss here in Section 5.6 when the b-analysis tools are developed.)

We finally mention the Fredholm context of the stability statement made after Proposition 5.27. Namely, the stability of the estimates implies the stability of the Fredholm statement, that is that if P fits into any of the Fredholm frameworks discussed above, then so does $\tilde{P}$ if $\tilde{P}$ is assumed to have real principal symbol and is sufficiently close to P in the pseudodifferential topology. In fact, it is not hard to show, see Section 2.7 of [48], that if the inverse P^{-1} exists, and one has a continuous family $w \mapsto \tilde{P}(w)$ of pseudodifferential operators with $\tilde{P}(w_0) = P$, then for w near w_0, $\tilde{P}(w)$ is invertible and the inverse depends continuously on w in the weak operator topology of $\mathcal{L}(\mathcal{Y}^{s-m+1,r-l+1}, \mathcal{X}^{s,r})$. Thus, our Fredholm framework is very stable in the context of pseudodifferential operators with real principal symbols. In particular, these remarks apply to the Klein–Gordon equation for Lorentzian metrics g which are close to the Minkowski metric in a symbolic sense, and provide a Fredholm framework even more generally than the setting of the Lorentzian scattering metrics of [2] which needed extra properties for the more precise asymptotic analysis presented there.

5.5. Conformally Compact Spaces

5.5.1. The Geometric Problem

We now show how the microlocal results of the previous sections give the meromorphic extension of the resolvent of the Laplacian on even conformally compact (asymptotically hyperbolic) spaces. Overall we follow [47] closely, but we de-emphasize the estimates outside strips (in the "physical half plane," $\operatorname{Im}\sigma > 0$). This has the benefit that conjugation by a σ-dependent power of $1+\mu$ becomes irrelevant, and thus can be dropped, simplifying the connection between the conformally compact and the extended problems. The downside is that while one obtains a Fredholm family whether the geometry is trapping or not, its invertibility at one point (thus the meromorphy of the inverse) follows only in the non-trapping setting when the large σ (in a strip) estimates hold. Thus, at the end of the section I present the additional conjugation, giving the full Mazzeo–Melrose and Guillarmou [29, 15] result in the setting of even metrics. In addition, I also refer to Zworski's expository article [57] for a particularly accessible treatment of part of the framework discussed below.

We start by recalling the definition of manifolds with *even* conformally compact metrics. These are Riemannian metrics g_0 on the interior of an n-dimensional compact manifold with boundary X_0 such that near the boundary Y, with a product decomposition nearby and a defining function x, they are of the form

$$g_0 = \frac{dx^2 + h}{x^2} \tag{5.67}$$

where h is a family of metrics on $Y = \partial X_0$ depending on x in an even manner, that is only even powers of x show up in the Taylor series, that is $h = h(x^2, y, dy)$ with smooth dependence on x^2. (There is a much more natural way to phrase the evenness condition due to Guillarmou [15].) It is convenient to take x to be a globally defined boundary defining function. Then the dual metric is

$$G_0 = x^2(\partial_x^2 + H),$$

with H the dual metric family of h (depending on x as a parameter), and

$$|dg_0| = \sqrt{|\det g_0|}\, dx\, dy = x^{-n}\sqrt{|\det h|}\, dx\, dy$$

so

$$\Delta_{g_0} = (xD_x)^2 + i(n - 1 + x^2\gamma)(xD_x) + x^2\Delta_h, \tag{5.68}$$

with γ even, and Δ_h the x-dependent family of Laplacians of h on Y. Below we consider the spectral family

$$\Delta_{g_0} - \frac{(n-1)^2}{4} - \sigma^2$$

of the Laplacian.

We start by noting that, regardless of the evenness of h as a function of x, g_0 is a complete Riemannian metric on X_0, and thus we have, writing $L_0^2(X_0)$ for the L^2 space of g_0, due to Chernoff's theorem:

Proposition 5.29 *The operator* $\Delta_{g_0} - \frac{(n-1)^2}{4}$ *is essentially self-adjoint on* $\dot{C}^\infty(X_0)$, *and thus* $\mathcal{R}(\sigma) : (\Delta_{g_0} - \frac{(n-1)^2}{4} - \sigma^2)^{-1} : L_0^2(X_0) \to L_0^2(X_0)$, $\operatorname{Im}\sigma > 0$, $\sigma \notin i[0, \frac{n-1}{2}]$.

As an aside, the natural Sobolev spaces for the analysis of Δ_{g_0} are those corresponding to vector fields of finite length with respect to g_0 with appropriate smoothness properties. Thus, letting $\mathcal{V}_0(X_0) = x\mathcal{V}(X_0)$ (recall that x is a boundary defining function), for integer $s \geq 0$,

$$H_0^s(X_0) = \{u \in L_0^2(X_0) : \forall k \leq s,\ V_1 \dots V_k u \in L_0^2(X_0),\ V_j \in \mathcal{V}_0(X_0)\},$$

where we again wrote $L_0^2(X_0)$ for the L^2 space of g_0. For negative integer s the spaces can be defined via dualization, and then for $s \in \mathbb{R}$ by interpolation. An alternative way of defining $H_0^s(X_0)$ is via the 0-pseudodifferential algebra developed by Mazzeo and Melrose in [29]. In any case, the domain of Δ_{g_0} then is $H_0^2(X_0)$, and for σ as indicated above,

$$\mathcal{R}(\sigma) : \left(\Delta_{g_0} - \frac{(n-1)^2}{4} - \sigma^2\right)^{-1} : H_0^s(X_0) \to H_0^{s-2}(X_0),$$

for all $s \in \mathbb{R}$.

The question one asks is if there is a meromorphic extension of $\mathcal{R}(\sigma)$ to $\mathbb{C}$. This was essentially proved by Mazzeo and Melrose [29], with an important improvement by Guillarmou [15] regarding the potential essential singularities:

Theorem 5.5 ([29, 15]) *Without assuming that h is even, the operator $\mathcal{R}(\sigma)$, given by $(\Delta_{g_0} - \frac{(n-1)^2}{4} - \sigma^2)^{-1}$ for $\operatorname{Im}\sigma > (n-1)/2$, has a meromorphic continuation to $\mathbb{C} \setminus \{-i(k+1/2),\ k \in \mathbb{Z},\ k \geq 0\}$, that is after one removes the pure imaginary half-integer essential singularities, as an operator $C_c^\infty(X_0^\circ) \to C^\infty(X_0^\circ)$, thus its Schwartz kernel continues meromorphically as a distribution.*

Further, if h is even, then actually there are no essential singularities, that is the analytic continuation is meromorphic on $\mathbb{C}$.

The goal of this section is to prove this theorem, and to provide uniform control for $\mathcal{R}(\sigma)$ as $|\sigma| \to \infty$ in strips $|\operatorname{Im}\sigma| < C$. The latter necessitates a rather different approach to that of Mazzeo, Melrose, and Guillarmou, based on extension of the operator across ∂X_0 in an appropriate sense and using the microlocal machinery that we have developed.

A particularly interesting point is the following: Mazzeo and Melrose [29] in fact show that for $f \in \dot{C}^\infty(X_0)$, that is for f vanishing to infinite order at ∂X_0, $\mathcal{R}(\sigma)f$ is of the form

$$\mathcal{R}(\sigma)f = x^{\frac{n-1}{2} - i\sigma} u,\ u \in C^\infty(X_0),$$

if $\operatorname{Im}\sigma > 0$ and σ is not a pole of $\mathcal{R}(.)$, and indeed for arbitrary σ which is not a pole of $\mathcal{R}(.)$ provided $2\sigma \notin i\mathbb{Z}$. (In the case of integer coincidences, $2\sigma \in i\mathbb{Z}$, one has additional logarithmic terms; if g is even then the integer coincidence condition is $\sigma \in i\mathbb{Z}$.) Note that if $\operatorname{Im}\sigma > 0$, then $\mathcal{R}(\sigma)f \in L^2$ as $\operatorname{Re}(-i\sigma) > 0$, but not if $\operatorname{Im}\sigma \leq 0$. Thus, for σ with $\operatorname{Im}\sigma > 0$, $u_\pm = \mathcal{R}(\pm\sigma)f$ both solve

$$\left(\Delta_g - \frac{(n-1)^2}{4} - \sigma^2\right) u_\pm = f;$$

the difference is that u_+ is the unique L^2-solution. This shows the difficulty of studying the analytic continuation because u_- cannot be so easily characterized without the full asymptotics: one cannot say that "the L^2-term is missing" since it is dominated by the growing asymptotic, $x^{(n-1)/2+i\sigma}\mathcal{C}^\infty(X_0)$, and further if $\operatorname{Im}\sigma$ changes sign the role of the two asymptotics (which is L^2 and which is not) reverses.

In the method we develop, the way of distinguishing between u_+ and u_- is rather different. We extend a modification of the operator to an extension X of the manifold X_0 endowed with the "even smooth structure" (only even functions on X_0 are considered smooth) across the boundary, and we do the same to $u_\pm$ after multiplying $u_\pm$ by $x^{-(n-1)/2+i\sigma}$. Then $x^{-(n-1)/2+i\sigma}u_+$ extends to be $\mathcal{C}^\infty$ across ∂X_0, but $x^{-(n-1)/2+i\sigma}u_-$ does not. Thus, in the extended picture we develop, the way of distinguishing between solutions is smoothness or lack thereof, *regardless* of the sign of $\operatorname{Im}\sigma$, that is smoothness rather than decay. Ultimately this allows for obtaining the Fredholm framework that plays a crucial role in the large parameter estimates.

Before proceeding with the analytic discussion we mention a closely related geometric problem that plays a major role below at least indirectly. Even asymptotically de Sitter spaces are manifolds with boundary $\tilde{X}_0$ (with boundary defining function $\tilde{x}$) equipped with a Lorentzian metric $\tilde{g}_0$ in the interior for which a neighborhood of the boundary has the form

$$\tilde{g}_0 = \frac{d\tilde{x}^2 - \tilde{h}}{\tilde{x}^2},$$

with $\tilde{h} = \tilde{h}(\tilde{x}^2, .)$ a family of metrics on $Y = \partial\tilde{X}_0$, which further satisfy that $Y = Y_+ \cup Y_-$ (disjoint union of connected components) and all null-geodesics of $\tilde{g}_0$ tend to either Y_+ in the forward direction along the Hamilton flow and to Y_- in the backward direction, or vice versa. While not apparent from the definition, these spaces are necessarily of the form $Y_+ \times I$, $I = [-1, 1]$ a compact interval, see [53] and the arguments of Geroch [13], and indeed the spacetime is globally hyperbolic, so if the (time-like) variable on I is denoted by T, with $T = 1 - \tilde{x}$ near Y_+, $T = -1 + \tilde{x}$ near Y_-, then the Cauchy problem at $T = T_0 \in (-1, 1)$ for the Klein–Gordon equation

$$\left(\Box_{\tilde{g}_0} - \frac{(n-1)^2}{4} - \sigma^2\right) u = f$$

is well-posed. Further, for $\mathcal{C}^\infty$ Cauchy data, the solution has an asymptotic expansion at $T = \pm 1$, as shown in [53]:

$$u = \tilde{x}^{\frac{n-1}{2}+i\sigma} u_+ + \tilde{x}^{\frac{n-1}{2}-i\sigma} u_-,\ u_\pm \in \mathcal{C}^\infty(\tilde{X}_0),$$

and indeed $u_\pm$ are even, as long as $\sigma \notin i\mathbb{Z}$ (there are potential logarithmic terms otherwise as for the asymptotically hyperbolic case above).

Now notice that if g is even asymptotically hyperbolic, and one replaces x by ix formally (called Wick rotation), one obtains

$$\frac{dx^2 - h(-x^2, y, dy)}{x^2},$$

which is asymptotically de Sitter, and vice versa. Here of course h needs to be extended smoothly to negative values of the argument. One way to think of the analytic constructions in Section 5.5.3 is that they give a precise setting in which Wick rotations are meaningful. Notice that if h is not even, but rather has the form $h'(x, y, dy)$, one would get $h'(ix, y, dy)$, which would require an analytic continuation, rather than a smooth extension, to make sense.

5.5.2. Large Parameter and Semiclassical Pseudodifferential Operators

As discussed above, in addition to working with finite σ, or σ in a compact set, we also want to consider $\sigma \to \infty$, mostly in strips, with $|\operatorname{Im}\sigma|$ bounded. In that case we should consider σ as a "large parameter." In general, on $\mathbb{R}^n$, the large parameter setting just means that one has a family of operators given by symbols a on $\mathbb{R}^n_z \times \mathbb{R}^n_{z'} \times \mathbb{R}^n_\zeta \times \Omega_\sigma$, $\Omega \subset \mathbb{C}$, satisfying, for example in the scattering setting, estimates

$$|D_z^\alpha D_{z'}^\beta D_\zeta^\gamma D_\sigma^k a| \le C_{\alpha\beta\gamma,k} \langle z\rangle^{\ell_1-|\alpha|} \langle z'\rangle^{\ell_2-|\beta|} \langle(\zeta,\sigma)\rangle^{m-|\gamma|-k},$$

that is there is joint symbolic behavior in (ζ, σ), with differentiation in either giving rise to joint decay. This is natural, as a typical way such a family of operators arises is by Mellin transforming the normal operator of a b-operator; then σ is the b-dual of the boundary defining function. For instance, starting with the elliptic b-operator on $[0,\infty)_x \times X$

$$L = (xD_x)^2 + \Delta_g,$$

with g a Riemannian metric on X, the Mellin transform gives rise to the family $\sigma^2 + \Delta_g$, and the (joint symbolic) large parameter behavior is a consequence of the fact that the b-principal symbol of L is a symbol (concretely a quadratic polynomial).

Since σ is simply a parameter as far as the action of pseudodifferential operators is concerned, in order to have a well-behaved algebra (so that e.g. left and right quantization are equivalent), one simply needs to make sure that the various asymptotic expansions are now asymptotic in this large parameter sense (i.e. jointly in (ζ,σ)), which is straightforward to check. Thus, the composition of order (m,ℓ) and (m',ℓ') large parameter operators is order $(m+m',\ell+\ell')$ as a large parameter operator, with principal symbol given by the product of the principal symbols, and so on Notice that the natural compactification on which the symbols are well behaved is not $\overline{\mathbb{R}^n_z}\times\overline{\mathbb{R}^n_\zeta}\times\overline{\mathbb{C}}$ (with the image of Ω inserted into the last factor), but $\overline{\mathbb{R}^n_z}\times\overline{\mathbb{R}^n_\zeta\times\mathbb{C}}$, that is (ζ,σ) are jointly compactified. Correspondingly, in the manifold setting, one needs to take $T^*X\times\mathbb{C}$, best thought of as the vector bundle $(T^*\oplus\mathbb{C})X$ over X, and then radially compactifying the fibers to obtain $\overline{T^*\oplus\mathbb{C}}X$.

Below we are interested in the setting of compact manifolds, which means that the behavior as $|z|\to\infty$ is irrelevant, so we could just as well use Ψ_∞-type estimates in our definition of the large parameter algebra. To make things concrete for differential operators, if

$$P(\sigma)=\sum_{|\alpha|+|\beta|\le m} a_\alpha(z)\sigma^\beta D_z^\alpha$$

is an order m differential operator depending on a large parameter σ, then the large-parameter symbol is denoted by

$$\sigma_{\mathrm{full}}(P(\sigma))=\sum_{|\alpha|+|\beta|=m} a_\alpha(z)\sigma^\beta\zeta^\alpha.$$

It is often convenient to relax the requirements a bit on the joint symbolic behavior, and consider the semiclassical operator algebra, which we at first simply pull out of a hat. (We refer to [58] for a thorough treatment.) We adopt the convention that $\hbar$ denotes semiclassical objects, while h is the actual semiclassical parameter. This algebra, $\cup_{m,\ell}\Psi_\hbar^{m,\ell}(\mathbb{R}^n)$, is given by

$$A_h=\mathrm{Op}_\hbar(a);\ \mathrm{Op}_\hbar(a)u(z)=(2\pi h)^{-n}\int_{\mathbb{R}^n\times\mathbb{R}^n} e^{i(z-z')\cdot\zeta/h}a(z,\zeta,h)\,u(z')\,d\zeta\,dz',$$
$$u\in\mathcal{S}(\mathbb{R}^n),\ a\in C^\infty([0,1)_h;S^{m,\ell}(\mathbb{R}^n_z;\mathbb{R}^n_\zeta));\tag{5.69}$$

its classical subalgebra, $\Psi_{\hbar,\mathrm{cl}}(\mathbb{R}^n)$ corresponds to $a\in C^\infty([0,1)_h;S^{m,\ell}_{\mathrm{cl}}(\mathbb{R}^n_z;\mathbb{R}^n_\zeta))$. More generally, we write

$$I_\hbar(a)u(z)=(2\pi h)^{-n}\int_{\mathbb{R}^n} e^{i(z-z')\cdot\zeta/h}a(z,z',\zeta,h)\,u(z')\,d\zeta\,dz',$$

$a \in \mathcal{C}^\infty([0,1)_h; S^{m,\ell_1,\ell_2}(\mathbb{R}^n_z, \mathbb{R}^n_{z'}; \mathbb{R}^n_\zeta))$. A straightforward computation gives, with $\iota : \mathbb{R}^n_z \times \mathbb{R}^n_\zeta \to \mathbb{R}^n_z \times \mathbb{R}^n_{z'} \times \mathbb{R}^n_\zeta$ the inclusion map as the diagonal ($z = z'$) in the first two factors, the left reduction formula

$$a_L \sim \sum_\alpha \frac{i^{|\alpha|}}{\alpha!} \iota^* D^\alpha_{z'} (hD_\zeta)^\alpha a, \tag{5.70}$$

and the right reduction formula

$$a_R \sim \sum_\alpha \frac{(-i)^{|\alpha|}}{\alpha!} \iota^* D^\alpha_z (hD_\zeta)^\alpha a,$$

where the factors of h arise in front of D_ζ since one has

$$I_\hbar((z-z')^\alpha a) = I_\hbar((hD_\zeta)^\alpha a)$$

as

$$(z-z')e^{i(z-z')\cdot\zeta/h} = hD_\zeta e^{i(z-z')\cdot\zeta}.$$

This provides that the αth term in (5.70) is in $h^{|\alpha|} S^{m-|\alpha|,\ell_1+\ell_2-|\alpha|}$, that is the terms not only become lower order as symbols, but also have higher order vanishing as $h \to 0$.

The semiclassical principal symbol is now $\sigma_{\hbar,m,\ell}(A) = a|_{h=0} \in S^{m,\ell}(\mathbb{R}^n \times \mathbb{R}^n)$; there is still the standard principal symbol $[a] \in \mathcal{C}^\infty([0,1)_h; S^{m,\ell}/S^{m-1,\ell-1})$ which is now a function depending on the parameter h. As usual, the classical setting is best encoded in terms of a compactification (or bordification if $h = 1$ is not added!)

$$[0,1)_h \times \overline{\mathbb{R}^n_z} \times \overline{\mathbb{R}^n_\zeta},$$

which has now three boundary hypersurfaces:

$$\{0\}_h \times \overline{\mathbb{R}^n_z} \times \overline{\mathbb{R}^n_\zeta},$$

carrying the semiclassical principal symbol,

$$[0,1)_h \times \partial\overline{\mathbb{R}^n_z} \times \overline{\mathbb{R}^n_\zeta},$$

carrying the scattering symbol at base-infinity (which is irrelevant when we transfer to compact manifolds without boundary) and

$$[0,1)_h \times \overline{\mathbb{R}^n_z} \times \partial\overline{\mathbb{R}^n_\zeta},$$

carrying the usual principal symbol at fiber infinity.

In the setting of a general manifold X, $\mathbb{R}^n \times \mathbb{R}^n$ is replaced by T^*X. Correspondingly, $\mathrm{WF}'_\hbar(A)$ and $\mathrm{Ell}_\hbar(A)$ are subsets of T^*X. We can add an extra parameter $\lambda \in O$, so $a \in \mathcal{C}^\infty([0,1)_h; S^m(\mathbb{R}^n \times O; \mathbb{R}^n_\zeta))$; then in the invariant setting the principal symbol is $a|_{h=0} \in S^m(T^*X \times O)$.

In order to motivate the definition a posteriori, notice that differential operators now take the form

$$A_{h,\lambda} = \sum_{|\alpha|\leq m} a_\alpha(z,\lambda;h)(hD_z)^\alpha. \tag{5.71}$$

Indeed, if a is actually a polynomial in ζ, depending on a parameter λ, in (5.69), namely

$$a_{h,\lambda} = \sum_{|\alpha|\leq m} a_\alpha(z,\lambda;h)\zeta^\alpha,$$

then letting $\tilde{\zeta} = \zeta/h$, we have

$$\begin{aligned}(\mathrm{Op}_\hbar(a)u)(z) &= (2\pi)^{-n}\int_{\mathbb{R}^n\times\mathbb{R}^n} e^{i(z-z')\cdot\tilde{\zeta}} \sum_{|\alpha|\leq m} a_\alpha(z,\lambda;h)(h\tilde{\zeta})^\alpha u(z')\,dz'\,d\tilde{\zeta}\\ &= \sum_{|\alpha|\leq m} a_\alpha(z)((hD_z)^\alpha u)(z).\end{aligned}$$

The two principal symbols (ignoring base-infinity) are the standard one (but taking into account the semiclassical degeneration, i.e. based on $(hD_z)^\alpha$ rather than D_z^α), which depends on h and is homogeneous, and the semiclassical one, which is at $h = 0$, and is not homogeneous:

$$\begin{aligned}\sigma_m(A_{h,\lambda}) &= \sum_{|\alpha|=m} a_\alpha(z,\lambda;h)\zeta^\alpha,\\ \sigma_\hbar(A_{h,\lambda}) &= \sum_{|\alpha|\leq m} a_\alpha(z,\lambda;0)\zeta^\alpha.\end{aligned}$$

However, the restriction of $\sigma_m(A_{h,\lambda})$ to $h = 0$ is the principal symbol of $\sigma_\hbar(A_{h,\lambda})$. In the special case in which $\sigma_m(A_{h,\lambda})$ is independent of h (which is true in the setting considered below), one can simply regard the usual principal symbol as the principal part of the semiclassical symbol.

In terms of the fiber compactification of T^*X, in the semiclassical context one considers $\overline{T^*}X \times [0,1)$ and notes that "classical" semiclassical operators of order 0 are given locally by $\mathrm{Op}_\hbar(a)$ with a extending to be smooth up to the boundaries of this space, with the semiclassical symbol given by restriction to $\overline{T^*}X \times \{0\}$, and the standard symbol given by restriction to $S^*X \times [0,1)$. Thus, the claim regarding the limit of the semiclassical symbol at infinity is simply a matching statement of the two symbols at the corner $S^*X \times \{0\}$ in this compactified picture.

We can convert a large parameter operator into a semiclassical one (but not conversely) as follows. If $P(\sigma) = \sum_{|\alpha|+|\beta|\le m} a_\alpha(z)\sigma^\beta D_z^\alpha$ is an order m differential operator depending on a large parameter σ, then letting $\sigma = h^{-1}\lambda$, where $h^{-1} \sim \sigma$ (e.g. one may take $h^{-1} = |\sigma|$, but this is often too restrictive), so λ is in a compact set, and where we restrict to $|\sigma| > 1$, say,

$$P_{h,\lambda} = h^m P(\sigma) = \sum_{|\alpha|+|\beta|\le m} h^{m-|\alpha|-|\beta|} a_\alpha(z)\lambda^\beta (hD_z)^\alpha$$

is a semiclassical differential operator with semiclassical symbol

$$\sigma_h(P_{h,\lambda}) = \sum_{|\alpha|+|\beta|=m} a_\alpha(z)\lambda^\beta \zeta^\alpha.$$

Note that the full large-parameter symbol and the semiclassical symbol are "the same," that is they are simply related to each other.

The semiclassical Sobolev norms are defined analogously to the standard ones. Namely, on $\mathbb{R}^n$, they are defined for h-dependent families of elements of the standard Sobolev space as follows. For $s \in \mathbb{R}$, $u_h \in H^s$, $h \in (0, h_0)$,

$$\|u\|_{H_h^s} = \|\mathcal{F}_h^{-1}(\langle . \rangle^s (\mathcal{F}_h u)(.))\|_{L^2} = (2\pi)^{-n/2} \|\langle . \rangle^s (\mathcal{F}_h u)(.)\|_{L^2},$$

where $\mathcal{F}_h$ is the semiclassical Fourier transform

$$(\mathcal{F}_h u)(\zeta) = h^{-n/2} \int_{\mathbb{R}^n} e^{-iz\cdot\zeta/h} u(z)\, dz,$$

with inverse

$$(\mathcal{F}_h^{-1} v)(z) = (2\pi)^{-n} h^{-n/2} \int_{\mathbb{R}^n} e^{iz\cdot\zeta/h} v(\zeta)\, d\zeta.$$

Thus, for $s \ge 0$ integer, the squared H_h^s-norm is equivalent to

$$\sum_{|\alpha|\le s} \|(hD)^\alpha u\|_{L^2}^2.$$

Thus, the semiclassical norms are very similar to the standard ones, but the derivative estimates become weaker as $h \to 0$ in that each derivative comes with a factor of h. The weighted norms are then

$$\|u\|_{H_h^{s,r}} = \|\langle z \rangle^r u\|_{H_h^s}.$$

We say that a family $u = (u_h)$, $h \in (0, h_0)$, is in H_h^s if $\|u_h\|_{H_h^s}$ is uniformly bounded for $h \in (0, h_0)$. Variable order semiclassical Sobolev spaces as well as semiclassical Sobolev spaces on manifolds can be introduced analogously to the standard case.

There are complete analogues of the microlocal results discussed so far in the semiclassical setting. For instance, the analogue of Proposition 5.24 is

Proposition 5.30 *For $\alpha_0 \in \overline{\mathbb{R}^n} \times \overline{\mathbb{R}^n}$ with $\mathsf{H}_{p,m,\ell}$ non-vanishing at α_0. Then α_0 has a neighborhood O in $\overline{\mathbb{R}^n} \times \overline{\mathbb{R}^n}$ on which $\mathsf{H}_{p,m,\ell}$ is non-vanishing, and if $\alpha \in O$, γ the integral curve of $\mathsf{H}_{p,m,\ell}$ through α with $\gamma(0) = \alpha$, and $\gamma(t) = \beta$ for some $t < 0$ such that $\gamma([t,0]) \subset O$, and further if we are given a neighborhood U_2 of β and U_1 of $\gamma([t,0])$ contained in O, $Q_1, Q_2 \in \Psi_\hbar^{0,0}$ such that Q_1 is elliptic on U_1, Q_2 is elliptic on U_2, then there exists $Q_3 \in \Psi_\hbar^{0,0}$ elliptic on $\gamma([t,0])$ such that the following holds. If $Q_1 u \in H_\hbar^{s-1/2,r-1/2}$, $Q_1 P u \in H_\hbar^{s-m+1,r-\ell+1}$, $Q_2 u \in H_\hbar^{s,r}$ then $Q_3 u \in H_\hbar^{s,r}$ and for all M, N, K there are $C > 0$, $h_0 > 0$ such that*

$$\|Q_3 u\|_{H_\hbar^{s,r}} \leq C(\|Q_2 u\|_{H_\hbar^{s,r}} + h^{-1}\|Q_1 P u\|_{H_\hbar^{s-m+1,r-\ell+1}} + h\|Q_1 u\|_{H_\hbar^{s-1/2,r-1/2}} + h^K \|u\|_{H_\hbar^{M,N}}).$$

The analogous result also holds for $t > 0$.

The proof is completely analogous to that of Proposition 5.24, with the point being that the commutator of two operators has an additional h vanishing relative to the product, hence the h^{-1} in front of Pu in the estimate, while the error term has an additional order of vanishing, hence the h in front of the $H_\hbar^{s-1/2,r-1/2}$ norm of $Q_1 u$.

This in turn gives rise to

Theorem 5.6 *Suppose that $P \in \Psi_{\hbar,\mathrm{cl}}^{m,\ell}$ with real principal symbol p. Suppose that $B, G, Q \in \Psi_\hbar^{0,0}$, $\mathrm{WF}'_\hbar(B) \subset \mathrm{Ell}_\hbar(G)$ and for every $\alpha \in \mathrm{WF}'_\hbar(B) \cap \mathrm{Char}_\hbar(P)$, there is a point $\alpha' = \gamma(t')$ on the bicharacteristic γ through α with $\gamma(0) = \alpha$ such that $\alpha' \in \mathrm{Ell}_\hbar(Q)$ and such that for $t \in [0,t']$ (or $t \in [t',0]$ if $t' < 0$), $\gamma(t) \in \mathrm{Ell}_\hbar(G)$. Then for any M, N, K, there is $C > 0$ such that if $Qu \in H_\hbar^{s,r}$, $GPu \in H_\hbar^{s-m+1,r-\ell+1}$ then $Bu \in H_\hbar^{s,r}$ and*

$$\|Bu\|_{H_\hbar^{s,r}} \leq C(\|Qu\|_{H_\hbar^{s,r}} + h^{-1}\|GPu\|_{H_\hbar^{s-m+1,r-\ell+1}} + h^K \|u\|_{H_\hbar^{M,N}}).$$

The analogous conclusion also holds in the variable order setting if either s, r are non-increasing along the Hamilton flow and $t' < 0$ or s, r are non-decreasing along the Hamilton flow and $t' > 0$.

The estimates involving complex absorption and radial points change analogously, for similar reasons.

5.5.3. From the Laplacian to the Extended Operator

We show now that if we change the smooth structure on X_0 by declaring that only even functions of x are smooth, that is introducing $\mu = x^2$ as the boundary defining function, then after a suitable conjugation and division by a vanishing factor the resulting operator smoothly and non-degenerately continues across the boundary, that is continues to $X_{-\delta_0} = (-\delta_0, 0)_\mu \times Y \sqcup X_{0,\mathrm{even}}$, where $X_{0,\mathrm{even}}$ is the manifold X_0 with the new smooth structure. At the level of the principal symbol, that is the dual metric, the conjugation is irrelevant, so we can easily see what happens: changing to coordinates (μ, y), $\mu = x^2$, as $x\partial_x = 2\mu\partial_\mu$,

$$G_0 = 4\mu^2\partial_\mu^2 + \mu H = \mu(4\mu\partial_\mu^2 + H),$$

so after dividing by μ, we obtain

$$\mu^{-1}G_0 = 4\mu\partial_\mu^2 + H.$$

This is a quadratic form that is positive definite for $\mu > 0$, is Lorentzian for $\mu < 0$, and has a transition at $\mu = 0$ that as we shall see involves radial points.

To see that the full spectral family of the Laplacian is well behaved, first, changing to coordinates (μ, y), $\mu = x^2$, we obtain

$$\Delta_{g_0} = 4(\mu D_\mu)^2 + 2i(n - 1 + \mu\gamma)(\mu D_\mu) + \mu\Delta_h. \tag{5.72}$$

Now we conjugate by $\mu^{-i\sigma/2+(n+1)/4}$ to obtain

$$\begin{aligned}
&\mu^{i\sigma/2-(n+1)/4}\Big(\Delta_{g_0} - \frac{(n-1)^2}{4} - \sigma^2\Big)\mu^{-i\sigma/2+(n+1)/4} \\
&= 4(\mu D_\mu - \sigma/2 - i(n+1)/4)^2 + 2i(n - 1 + \mu\gamma)\big(\mu D_\mu - \sigma/2 - i(n+1)/4\big) \\
&\qquad + \mu\Delta_h - \frac{(n-1)^2}{4} - \sigma^2 \\
&= 4(\mu D_\mu)^2 - 4\sigma(\mu D_\mu) + \mu\Delta_h - 4i(\mu D_\mu) + 2i\sigma - 1 \\
&\qquad + 2i\mu\gamma\big(\mu D_\mu - \sigma/2 - i(n+1)/4\big).
\end{aligned}$$

Next we multiply by $\mu^{-1/2}$ from both sides to obtain

$$\begin{aligned}
&\mu^{-1/2}\mu^{i\sigma/2-(n+1)/4}\Big(\Delta_{g_0} - \frac{(n-1)^2}{4} - \sigma^2\Big)\mu^{-i\sigma/2+(n+1)/4}\mu^{-1/2} \\
&= 4\mu D_\mu^2 - \mu^{-1} - 4\sigma D_\mu - 2i\sigma\mu^{-1} + \Delta_h - 4iD_\mu + 2\mu^{-1} + 2i\sigma\mu^{-1} - \mu^{-1} \\
&\qquad + 2i\gamma\big(\mu D_\mu - \sigma/2 - i(n-1)/4\big) \\
&= 4\mu D_\mu^2 - 4\sigma D_\mu + \Delta_h - 4iD_\mu + 2i\gamma\big(\mu D_\mu - \sigma/2 - i(n-1)/4\big).
\end{aligned} \tag{5.73}$$

This operator is in $\mathrm{Diff}^2(X_{0,\mathrm{even}})$ and now can be continued smoothly across the boundary to an operator $\tilde{P}_\sigma$, by extending h and γ in an arbitrary smooth manner. This form suffices for analyzing the problem for σ in a compact set, or indeed for σ going to infinity in a strip near the reals. Later we discuss a modification that is necessary for semiclassical ellipticity when σ is far away from the reals in $\mathrm{Im}\,\sigma > 0$. Since the details of the operator (5.73) do not matter much, it is convenient to write

$$\begin{aligned}\tilde{P}_\sigma = 4(1+\tilde{a}_1)\mu D_\mu^2 - 4(1+\tilde{a}_2)\sigma D_\mu - \tilde{a}_3\sigma^2 + \Delta_h \\ - 4iD_\mu + \tilde{b}_1\mu D_\mu + \tilde{b}_2\sigma + \tilde{c}_1\end{aligned} \tag{5.74}$$

with $\tilde{a}_j$ smooth, real, and vanishing at $\mu = 0$, $\tilde{b}_j$ and $\tilde{c}_1$ smooth. In fact, we have $\tilde{a}_1 \equiv 0$, but it is sometimes convenient to have more flexibility in the form of the operator since this means that we do not need to start from the relatively rigid form (5.68). For the purposes of finite σ behavior this can be further simplified to

$$\tilde{P}_\sigma = 4(1+\tilde{a})\mu D_\mu^2 - 4(\sigma + i + \mu\tilde{b}(\sigma))D_\mu + \Delta_h + \tilde{c}(\sigma), \tag{5.75}$$

with $\tilde{a}$ smooth, real, and vanishing at $\mu = 0$, independent of σ, $\tilde{b}$, $\tilde{c}$ being smooth, including in σ-dependence, which in fact is holomorphic.

Writing covectors as

$$\xi\, d\mu + \eta\, dy,$$

the principal symbol of $\tilde{P}_\sigma \in \mathrm{Diff}^2(X_{-\delta_0})$, including in the high energy sense ($\sigma \to \infty$), is

$$\tilde{p}_{\mathrm{full}} = 4(1+\tilde{a}_1)\mu\xi^2 - 4(1+\tilde{a}_2)\sigma\xi - \tilde{a}_3\sigma^2 + |\eta|^2_{\mu,y}, \tag{5.76}$$

and is real for σ real. The Hamilton vector field is

$$\begin{aligned}\mathsf{H}_{\tilde{p}_{\mathrm{full}}} &= 4(2(1+\tilde{a}_1)\mu\xi - (1+\tilde{a}_2)\sigma)\partial_\mu + \tilde{\mathsf{H}}_{|\eta|^2_{\mu,y}} \\ &\quad - \left(4\left(1+\tilde{a}_1+\mu\frac{\partial\tilde{a}_1}{\partial\mu}\right)\xi^2 - 4\frac{\partial\tilde{a}_2}{\partial\mu}\sigma\xi + \frac{\partial\tilde{a}_3}{\partial\mu}\sigma^2 + \frac{\partial|\eta|^2_{\mu,y}}{\partial\mu}\right)\partial_\xi \\ &\quad - \left(4\frac{\partial\tilde{a}_1}{\partial y}\mu\xi^2 - 4\frac{\partial\tilde{a}_2}{\partial y}\sigma\xi - \frac{\partial\tilde{a}_3}{\partial y}\sigma^2\right)\partial_\eta,\end{aligned} \tag{5.77}$$

where $\tilde{\mathsf{H}}$ indicates that this is the Hamilton vector field in T^*Y, that is with μ considered a parameter. Correspondingly, the standard, "classical," principal symbol is

$$\tilde{p} = \sigma_2(\tilde{P}_\sigma) = 4(1+\tilde{a}_1)\mu\xi^2 + |\eta|^2_{\mu,y}, \tag{5.78}$$

which is real, independent of σ, while the Hamilton vector field is

$$\begin{aligned}\mathsf{H}_{\tilde{p}} &= 8(1+\tilde{a}_1)\mu\xi\partial_\mu + \tilde{\mathsf{H}}_{|\eta|^2_{\mu,y}} \\ &- \left(4\left(1+\tilde{a}_1+\mu\frac{\partial\tilde{a}_1}{\partial\mu}\right)\xi^2 + \frac{\partial|\eta|^2_{\mu,y}}{\partial\mu}\right)\partial_\xi - 4\frac{\partial\tilde{a}_1}{\partial y}\mu\xi^2\partial_\eta.\end{aligned} \tag{5.79}$$

In order to proceed in computing the principal symbol of $\tilde{P}_\sigma - \tilde{P}_\sigma^*$, it is useful to keep in mind that as $\Delta_{g_0} - \sigma^2 - (n-1)^2/4$ is formally self-adjoint relative to the metric density $|dg_0|$ for σ real, so the same holds for $\mu^{-1/2}(\Delta_{g_0} - \sigma^2 - (n-1)^2/4)\mu^{-1/2}$ (as μ is real), and indeed for its conjugate by $\mu^{-i\sigma/2}$ for σ real since this is merely unitary conjugation. As for f real, A formally self-adjoint relative to $|dg_0|$, $f^{-1}Af$ is formally self-adjoint relative to $f^2|dg_0|$, we then deduce that for σ real, $\tilde{P}_\sigma$ is formally self-adjoint relative to

$$\mu^{(n+1)/2}|dg_0| = \frac{1}{2}|dh|\,|d\mu|,$$

as $x^{-n}\,dx = \frac{1}{2}\mu^{-(n+1)/2}\,d\mu$. Note that $\mu^{(n+1)/2}|dg_0|$ thus extends to a C^∞ density to $X_{-\delta_0}$, and we deduce that with respect to the extended density, $\sigma_1(\frac{1}{2i}(\tilde{P}_\sigma - \tilde{P}_\sigma^*))|_{\mu\geq 0}$ vanishes when $\sigma \in \mathbb{R}$. Since in general $\tilde{P}_\sigma - \tilde{P}_{\mathrm{Re}\,\sigma}$ differs from $-4i(1+a_2)\,\mathrm{Im}\,\sigma D_\mu$ by a zeroth order operator, we conclude that

$$\sigma_1\left(\frac{1}{2i}(\tilde{P}_\sigma - \tilde{P}_\sigma^*)\right)\Big|_{\mu=0} = -4(\mathrm{Im}\,\sigma)\xi. \tag{5.80}$$

In fact, it is not important that $\tilde{P}_\sigma$ is related to $\Delta_{g_0} - \sigma^2 - (n-1)^2/4$ in this manner as shown by an explicit computation, or simply by noting that the coefficients $\tilde{a}_j$ do not affect the result of this computation.

In the interior, away from the region of validity of the product decomposition (5.67) (where we had no requirements so far on μ), one can choose μ arbitrarily, without affecting our arguments. Indeed, the classical principal symbol is unaffected by the choice, and for the semiclassical principal symbol, for real z, with $\sigma = z/h$, one simply has a conjugation by $e^{-i\sigma\phi}$ where $e^\phi = \mu^{1/2}$, which does not affect any of the relevant properties.

While we have not considered vector bundles over X_0, for instance for the Laplacian on the differential form bundles, it is straightforward to check that slightly changing the power of μ in the conjugation the resulting operator extends smoothly across ∂X_0, has scalar principal symbol of the form (5.76), and the principal symbol of $\frac{1}{2i}(\tilde{P}_\sigma - \tilde{P}_\sigma^*)$, which plays a role below, is also as in the scalar setting, so all the results in fact go through; see [46].

Notice that if one starts with an even asymptotically de Sitter space $(\tilde{X}_0, \tilde{g}_0)$, with boundary defining function $\tilde{x}$, and considers

$$\Box_{\tilde{g}_0} - \frac{(n-1)^2}{4} - \sigma^2,$$

the same construction gives rise to a family of operators $\tilde{P}'_\sigma$ on $\tilde{X}$, the extension of $\tilde{X}_{0,\text{even}}$ across its boundary. Namely, with $\tilde{\mu} = \tilde{x}^2$,

$$\begin{aligned} &\tilde{\mu}^{-1/2}\tilde{\mu}^{i\sigma/2-(n+1)/4}\left(\Box_{g_0} - \frac{(n-1)^2}{4} - \sigma^2\right)\tilde{\mu}^{-i\sigma/2+(n+1)/4}\tilde{\mu}^{-1/2} \\ &\quad = 4\tilde{\mu}D_{\tilde{\mu}}^2 - 4\sigma D_{\tilde{\mu}} - \Delta_{\tilde{h}} - 4iD_{\tilde{\mu}} + 2i\gamma\left(\tilde{\mu}D_{\tilde{\mu}} - \sigma/2 - i(n-1)/4\right), \end{aligned} \tag{5.81}$$

and again one just needs to extend h and γ in an arbitrary smooth manner to obtain the operator $\tilde{P}'_\sigma$. Thus, $\tilde{\mu} > 0$ is the Lorentzian region and $\tilde{\mu} < 0$ the Riemannian one.

Furthermore, if one has a fixed tangential metric h defined near $\mu = 0$ in $\mathbb{R}$, the two constructions are closely related in the sense that with $\tilde{h}(\tilde{\mu}, y, dy) = h(-\tilde{\mu}, y, dy)$, $\tilde{P}_\sigma$ can be chosen the same as the pull-back of $\tilde{P}'_\sigma$ by the map $\mu \mapsto -\mu = \tilde{\mu}$, or in other words $\tilde{P}_\sigma$ and the pull-back of $\tilde{P}'_\sigma$ are extensions of each other (from $\mu > 0$, resp. $\mu < 0$).

5.5.4. Local Dynamics Near the Radial Set

First, $\tilde{p}$ does not vanish in $\mu > 0$, and in general $\tilde{p} = 0$ and $H_{\tilde{p}} = 0$ at a point imply $\xi \neq 0$, $4(1+\tilde{a}_1)\mu = -\xi^{-2}|\eta|^2_{\mu,y}$ together with $\mathsf{H}_{|\eta|^2_{\mu,y}} = 0$, so $\eta = 0$, so $\mu = 0$, that is the point is in

$$N^*S \setminus o = \{(\mu, y, \xi, \eta) : \ \mu = 0, \ \eta = 0, \ \xi \neq 0\}, \qquad S = \{\mu = 0\},$$

so $S \subset X_{-\delta_0}$ can be identified with $Y = \partial X_0 (= \partial X_{0,\text{even}})$; but then

$$\mathsf{H}_{\tilde{p}}|_{N^*S} = -4\xi^2\partial_\xi, \tag{5.82}$$

which is non-zero at $N^*S \setminus o$, a contradiction. Thus, if $\tilde{p} = 0$ then $\mathsf{H}_{\tilde{p}}$ never vanishes, and thus $d\tilde{p}$ never vanishes; in particular $d\tilde{p} = 4\xi^2\, d\mu$ at N^*S. Thus, the characteristic set $\Sigma = \{\tilde{p} = 0\}$ is smooth. Note that $\tilde{p} = 0$ at $N^*S \setminus o$ indeed, and $\mathsf{H}_{\tilde{p}}$ is radial there in view of (5.82). We further divide

$$N^*S \setminus o = \Lambda_+ \cup \Lambda_-, \qquad \Lambda_\pm = N^*S \cap \{\pm\xi > 0\}.$$

Let $L_\pm$ be the image of $\Lambda_\pm$ in $S^*X_{-\delta_0}$. Next we analyze the Hamilton flow at $\Lambda_\pm$. First,

$$\mathsf{H}_{\tilde{p}}|\eta|^2_{\mu,y} = 8(1+\tilde{a}_1)\mu\xi\,\partial_\mu|\eta|^2_{\mu,y} - 4\mu\xi^2\frac{\partial\tilde{a}_1}{\partial y}\cdot_h\eta \tag{5.83}$$

and

$$\mathsf{H}_{\tilde{p}}\mu = 8(1+\tilde{a}_1)\xi\mu. \tag{5.84}$$

In terms of linearizing the flow at N^*S, $\tilde{p}$ and μ are equivalent as $d\tilde{p} = 4\xi^2\,d\mu$ there, so one can simply use $\hat{p} = p/|\xi|^2$ (which is homogeneous of degree 0, like μ), in place of μ. Finally,

$$\mathsf{H}_{\tilde{p}}|\xi| = -4\,\mathrm{sgn}(\xi) + \tilde{b}, \tag{5.85}$$

with $\tilde{b}$ vanishing at $\Lambda_\pm$.

It is convenient to rehomogenize (5.83) in terms of $\hat{\eta} = \eta/|\xi|$. This can be phrased more invariantly by working with $S^*X_{-\delta_0} = (T^*X_{-\delta_0}\setminus o)/\mathbb{R}^+$ (Figure 5.8). Let $L_\pm$ be the image of $\Lambda_\pm$ in $S^*X_{-\delta_0}$. Homogeneous degree zero functions on $T^*X_{-\delta_0}\setminus o$, such as $\hat{p}$, can be regarded as functions on $S^*X_{-\delta_0}$. For semiclassical purposes, it is best to consider $S^*X_{-\delta_0}$ as the boundary at fiber infinity of the fiber-radial compactification $\overline{T}^*X_{-\delta_0}$ of $T^*X_{-\delta_0}$. Then at fiber infinity near N^*S, we can take $(|\xi|^{-1},\hat{\eta})$ as (projective, rather than polar) coordinates on the fibers of the cotangent bundle, with $\tilde{\rho} = |\xi|^{-1}$ defining $S^*X_{-\delta_0}$ in $\overline{T}^*X_{-\delta_0}$. Then $W = |\xi|^{-1}\mathsf{H}_{\tilde{p}}$ is a $\mathcal{C}^\infty$ vector field in this region and

$$|\xi|^{-1}\mathsf{H}_{\tilde{p}}|\hat{\eta}|^2_{\mu,y} = 2|\hat{\eta}|^2_{\mu,y}\mathsf{H}_{\tilde{p}}|\xi|^{-1} + |\xi|^{-3}\mathsf{H}_{\tilde{p}}|\eta|^2_{\mu,y} = 8(\mathrm{sgn}\,\xi)|\hat{\eta}|^2 + \tilde{a}, \tag{5.86}$$

where $\tilde{a}$ vanishes cubically at N^*S. In similar notation we have

$$\mathsf{H}_{\tilde{p}}\tilde{\rho} = 4\,\mathrm{sgn}(\xi) + \tilde{a}', \qquad \tilde{\rho} = |\xi|^{-1}, \tag{5.87}$$

and

$$|\xi|^{-1}\mathsf{H}_{\tilde{p}}\mu = 8(\mathrm{sgn}\,\xi)\mu + \tilde{a}'', \tag{5.88}$$

with $\tilde{a}'$ smooth (indeed, homogeneous degree zero without the compactification) vanishing at N^*S, and $\tilde{a}''$ is also smooth, vanishing quadratically at N^*S. As the vanishing of $\hat{\eta}, |\xi|^{-1}$ and μ defines ∂N^*S, we conclude that $L_- = \partial\Lambda_-$ is a sink, while $L_+ = \partial\Lambda_+$ is a source, in the sense that all nearby bicharacteristics (in fact, including semiclassical (null) bicharacteristics, since $\mathsf{H}_{\tilde{p}}|\xi|^{-1}$ contains the additional information needed; see (5.97)) converge to $L_\pm$ as the parameter along the bicharacteristic goes to $\mp\infty$. In particular, the quadratic defining function of $L_\pm$ in Σ given by

$$\rho_0 = \widehat{p_\partial}, \text{ where } \widehat{p_\partial} = |\hat{\eta}|^2,$$

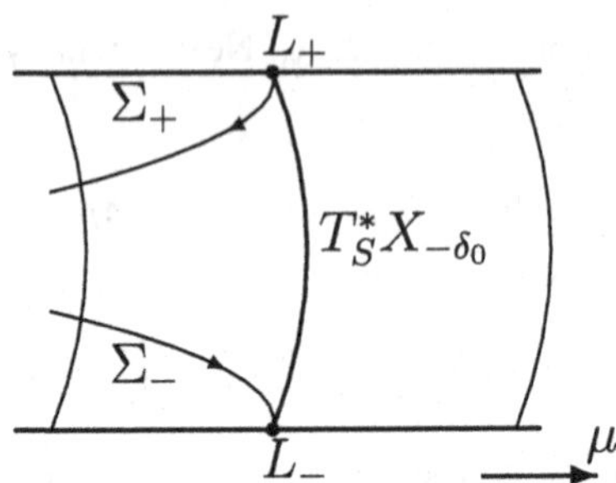

Figure 5.8. The cotangent bundle of $X_{-\delta_0}$ near $S = \{\mu = 0\}$, which is drawn in a fiber-radially compactified view. The boundary of the fiber compactification is the cosphere bundle $S^*X_{-\delta_0}$; it is the surface of the cylinder shown. $\Sigma_\pm$ are the components of the (classical) characteristic set containing $L_\pm$. They lie in $\mu \le 0$, only meeting $S^*_S X_{-\delta_0}$ at $L_\pm$. Semiclassically, that is in the interior of $\overline{T}^* X_{-\delta_0}$, for $z = h^{-1}\sigma > 0$, only the component of the semiclassical characteristic set containing L_+ can enter $\mu > 0$. This is reversed for $z < 0$

satisfies, on Σ,

$$(\operatorname{sgn}\xi) W\rho_0 \ge 8\rho_0 + \mathcal{O}(\rho_0^{3/2}). \tag{5.89}$$

We also need information on the principal symbol of $\frac{1}{2i}(\tilde{P}_\sigma - \tilde{P}_\sigma^*)$ at the radial points. At $L_\pm$ this is given by

$$\sigma_1\left(\frac{1}{2i}(\tilde{P}_\sigma - \tilde{P}_\sigma^*)\right)|_{N^*S} = -(4\operatorname{sgn}(\xi))\operatorname{Im}\sigma\,|\xi|; \tag{5.90}$$

here $(4\operatorname{sgn}(\xi))$ is pulled out due to (5.87), namely its size relative to $\mathsf{H}_{\tilde{p}}|\xi|^{-1}$ matters. This corresponds to the fact that $(\mu \pm i0)^{i\sigma}$, which are Lagrangian distributions associated with $\Lambda_\pm$, solve the PDE (5.101) modulo an error that is two orders lower than what one might a priori expect, that is $\tilde{P}_\sigma(\mu \pm i0)^{i\sigma} \in (\mu \pm i0)^{i\sigma}\mathcal{C}^\infty(X_{-\delta_0})$. Note that $\tilde{P}_\sigma$ is second order, so one should lose two orders a priori, that is get an element of $(\mu \pm i0)^{i\sigma-2}\mathcal{C}^\infty(X_{-\delta_0})$; the characteristic nature of $\Lambda_\pm$ reduces the loss to 1, and the particular choice of exponent eliminates the loss. This has much in common with $e^{i\lambda/x}x^{(n-1)/2}$ being an approximate solution in asymptotically Euclidean scattering.

5.5.5. Global Behavior of the Characteristic Set

By (5.78), points with $\xi = 0$ cannot lie in the characteristic set. Thus, with

$$\Sigma_\pm = \Sigma \cap \{\pm\xi > 0\},$$

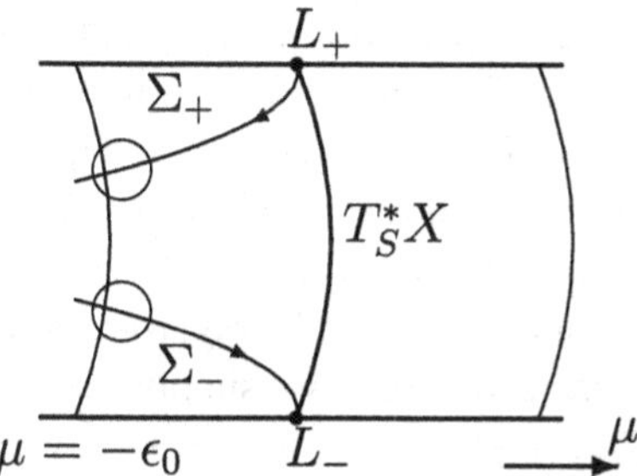

Figure 5.9. The cotangent bundle near $S = \{\mu = 0\}$, which is drawn in a fiber-radially compactified view, as in Figure 5.8. The circles on the left show the support of q; it has opposite signs on the two disks corresponding to the opposite directions of propagation relative to the Hamilton vector field

$\Sigma = \Sigma_+ \cup \Sigma_-$ and $\Lambda_\pm \subset \Sigma_\pm$. Further, the characteristic set lies in $\mu \leq 0$, and intersects $\mu = 0$ only in $\Lambda_\pm$.

Moreover, as $\mathsf{H}_{\tilde{p}}\mu = 8(1 + \tilde{a}_1)\xi\mu$ and $\xi \neq 0$ on Σ, and μ only vanishes at $\Lambda_+ \cup \Lambda_-$ there, for $\epsilon_0 > 0$ sufficiently small the C^∞ function μ provides a negative global escape function on $\mu \geq -\epsilon_0$ which is decreasing on Σ_+, increasing on Σ_-. Correspondingly, bicharacteristics in Σ_- travel from $\mu = -\epsilon_0$ to L_-, while in Σ_+ they travel from L_+ to $\mu = -\epsilon_0$.

5.5.6. Complex Absorption

The final step of fitting $\tilde{P}_\sigma$ into our general microlocal framework is moving the problem to a compact manifold and adding a complex absorbing second order operator. We thus consider a compact manifold without boundary X for which $X_{\mu_0} = \{\mu > \mu_0\}$, $\mu_0 = -\epsilon_0 < 0$, with $\epsilon_0 > 0$ as above, and which is identified as an open subset with smooth boundary; it is convenient to take X to be the double of X_{μ_0}, so there are two copies of $X_{0,\text{even}}$ in X.

In the case of hyperbolic space, this doubling process can be realized from the perspective of $(n+1)$-dimensional Minkowski space as discussed in [46, 54]. Then the Poincaré model shows up in two copies, namely in the interior of the future and past light cone inside the sphere at infinity, while de Sitter space shows up as the "equatorial belt," that is the exterior of the light cone at the sphere at infinity. One can take the Minkowski equatorial plane, $t = 0$, as $\mu = \mu_0$, and place the complex absorption there, thereby decoupling the future and past hemispheres.

At first we consider the bounded σ problem; in this case the potential trapping for g_0 is irrelevant.

We then introduce a "complex absorption" operator $Q_\sigma \in \Psi^2_{\mathrm{cl}}(X)$ with real principal symbol q supported in, say, $\mu < -\epsilon_1$, with the Schwartz kernel also supported in the corresponding region (i.e. in both factors on the product space this condition holds on the support) such that $\tilde{p} \pm iq$ is elliptic near ∂X_{μ_0}, that is near $\mu = \mu_0$, and which satisfies that $\pm q \geq 0$ near $\Sigma_\pm$ (Figure 5.9). This can easily be done since $\Sigma_\pm$ are disjoint, and away from these $\tilde{p}$ is elliptic, hence so is $\tilde{p} \pm iq$ regardless of the choice of q; we simply need to ensure q has support sufficiently close to $\Sigma_\pm$, elliptic on $\Sigma_\pm$ at $\mu = -\epsilon_0$, with the appropriate sign near $\Sigma_\pm$. Having done this, we extend $\tilde{p}$ and q to X in such a way that $\tilde{p} \pm iq$ are elliptic near ∂X_{μ_0}; the region we added is thus irrelevant at the level of bicharacteristic dynamics (of $\tilde{p}$) in so far as it is decoupled from the dynamics in X_0, and indeed also for analysis as we see shortly (in so far as we have two essentially decoupled copies of the same problem). This is accomplished, for instance, by using the doubling construction to define $\tilde{p}$ on $X \setminus X_{\mu_0}$ (in a smooth fashion at ∂X_{μ_0}, as can be easily arranged; the holomorphic dependence of $\tilde{P}_\sigma$ on σ is still easily preserved), and then, noting that the characteristic set of $\tilde{p}$ still has two connected components, making q elliptic on the characteristic set of $\tilde{p}$ near ∂X_{μ_0}, with the same sign in each component as near ∂X_{μ_0}. (An alternative would be to make q elliptic on the characteristic set of $\tilde{p}$ near $X \setminus X_{\mu_0}$; it is just slightly more complicated to write down such a q when the high energy behavior is taken into account. With the present choice, due to the doubling, there are essentially two copies of the problem on X_0: the original and the one from the doubling.) Finally we take Q_σ to be any operator with principal symbol q with Schwartz kernel satisfying the desired support conditions and which depends on σ holomorphically. For instance, we may take q and Q_σ to be $\pm$ a square, as convenient for the complex absorption discussion. We may even choose Q_σ to be independent of σ so Q_σ is indeed holomorphic.

In view of Section 5.5.5 we have arranged the following. For $\alpha \in S^*X \cap \Sigma$, let $\gamma_+(\alpha)$, resp. $\gamma_-(\alpha)$, denote the image of the forward, resp. backward, half-bicharacteristic of $\tilde{p}$ from α. We write $\gamma_\pm(\alpha) \to L_\pm$ (and say $\gamma_\pm(\alpha)$ tends to $L_\pm$) if given any neighborhood O of $L_\pm$, $\gamma_\pm(\alpha) \cap O \neq \emptyset$; by the source/sink property this implies that the points on the curve are in O for sufficiently large (in absolute value) parameter values. Then, with $\mathrm{Ell}(Q_\sigma)$ denoting the elliptic set of Q_σ,

$$\begin{aligned} &\alpha \in \Sigma_- \setminus L_- \Rightarrow \gamma_+(\alpha) \to L_- \text{ and } \gamma_-(\alpha) \cap \mathrm{Ell}(Q_\sigma) \neq \emptyset, \\ &\qquad \alpha \in \Sigma_+ \setminus L_+ \Rightarrow \gamma_-(\alpha) \to L_+ \text{ and } \gamma_+(\alpha) \cap \mathrm{Ell}(Q_\sigma) \neq \emptyset. \end{aligned} \tag{5.91}$$

That is, all forward and backward half-(null) bicharacteristics of $\tilde{P}_\sigma$ either enter the elliptic set of Q_σ, or go to $\Lambda_\pm$, that is $L_\pm$ in S^*X. The point of the arrangements regarding Q_σ and the flow is that we are able to propagate estimates forward near where $q \geq 0$, backward near where $q \leq 0$, so by our hypotheses we can always propagate estimates for $\tilde{P}_\sigma - iQ_\sigma$ from $\Lambda_\pm$ towards the elliptic set of Q_σ. On the other hand, for $\tilde{P}_\sigma^* + iQ_\sigma^*$, we can propagate estimates from the elliptic set of Q_σ towards $\Lambda_\pm$. This behavior of $\tilde{P}_\sigma - iQ_\sigma$ vs. $\tilde{P}_\sigma^* + iQ_\sigma^*$ is important for duality reasons as discussed in Section 5.4.8.

An alternative to the complex absorption would be simply adding a boundary at $\mu = \mu_0$; this is easy to do since this is a space-like hypersurface. This is less convenient from the perspective of microlocal analysis, but is straightforward to do, and is indeed very useful for non-linear problems, see [26, 24, 22].

5.5.7. Fredholm Theory

An immediate consequence of (5.91) is

Proposition 5.31 *Let $r \in \mathbb{R}$, and let $s \geq \frac{1}{2} + r$. Let $\Omega_r = \{\sigma \in \mathbb{C} \colon \operatorname{Im} \sigma > -r\}$, and let*

$$\mathcal{X}^s = \{u \in H^s(X) : (\tilde{P}_\sigma - iQ_\sigma) \in H^{s-1}(X)\},\ \mathcal{Y}^{s-1} = H^{s-1}(X).$$

For $\sigma \in \mathbb{C}$, the operator $\tilde{P}_\sigma - iQ_\sigma : \mathcal{X}^s \to \mathcal{Y}^{s-1}$ is Fredholm, and $\Omega_r \ni \sigma \mapsto \tilde{P}_\sigma - iQ_\sigma \in \mathcal{L}(\mathcal{X}^s, \mathcal{Y}^{s-1})$ is a holomorphic Fredholm family. Furthermore, $\operatorname{Ker}(\tilde{P}_\sigma - iQ_\sigma) \subset C^\infty(X)$ for all σ, while $\operatorname{Ker}(\tilde{P}_\sigma^* + iQ_\sigma^*) \subset H^{s'}(X)$ for all $s' < \frac{1}{2} - \operatorname{Im} \sigma$.

Remark 5.4 Note here that the principal symbol of $\tilde{P}_\sigma - iQ_\sigma$ is $\tilde{p} - iq$, independent of σ. On the other hand, $\mathcal{X}^s$ only depends on the principal symbol of $\tilde{P}_\sigma - iQ_\sigma$, since elements of $\operatorname{Diff}^1(X)$ map $H^s(X)$ to $H^{s-1}(X)$. Thus, any fixed value σ_0 is σ and can be used to define $\mathcal{X}^s$, and thus the meromorphic statement actually makes sense as a family of maps between fixed spaces.

Proof For the proof we simply need to observe the non-trapping statement (5.91) together with the threshold regularity at the radial points given by $1/2 - \operatorname{Im} \sigma$ due to (5.90) and that the order of the operator $\tilde{P}_\sigma - iQ_\sigma$ is $m = 2$. Thus, Section 5.4.8 is applicable for each fixed σ, and indeed in the analytic Fredholm sense in the indicated half-spaces of σ. The regularity statement for $\operatorname{Ker}(\tilde{P}_\sigma - iQ_\sigma)$ is a consequence of propagation of singularities, since $s > \frac{1}{2} + r$ ensures that the regularity of elements of $\mathcal{X}^s$ is above the threshold at the radial set, and so $(\tilde{P}_\sigma - iQ_\sigma)u = 0$ implies $\operatorname{WF}(u) \cap (L_+ \cup L_-) = \emptyset$, and then by (5.91), $\operatorname{WF}(u) = \emptyset$. Similarly, the regularity statement for $\operatorname{Ker}(\tilde{P}_\sigma^* + iQ_\sigma^*)$ follows from propagation of singularities, this time from $\operatorname{Ell}(Q_\sigma)$, taking into account that s' is below the threshold regularity. □

What we do not have at this point is that $\tilde{P}_\sigma - iQ_\sigma$ is invertible for any fixed value σ_0 of σ; if it is we conclude that

$$(\tilde{P}_\sigma - iQ_\sigma)^{-1} : \mathcal{Y}^{s-1} \to \mathcal{X}^s$$

is a meromorphic family. Note that for any fixed σ invertibility is independent of the choice of s satisfying $s > \frac{1}{2} - \operatorname{Im}\sigma$ in view of the regularity of the kernel of $\tilde{P}_\sigma - iQ_\sigma$ and its adjoint, so *one does not need to specify s explicitly as long as it is understood that it satisfies the constraint.*

Correspondingly we have:

Proposition 5.32 *Suppose that (X_0, g_0) is an n-dimensional manifold with boundary with an even conformally compact metric and boundary defining function x. Let $X_{0,\mathrm{even}}$ denote the even version of X_0, that is with the boundary defining function replaced by its square with respect to a decomposition in which g_0 is even, and let $\tilde{P}_\sigma$ be the extended operator in the sense of Section 5.5.3, and Q_σ be the complex absorption in the sense of Section 5.5.6. Finally suppose $\tilde{P}_\sigma - iQ_\sigma : \mathcal{X}^s \to \mathcal{Y}^{s-1}$ is invertible for some $\sigma_0 \in \Omega_r$, s, r, Ω_r, $\mathcal{X}^s, \mathcal{Y}^{s-1}$ as in Proposition 5.31.*

Then $(\tilde{P}_\sigma - iQ_\sigma)^{-1}$ is a meromorphic family in Ω_r; further the inverse of

$$\Delta_{g_0} - \left(\frac{n-1}{2}\right)^2 - \sigma^2,$$

written as $\mathcal{R}(\sigma) : L^2 \to L^2$, has a meromorphic continuation from $\operatorname{Im}\sigma \gg 0$ *to Ω_r,*

$$\mathcal{R}(\sigma) : \dot{C}^\infty(X_0) \to C^{-\infty}(X_0),$$

with poles with finite rank residues, and for $f \in \dot{C}^\infty(X_0)$,

$$\mathcal{R}(\sigma)f = x^{-i\sigma} x^{(n+1)/2} x^{-1} R_s (\tilde{P}_\sigma - iQ_\sigma)^{-1} E_{s-1} x^{i\sigma} x^{-(n+1)/2} x^{-1} f, \quad (5.92)$$

where

$$E_s : H^s(X_{0,\mathrm{even}}) \to H^s(X)$$

is a continuous extension operator, $R_s : H^s(X) \to H^s(X_{0,\mathrm{even}})$ the restriction map. Furthermore, for $s \geq \frac{1}{2} + r$, $\sigma \in \Omega_r$ not a pole of $\mathcal{R}(.)$,

$$\|x^{-(n-1)/2+i\sigma}\mathcal{R}(\sigma)f\|_{H^s(X_{0,\mathrm{even}})} \leq C\|x^{-(n+3)/2+i\sigma} f\|_{H^{s-1}(X_{0,\mathrm{even}})}, \quad (5.93)$$

where C depends on σ. If f is supported in X_0°, the $s-1$ norm on f can be replaced by the $s-2$ norm.

Proof By Proposition 5.31, $(\tilde{P}_\sigma - iQ_\sigma)^{-1}$ is meromorphic on Ω_r with values in $\mathcal{L}(\mathcal{Y}^{s-1}, \mathcal{X}^s)$. On the other hand, by self-adjointness and positivity of Δ_{g_0} and as $\dot{C}^\infty(X_0)$ is in its domain,

$$\left(\Delta_{g_0} - \sigma^2 - \left(\frac{n-1}{2}\right)^2\right) u = f \in \dot{C}^\infty(X_0)$$

has a unique solution $u = \mathcal{R}(\sigma) f \in L^2(X_0, |dg_0|)$ when $\operatorname{Im}\sigma \gg 0$. On the other hand, with $\tilde{f}_0 = x^{i\sigma-(n+1)/2-1} f$ in $\mu \geq 0$, and $\tilde{f}_0$ still vanishes to infinite order at $\mu = 0$, let $\tilde{f}$ be an arbitrary smooth extension of $\tilde{f}_0$ to the compact manifold X on which $\tilde{P}_\sigma - iQ_\sigma$ is defined; for instance we can take $\tilde{f} = E_{s-1}\tilde{f}_0$. Let $\tilde{u} = (\tilde{P}_\sigma - iQ_\sigma)^{-1}\tilde{f}$; this satisfies $(\tilde{P}_\sigma - iQ_\sigma)\tilde{u} = \tilde{f}$ and $\tilde{u} \in C^\infty(X)$. Thus, $u' = x^{-i\sigma+(n+1)/2-1} R_s \tilde{u}$ satisfies $u' \in x^{(n-1)/2} x^{-i\sigma} C^\infty(X_0)$, and

$$\left(\Delta_{g_0} - \sigma^2 - \left(\frac{n-1}{2}\right)^2\right) u' = f$$

by (5.101) and (5.110) (as Q_σ is supported in $\mu < 0$). Since $u' \in L^2(X_0, |dg_0|)$ for $\operatorname{Im}\sigma > 0$, by the aforementioned uniqueness, $u = u'$.

While, for the sake of simplicity, Q_σ is constructed in Section 5.5.6 in such a manner that it is not holomorphic in all of $\operatorname{Im}\sigma > -C$ due to a cut in the upper half-plane, this cut can be moved outside any fixed compact subset, so taking into account that $\mathcal{R}(\sigma)$ is independent of the choice of Q_σ, the theorem follows immediately from our earlier microlocal results. □

Thus, in order to prove the theorem of Mazzeo and Melrose [29] and Guillarmou [15], it remains to show the invertibility of $\tilde{P}_\sigma - iQ_\sigma$ for some σ_0. We do this by considering the high energy asymptotics of the $\tilde{P}_\sigma - iQ_\sigma$, and as a byproduct we obtain high energy resolvent estimates on even asymptotically hyperbolic spaces.

We now proceed to make some remarks under this invertibility hypothesis at some σ_0. Our proof of Proposition 5.32 implies that every pole of $\mathcal{R}(\sigma)$ is a pole of $(\tilde{P}_\sigma - iQ_\sigma)^{-1}$ (for otherwise (5.92) would show $\mathcal{R}(\sigma)$ does not have a pole either), but it is possible for $(\tilde{P}_\sigma - iQ_\sigma)^{-1}$ to have poles which are not poles of $\mathcal{R}(\sigma)$. However, in the latter case, the Laurent coefficients of $(\tilde{P}_\sigma - iQ_\sigma)^{-1}$ would be annihilated by multiplication by R_s from the left, that is the resonant states (which are smooth) would be supported in $\mu \leq 0$, in particular vanish to infinite order at $\mu = 0$.

In fact, a stronger statement can be made: as remarked at the end of Section 5.5.3, in $\mu < 0$, $\tilde{P}_\sigma$ is a conjugate (times a power of μ) of a Klein–Gordon-type operator on n-dimensional de Sitter space with $\mu = 0$ being the boundary (i.e. where time goes to infinity). Thus, if σ is not a pole of $\mathcal{R}(\sigma)$ and $(\tilde{P}_\sigma - iQ_\sigma)\tilde{u} = 0$ then one would have a solution u of this Klein–Gordon-type equation near $\mu = 0$, that is infinity, that rapidly vanishes at infinity. As shown in [53], one can use a Carleman-type estimate to show that this cannot happen; alternatively one can use an argument due to Zworski [57, Lemma 1] which makes the usual energy estimate quantitative in moving the initial slice $\mu = -\epsilon$, $\epsilon \to 0$. Thus, if Q_σ is supported in $\mu < c$, $c < 0$, then $\tilde{u}$ is also supported in $\mu < c$. This argument can be iterated for Laurent coefficients of higher order poles; their range (which is finite dimensional) contains only functions supported in $\mu < c$.

Remark 5.5 We now return to our previous remarks regarding the fact that our solution disallows the conormal singularities $(\mu \pm i0)^{i\sigma}$ from the perspective of conformally compact spaces of dimension n. Recalling that $\mu = x^2$, the two indicial roots on these spaces correspond to the asymptotics $\mu^{\pm i\sigma/2+(n-1)/4}$ in $\mu > 0$. Thus for the operator

$$\mu^{-1/2}\mu^{i\sigma/2-(n+1)/4}\left(\Delta_{g_0} - \frac{(n-1)^2}{4} - \sigma^2\right)\mu^{-i\sigma/2+(n+1)/4}\mu^{-1/2},$$

or indeed $\tilde{P}_\sigma$, they correspond to

$$\left(\mu^{-i\sigma/2+(n+1)/4}\mu^{-1/2}\right)^{-1}\mu^{\pm i\sigma/2+(n-1)/4} = \mu^{i\sigma/2\pm i\sigma/2}.$$

Here the indicial root $\mu^0 = 1$ corresponds to the smooth solutions we construct for $\tilde{P}_\sigma$, while $\mu^{i\sigma}$ corresponds to the conormal behavior we rule out. Back to the original Laplacian, thus, $\mu^{-i\sigma/2+(n-1)/4}$ is the allowed asymptotic and $\mu^{i\sigma/2+(n-1)/4}$ is the disallowed one. Notice that $\operatorname{Re} i\sigma = -\operatorname{Im}\sigma$, so the disallowed solution is growing at $\mu = 0$ relative to the allowed one, as expected in the physical half-plane, and the behavior reverses when $\operatorname{Im}\sigma < 0$. Thus, in the original asymptotically hyperbolic picture one has to distinguish two different rates of growth, whose relative size changes. On the other hand, in our approach, we rule out the singular solution and allow the non-singular (smooth one), so there is no change in behavior at all for the analytic continuation.

5.5.8. High Energy, or Semiclassical, Asymptotics

We are thus interested in the high energy behavior, as $|\sigma| \to \infty$. For the associated semiclassical problem one obtains a family of operators

$$\tilde{P}_{h,z} = h^2\tilde{P}_{h^{-1}z},$$

with $h = |\sigma|^{-1}$, and z corresponding to $\sigma/|\sigma|$ in the unit circle in $\mathbb{C}$. Then the semiclassical principal symbol $\tilde{p}_{\hbar,z}$ of $\tilde{P}_{h,z}$ is a function on $T^*X_{-\delta_0}$, whose asymptotics at fiber infinity of $T^*X_{-\delta_0}$ is given by the classical principal symbol $\tilde{p}$. We are interested in $\operatorname{Im}\sigma \geq -C$, which in semiclassical notation corresponds to $\operatorname{Im} z \geq -Ch$. However, the present form of $\tilde{P}_\sigma$ is only suitable for analysis in the strip $|\operatorname{Im}\sigma| \leq C$, that is $|\operatorname{Im} z| \leq Ch$; the more general case is taken up in Section 5.5.11.

It is often convenient to think of $\tilde{p}_{\hbar,z}$, and its rescaled Hamilton vector field, as objects on $\overline{T}^* X_{-\delta_0}$. Thus,

$$\tilde{p}_{\hbar,z} = \sigma_{2,\hbar}(\tilde{P}_{\hbar,z}) = 4(1+\tilde{a}_1)\mu\xi^2 - 4(1+\tilde{a}_2)z\xi - \tilde{a}_3 z^2 + |\eta|^2_{\mu,y}; \tag{5.94}$$

when $|\operatorname{Im} z| \leq Ch$, one can simply replace z by $\operatorname{Re} z$ since one obtains an equivalent representative of the semiclassical principal symbol.

Explicitly, if we introduce for instance

$$(\mu, y, \nu, \hat{\eta}), \qquad \nu = |\xi|^{-1},\ \hat{\eta} = \eta/|\xi|, \tag{5.95}$$

as valid projective coordinates in a (large!) neighborhood of $L_\pm$ in $\overline{T}^* X_{-\delta_0}$, then

$$\nu^2 \tilde{p}_{\hbar,z} = 4(1+\tilde{a}_1)\mu - 4(1+\tilde{a}_2)(\operatorname{sgn}\xi)z\nu - \tilde{a}_3 z^2\nu^2 + |\hat{\eta}|^2_{y,\mu}.$$

We assume that $\operatorname{Re} z > 0$ for the sake of definiteness. Observe that the semiclassical Hamilton vector field is

$$\begin{aligned}
\mathsf{H}_{\tilde{p}_{\hbar,z}} &= 4(2(1+\tilde{a}_1)\mu\xi - (1+\tilde{a}_2)z)\partial_\mu + \tilde{\mathsf{H}}_{|\eta|^2_{\mu,y}} \\
&\quad - \left(4\left(1+\tilde{a}_1+\mu\frac{\partial\tilde{a}_1}{\partial\mu}\right)\xi^2 - 4\frac{\partial\tilde{a}_2}{\partial\mu}z\xi + \frac{\partial\tilde{a}_3}{\partial\mu}z^2 + \frac{\partial|\eta|^2_{\mu,y}}{\partial\mu}\right)\partial_\xi \\
&\quad - \left(4\frac{\partial\tilde{a}_1}{\partial y}\mu\xi^2 - 4\frac{\partial\tilde{a}_2}{\partial y}z\xi - \frac{\partial\tilde{a}_3}{\partial y}z^2\right)\partial_\eta;
\end{aligned} \tag{5.96}$$

here we are concerned about z real. Near $S^*X_{-\delta_0} = \partial\overline{T}^* X_{-\delta_0}$, using the coordinates (5.95) (which are valid near the characteristic set)

$$\begin{aligned}
W_\hbar = \nu\mathsf{H}_{\tilde{p}_{\hbar,z}} &= 4(2(1+\tilde{a}_1)\mu(\operatorname{sgn}\xi) - (1+\tilde{a}_2)z\nu)\partial_\mu + \nu\tilde{\mathsf{H}}_{|\eta|^2_{\mu,y}} \\
&\quad + (\operatorname{sgn}\xi)\left(4\left(1+\tilde{a}_1+\mu\frac{\partial\tilde{a}_1}{\partial\mu}\right) - 4\frac{\partial\tilde{a}_2}{\partial\mu}z(\operatorname{sgn}\xi)\nu\right. \\
&\quad \left. + \frac{\partial\tilde{a}_3}{\partial\mu}z^2\nu^2 + \frac{\partial|\hat{\eta}|^2_{\mu,y}}{\partial\mu}\right)(\nu\partial_\nu + \hat{\eta}\partial_{\hat{\eta}}) \\
&\quad - \left(4\frac{\partial\tilde{a}_1}{\partial y}\mu - 4(\operatorname{sgn}\xi)\frac{\partial\tilde{a}_2}{\partial y}z\nu - \frac{\partial\tilde{a}_3}{\partial y}z^2\nu^2\right)\partial_{\hat{\eta}},
\end{aligned} \tag{5.97}$$

with $\nu\tilde{\mathsf{H}}_{|\eta|^2_{\mu,y}} = \sum_{ij} H_{ij}\hat{\eta}_i\partial_{y_j} - \sum_{ijk} \frac{\partial H_{ij}}{\partial y_k}\hat{\eta}_i\hat{\eta}_j\partial_{\hat{\eta}_k}$ smooth. Thus, $W_\hbar$ is a smooth vector field on the compactified cotangent bundle, $\overline{T}^*X_{-\delta_0}$ which is tangent to its boundary, $S^*X_{-\delta_0}$, and $W_\hbar - W = \nu W^\sharp$ (with W considered as a homogeneous degree zero vector field) with $W^\sharp$ smooth and tangent to $S^*X_{-\delta_0}$. In particular, by (5.87) and (5.89), using the fact that $\tilde{\rho}^2 + \rho_0$ is a quadratic defining function of $L_\pm$,

$$(\operatorname{sgn}\xi)W_\hbar(\tilde{\rho}^2 + \rho_0) \geq 8(\tilde{\rho}^2 + \rho_0) - \mathcal{O}((\tilde{\rho}^2 + \rho_0)^{3/2})$$

shows that there is $\epsilon_1 > 0$ such that in $\tilde{\rho}^2 + \rho_0 \leq \epsilon_1$, $\xi > 0$, $\tilde{\rho}^2 + \rho_0$ is strictly increasing along the Hamilton flow except at L_+ (where it is constant), while in $\tilde{\rho}^2 + \rho_0 \leq \epsilon_1$, $\xi < 0$, $\tilde{\rho}^2 + \rho_0$ is strictly decreasing along the Hamilton flow except at L_-. Indeed, all null-bicharacteristics in this neighborhood of $L_\pm$ except the constant ones at $L_\pm$ tend to $L_\pm$ in one direction and to $\tilde{\rho}^2 + \rho_0 = \epsilon_1$ in the other direction. In particular, the local semiclassical bicharacteristic geometry near $L_\pm$ consists of sources/sinks, and the semiclassical versions of the radial point estimates are applicable.

In order to proceed, we need to understand the global structure of the bicharacteristics. Namely, we would like to be able to say that if the original metric is non-trapping, that is for all points in $T^*X_0^\circ$, the projection to X_0 of the bicharacteristic through that point tends to ∂X_0 in both the forward and backward directions, then in fact with suitable complex absorption near $\mu = -\epsilon_0$, the problem is globally non-trapping in the sense that all semiclassical bicharacteristics except those contained in $L_+ \cup L_-$ either reach the complex absorption in finite time or tend to the radial set $L_+ \cup L_-$. Since there is an asymmetry between $L_\pm$, namely one of them is a source, another is a sink, and correspondingly there is a sign difference for the complex absorption along the bicharacteristics, we need to be able to separate these, which is done by observing that, semiclassically, for $\operatorname{Im} z = \mathcal{O}(h)$, that is z almost real, the characteristic set can be divided into two components $\Sigma_{\hbar,\pm}$, with $L_\pm$ in different components.

Indeed, consider the hypersurface given by

$$\xi = -\operatorname{Re} z/4,$$

that is $\operatorname{sgn}\xi = -\nu \operatorname{Re} z/4$ in coordinates valid near fiber infinity away from $\xi = 0$, on which, by (5.94), $\tilde{p}_{\hbar,z}$ cannot vanish where $|\mu|$ is small as it is

$$\tilde{p}_{\hbar,z} = \left(\frac{1}{4}(1+\tilde{a}_1)\mu + 1 + \tilde{a}_2 - \tilde{a}_3\right)(\operatorname{Re} z)^2 + |\eta|^2 > 0,$$

and $\tilde{a}_2, \tilde{a}_3$ vanish at $\mu = 0$. Moreover, if $\xi \leq -\operatorname{Re} z/4$ and $\mu > 0$ then $\tilde{p}_{\hbar,z}$ is a decreasing function of ξ, so it cannot vanish for $\xi < -\operatorname{Re} z/4$ either; if

$\mu = 0$, the only place where it can vanish in $\xi < -\operatorname{Re} z/4$ is at fiber infinity. Thus, for small $\epsilon_2 > 0$, in $|\mu| < \epsilon_2$ the characteristic set $\Sigma_\hbar$ can be written as the disjoint union of two non-empty components:

$$\Sigma_\hbar \cap \{|\mu| < \epsilon_2\} = (\Sigma_{\hbar,+} \cap \{|\mu| < \epsilon_2\}) \cup (\Sigma_{\hbar,-} \cap \{|\mu| < \epsilon_2\}),$$
$$\Sigma_{\hbar,\pm} \cap \{|\mu| < \epsilon_2\} = \Sigma_\hbar \cap \{\pm(\xi + \operatorname{Re} z/4) > 0\} \cap \{|\mu| < \epsilon_2\},$$

and

$$\Sigma_{\hbar,-} \cap \{0 < \mu < \epsilon_2\} = \emptyset.$$

We finally need more information about the global semiclassical dynamics.

Lemma 5.14 *There exists $\epsilon_0 > 0$ such that the following holds. Suppose $\operatorname{Re} z > 0$. Then only $\Sigma_{\hbar,+}$ enters $\mu > 0$, and all semiclassical null-bicharacteristics in $(\Sigma_{\hbar,+} \setminus L_+) \cap \{-\epsilon_0 \le \mu \le \epsilon_0\}$ go to either L_+ or to $\mu = \epsilon_0$ in the backward direction and to $\mu = \epsilon_0$ or $\mu = -\epsilon_0$ in the forward direction, with the limit being at $\mu = \epsilon_0$ in at most one of the cases, while all semiclassical null-bicharacteristics in $(\Sigma_{\hbar,-} \setminus L_-) \cap \{-\epsilon_0 \le \mu \le \epsilon_0\}$ go to L_- in the forward direction and to $\mu = -\epsilon_0$ in the backward direction.*

For $\operatorname{Re} z < 0$ the analogous conclusion holds with the $\mu = \epsilon_0$ possibilities added to the case of $\Sigma_{\hbar,-}$ instead of $\Sigma_{\hbar,+}$.

Proof Let $\epsilon_1 > 0$ be as above, so in $\tilde\rho^2 + \rho_0 \le \epsilon_1$, $\xi > 0$, $\tilde\rho^2 + \rho_0$ is strictly increasing along the Hamilton flow except at L_+ (where it is constant), while in $\tilde\rho^2 + \rho_0 \le \epsilon_1$, $\xi < 0$, $\tilde\rho^2 + \rho_0$ is strictly decreasing along the Hamilton flow except at L_-.

Choosing $\epsilon_0' > 0$ sufficiently small, the characteristic set in $\overline{T}^*X_{-\delta_0} \cap \{-\epsilon_0' \le \mu \le \epsilon_0'\}$ is disjoint from $S^*X_{-\delta_0} \setminus \{\tilde\rho^2 + \rho_0 \le \epsilon_1\}$, and indeed only contains points in $\Sigma_{\hbar,+}$ as $\operatorname{Re} z > 0$. Since $\mathsf{H}_{p_{\hbar,z}}\mu = 4(2(1+\tilde a_1)\mu\xi - (1+\tilde a_2)z)$, it is negative on $\overline{T}^*_{\{\mu=0\}}X_{-\delta_0} \setminus S^*X_{-\delta_0}$. In particular, there is a neighborhood U of $\mu = 0$ in $\Sigma_{\hbar,+} \setminus S^*X_{-\delta_0}$ on which the same sign is preserved; since the characteristic set in $\overline{T}^*X_{-\delta_0} \setminus \{\tilde\rho^2 + \rho_0 < \epsilon_1\}$ is compact, and is indeed a subset of $T^*X_{-\delta_0} \setminus \{\tilde\rho^2 + \rho_0 < \epsilon_1\}$, we deduce that $|\mu|$ is bounded below on $\Sigma \setminus (U \cup \{\tilde\rho^2 + \rho_0 < \epsilon_1\})$, say $|\mu| \ge \epsilon_0'' > 0$ there, so with $\epsilon_0 = \min(\epsilon_0', \epsilon_0'')$, $\mathsf{H}_{p_{\hbar,z}}\mu < 0$ on $\Sigma_{\hbar,+} \cap \{-\epsilon_0 \le \mu \le \epsilon_0\} \setminus \{\tilde\rho^2 + \rho_0^2 < \epsilon_1\}$. As $\mathsf{H}_{p_{\hbar,z}}\mu < 0$ at $\mu = 0$, bicharacteristics can only cross $\mu = 0$ in the outward direction.

Thus, if γ is a bicharacteristic in $\Sigma_{\hbar,+}$, there are two possibilities. If γ is disjoint from $\{\tilde\rho^2 + \rho_0 < \epsilon_1\}$, it has to go to $\mu = \epsilon_0$ in the backward direction and to $\mu = -\epsilon_0$ in the forward direction. If γ has a point in $\{\tilde\rho^2 + \rho_0 < \epsilon_1\}$, then it has to go to L_+ in the backward direction and to $\tilde\rho^2 + \rho_0 = \epsilon_1$ in the forward direction; if $|\mu| \ge \epsilon_0$ by the time $\tilde\rho^2 + \rho_0 = \epsilon_1$ is reached, the result is proved, and otherwise $\mathsf{H}_{p_{\hbar,z}}\mu < 0$ in $\tilde{\rho}^2 + \rho_0 \geq \epsilon_1$, $|\mu| \leq \epsilon_0$, shows that the bicharacteristic goes to $\mu = -\epsilon_0$ in the forward direction.

If γ is a bicharacteristic in $\Sigma_{\hbar,-}$, only the second possibility exists, and the bicharacteristic cannot leave $\{\tilde{\rho}^2+\rho_0 < \epsilon_1\}$ in $|\mu| \leq \epsilon_0$, so it reaches $\mu = -\epsilon_0$ in the backward direction (as the characteristic set is in $\mu \leq 0$). □

If we assume that g_0 is a non-trapping metric we have the following stronger conclusion:

Lemma 5.15 *Suppose g_0 is non-trapping, that is all the bicharacteristics of g_0 in $T^*X_0^\circ \setminus o$ tend to ∂X_0 in both the forward and the backward directions. Then for sufficiently small $\epsilon_0 > 0$, and for* $\operatorname{Re} z > 0$, *any bicharacteristic in $\Sigma_{\hbar,+}$ in $-\epsilon_0 \leq \mu$ has to go to L_+ in the backward direction, and to $\mu = -\epsilon_0$ in the forward direction (with the exception of the constant bicharacteristics at L_+), while in $\Sigma_{\hbar,-}$, all bicharacteristics in $-\epsilon_0 \leq \mu$ lie in $-\epsilon_0 \leq \mu \leq 0$, and go to L_- in the forward direction and to $\mu = -\epsilon_0$ in the backward direction (with the exception of the constant bicharacteristics at L_-).*

Proof The conclusion follows from Lemma 5.14 in the case of $\Sigma_{\hbar,-}$, so it suffices to consider $\Sigma_{\hbar,+}$.

We first observe that by the non-trapping hypothesis all bicharacteristics in $\Sigma_{\hbar,+} \cap \{\mu \geq -\epsilon_0\}$ enter $-\epsilon_0 \leq \mu \leq \epsilon_0$. Now, for a bicharacteristic with a point in $-\epsilon_0 \leq \mu \leq \epsilon_0$, the conclusion of this lemma holds by Lemma 5.14 unless the $\mu = \epsilon_0$ possibility holds in its conclusion either in the forward or backward direction. Suppose for instance this $\mu = \epsilon_0$ possibility holds in the backward direction. Then by the non-trapping hypothesis, the backward bicharacteristic segment continued from $\mu = \epsilon_0$ enters $\mu < \epsilon_0$ again, and then applying Lemma 5.14 it must approach L_+ (otherwise the local segment in $|\mu| \leq \epsilon_0$ would have both endpoints at $|\mu| = \epsilon_0$). Similarly, if the $\mu = \epsilon_0$ possibility holds in the forward direction, the analogous argument shows the further forward extend bicharacteristic in fact reaches $\mu = -\epsilon_0$. This completes the proof. □

For applications to gluing constructions as in [5] it is also useful to remark that μ is bicharacteristically strictly convex (in the sense that the super-level sets are convex) for $\mu > 0$ small. More precisely, for sufficiently small $\epsilon_0 > 0$, and for $\alpha \in T^*X_0$,

$$0 < \mu(\alpha) < \epsilon_0,\ \tilde{p}_{\hbar,z}(\alpha) = 0 \text{ and } (\mathsf{H}_{\tilde{p}_{\hbar,z}}\mu)(\alpha) = 0 \Rightarrow (\mathsf{H}^2_{\tilde{p}_{\hbar,z}}\mu)(\alpha) < 0. \tag{5.98}$$

Indeed, as $\mathsf{H}_{\tilde{p}_{\hbar,z}}\mu = 4(2(1 + \tilde{a}_1)\mu\xi - (1 + \tilde{a}_2)z)$, the hypotheses imply $z = 2(1 + \tilde{a}_1)(1 + \tilde{a}_2)^{-1}\mu\xi$ and

$$
\begin{aligned}
0 &= \tilde{p}_{\hbar,z} \\
&= 4(1+\tilde{a}_1)\mu\xi^2 - 8(1+\tilde{a}_1)\mu\xi^2 - 4(1+\tilde{a}_1)^2(1+\tilde{a}_2)^{-2}\tilde{a}_3\mu^2\xi^2 + |\eta|^2_{\mu,y} \\
&= -4(1+\tilde{a}_1)\mu\xi^2 - 4(1+\tilde{a}_1)^2(1+\tilde{a}_2)^{-2}\tilde{a}_3\mu^2\xi^2 + |\eta|^2_{\mu,y},
\end{aligned}
$$

so $|\eta|^2_{\mu,y} = 4(1+b)\mu\xi^2$, with b vanishing at $\mu = 0$. Thus, at points where $\mathsf{H}_{\tilde{p}_{\hbar,z}}\mu$ vanishes, writing $\tilde{a}_j = \mu\hat{a}_j$,

$$
\begin{aligned}
\mathsf{H}^2_{\tilde{p}_{\hbar,z}}\mu &= 8(1+\tilde{a}_1)\mu\mathsf{H}_{\tilde{p}_{\hbar,z}}\xi + 8\mu^2\xi\mathsf{H}_{\tilde{p}_{\hbar,z}}\hat{a}_1 - 4z\mu\mathsf{H}_{\tilde{p}_{\hbar,z}}\hat{a}_2 \\
&= 8(1+\tilde{a}_1)\mu\mathsf{H}_{\tilde{p}_{\hbar,z}}\xi + \mathcal{O}(\mu^2\xi^2).
\end{aligned} \tag{5.99}
$$

Now

$$
\mathsf{H}_{\tilde{p}_{\hbar,z}}\xi = -\left(4\left(1+\tilde{a}_1+\mu\frac{\partial\tilde{a}_1}{\partial\mu}\right)\xi^2 - 4\frac{\partial\tilde{a}_2}{\partial\mu}z\xi + \frac{\partial\tilde{a}_3}{\partial\mu}z^2 + \frac{\partial|\eta|^2_{\mu,y}}{\partial\mu}\right).
$$

Since $z\xi$ is $\mathcal{O}(\mu\xi^2)$ due to $\mathsf{H}_{\tilde{p}_{\hbar,z}}\mu = 0$, z^2 is $\mathcal{O}(\mu^2\xi^2)$ for the same reason, and $|\eta|^2$ and $\partial_\mu|\eta|^2$ are $\mathcal{O}(\mu\xi^2)$ due to $\tilde{p}_{\hbar,z} = 0$, we deduce that $\mathsf{H}_{\tilde{p}_{\hbar,z}}\xi < 0$ for sufficiently small $|\mu|$, so (5.99) implies (5.98). Thus, μ can be used for gluing constructions.

5.5.9. Complex Absorption at High Energies

For the semiclassical problem, when z is almost real we need to increase the requirements on Q_σ, and what we need to do depends on whether g_0 is non-trapping.

If g_0 is non-trapping, we choose Q_σ such that $h^2 Q_{h^{-1}z} \in \Psi^2_{\hbar,\mathrm{cl}}(X)$ with semiclassical principal symbol $q_{\hbar,z}$, and in addition to the above requirement for the classical symbol, we need semiclassical ellipticity near $\mu = \mu_0$, that is $\tilde{p}_{\hbar,z} - iq_{\hbar,z}$ and its complex conjugate are elliptic near ∂X_{μ_0}, that is near $\mu = \mu_0$, and which satisfies that for z real $\pm q_{\hbar,z} \geq 0$ on $\Sigma_{\hbar,\pm}$. Again, we extend $\tilde{P}_\sigma$ and Q_σ to X in such a way that $\tilde{p} - iq$ and $\tilde{p}_{\hbar,z} - iq_{\hbar,z}$ (and thus their complex conjugates) are elliptic near ∂X_{μ_0}; the region we added is thus irrelevant. This is straightforward to arrange if one ignores the fact that one wants Q_σ to be holomorphic: one easily constructs a function $q_{\hbar,z}$ on T^*X (taking into account the disjointness of $\Sigma_{\hbar,\pm}$), and defines $Q_{h^{-1}z}$ to be h^{-2} times the semiclassical quantization of $q_{\hbar,z}$ (or any other operator with the same semiclassical and standard principal symbols). However, it is not hard to have an example of a holomorphic family Q_σ in a strip at least: in view of (5.112) we can take (with $C > 0$ sufficiently large, depending on the strip in $\mathbb{C}$ we are working in)

$$
q_{\hbar,z} = (4\xi + z)(\xi^2 + |\eta|^2 + z^2 + C^2h^2)^{1/2}\chi(\mu),
$$

where $\chi \geq 0$ is supported near μ_0; the corresponding full symbol is

$$\sigma_{\mathrm{full}}(Q_\sigma) = (4\xi + \sigma)(\xi^2 + |\eta|^2 + \sigma^2)^{1/2}\chi(\mu),$$

and Q_σ is taken as a quantization of this full symbol. Here the square root is defined on $\mathbb{C} \setminus [0, -\infty)$, with the real part of the result being positive, and correspondingly $q_{\hbar,z}$ is defined and is holomorphic away from $h^{-1}z \in \pm i[C, +\infty)$. We remark for future use that $\xi^2+|\eta|^2+\sigma^2$ is an elliptic symbol in $(\xi, \eta, \operatorname{Re}\sigma, \operatorname{Im}\sigma)$ in the larger region $|\operatorname{Im}\sigma| < C'|\operatorname{Re}\sigma|$, so the corresponding statement also holds for its square root.

If g_0 is trapping, we need to add complex absorption inside X_0 as well, at $\mu = \epsilon_0$, so in addition to Q_σ being supported in $\mu < -\epsilon_0/2$, we add Q'_σ supported in $\mu > \epsilon_0/2$, but we require in addition to the other classical requirements that $\tilde{p}_{\hbar,z} - i(q_{\hbar,z} + q'_{\hbar,z})$ and its complex conjugate are elliptic near $\mu = \pm\epsilon_0$, and which satisfies $\pm q_{\hbar,z} \geq 0$ on $\Sigma_{\hbar,\pm}$. This can be achieved as above for μ near μ_0. Again, we extend $\tilde{P}_\sigma$ and Q_σ to X in such a way that $\tilde{p} - iq$ and $\tilde{p}_{\hbar,z} - iq_{\hbar,z}$ (and thus their complex conjugates) are elliptic near ∂X_{μ_0}.

In either of these semiclassical cases we have arranged that for sufficiently small $\delta_0 > 0$, $\tilde{p}_{\hbar,z} - iq_{\hbar,z}$ and its complex conjugate are *semiclassically non-trapping* for $|\operatorname{Im} z| < \delta_0$, namely the bicharacteristics from any point in $\Sigma_\hbar \setminus (L_+ \cup L_-)$ flow to $\operatorname{Ell}(q_{\hbar,z}) \cup L_-$ (i.e. either enter $\operatorname{Ell}(q_{\hbar,z})$ at some finite time, or tend to L_-) in the forward direction, and to $\operatorname{Ell}(q_{\hbar,z}) \cup L_+$ in the backward direction. Here $\delta_0 > 0$ arises from the particularly simple choice of $q_{\hbar,z}$ for which semiclassical ellipticity is easy to check for $\operatorname{Im} z > 0$ (bounded away from 0) and small; a more careful analysis would give a specific value of δ_0, and a more careful choice of $q_{\hbar,z}$ would give a better result.

5.5.10. Meromorphic Continuation of the Resolvent

We now state our results in the original conformally compact setting. Without the non-trapping estimate, these are a special case of a result of Mazzeo and Melrose [29], with improvements by Guillarmou [15], with "special" meaning that evenness is assumed. If the space is asymptotic to actual hyperbolic space, the non-trapping estimate is a stronger version of the estimate of Melrose, Sá Barreto, and me [32], where it is shown by a parametrix construction; here conformal infinity can have arbitrary geometry. The point is thus that, first, we do not need the machinery of the zero calculus here, second, we do have non-trapping high energy estimates in general (and without a parametrix construction), and third, we add the semiclassically outgoing property which is useful for resolvent gluing, including for proving non-trapping bounds microlocally away from trapping, provided the latter is mild, as shown by

Datchev and Vasy [5]. Since the original work in [48, 47] (somewhat weaker, but the full result should follow with some effort), high energy estimates have been obtained by zero calculus parametrix methods by Chen and Hassell [4] for real σ and by Wang [55] in general.

Theorem 5.7 *Suppose that (X_0, g_0) is an n-dimensional non-trapping manifold with boundary with an even conformally compact metric and boundary defining function x. Let $X_{0,\text{even}}$ denote the even version of X_0, that is with the boundary defining function replaced by its square with respect to a decomposition in which g_0 is even. Then the inverse of*

$$\Delta_{g_0} - \left(\frac{n-1}{2}\right)^2 - \sigma^2,$$

written as $\mathcal{R}(\sigma) : L^2 \to L^2$, has a meromorphic continuation from $\operatorname{Im}\sigma \gg 0$ to $\mathbb{C}$,

$$\mathcal{R}(\sigma) : \dot{C}^\infty(X_0) \to C^{-\infty}(X_0),$$

with poles with finite rank residues. Furthermore, non-trapping estimates hold in every region $|\operatorname{Im}\sigma| < r$, $|\operatorname{Re}\sigma| \gg 0$: for $s > \frac{1}{2} + r$,

$$\|x^{-(n-1)/2}x^{i\sigma}\mathcal{R}(\sigma)f\|_{H^s_{|\sigma|^{-1}}(X_{0,\text{even}})} \leq \tilde{C}|\sigma|^{-1}\|x^{-(n+3)/2}x^{i\sigma}f\|_{H^{s-1}_{|\sigma|^{-1}}(X_{0,\text{even}})}. \tag{5.100}$$

If f is supported in X_0°, the $s-1$ norm on f can be replaced by the $s-2$ norm.

Furthermore, for $\operatorname{Re} z > 0$, $\operatorname{Im} z = \mathcal{O}(h)$, the resolvent $\mathcal{R}(h^{-1}z)$ is semiclassically outgoing with a loss of h^{-1} *in the sense that if f has compact support in X_0°, $\alpha \in T^*X$ is in the semiclassical characteristic set and if* $\mathrm{WF}_\hbar^{s-1}(f)$ *is disjoint from the backward bicharacteristic from α, then $\alpha \notin$* $\mathrm{WF}_\hbar^{s}(h\mathcal{R}(h^{-1}z)f)$.

We remark that, although in order to go through without changes our methods require the evenness property, it is not hard to deduce more restricted results without this. Essentially one would have operators with coefficients that have a conormal singularity at the event horizon; as long as this is sufficiently mild relative to what is required for the analysis, it does not affect the results. The problems arise for the analytic continuation, when one needs strong function spaces (H^s with s large); these are not preserved when one multiplies by the singular coefficients.

Proof First note that semiclassical estimates hold for u in terms of $(\tilde{P}_{h,z} - iQ_{h,z})u$ plus a positive power of h times a weak norm of u by the same arguments as in the standard setting, using Lemma 5.15 to control u by propagating estimates within $\Sigma_\hbar$ from the radial sets $L_+ \cup L_-$: for all r, R, for $\{(h,z) : -r < h^{-1} \operatorname{Im} z < R\}$, for $s > \frac{1}{2} + r$, and for all $k > 0, N$,

$$\|u\|_{H_\hbar^s} \le C(h^{-1}\|(\tilde{P}_{h,z} - iQ_{h,z})u\|_{H_\hbar^{s-1}} + h^k\|u\|_{H_\hbar^N}).$$

Similarly, semiclassical estimates hold for v in terms of $(\tilde{P}^*_{h,z} + iQ^*_{h,z})v$ plus a positive power of h times a weak norm of v, this time using Lemma 5.15 to control u by propagating estimates within $\Sigma_\hbar$ from the complex absorption $\mathrm{Ell}_\hbar(Q_{h,z})$. This gives that in fact $\tilde{P}_{h,z} - iQ_{h,z}$ is invertible for sufficiently small h.

Now, Proposition 5.32 gives the non-high energy part of the statement provided $\tilde{P}_\sigma - iQ_\sigma$ is invertible for *some* σ; that this is the case follows from the semiclassical behavior in strips $|\operatorname{Im}\sigma| < C$, since for large $\operatorname{Re}\sigma$ these guarantee invertibility. Further, the semiclassical estimates also yield high energy bounds for $(\tilde{P}_\sigma - iQ_\sigma)^{-1}$. By (5.92) this gives the desired high energy estimates for $\mathcal{R}(\sigma)$ in strips, as stated. □

5.5.11. High Energy Estimates in $\operatorname{Im}\sigma > 0$ Farther from the Real Axis

It is instructive to modify our construction of $\tilde{P}_\sigma$ so that the resulting operator is semiclassically elliptic when σ is away from the reals. We achieve this via conjugation by an additional power of a smooth function, with the exponent depending on σ. The latter would make no difference even semiclassically in the real regime as it is conjugation by an elliptic semiclassical Fourier integral operator, which simply rearranges the cotangent bundle. However, in the non-real regime (where we would like ellipticity) it does matter; $\tilde{P}_\sigma$ is not semiclassically elliptic at the zero section. So finally we conjugate by $(1+\mu)^{i\sigma/4}$ to obtain

$$\begin{aligned} P_\sigma = 4(1+a_1)\mu D_\mu^2 - 4(1+a_2)\sigma D_\mu - (1+a_3)\sigma^2 + \Delta_h \\ - 4iD_\mu + b_1\mu D_\mu + b_2\sigma + c_1 \end{aligned} \tag{5.101}$$

with a_j smooth, real, and vanishing at $\mu = 0$, b_j and c_1 smooth. In fact, we have $a_1 \equiv 0$, but it is sometimes convenient to have more flexibility in the form of the operator since this means that we do not need to start from the relatively rigid form (5.68). Note that the only change compared to (5.74) is in the coefficient of σ^2; this is exactly what is needed for the far-from-real-axis ellipticity.

Now the full (including high energy) principal symbol of $P_\sigma \in \mathrm{Diff}^2(X_{-\delta_0})$ is

$$p_{\mathrm{full}} = 4(1+a_1)\mu\xi^2 - 4(1+a_2)\sigma\xi - (1+a_3)\sigma^2 + |\eta|^2_{\mu,y}. \tag{5.102}$$

The Hamilton vector field is

$$\begin{aligned}\mathsf{H}_{p_{\rm full}} &= 4(2(1+a_1)\mu\xi - (1+a_2)\sigma)\partial_\mu + \tilde{\mathsf{H}}_{|\eta|^2_{\mu,y}}\\ &\quad - \left(4\left(1+a_1+\mu\frac{\partial a_1}{\partial\mu}\right)\xi^2 - 4\frac{\partial a_2}{\partial\mu}\sigma\xi + \frac{\partial a_3}{\partial\mu}\sigma^2 + \frac{\partial|\eta|^2_{\mu,y}}{\partial\mu}\right)\partial_\xi\\ &\quad - \left(4\frac{\partial a_1}{\partial y}\mu\xi^2 - 4\frac{\partial a_2}{\partial y}\sigma\xi - \frac{\partial a_3}{\partial y}\sigma^2\right)\partial_\eta, \end{aligned} \tag{5.103}$$

where $\tilde{\mathsf{H}}$ still indicates that this is the Hamilton vector field in T^*Y, that is with μ considered a parameter. Correspondingly, essentially unchanged relative to $\tilde{P}_\sigma$, the standard, "classical," principal symbol is

$$p = \sigma_2(P_\sigma) = 4(1+a_1)\mu\xi^2 + |\eta|^2_{\mu,y}, \tag{5.104}$$

which is real, independent of σ, while the Hamilton vector field is

$$\begin{aligned}\mathsf{H}_p &= 8(1+a_1)\mu\xi\partial_\mu + \tilde{\mathsf{H}}_{|\eta|^2_{\mu,y}}\\ &\quad - \left(4\left(1+a_1+\mu\frac{\partial a_1}{\partial\mu}\right)\xi^2 + \frac{\partial|\eta|^2_{\mu,y}}{\partial\mu}\right)\partial_\xi - 4\frac{\partial a_1}{\partial y}\mu\xi^2\partial_\eta.\end{aligned} \tag{5.105}$$

We still have the fact that as $\Delta_{g_0} - \sigma^2 - (n-1)^2/4$ is formally self-adjoint relative to the metric density $|dg_0|$ for σ real, the same holds for the conjugate of $\mu^{-1/2}(\Delta_{g_0} - \sigma^2 - (n-1)^2/4)\mu^{-1/2}$ by $\mu^{-i\sigma/2}(1+\mu)^{i\sigma/4}$ for σ real since this is merely unitary conjugation. This gives finally that for σ real, P_σ is formally self-adjoint relative to

$$\mu^{(n+1)/2}|dg_0| = \frac{1}{2}|dh|\,|d\mu|,$$

as $x^{-n}\,dx = \frac{1}{2}\mu^{-(n+1)/2}\,d\mu$. We deduce that with respect to the extension of the density $\mu^{(n+1)/2}|dg_0|$ to $X_{-\delta_0}$, $\sigma_1(\frac{1}{2i}(P_\sigma - P_\sigma^*))|_{\mu\geq 0}$ vanishes when $\sigma \in \mathbb{R}$. Since in general $P_\sigma - P_{\mathrm{Re}\,\sigma}$ differs from $-4i(1+a_2)\,\mathrm{Im}\,\sigma D_\mu$ by a zeroth order operator, we conclude that

$$\sigma_1\left(\frac{1}{2i}(P_\sigma - P_\sigma^*)\right)\Big|_{\mu=0} = -4(\mathrm{Im}\,\sigma)\xi. \tag{5.106}$$

We now need to choose μ appropriately in the interior of X_0 far from ∂X_0; this choice matters greatly for the far-from-real-axis ellipticity. Thus, near $\mu = 0$, but μ bounded away from 0, the only semiclassically non-trivial action we have done was to conjugate the operator by $e^{-i\sigma\phi}$ where $e^\phi = \mu^{1/2}(1+\mu)^{-1/4}$; we need to extend ϕ into the interior. But the semiclassical principal symbol of the conjugated operator is, with $\sigma = z/h$,

$$(\zeta - z\,d\phi, \zeta - z\,d\phi)_{G_0} - z^2 = |\zeta|^2_{G_0} - 2z(\zeta, d\phi)_{G_0} - (1 - |d\phi|^2_{G_0})z^2. \quad (5.107)$$

For z non-real this is elliptic if $|d\phi|_{G_0} < 1$. Indeed, if (5.107) vanishes then from the vanishing imaginary part we get

$$2\,\mathrm{Im}\,z((\zeta, d\phi)_{G_0} + (1 - |d\phi|^2_{G_0})\,\mathrm{Re}\,z) = 0, \quad (5.108)$$

and then the real part is

$$\begin{aligned} &|\zeta|^2_{G_0} - 2\,\mathrm{Re}\,z(\zeta, d\phi)_{G_0} - (1 - |d\phi|^2_{G_0})((\mathrm{Re}\,z)^2 - (\mathrm{Im}\,z)^2) \\ &= |\zeta|^2_{G_0} + (1 - |d\phi|^2_{G_0})((\mathrm{Re}\,z)^2 + (\mathrm{Im}\,z)^2), \end{aligned} \quad (5.109)$$

which cannot vanish if $|d\phi|_{G_0} < 1$. But, reading off the dual metric from the principal symbol of (5.72),

$$\frac{1}{4}\left|d(\log\mu - \frac{1}{2}\log(1+\mu))\right|^2_{G_0} = \left(1 - \frac{\mu}{2(1+\mu)}\right)^2 < 1$$

for $\mu > 0$, with a strict bound as long as μ is bounded away from 0. Correspondingly, $\mu^{1/2}(1+\mu)^{-1/4}$ can be extended to a function e^{ϕ} on all of X_0 so that semiclassical ellipticity for z away from the reals is preserved, and we may even require that ϕ is constant on a fixed (but arbitrarily large) compact subset of X_0°. Then, after conjugation by $e^{-i\sigma\phi}$,

$$P_{h,z} = e^{iz\phi/h}\mu^{-(n+1)/4-1/2}(h^2\Delta_{g_0} - z)\mu^{(n+1)/4-1/2}e^{-iz\phi/h} \quad (5.110)$$

is semiclassically elliptic in $\mu > 0$ (as well as in $\mu \leq 0$, μ near 0, where this is already guaranteed), as desired.

We again want to think of $p_{\hbar,z}$, and its rescaled Hamilton vector field, as objects on $\overline{T}^*X_{-\delta_0}$. Thus,

$$p_{\hbar,z} = \sigma_{2,\hbar}(P_{h,z}) = 4(1+a_1)\mu\xi^2 - 4(1+a_2)z\xi - (1+a_3)z^2 + |\eta|^2_{\mu,y}, \quad (5.111)$$

so

$$\mathrm{Im}\,p_{\hbar,z} = -2\,\mathrm{Im}\,z(2(1+a_2)\xi + (1+a_3)\,\mathrm{Re}\,z). \quad (5.112)$$

In particular, for z non-real, $\mathrm{Im}\,p_{\hbar,z} = 0$ implies $2(1+a_2)\xi + (1+a_3)\,\mathrm{Re}\,z = 0$, so

$$\begin{aligned} \mathrm{Re}\,p_{\hbar,z} = ((1+a_1)(1+a_3)^2(1+a_2)^{-2}\mu + (1+2a_2)(1+a_3))(\mathrm{Re}\,z)^2 \\ + (1+a_3)(\mathrm{Im}\,z)^2 + |\eta|^2_{\mu,y} > 0 \end{aligned} \quad (5.113)$$

near $\mu = 0$, that is $p_{\hbar,z}$ is semiclassically elliptic on $T^*X_{-\delta_0}$, but *not* at fiber infinity, that is at $S^*X_{-\delta_0}$ (standard ellipticity is lost only in $\mu \le 0$, of course). In $\mu > 0$ we have semiclassical ellipticity (and automatically classical ellipticity) by our choice of ϕ following (5.107). Explicitly, if we introduce for instance

$$(\mu, y, \nu, \hat{\eta}), \qquad \nu = |\xi|^{-1},\ \hat{\eta} = \eta/|\xi|, \tag{5.114}$$

as valid projective coordinates in a (large!) neighborhood of $L_\pm$ in $\overline{T}^*X_{-\delta_0}$, then

$$\nu^2 p_{\hbar,z} = 4(1+a_1)\mu - 4(1+a_2)(\operatorname{sgn}\xi)z\nu - (1+a_3)z^2\nu^2 + |\hat{\eta}|^2_{y,\mu}$$

so

$$\nu^2 \operatorname{Im} p_{\hbar,z} = -4(1+a_2)(\operatorname{sgn}\xi)\nu \operatorname{Im} z - 2(1+a_3)\nu^2 \operatorname{Re} z \operatorname{Im} z$$

which automatically vanishes at $\nu = 0$, that is at $S^*X_{-\delta_0}$. Thus, for σ large and pure imaginary, the semiclassical problem adds no complexity to the "classical" quantum problem, but of course it does not simplify it. In fact, we need somewhat more information at the characteristic set, which is thus at $\nu = 0$ when $\operatorname{Im} z$ is bounded away from 0:

$$\begin{aligned} &\nu \text{ small},\ \operatorname{Im} z \ge 0 \Rightarrow (\operatorname{sgn}\xi)\operatorname{Im} p_{\hbar,z} \le 0 \Rightarrow \pm \operatorname{Im} p_{\hbar,z} \le 0 \text{ near } \Sigma_{\hbar,\pm},\\ &\nu \text{ small},\ \operatorname{Im} z \le 0 \Rightarrow (\operatorname{sgn}\xi)\operatorname{Im} p_{\hbar,z} \ge 0 \Rightarrow \pm \operatorname{Im} p_{\hbar,z} \ge 0 \text{ near } \Sigma_{\hbar,\pm}, \end{aligned} \tag{5.115}$$

which means that for $P_{h,z}$ with $\operatorname{Im} z > 0$ one can propagate estimates forwards along the bicharacteristics where $\xi > 0$ (in particular, away from L_+, as the latter is a source) and backwards where $\xi < 0$ (in particular, away from L_-, as the latter is a sink), while for $P^*_{h,z}$ the directions are reversed since its semiclassical symbol is $\overline{p_{\hbar,z}}$. The directions are also reversed if $\operatorname{Im} z$ switches sign. This is important because it gives invertibility for $z = i$ (corresponding to $\operatorname{Im}\sigma$ large positive, i.e. the physical half-plane), but does not give invertibility for $z = -i$ negative.

We now give a different perspective on what we have shown already (for $\tilde{P}_\sigma$, but as it is related to P_σ by conjugation by a smooth function, this distinction does not matter), namely that semiclassically, for z almost real (i.e. when z is not bounded away from the reals; we are not fixing z as we let h vary!), when the operator is not semiclassically elliptic on $T^*X_{-\delta_0}$ as mentioned above, the characteristic set can be divided into two components $\Sigma_{\hbar,\pm}$, with $L_\pm$ in different components. The vanishing of the factor following $\operatorname{Im} z$ in (5.112) gives a hypersurface that separates $\Sigma_\hbar$ into two parts. Indeed, this is the hypersurface given by

$$2(1 + a_2)\xi + (1 + a_3)\operatorname{Re} z = 0,$$

on which, by (5.113), $\operatorname{Re} p_{\hbar,z}$ cannot vanish, so

$$\Sigma_\hbar = \Sigma_{\hbar,+} \cup \Sigma_{\hbar,-}, \qquad \Sigma_{\hbar,\pm} = \Sigma_\hbar \cap \{\pm(2(1 + a_2)\xi + (1 + a_3)\operatorname{Re} z) > 0\}.$$

Farther in $\mu > 0$, the hypersurface is given, due to (5.108), by

$$(\zeta, d\phi)_{G_0} + (1 - |d\phi|^2_{G_0})\operatorname{Re} z = 0,$$

and on it, by (5.109), the real part is $|\zeta|^2_{G_0} + (1 - |d\phi|^2_{G_0})((\operatorname{Re} z)^2 + (\operatorname{Im} z)^2) > 0$; correspondingly

$$\Sigma_\hbar = \Sigma_{\hbar,+} \cup \Sigma_{\hbar,-}, \qquad \Sigma_{\hbar,\pm} = \Sigma_\hbar \cap \{\pm((\zeta, d\phi)_{G_0} + (1 - |d\phi|^2_{G_0})\operatorname{Re} z) > 0\}.$$

In fact, more generally, the real part is

$$\begin{aligned} &|\zeta|^2_{G_0} - 2\operatorname{Re} z(\zeta, d\phi)_{G_0} - (1 - |d\phi|^2_{G_0})((\operatorname{Re} z)^2 - (\operatorname{Im} z)^2) \\ &= |\zeta|^2_{G_0} - 2\operatorname{Re} z((\zeta, d\phi)_{G_0} + (1 - |d\phi|^2_{G_0})\operatorname{Re} z) \\ &\quad + (1 - |d\phi|^2_{G_0})((\operatorname{Re} z)^2 + (\operatorname{Im} z)^2), \end{aligned}$$

so for $\pm \operatorname{Re} z > 0$, $\mp((\zeta, d\phi)_{G_0} + (1 - |d\phi|^2_{G_0})\operatorname{Re} z) > 0$ implies that $p_{\hbar,z}$ does not vanish. Correspondingly, only one of the two components of $\Sigma_{\hbar,\pm}$ enter $\mu > 0$, namely for $\operatorname{Re} z > 0$, only $\Sigma_{\hbar,+}$ enters, while for $\operatorname{Re} z < 0$, only $\Sigma_{\hbar,-}$ enters.

This completes the characteristic set and bicharacteristic geometry discussion even for $\operatorname{Im}\sigma$ large. Correspondingly, we obtain high energy estimates in cones $c_1|\operatorname{Re}\sigma| < \operatorname{Im}\sigma < c_2|\operatorname{Re}\sigma|$, $0 < c_1 < c_2$, where the limitation given by c_2 arises due to the construction of the complex absorption in Section 5.5.9. This then yields the invertibility of $P_\sigma - iQ_\sigma$ for large $|\sigma|$ in these cones. This shows that the invertibility assumption of Proposition 5.32 is satisfied, and thus we have the meromorphic continuation theorem in general:

Theorem 5.8 (Special case of the theorem of Mazzeo and Melrose [29] and Guillarmou [15].) *Suppose that (X_0, g_0) is an n-dimensional manifold with boundary with an even conformally compact metric and boundary defining function x. Let $X_{0,\mathrm{even}}$ denote the even version of X_0, that is with the boundary defining function replaced by its square with respect to a decomposition in which g_0 is even, and let P_σ be the extended operator in the sense of Section 5.5.3, and Q_σ be the complex absorption in the sense of Section 5.5.6. Let s, r, Ω_r be as in Proposition 5.31.*

Then $(P_\sigma - iQ_\sigma)^{-1}$ is a meromorphic family in Ω_r, further the inverse of

$$\Delta_{g_0} - \left(\frac{n-1}{2}\right)^2 - \sigma^2,$$

written as $\mathcal{R}(\sigma) : L^2 \to L^2$, has a meromorphic continuation from $\operatorname{Im}\sigma \gg 0$ *to Ω_r,*

$$\mathcal{R}(\sigma) : \dot{C}^\infty(X_0) \to C^{-\infty}(X_0),$$

with poles with finite rank residues, and for $f \in \dot{C}^\infty(X_0)$,

$$\mathcal{R}(\sigma)f = x^{-i\sigma+(n-1)/2} R_s(P_\sigma - iQ_\sigma)^{-1} E_{s-1} x^{i\sigma-(n-3)/2} f, \tag{5.116}$$

where

$$E_s : H^s(X_{0,\mathrm{even}}) \to H^s(X)$$

is a continuous extension operator, $R_s : H^s(X) \to H^s(X_{0,\mathrm{even}})$ the restriction map. Furthermore, for $s > \frac{1}{2} + r$, $\sigma \in \Omega_r$ not a pole of $\mathcal{R}(.)$,

$$\|x^{-(n-1)/2+i\sigma}\mathcal{R}(\sigma)f\|_{H^s(X_{0,\mathrm{even}})} \leq C\|x^{-(n+3)/2+i\sigma}f\|_{H^{s-1}(X_{0,\mathrm{even}})}, \tag{5.117}$$

where C depends on σ. If f is supported in X_0°, the $s-1$ norm on f can be replaced by the $s-2$ norm.

5.6. Microlocal Analysis in the b-Setting

We now consider the geometric settings of b- and (complete) cusp analysis. For a much more detailed description of the former, including parametrix constructions for resolvents of elliptic operators, I refer to [34] and to [14]; here we proceed via estimates as these generalize easily to non-elliptic settings.

Recall that $\mathcal{V}_b(M)$ is the set of smooth vector fields on a manifold with boundary of corners; this does not depend on any choices. Fixing a boundary defining function x modulo $x^2C^\infty(M)$, we also let $\mathcal{V}_{cu}(M)$ be the set of vector fields V tangent to ∂M such that $Vx \in x^2C^\infty(M)$. This means that $V = b_0(x\partial_x) + \sum b_j\partial y_j$ and $Vx = b_0x \in x^2C^\infty(M)$, so $b_0 \in xC^\infty(M)$, so

$$V = a_0(x^2\partial_x) + \sum a_j\partial_{y_j},\ a_j \in C^\infty(M).$$

These two structures are closely connected to each other (Figure 5.10). Indeed, given a manifold with boundary with a C^∞ defining function x, choose a local product decomposition near ∂M, so a neighborhood U of ∂M is identified with $[0,\epsilon)_x \times \partial M$; one gets local coordinates on this space via local coordinates y_j on ∂M. If we now introduce $\rho(x) = e^{-1/x}$, $x > 0$, $\rho(0) = 0$, the map $\rho \times \mathrm{id} : [0,\epsilon) \times \partial M \to [0, e^{-1/\epsilon}) \times \partial M$ is C^∞, and the push forward of a cusp vector field V is

$$a_0\rho\partial_\rho + \sum a_j\partial_{y_j}.$$

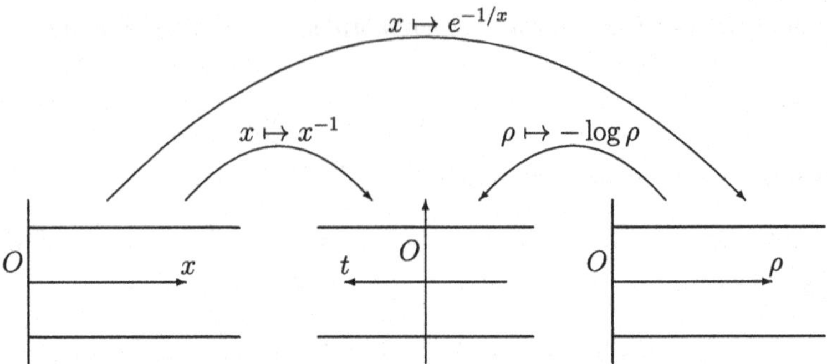

Figure 5.10. The relationship between the b- (right), cusp (left), and cylindrical (center) structures. The vertical direction is that of $O \subset \mathbb{R}^{n-1}$, that is the y-variables

Note that the a_j here need not be smooth after the push-forward: they are smooth functions of (x, y), $y \in \partial M$, that is of $(-\frac{1}{\log \rho}, y)$, which is a much weaker statement than smoothness in ρ.

Conversely, starting with $\mathcal{V}_b(M)$, with a boundary defining function ρ and a local product decomposition of a neighborhood U of ∂M as $[0, \epsilon)_\rho \times \partial M$, we can let $x = -1/(\log \rho)$, which then gives rise to a cusp structure. The push-forward of a $\mathcal{C}^\infty$ b-vector field is then a $\mathcal{C}^\infty$ cusp vector field, since smoothness in ρ implies smoothness in x.

Letting $t = x^{-1}$, we have $t = -\log \rho$, and the identification of $(0, \epsilon)_\rho \times \partial M$ is with $(-\log \epsilon, +\infty)_t \times \partial M$ (the boundary is pushed out to infinity). Then the push-forward of $a_0 \rho \partial_\rho + \sum a_j \partial_{y_j}$ is $-a_0 \partial_t + \sum a_j \partial_{y_j}$, and a coordinate chart O in ∂M with coordinates y gives an identification of $(0, \epsilon) \times O$ with a cylinder $(-\log \epsilon, +\infty)_t \times O \subset \mathbb{R} \times \mathbb{R}^{n-1} = \mathbb{R}^n$. Now, the smoothness of the coefficients a_j in $[0, \epsilon) \times O$ is equivalent to the statement that $\partial_\rho^k \partial_y^\alpha a_j$ is bounded for all k, α on $(0, \epsilon) \times O$ (with $a_j(0, y) = \lim_{\rho \to 0} a_j(\rho, y)$, and the limit automatically existing under these conditions), and thus (as the push-forward of $\partial_\rho = \rho^{-1}(\rho \partial_\rho)$ is $-e^t \partial_t$) to the statement that $(-e^t \partial_t)^k \partial_y^\alpha a_j$ is bounded on $(-\log \epsilon, +\infty)_t \times O$. This is a rather stringent requirement on the t derivatives.

A weaker, and more convenient, requirement on the a_j is *conormality*. In general, on a manifold with boundary with a boundary defining function ρ, one says that a function a is conormal to ∂M of order s and writes $a \in \mathcal{A}^s$, if $\rho^{-s} V_1 \ldots V_\ell a$ is bounded for all ℓ and all $V_j \in \mathcal{V}_b(M)$. Note that such a statement is independent of the choice of ρ (since it is unchanged by inserting a positive smooth factor into ρ), and locally, in coordinates (ρ, y), it is equivalent to $\rho^{-s}(\rho \partial_\rho)^k \partial_y^\alpha a$ being bounded for all k, α, as can be seen (in the less trivial direction) by expanding the V_j in terms of $\rho \partial_\rho$ and ∂_{y_j} with smooth coefficients, and using the product rule iteratively. Notice that this translates to $e^{st}(-\partial_t)^k\partial_y^\alpha a$ being bounded for all k, α, that is for $s = 0$ simply to the boundedness of all (standard, constant coefficient) derivatives in the cylindrical region.

This means that conormal b-vector fields, that is elements of $\mathcal{A}^0 \otimes_{C^\infty} \mathcal{V}_b(M)$, that is vector fields of the form

$$a_0(\rho\partial_\rho) + \sum a_j\partial_{y_j},\ a_j \in \mathcal{A}^0,$$

are exactly vector fields in the cylindrical region of $\mathbb{R}^n$ with coefficients that are bounded with all derivatives; if they have support inside $[t_0, +\infty) \times K$, $K \subset O$ compact, $t_0 > -\log\epsilon$, as one would arrange when using a partition of unity on M, then they can also be regarded as vector fields on all of $\mathbb{R}^n$ with coefficients bounded with all derivatives. But this means that they are elements of the $\Psi^1_\infty = \Psi^{1,0}_\infty$ space, and differential operators based on them are elements of $\Psi^m_\infty = \Psi^{m,0}_\infty$. Correspondingly, the analysis of *conormal* b-differential operators, that is elements of $\mathcal{A}^0\mathrm{Diff}_b(M)$, is locally equivalent to the analysis of Ψ^m_∞.

Notice that $\mathrm{Diff}_b(M) \subset \mathcal{A}^0\mathrm{Diff}_b(M)$, so this in particular includes at least aspects of the $\mathrm{Diff}_b(M)$-analysis (which however has stronger aspects as well). In fact, what matters for us most right now is that if $P \in \mathrm{Diff}_b(M)$ then one can write in the local coordinate decomposition $P = \sum_{j=0}^{N-1} \rho^j P_j + \rho^N \tilde{P}$, where $P_j \in \mathrm{Diff}_b(M)$ have coefficients independent of ρ, and $\tilde{P} \in \mathrm{Diff}_b(M)$. Correspondingly, translated to $\mathbb{R}^n$, we have

$$P = \sum_{j=0}^{N-1} e^{-jt}P_j + e^{-Nt}\tilde{P},$$

with $\tilde{P} \in \Psi_\infty$, and P_j having t-independent coefficients.

Turning to the cusp vector fields, as $t = 1/x$, the bordification of $(\epsilon', \infty)_t$ as $[0, 1/\epsilon')$ via $t \mapsto 1/t$ simply undoes the change of variables $x \mapsto t$, so elements of $C^\infty(\tilde{M})$ are locally equivalently in $C^\infty(\overline{\mathbb{R}_t} \times O_y)$.

In order to do the analysis on $\mathbb{R}^n$, it is more convenient to bordify all of $\mathbb{R}^n$, not just a cylindrical subset, as $\overline{\mathbb{R}_t} \times \mathbb{R}^{n-1}_y$, with the bordification of the first factor given by $t \mapsto t^{-1}$ in the cusp setting, and $t \mapsto e^{-t}$ in the b-setting; in the latter case we write the compactification as $\overline{\mathbb{R}_{\exp}} \times \mathbb{R}^{n-1}$. Then for $P \in \mathrm{Diff}^m_b(M)$, one has a principal symbol in $S^m_\infty(\mathbb{R}^n; \mathbb{R}^n)$ in fact lying in $C^\infty(\overline{\mathbb{R}_{\exp}} \times \mathbb{R}^{n-1}; S^m(\mathbb{R}^n))$. Correspondingly, it is convenient to regard $\overline{\mathbb{R}_{\exp}} \times \mathbb{R}^{n-1} \times \overline{\mathbb{R}^n}$ as the locus of b-microlocal analysis, with $\sigma_{b,m}(P)$ defined on this space. Thus, one defines the elliptic set $\mathrm{Ell}(P)$, the characteristic set $\mathrm{Char}(P)$, and the operator wavefront set $\mathrm{WF}'_b(P)$ as subsets of this space. Similarly, for cusp analysis the locus of microlocalization is $\overline{\mathbb{R}} \times \mathbb{R}^{n-1}$ where the lack of the subscript exp means that the reciprocal map $t \mapsto t^{-1}$ is used to construct the bordification. Since the connection between these two compactifications is $\rho = e^{-1/x}$, resp. $x = -1/(\log \rho)$, the two smooth structures differ, with the extension of the identity map $\mathbb{R} \to \mathbb{R}$ to $\overline{\mathbb{R}} \to \overline{\mathbb{R}_{\exp}}$ being a C^∞ map, but its inverse is only continuous, not differentiable, that is $\overline{\mathbb{R}} \times \mathbb{R}^{n-1}$ and $\overline{\mathbb{R}_{\exp}} \times \mathbb{R}^{n-1}$ are naturally homeomorphic (in the sense of the extension of the identity map on $\mathbb{R}^n$), that is the same as topological manifolds, but not diffeomorphic, though the identity map is (in the sense of "extends to") C^∞ as a map $\overline{\mathbb{R}} \times \mathbb{R}^{n-1} \to \overline{\mathbb{R}_{\exp}} \times \mathbb{R}^{n-1}$.

Concretely then define the *local cusp algebra* Ψ^m_{lcu}, resp. the *extended local b-algebra* Ψ^m_{elb}, as quantizations of symbols $a \in S^m_{\mathrm{lcu}}(\overline{\mathbb{R}} \times \mathbb{R}^{n-1}; \mathbb{R}^n)$, resp. $a \in S^m_{\mathrm{elb}}(\overline{\mathbb{R}} \times \mathbb{R}^{n-1}; \mathbb{R}^n)$, where $a \in S^m_{\mathrm{lcu}}(\overline{\mathbb{R}} \times \mathbb{R}^{n-1}; \mathbb{R}^n)$ if

$$|(t^2\partial_t)^k \partial_y^\beta \partial_\zeta^\alpha a| \leq C_{\alpha\beta,k} \langle \zeta \rangle^m,$$

and $a \in S^m_{\mathrm{elb}}(\overline{\mathbb{R}} \times \mathbb{R}^{n-1}; \mathbb{R}^n)$ if

$$|(e^{\phi(t)}\partial_t)^k \partial_y^\beta \partial_\zeta^\alpha a| \leq C_{\alpha\beta,k} \langle \zeta \rangle^m,$$

where $\phi \in C^\infty(\mathbb{R})$ with $\phi(t) = |t|$ for $|t| \geq 1$. We also define the *extended local symbolic b-algebra* as Ψ^m_∞. This means that

$$S^m_{\mathrm{lcu}}(\overline{\mathbb{R}} \times \mathbb{R}^{n-1}; \mathbb{R}^n) = C^\infty_\infty(\overline{\mathbb{R}} \times \mathbb{R}^{n-1}; S^m(\mathbb{R}^n))$$

and

$$S^m_{\mathrm{elb}}(\overline{\mathbb{R}} \times \mathbb{R}^{n-1}; \mathbb{R}^n) = C^\infty_\infty(\overline{\mathbb{R}_{\exp}} \times \mathbb{R}^{n-1}; S^m(\mathbb{R}^n)),$$

where the subscript ∞ refers to uniform bounds in the non-compact variable y. The reason for the term "extended" is the slow off-diagonal decay that we discuss below; we in fact want to impose faster decay conditions.

The membership of Ψ^m_{elb} is equivalent to the statement that $P \in \Psi^m_\infty$ and for all N there exist $\mathbb{R}$-translation invariant operators $P_{j,\pm} \in \Psi^m_\infty$, $j = 0, 1, \ldots, N-1$, and $\tilde{P}_{N,\pm} \in \Psi^m_\infty$ such that

$$P = \sum_{j=0}^{N-1} e^{-j\phi_\pm(t)} P_{j,\pm} + e^{-N\phi_\pm(t)} \tilde{P}_{N,\pm}, \tag{5.118}$$

where $\phi_\pm(t) = \pm t$ for $\pm t > 1$, $\phi_\pm(t) = 0$ for $\pm t < -1$, and $\phi_\pm \in C^\infty(\mathbb{R})$; note that the $+$ sign is making a non-trivial statement over membership in Ψ^m_∞ only as $t \to +\infty$, while the $-$ sign is making a non-trivial statement as $t \to -\infty$. Thus, the $P_{j,\pm}$ are Taylor series of P at $t = \pm\infty$ in $\overline{\mathbb{R}_{\exp}}$.

In view of the algebra property of Ψ^m_∞, it is easy to check that the other spaces mentioned also form algebras. Further, as Ψ^m_∞ is bounded between weighted Sobolev spaces $H^{s,r} \to H^{s-m,r}$, so are its subalgebras. For the cusp algebra this is the best boundedness statement one would like (essentially because smoothness in x rather than in $\rho = e^{-1/x}$ is considered from the perspective described at the beginning of this section), but in the b-setting one would like to improve this to boundedness between exponentially weighted Sobolev spaces:

$$H_{\mathrm{b}}^{s,r} = e^{-r\phi} H^s(\mathbb{R}^n).$$

The obstacle for this is that the Schwartz kernels of elements of Ψ_∞ only decay as $C_N \langle z - z' \rangle^{-N}$; to have action on the exponentially weighted spaces, with exponential weight in t, one needs superexponential decay. Note that if $A \in \Psi_\infty^m$, and K_A is its Schwartz kernel, then with $\psi \in C_c^\infty(\mathbb{R})$, identically 1 near 0,

$$K_R = (1 - \psi(t - t'))K_A$$

defines an element R of $\Psi_\infty^{-\infty}$, that is if we let $K_B = \psi(t - t')K_A$, then $R = A - B \in \Psi_\infty^{-\infty}$, and the Schwartz kernel of B is supported in a region where $|t - t'| < C$. Correspondingly, for any r,

$$e^{-r\phi} B e^{r\phi} \in \Psi_\infty^m,$$

since it has Schwartz kernel

$$e^{-r(\phi(t) - \phi(t'))} K_B,$$

and in view of the boundedness of $t - t'$, $\phi(t) - \phi(t')$ is also bounded, with all derivatives. Now, we do not actually need compact supports in $t - t'$, only superexponential decay, so it is natural to consider operators R' with C^∞ Schwartz kernel $K_{R'}$ satisfying estimates

$$|\partial_t^k \partial_{t'}^l \partial_y^\alpha \partial_{y'}^\beta K_{R'}(t, y, t', y')| \leq C_{kl\alpha\beta NM} \langle y - y' \rangle^{-N} e^{-M\phi(t-t')} \tag{5.119}$$

for all $\alpha, \beta, k, l, M, N$. These are of course elements of $\Psi_\infty^{-\infty}$, whose elements have similar estimates, except the decay is superpolynomial in $t - t'$.

Definition 5.15 The *local symbolic b-algebra* $\cup_m \Psi_{\mathrm{lbc}}^m$ consists of operators L of the form $L = B + R'$, with $K_B = \psi(t - t')K_A$, $A \in \Psi_\infty^m$, $\psi \in C_c^\infty(\mathbb{R})$ identically 1 near 0, while R' is as in (5.119).

The *local b-algebra* $\cup_m \Psi_{\mathrm{lb}}^m$ consists of operators $L \in \Psi_{\mathrm{lbc}}^m$ of the form $B+R'$, with $K_B = \psi(t - t')K_A$, $A \in \Psi_{\mathrm{elb}}^m$, $\psi \in C_c^\infty(\mathbb{R})$ identically 1 near 0, while R' in addition to (5.119) has an expansion

$$R' = \sum_{j=0}^{N-1} e^{-j\phi_\pm(t)} R'_{j,\pm} + e^{-N\phi_\pm(t)} \tilde{R}'_{N,\pm}, \tag{5.120}$$

where $R'_{j,\pm}$ are $\mathbb{R}$-translation invariant (i.e. the Schwartz kernels depend on t, t' only via $t - t'$) satisfying the estimates (5.119) and $\tilde{R}'_{N,\pm}$ also satisfies (5.119). Thus, both B and R' have expansions in terms of translation-invariant operators as in (5.118).

Notice that for R' as in (5.119), for any r,

$$e^{-r\phi}R'e^{r\phi} \in \Psi_\infty^{-\infty},$$

since for $\pm r \geq 0$,

$$e^{-r(\phi(t)-\phi(t'))\mp r\phi(t-t')}$$

is bounded: it is essentially $e^{-r(|t|-|t'|\pm|t-t'|)}$, and $|t| - |t'| + |t - t'| \geq 0$ (the case $r \geq 0$), while $|t| - |t'| - |t - t'| \leq 0$ (the case $r \leq 0$). In particular, such R' is bounded on $H_{\mathrm{b}}^{s,r} \to H_{\mathrm{b}}^{s',r}$ for all s, s', r since this boundedness is equivalent to the boundedness of $e^{r\phi}R'e^{-r\phi} : H^s \to H^{s'}$. Similarly, for $B \in \Psi_\infty^m$ arising by localizing the Schwartz kernel of $A \in \Psi_\infty^m$ near the diagonal as above, $B : H_{\mathrm{b}}^{s,r} \to H_{\mathrm{b}}^{s-m,r}$, so we conclude that elements of Ψ_{lb}^m map $H_{\mathrm{b}}^{s,r} \to H_{\mathrm{b}}^{s-m,r}$ continuously for all s, r.

For $L \in \Psi_{\mathrm{lbc}}^m$ we can now define the principal symbol $\sigma_{\mathrm{b}}(L)$ and the wavefront set $\mathrm{WF}'_{\mathrm{b}}(L)$ as the corresponding object in the decomposition of Definition 5.15 for A. However, the subtlety is that we use a *different* notion from that of the standard case at $\partial\overline{\mathbb{R}^n}$. Indeed, the partial compactification of $\mathbb{R}^n$ we want to use is $\overline{\mathbb{R}_{\exp}} \times \mathbb{R}^{n-1}$ (the compactification is partial as $\mathbb{R}^{n-1}$ is non-compact, but this is not important here). For $A \in \Psi_\infty^m$ we thus take the standard notion of the principal symbol, by writing $A = q_L(a)$, and taking the class of a in $S_\infty^m/S_\infty^{m-1}$. However, the notion of the wavefront set we want is based on the notion of essential support in $\overline{\mathbb{R}_{\exp}} \times \mathbb{R}^{n-1} \times \overline{\mathbb{R}^n}$; thus a point $\alpha \in \overline{\mathbb{R}_{\exp}} \times \mathbb{R}^{n-1} \times \overline{\mathbb{R}^n}$ is not in $\mathrm{esssupp}_{\mathrm{b}}(a)$ if there exists a neighborhood U of α in $\overline{\mathbb{R}_{\exp}} \times \mathbb{R}^{n-1} \times \overline{\mathbb{R}^n}$ such that $a|_{U\cap(\mathbb{R}^n\times\mathbb{R}^n)}$ is in $S_\infty^{-\infty}$. Similarly, we define the elliptic and characteristic sets as subsets of $\overline{\mathbb{R}_{\exp}} \times \mathbb{R}^{n-1} \times \overline{\mathbb{R}^n}$. Note that given L, A is defined modulo $\Psi_\infty^{-\infty}$ since B is also so determined, and A and B differ by an element of $\Psi_\infty^{-\infty}$. Thus, $\mathrm{WF}'_{\mathrm{b}}(L)$ is a subset of $\overline{\mathbb{R}_{\exp}} \times \mathbb{R}^{n-1} \times \partial\overline{\mathbb{R}^n}$, and is well-defined as such.

In order to explain the notion of compactification we have used, recall that for $L \in \Psi_{\mathrm{lb}}^m$ we may regard the principal symbol as a function on $\overline{\mathbb{R}_{\exp}} \times \mathbb{R}^{n-1} \times \mathbb{R}^n$ which is symbolic in the last variable of order m, modulo such symbols of order $m - 1$. Furthermore, if L has a homogeneous principal symbol, for instance if it is classical, we may regard the principal symbol as a homogeneous degree m function on $\overline{\mathbb{R}_{\exp}} \times \mathbb{R}^{n-1} \times (\mathbb{R}^n \setminus \{0\})$. In particular, if $L \in \Psi_{\mathrm{lbc}}^0$ is classical of order 0 (in the sense that A is such, which property is again independent of the particular choice of A) then $\sigma_{\mathrm{b},0}(L)$ can be regarded as a C^∞ function on $\overline{\mathbb{R}_{\exp}} \times \mathbb{R}^{n-1} \times \partial\overline{\mathbb{R}^n}$.

We also need the notion of the wavefront set of a distribution. Since the principal symbol captures b-pseudodifferential operators modulo one order lower pseudodifferential operators, but without giving any additional decay properties, the b-wavefront set microlocally captures the regularity (differentiability) of distributions, but does not imply any additional decay properties beyond what they a priori possess, much like the case of $\mathrm{WF}_\infty^{m,\ell}$. Indeed, the difference between WF_b and WF_∞ is the space on which the wavefront set lives, reflecting the analogous difference between elliptic sets. Concretely:

Definition 5.16 Suppose $k, \ell, m \in \mathbb{R}$, $u \in H_\mathrm{b}^{k,\ell}$. For $\alpha \in \overline{\mathbb{R}_{\exp}} \times \mathbb{R}^{n-1} \times \mathbb{R}^n$, we say $\alpha \notin \mathrm{WF}_\mathrm{b}^{m,\ell}(u)$ if there exists $A \in \Psi_{\mathrm{lbc}}^{0,0}$ elliptic at α such that $Au \in H_\mathrm{b}^{m,\ell}$. We say that $\alpha \notin \mathrm{WF}_\mathrm{b}^{\infty,\ell}(u)$ if there exists $A \in \Psi_{\mathrm{lbc}}^{0,0}$ elliptic at α such that $Au \in H_\mathrm{b}^{\infty,\ell}$.

We also make the analogous definition for variable order spaces.

When $\ell = 0$, we sometimes drop the weight from the notation and simply write $\mathrm{WF}_\mathrm{b}^m(u)$.

Now, if $p \in S_{\mathrm{lcu}}^m$ is homogeneous of degree m (outside a neighborhood of the zero section) then writing $\zeta = (\tau, \eta)$,

$$H_p = \frac{\partial p}{\partial \tau}\frac{\partial}{\partial t} - \frac{\partial p}{\partial t}\frac{\partial}{\partial \tau} + \sum_j \left(\frac{\partial p}{\partial \eta_j}\frac{\partial}{\partial y_j} - \frac{\partial p}{\partial y_j}\frac{\partial}{\partial \eta_j} \right) \tag{5.121}$$

is homogeneous of degree $m-1$ away from o, and it is smooth on $\overline{\mathbb{R}} \times \mathbb{R}^{n-1} \times \mathbb{R}^n$, tangent to the boundary, $\partial\overline{\mathbb{R}} \times \mathbb{R}^{n-1} \times \mathbb{R}^n$. Note that $\partial_t = -x^2\partial_x$ shows that the first two terms even vanish as b-vector fields at this boundary, that is the lack of extra decay there arises from the y and η derivatives.

Similarly, if $p \in S_{\mathrm{elb}}^m$ is homogeneous of degree m (outside a neighborhood of the zero section) then H_p is homogeneous of degree $m-1$ away from o, and it is smooth on $\overline{\mathbb{R}_{\exp}} \times \mathbb{R}^{n-1} \times \mathbb{R}^n$, tangent to the boundary $\partial\overline{\mathbb{R}_{\exp}} \times \mathbb{R}^{n-1} \times \mathbb{R}^n$. Correspondingly, we define the b-Hamilton vector field

$$\mathsf{H}_{p,\mathrm{b},m} = \langle \zeta \rangle^{-m+1} H_p \in \mathcal{V}_\mathrm{b}(\overline{\mathbb{R}_{\exp}} \times \mathbb{R}^{n-1} \times \mathbb{R}^n),$$

well defined modulo $\rho_{\mathrm{fiber}}\mathcal{V}_\mathrm{b}(\overline{\mathbb{R}_{\exp}} \times \mathbb{R}^{n-1} \times \mathbb{R}^n)$. Now $\partial_t = -\rho\partial_\rho$, so unlike in the cusp setting none of the terms in (5.121) vanish in general at the boundary, $\rho = 0$, as b-vector fields. Then the analogues of our results from the scattering setting all go through, keeping in mind that we only have gains in differentiability.

Note that the additional content in the next propositions is only what happens at $\partial\overline{\mathbb{R}_{\exp}} \times \mathbb{R}^{n-1} \times \mathbb{R}^n$ in the homogeneous setting. First, the elliptic result is essentially the result for Ψ^m_∞, except that if $P \in \Psi^m_{\mathrm{lbc}}$ then one can also work with exponentially weighted spaces.

Proposition 5.33 *Suppose $P \in \Psi^m_{\mathrm{elbc}} = \Psi^m_\infty$ is elliptic at $\alpha \in \overline{\mathbb{R}_{\mathrm{exp}}} \times \mathbb{R}^{n-1} \times \partial\overline{\mathbb{R}^n}$. If $u \in H^{-N}$ for some N, $\alpha \notin \mathrm{WF}_{\mathrm{b}}^{s-m}(Pu)$, then $\alpha \notin \mathrm{WF}_{\mathrm{b}}^{s}(u)$.*

On the other hand, suppose $P \in \Psi^m_{\mathrm{lbc}}$ is elliptic at $\alpha \in \overline{\mathbb{R}_{\mathrm{exp}}} \times \mathbb{R}^{n-1} \times \partial\overline{\mathbb{R}^n}$. If $u \in H_{\mathrm{b}}^{-N,r}$ for some N, r, $\alpha \notin \mathrm{WF}_{\mathrm{b}}^{s-m,r}(Pu)$, then $\alpha \notin \mathrm{WF}_{\mathrm{b}}^{s,r}(u)$.

For the non-elliptic setting, we assume that P has an expansion at infinity, that is $P \in \Psi^m_{\mathrm{elb}}$; this assures a well-behaved Hamilton flow. In fact, one should think of the expansion of P as being analogous to the requirement of the operator possessing a homogeneous principal symbol in the standard microlocal propagation of the singularities theorem. Then the proof of the propagation of the singularities theorem presented in Section 5.4 goes through without essential changes; one just needs to keep in mind that $\overline{\mathbb{R}_{\mathrm{exp}}} \times \mathbb{R}^{n-1} \times \mathbb{R}^n$ is the location of the analysis.

Proposition 5.34 *Suppose $P \in \Psi^m_{\mathrm{elb}}$. If $u \in H^{-N}$ for some N, then $\mathrm{WF}_{\mathrm{b}}^{s}(u) \setminus \mathrm{WF}_{\mathrm{b}}^{s-m+1}(Pu)$ is a union of maximally extended bicharacteristics, that is integral curves of $\mathsf{H}_{p,\mathrm{b},m}$ in $\mathrm{Char}(P)$.*

On the other hand, suppose $P \in \Psi^m_{\mathrm{lb}}$. If $u \in H_{\mathrm{b}}^{-N,r}$ for some N, r, then $\mathrm{WF}_{\mathrm{b}}^{s,r}(u) \setminus \mathrm{WF}_{\mathrm{b}}^{s-m+1,r}(Pu)$ is a union of maximally extended bicharacteristics, that is integral curves of $\mathsf{H}_{p,\mathrm{b},m}$ in $\mathrm{Char}(P)$.

Proposition 5.35 *Suppose $P \in \Psi^m_{\mathrm{elb}}$, $L \subset \partial\overline{\mathbb{R}_{\mathrm{exp}}} \times \mathbb{R}^{n-1} \times \partial\overline{\mathbb{R}^n}$ is a compact smooth embedded submanifold, and*

$$\mp\mathsf{H}_{p,\mathrm{b},m}\rho_{\mathrm{fiber}} = \beta_0\rho_{\mathrm{fiber}},\ \beta_0|_L > 0,$$

$$\tilde{p} = \sigma_{\mathrm{b},m-1}\left(\frac{1}{2i}(P - P^*)\right)$$

satisfies

$$\tilde{p} = \pm\beta_0\tilde{\beta}\rho_{\mathrm{fiber}}^{1-m},$$

and there is a quadratic defining function ρ_1 of L in $\mathrm{Char}(P) \subset \overline{\mathbb{R}_{\mathrm{exp}}} \times \mathbb{R}^{n-1} \times \partial\overline{\mathbb{R}^n}$ such that

$$\mp\mathsf{H}_{p,\mathrm{b},m}\rho_1 = \beta_1\rho_1 + F_2 + F_3,\ \beta_1 > 0,$$

where $F_2 \geq 0$ and F_3 vanishes cubically at L.

(i) *Suppose that $s - (m-1)/2 + \tilde{\beta} > 0$ on L, and $s' - (m-1)/2 + \tilde{\beta} > 0$ on L. Then $u \in H^{-N}$ for some N, $L \cap \mathrm{WF}_{\mathrm{b}}^{s-m+1}(Pu) = \emptyset$, $\mathrm{WF}_{\mathrm{b}}^{s'}(u) \cap L = \emptyset$ imply $\mathrm{WF}_{\mathrm{b}}^{s}(u) \cap L = \emptyset$.*

(ii) *Suppose that* $s-(m-1)/2+\tilde\beta<0$ *on* L. *If* $u\in H^{-N}$ *and* L *has a neighborhood* U *such that* $(U\setminus L)\cap \mathrm{WF}_{\mathrm b}^s(u)=\emptyset$ *and* $U\cap \mathrm{WF}_{\mathrm b}^{s-m+1}(Pu)=\emptyset$, *then* $L\cap \mathrm{WF}_{\mathrm b}^s(u)=\emptyset$.

The exponentially weighted version is:

Proposition 5.36 *Suppose* $P\in\Psi^m_{\mathrm{lb}}$, $L\subset\partial\overline{\mathbb R_{\mathrm{exp}}}\times\mathbb R^{n-1}\times\partial\overline{\mathbb R^n}$ *is a compact smooth embedded submanifold, with*

$$\mp\mathsf H_{p,\mathrm b,m}\rho_{\mathrm{fiber}}=\beta_0\rho_{\mathrm{fiber}},\ \beta_0|_L>0,$$

and

$$\mp\mathsf H_{p,\mathrm b,m}\rho_{\mathrm{base}}=\beta_2\beta_0\rho_{\mathrm{base}},\ \beta_2|_L>0.$$

Suppose further that

$$\tilde p=\sigma_{\mathrm b,m-1}\left(\frac{1}{2i}(P-P^*)\right)$$

satisfies

$$\tilde p=\pm\beta_0\tilde\beta\rho_{\mathrm{fiber}}^{1-m},$$

and there is a quadratic defining function ρ_1 *of* L *in* $\mathrm{Char}(P)\subset\overline{\mathbb R_{\mathrm{exp}}}\times\mathbb R^{n-1}\times\partial\overline{\mathbb R^n}$ *such that*

$$\mp\mathsf H_{p,\mathrm b,m}\rho_1=\beta_1\rho_1+F_2+F_3,\ \beta_1>0,$$

where $F_2\geq 0$ *and* F_3 *vanishes cubically at* L.

(i) *Suppose that* $s+\beta_2 r-(m-1)/2+\tilde\beta>0$ *on* L, *and* $s'+\beta_2 r-(m-1)/2+\tilde\beta>0$ *on* L. *Then* $u\in H_{\mathrm b}^{-N,r}$ *for some* N,r, $L\cap\mathrm{WF}_{\mathrm b}^{s-m+1,r}(Pu)=\emptyset$, $\mathrm{WF}_{\mathrm b}^{s',r}(u)\cap L=\emptyset$ *imply* $\mathrm{WF}_{\mathrm b}^{s,r}(u)\cap L=\emptyset$.
(ii) *Suppose that* $s+\beta_2 r-(m-1)/2+\tilde\beta<0$ *on* L. *If* $u\in H^{-N,r}$ *and* L *has a neighborhood* U *such that* $(U\setminus L)\cap\mathrm{WF}_{\mathrm b}^{s,r}(u)=\emptyset$ *and* $U\cap\mathrm{WF}_{\mathrm b}^{s-m+1,r}(Pu)=\emptyset$, *then* $L\cap\mathrm{WF}_{\mathrm b}^{s,r}(u)=\emptyset$.

We eventually consider the case when L is a saddle point of the flow, but what is stated so far suffices for elliptic problems, as well as for the Minkowski analysis.

While often not strictly necessary, it is convenient to globalize the boundary pseudo-differential operators algebra to simplify various statements. The reason this is often not necessary is that the symbolic considerations above are almost all local, except that the radial set results are global within the radial set. Thus, as long as the radial sets can be thought to lie over a single coordinate chart, no globalization is needed. In fact, the radial set results themselves can be localized, as was shown by Haber and Vasy [17], but we do not pursue this here. On the other hand, the normal operator considerations are global, but for this we use the global Mellin transform on a collar neighborhood of the boundary of our manifold M, and thus one need not explicitly globalize the pseudo-differential operator algebra to deal with $\mathrm{Diff}_{\mathrm{b}}(M)$.

For the next definition recall that $\dot{C}^\infty(M)$ is the subset of $C^\infty(M)$ consisting of functions vanishing with all derivatives at ∂M.

Definition 5.17 Suppose M is a compact manifold with boundary. We define the symbolic (or conormal) b-algebra $\cup_m \Psi^m_{\mathrm{bc}}(M)$, in analogy with standard pseudodifferential operators on manifolds without boundary by requiring for $A \in \Psi^m_{\mathrm{bc}}(M)$ that

(i) $A : \dot{C}^\infty(M) \to \dot{C}^\infty(M)$ continuously (so in particular A has a distributional Schwartz kernel),
(ii) if U is any coordinate chart with $\Phi : U \to \tilde{U} \subset \mathbb{R}^n$ a diffeomorphism, then for $\chi \in C^\infty_c(U)$, $(\Phi^{-1})^*\chi A \chi \Phi^* \in \Psi^m_{\mathrm{lbc}}$,
(iii) and if $\chi, \phi \in C^\infty(M)$ have disjoint support in local coordinate charts U, resp. V, then $\chi A \phi$ has a Schwartz kernel which is conormal on $M \times M$ (i.e. has iterated regularity relative to b-vector fields) relative to bounded functions which decay rapidly as ρ/ρ' tends to 0 or $+\infty$, that is have bounds $C_N(\rho/\rho')^N$ in $\rho/\rho' < 1$, $C_N(\rho'/\rho)^N$ in $\rho/\rho' > 1$, where ρ stands for the pull-back of ρ from the first factor, while ρ' for the pull-back from the second factor.

Specifically, (iii) means (see Figure 5.11)

(i) if U, V are both disjoint from ∂M then $\chi A \phi$ has a C^∞ Schwartz kernel,
(ii) if U is disjoint from ∂M and V is not, or vice versa, then $\chi A \phi$ has a C^∞ Schwartz kernel on $M \times M$ vanishing to infinite order $M \times \partial M$, resp. $\partial M \times M$,
(iii) if both U and V intersect ∂M, with a product decomposition $[0, \epsilon)_\rho \times U_0$, resp. $[0, \epsilon)_\rho \times V_0$, and with coordinate maps Φ and Ψ then the Schwartz kernel $K_{\chi,\psi}$ of $(\Phi^{-1})^*\chi A \phi \Psi^*$ satisfies, with e^{-t} being the pull-back of ρ from the first factor, $e^{-t'}$ from the second factor, y, resp. y', coordinates on U_0, resp. V_0, estimates

$$|\partial_t^k \partial_{t'}^\ell \partial_y^\alpha \partial_{y'}^\beta K_{\chi,\psi}| \leq C_{kl\alpha\beta M} e^{-M|t-t'|}.$$

Notice that these are exactly the estimates (5.119), apart from the $y - y'$ factor, which would not make sense here as U_0, V_0 are unrelated coordinate charts and which is irrelevant even for related coordinate charts, as we have compact support in both y and y'.

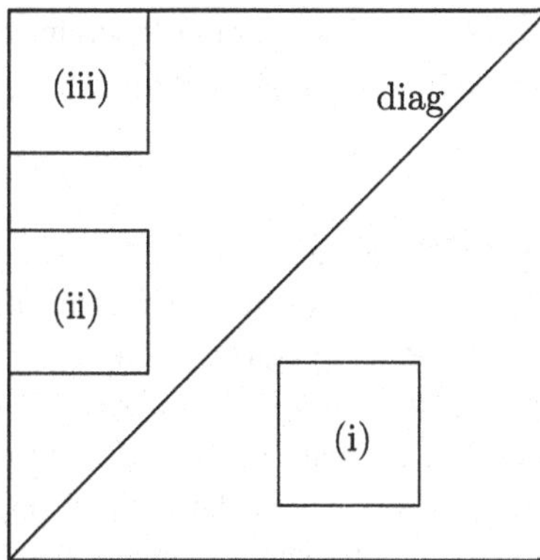

Figure 5.11. The product space $M \times M$, with the two factors being horizontal and vertical, respectively, and regions of the kind (i), (ii), and (iii) corresponding to product sets $U \times V$. Note that $U \cap V = \emptyset$ means that $(U \times V) \cap \mathrm{diag} = \emptyset$

Although we have not discussed geometric blow-ups, see [34] for a discussion in precisely this setting, these are exactly the statement that the Schwartz kernel of $\chi A \phi$ is conormal on $[[0,\epsilon)_\rho \times [0,\epsilon)_{\rho'}; \{0\} \times \{0\}] \times U_0 \times V_0$, relative to bounded functions with infinite order of vanishing at the lift of $\rho = 0$ and $\rho' = 0$: the various regions above describe various regions of this blow-up.

Note that the terms in part (iii) of the definition, corresponding to R' in the local algebra, are in $\Psi_{\mathrm{bc}}^{-\infty}(M)$, and thus do not affect the symbolic statements/estimates. Here we emphasize that elements of the global algebra $\Psi_{\mathrm{bc}}^{-\infty}(M)$ preserve the Sobolev spaces $H_{\mathrm{b}}^{s,r}(M)$, by the same arguments as above for R'. Correspondingly, the above propositions are all valid in the global algebra $\cup_m \Psi_{\mathrm{bc}}^m(M)$, provided one also assumes smoothness in the coefficients in the non-elliptic settings in the following sense.

Definition 5.18 The *smooth subalgebra* $\Psi_{\mathrm{b}}(M)$ of $\Psi_{\mathrm{bc}}^m(M)$ consists of elements P of $\Psi_{\mathrm{bc}}^m(M)$ for which there exist $P_j \in \Psi_{\mathrm{bc}}^m([0,\infty) \times \partial M)$ which are dilation invariant and such that for all k, using a collar neighborhood identification of a neighborhood of ∂M with $[0,\epsilon) \times \partial M$, $P - \sum_{j<k} P_j \in \rho^k \Psi_{\mathrm{bc}}^m(M)$.

The one-step nature is not very important here; more general expansions would also work. However, if one wants to ensure diffeomorphism invariance, one *must* include all powers $\rho^{\alpha+j}$, j a positive integer, if one includes ρ^α in the expansion.

First consider elliptic problems. Suppose that $L \in \Psi_{\mathrm{bc}}^m(M)$ where M is a compact manifold with boundary, and suppose that L is elliptic. Then, by the elliptic regularity (plus the closed graph theorem, or doing the estimate directly), for all $s, r \in \mathbb{R}$, and for all $s' \in \mathbb{R}$, there exists C such that

$$\|u\|_{H_b^{s,r}} \leq C(\|Lu\|_{H_b^{s-m,r}} + \|u\|_{H_b^{s',r}}), \tag{5.122}$$

and one also has a similar estimate for L^* as it is also elliptic. Since the inclusion $H_b^{s,r} \to H_b^{s',r}$ is non-compact, this is insufficient for Fredholm analysis. This is where the normal operators enter.

So let us assume that $L \in \Psi_b^m(M)$, that is L is *smooth* in the above sense.

Let $N(L) = L_0$ denote the normal operator of L; $\hat{L}(\sigma)$ be its conjugate by the Mellin transform (sometimes called the indicial operator). Then $\hat{L}(\sigma)$ is elliptic, even in the large parameter sense, when σ lies in strips $|\operatorname{Im}\sigma| < C$. Correspondingly, $\hat{L}(\sigma)^{-1}$ exists in such strips when $|\operatorname{Re}\sigma|$ is large, and thus $\hat{L}(\sigma)^{-1}$ is a meromorphic family in any such strip with finite rank poles. Choose r such that $-r$ is *not* the imaginary part of any of the poles. Then one has the estimate

$$\|v\|_{H^s} \leq C\|\hat{L}(\sigma)v\|_{H^{s-m}} \tag{5.123}$$

for compact regions of σ; notice that this is just the standard elliptic estimate with the usual weak error term $\|v\|_{H^{-N}}$ dropped due to invertibility, as a standard functional analysis argument allows one to do, see [27, Proof of Theorem 26.1.7] or [47, Section 4.3]. Furthermore, in strips $|\operatorname{Im}\sigma| < C'$, one has the semiclassical/large parameter estimates (recall $|\sigma|^{-m}\hat{L}(\sigma) = \hat{L}_{h,z}$)

$$\|v\|_{H^s_{|\sigma|^{-1}}} \leq C|\sigma|^{-m}\|\hat{L}(\sigma)v\|_{H^{s-m}_{|\sigma|^{-1}}}, \quad |\sigma| \gg 1. \tag{5.124}$$

Now, the Mellin transform is an isomorphism between $H_b^{s,r}([0,\infty) \times \partial M)$ and the large parameter Sobolev space, which for $s \geq 0$ integer corresponds to the squared norm

$$\sum_{|\alpha|+k \leq s} \int_{\operatorname{Im}\sigma=-r} \|\sigma^k V^\alpha v(\sigma)\|^2_{L^2(\partial M)}\, d\sigma,$$

where $V = (V_1, \ldots, V_l)$ are vector fields on ∂M spanning $\mathcal{V}(\partial M)$, with the general case following by interpolation and duality. Thus, for $v = v(\sigma) = (\mathcal{M}\tilde{u})(\sigma)$, (5.123)–(5.124) is exactly an estimate in $H_b^{s,r}$, if one performs the inverse Mellin transform integral on the contour line $\operatorname{Im}\sigma = -r$ (which, recall, is disjoint from poles). Thus, one has the estimate

$$\|\tilde{u}\|_{H_b^{s,r}} \leq C\|N(L)\tilde{u}\|_{H_b^{s-m,r}}.$$

Let $\chi \in C^\infty(M)$ be supported in a collar neighborhood of ∂M as above in which we have the stated expansion and the identification with $[0,\epsilon)_\rho \times \partial M$, also assume $\chi \equiv 1$ near ∂M. We apply our normal operator estimate with s replaced by s', and with $\tilde{u} = \chi u$, so:

$$\|u\|_{H_b^{s',r}} \le \|\chi u\|_{H_b^{s',r}} + \|(1-\chi)u\|_{H_b^{s',r}} \le C\|N(L)(\chi u)\|_{H_b^{s'-m,r}} + \|(1-\chi)u\|_{H_b^{s',r}}.$$

Since $1 - \chi$ is compactly supported in M°, for any r', $\|(1-\chi)u\|_{H_b^{s',r}} \le C\|u\|_{H_b^{s',r'}}$. On the other hand,

$$N(L)(\chi u) = [N(L), \chi]u + \chi Lu + \chi(N(L) - L)u$$

so, using $\chi(N(L) - L) \in \rho\Psi_b^m(M)$, $[N(L), \chi] \in \Psi_{bc}^{m-1}$ supported away from ∂M, for all r',

$$\|N(L)(\chi u)\|_{H_b^{s'-m,r}} \le C(\|u\|_{H_b^{s'-1,r'}} + \|Lu\|_{H_b^{s'-m,r}} + \|u\|_{H_b^{s',r-1}}).$$

Thus,

$$\|u\|_{H_b^{s',r}} \le C\|N(L)(\chi u)\|_{H_b^{s'-m,r}} + \|(1-\chi)u\|_{H_b^{s',r}} \le C(\|Lu\|_{H_b^{s'-m,r}} + \|u\|_{H_b^{s',r-1}}).$$

Combining with (5.122),

$$\begin{aligned}\|u\|_{H_b^{s,r}} &\le C(\|Lu\|_{H_b^{s-m,r}} + \|u\|_{H_b^{s',r}}) \\ &\le C(\|Lu\|_{H_b^{s-m,r}} + \|u\|_{H_b^{s',r-1}}).\end{aligned}$$

We now have the compact inclusion $H_b^{s,r} \to H_b^{s',r-1}$ that we desired.

A completely analogous argument for L^* gives for any $\tilde{s}, \tilde{s}', \tilde{r}$ with $-\tilde{r}$ not being the imaginary part of any pole of $\widehat{L^*}(\sigma)$,

$$\begin{aligned}\|u\|_{H_b^{\tilde{s},\tilde{r}}} &\le C(\|L^*u\|_{H_b^{\tilde{s}-m,\tilde{r}}} + \|u\|_{H_b^{\tilde{s}',\tilde{r}}}) \\ &\le C(\|L^*u\|_{H_b^{\tilde{s}-m,\tilde{r}}} + \|u\|_{H_b^{\tilde{s}',\tilde{r}-1}}).\end{aligned}$$

Note that for a product-type b-metric used to define the inner product $\widehat{L^*}(\sigma) = (\hat{L}(\overline{\sigma}))^*$ as shown by a simple computation, so the requirement on $\tilde{r}$ is that $\tilde{r}$ is not the imaginary part of any pole of $\hat{L}(\sigma)$.

Now take $\tilde{r} = -r$, $\tilde{s} = m - s$, so $\tilde{s} - m = -s$, and notice the expected duality between the spaces in the estimates for L and L^*, as before. This yields that

$$L : H_b^{s,r} \to H_b^{s-m,r}$$

is Fredholm. Notice that this is not quite an invertibility statement, though it gives an inverse on a finite codimensional subspace.

For the Laplacian Δ_g of a Riemannian b-metric g, if g is product type at ∂M, that is

$$g = \frac{d\rho^2}{\rho^2} + h + g_1,\ g_1 \in \rho C^\infty(M; {}^{\mathrm{b}}T^*M \otimes_s {}^{\mathrm{b}}T^*M), \tag{5.125}$$

h a metric on ∂M, extended via the local collar neighborhood identification, and $L = \Delta_g - \lambda$, $N(L) = (\rho D_\rho)^2 + \Delta_h - \lambda$, so $\hat{N}(L) = \Delta_h + \sigma^2 - \lambda$. In particular, this is invertible as long as $\lambda - \sigma^2$ is not an eigenvalue of Δ_h, in particular if $\lambda - \sigma^2 \notin [0, \infty)$. So for $\lambda \in \mathbb{C} \setminus [0, \infty)$, $\sigma \in \mathbb{R}$, one has the desired invertibility, which then gives the Fredholm statement for L as a map $H_{\mathrm{b}}^{s,0} \to H_{\mathrm{b}}^{s-2,0}$, as well as for $H_{\mathrm{b}}^{s,r} \to H_{\mathrm{b}}^{s-2,r}$ for $|r|$ small. Since any element of $\operatorname{Ker} L$ and $\operatorname{Ker} L^*$ would then lie in $H_{\mathrm{b}}^{\infty,r}$ for some $r > 0$, it follows from the symmetry of L that these kernels are trivial, thus the operator is invertible. We state this as a theorem:

Theorem 5.9 *Let g be a Riemannian b-metric of the type* (5.125)*, and let λ_j, $j \geq 1$, be the eigenvalues of the boundary Laplacian Δ_h. For $\lambda \in \mathbb{C} \setminus [0, \infty)$, $\Delta_g - \lambda$ is invertible as a map $H_{\mathrm{b}}^{s,r} \to H_{\mathrm{b}}^{s-2,r}$ for $|r| < \operatorname{Im} \sqrt{\lambda}$.*

Furthermore, with $\sigma_j = \sqrt{\lambda - \lambda_j}$, with the square root being in the lower half-plane, for $r' > 0$ such that $r' \neq \sigma_j$ for any j, if $f \in H_{\mathrm{b}}^{s-2,r'}$ then there exist $a_{j,k,\ell} \in C^\infty(\partial M)$, $j \in \mathbb{N}^+$, $k \in \mathbb{N}$, $\mathbb{N} \ni \ell \leq \ell_{j,k}$ such that

$$\begin{aligned}(\Delta_g - \lambda)^{-1} f|_{\rho<\epsilon/2} \\ = \sum_{j:0<-\operatorname{Im}\sigma_j<r'} \sum_{k:-\operatorname{Im}\sigma_j+k<r'} \sum_{\ell\leq\ell_{j,k}} \rho^{i\sigma_j+k}(-\log\rho)^\ell a_{j,k,\ell} + u',\ u' \in H_{\mathrm{b}}^{s,r'}.\end{aligned}$$

Remark 5.6 Note that if $\operatorname{Im} \sigma_j$ becomes more negative, then $\rho^{i\sigma_j}$ vanishes faster as $\rho \to 0$.

Proof The first part is already proved, taking into account that $\lambda_j \geq 0$ implies that $\operatorname{Im} \sqrt{\lambda - \lambda_j} \geq \operatorname{Im} \sqrt{\lambda}$ if both square roots are taken to have positive imaginary parts, so the poles σ_j, which satisfy $\sigma_j^2 + \lambda_j = \lambda$, lie at least distance $\operatorname{Im} \sqrt{\lambda}$ from the real axis.

For the second half, note that we have $u = (\Delta_g - \lambda)^{-1} f \in H_{\mathrm{b}}^{s,0}$. Thus, if $\chi \equiv 1$ where $\rho \leq \epsilon/2$, supported in $\rho \leq \epsilon$, then

$$N(\Delta_g - \lambda)(\chi u) = f + [\Delta_g, \chi]u - (\Delta_g - \lambda - N(\Delta_g - \lambda))(\chi u),$$

and $[\Delta_g, \chi]$ and thus $[\Delta_g, \chi]u$ are compactly supported, thus in $H_{\mathrm{b}}^{s-1,\infty}$, while $(\Delta_g - \lambda - N(\Delta_g - \lambda))(\chi u) \in H_{\mathrm{b}}^{s-2,1}$. Thus,

$$N(\Delta_g - \lambda)(\chi u) = f_1 \in H_{\mathrm{b}}^{s-2,1}.$$

Mellin transforming both sides we obtain holomorphic functions in $\operatorname{Im}\sigma > 0$ with a continuous extension to $\operatorname{Im}\sigma \geq 0$ in the Sobolev sense discussed above. Moreover the right hand side is actually holomorphic in $\operatorname{Im}\sigma > -1$ with continuous extension to $\operatorname{Im}\sigma \geq -1$. On the other hand, $\hat{N}(\Delta_g - \lambda)^{-1}$ is meromorphic in $\mathbb{C}$, and thus $(\hat{N}(\Delta_g-\lambda)(\sigma))^{-1}(\mathcal{M}f_1)(\sigma)$ is such in $\operatorname{Im}\sigma > -1$ (with continuous extension to $\operatorname{Im}\sigma \geq -1$ if there is no resonance with imaginary part $= -1$). Letting $r_1 \in (0, 1]$ (so we may assume r_1 is arbitrarily close to 1) be such that no resonance lies in $\operatorname{Im}\sigma = -r_1$, we have by Cauchy's theorem (crucially using the high energy estimates for $(\hat{N}(\Delta_g - \lambda)(\sigma))^{-1}$ to ensure that there are no issues at infinity for the contour shifting)

$$\begin{aligned}
\rho u &= (2\pi)^{-1} \int_{\operatorname{Im}\sigma=0} \rho^{i\sigma} \hat{N}(\Delta_g - \lambda)(\sigma)^{-1}(\mathcal{M}f_1)(\sigma)\, d\sigma \\
&= (2\pi)^{-1} \int_{\operatorname{Im}\sigma=-r_1} \rho^{i\sigma} \hat{N}(\Delta_g - \lambda)(\sigma)^{-1}(\mathcal{M}f_1)(\sigma)\, d\sigma \\
&\quad + \sum_{j:0<-\operatorname{Im}\sigma_j\leq r_1} \sum_{\ell} a_{j,0,\ell}\rho^{i\sigma_j}(-\log\rho)^{\ell} a_{j,0,\ell},
\end{aligned}$$

and the first term is in H_{b}^{s,r_1}. This proves the theorem if $r' < 1$.

In general, we iterate, using the already obtained partial expansion for ρu. This gives that

$$(\Delta_g - \lambda - N(\Delta_g - \lambda))\chi u = f_2 + \sum_{j:0<-\operatorname{Im}\sigma_j\leq r_1} \sum_{\ell} a_{j,0,\ell}\rho^{i\sigma_j+1}(-\log\rho)^{\ell} a_{j,0,\ell},$$

where now $f_2 \in H_{\mathrm{b}}^{s-2,r_1+1}$, and correspondingly its Mellin transform is meromorphic in $\operatorname{Im}\sigma > -r_1$ with poles at $\sigma_j - i$ with $\operatorname{Im}\sigma_j > -r_1$. Taking r_2 now with $r_2 \in (r_1, r_1 + 1]$ and with no resonance on $\operatorname{Im}\sigma_j = -r_2$ (so again r_2 can be arbitrarily close to $r_1 + 1$) we repeat the above argument, except that Cauchy's theorem now gives contributions from the positive integer shifted poles $\sigma_j - i$, $\operatorname{Im}\sigma_j > -r_1$, in addition to the poles σ_j with $-\operatorname{Im}\sigma_j \in (r_1, r_2)$. This gives the statement of the theorem if $r' < 2$; in general it follows by induction. □

Note that the same argument gives in general the asymptotic expansion of approximate solutions, with "approximate" understood in the sense that $(\Delta_g - \lambda)u$ decays more than u does:

Proposition 5.37 *Suppose $u \in H_{\mathrm{b}}^{s,r}$, $r \in \mathbb{R}$, and $(\Delta_g - \lambda)u = f \in H_{\mathrm{b}}^{s-2,r'}$, $r' > r$ and $r' \neq -\operatorname{Im}\sigma_{j,\pm}$ for any j. Let $\sigma_{j,\pm} = \pm\sqrt{\lambda - \lambda_j}$ be the resonances. Then*

$$u|_{\rho<\epsilon/2} = \sum_{j,\pm:r\leq -\operatorname{Im}\sigma_{j,\pm}<r'} \sum_{k:\operatorname{Im}\sigma_{j,\pm}+k<r'} \sum_{\ell\leq \ell_{j,k}} \rho^{i\sigma_{j,\pm}+k}(-\log\rho)^{\ell} a_{j,k,\ell}$$
$$+ u',\ u' \in H_{\mathrm{b}}^{s,r'}.$$

Indeed, the same expansion works whenever one has a meromorphic inverse normal operator family with large parameter estimates.

For a Riemannian scattering metric g on an n-dimensional space, one computes quite easily that

$$L = \rho^{-(n-2)/2}\rho^{-2}\Delta_g\rho^{(n-2)/2} \in \mathrm{Diff}_{\mathrm{b}}^2$$

has normal operator

$$N(L) = (\rho D_\rho)^2 + \Delta_h + \frac{(n-2)^2}{4},$$

so

$$\hat{L}(\sigma) = \Delta_h + \sigma^2 + \frac{(n-2)^2}{4}.$$

Thus the resonances σ satisfy $\sigma^2 + \frac{(n-2)^2}{4} = -\lambda_j$, where $\lambda_j \geq 0$ are the eigenvalues of Δ_h. In particular, for $n \geq 3$, there are no poles of $\hat{L}(\sigma)^{-1}$ in the strip $|\operatorname{Im}\sigma| < (n-2)/2$, so for any $|r| < (n-2)/2$,

$$L : H_{\mathrm{b}}^{s,r} \to H_{\mathrm{b}}^{s-2,r}$$

is Fredholm. Moreover, any element of $\operatorname{Ker} L$ or $\operatorname{Ker} L^*$ would necessarily be in $H_{\mathrm{b}}^{s',r'}$ for all s' and for all $r' < (n-2)/2$. This means for $u \in \operatorname{Ker} L$ that $\rho^{(n-2)/2}u \in \operatorname{Ker}\Delta_g$, and $\rho^{(n-2)/2}u \in H_{\mathrm{b}}^{s',r''}$ for all $r'' < n-2$. Now, in view of $\rho^2 g$ being a b-metric, the L^2 space of the metric, L^2_g, which is the scattering L^2 space, L^2_{sc}, is $\rho^{n/2}L^2_{\mathrm{b}}$. Thus, for $n-2 > n/2$, that is $n > 4$, such a u is in L^2_g, with indeed all b-derivatives being in this space (so the scattering derivatives are in better spaces). This suffices to use that $\Delta_g = d_g^* d$ to conclude that $\|du\|^2 = 0$, so $du = 0$, and thus $u = 0$. As a similar argument also applies to L^* (which, recall, was the adjoint with respect to a b-metric, such as $\rho^2 g$). This yields that Δ_g is invertible as a map

$$\Delta_g : H_{\mathrm{b}}^{s,r+(n-2)/2} \to H_{\mathrm{b}}^{s-2,r+(n+2)/2}$$

if $r = 0$, and thus if $|r| < (n-2)/2$ by the Fredholm property. One also gets an expansion of solutions of $\Delta_g u = f$, with, say, $f \in \dot{C}^\infty$, in terms of the resonances of $\hat{N}(L)$.

The cases $n = 3, 4$ require just a bit more work. Indeed, with λ_j still denoting the eigenvalues of Δ_h, the resonances in the lower half-plane are

$$\sigma_j = -i\sqrt{\lambda_j + (n-2)^2/4},$$

the square root being the positive one. Thus for any r, any element of $\operatorname{Ker} L$ (or $\operatorname{Ker} L^*$) on $H_{\mathrm{b}}^{s,r}$ with $|r| < (n-2)/2$ has an asymptotic expansion

$$u|_{\rho<\epsilon/2} = \sum_{j:0<-\operatorname{Im}\sigma_j<r} \; \sum_{k:\operatorname{Im}\sigma_j+k<r} \; \sum_{\ell\leq\ell_{j,k}} \rho^{i\sigma_j+k}(-\log\rho)^{\ell} a_{j,k,\ell} + u',\ u' \in H_{\mathrm{b}}^{s,r}.$$

Thus, $\rho^{(n-2)/2}u \in \operatorname{Ker}\Delta_g$ is C^∞ on M°, and it tends to 0 at infinity, so it vanishes by the maximum principle. Thus, in all dimensions ≥ 3, Δ_g is invertible as a map $H_{\mathrm{b}}^{s,r+(n-2)/2} \to H_{\mathrm{b}}^{s-2,r+(n+2)/2}$ for $|r| < (n-2)/2$. We now state this as a theorem:

Theorem 5.10 *Let g be a Riemannian scattering metric on M,* $\dim M \geq 3$, *and let λ_j, $j \geq 1$, be the eigenvalues of the boundary Laplacian Δ_h. Then Δ_g is invertible as a map $H_{\mathrm{b}}^{s,r+(n-2)/2} \to H_{\mathrm{b}}^{s-2,r+(n+2)/2}$ for $|r| < \frac{n-2}{2}$.*

Furthermore, with $\sigma_j = -i\sqrt{\lambda_j + (n-2)^2/4}$, with the square root being the positive one, for $r' > 0$ such that $r' \neq \sigma_{j,+}$ for any j, if $f \in H_{\mathrm{b}}^{s-2,r'}$ then there exist $a_{j,k,\ell} \in C^\infty(\partial M)$, $j \in \mathbb{N}^+$, $k \in \mathbb{N}$, $\mathbb{N} \ni \ell \leq \ell_{j,k}$ such that

$$\Delta_g^{-1} f|_{\rho<\epsilon/2} = \sum_{j:0<-\operatorname{Im}\sigma_j<r'} \; \sum_{k:\operatorname{Im}\sigma_j+k<r'} \; \sum_{\ell\leq\ell_{j,k}} \rho^{i\sigma_j+k+(n-2)/2}(-\log\rho)^{\ell} a_{j,k,\ell} + u',$$

where $u' \in H_{\mathrm{b}}^{s,r'+(n-2)/2}$.

The Fredholm argument goes through with minor changes if L is non-elliptic:

Theorem 5.11 *Suppose $P \in \operatorname{Diff}_{\mathrm{b}}^m(M)$, with real principal symbol, and the bicharacteristic flow is non-trapping in the sense that all the (null) bicharacteristics, apart from the ones contained in the radial sets, tend to radial points of the kind described above (sources/sinks) in both the forward and in the backward direction along the Hamilton flow.*

Let $r \in \mathbb{R}$. Suppose also that in each component of the characteristic set we choose either the sources or the sinks as the locations to propagate estimates from by demanding that s is larger than a threshold regularity, and demand that at the other radial set in each component of the characteristic set s is smaller than the threshold regularity, where s is a variable order monotone along the bicharacteristics. Suppose that $\hat{L}(\sigma)$ in invertible on the induced spaces on ∂M when $\operatorname{Im}\sigma = -r$. Then

$$L : \mathcal{X}^{s,r} \to \mathcal{Y}^{s-m+1,r}$$

is Fredholm with

$$\mathcal{Y}^{s',r'} = H_{\mathrm{b}}^{s',r'},\ \mathcal{X}^{s',r'} = \{u \in H_{\mathrm{b}}^{s',r'} : Lu \in H_{\mathrm{b}}^{s'-m+1,r'}\}.$$

In particular, this applies to

$$L = \rho^{-(n-2)/2}\rho^{-2}\Box_g\rho^{(n-2)/2} \in \mathrm{Diff}_{\mathrm{b}}^2(M)$$

for non-trapping Lorentzian scattering metrics g (generalizing spaces asymptotic to Minkowski space), as shown by Baskin, Vasy, and Wunsch [2], extended to a long-range setting in [1]. Lorentzian scattering metrics are symmetric bilinear forms (non-positive definite inner products) of signature $(1, n-1)$ on ${}^{\mathrm{sc}}TM$, with ${}^{\mathrm{sc}}TM$ having been discussed in Section 5.3.11, which have a certain form near ∂M generalizing that of Minkowski space, see [2, Definition 3.1]. The main point of this form is that the radial set for the Hamilton flow is the conormal bundle of a codimension 1 submanifold S of ∂M, given by the zero set of a function v, with a specific structure of the linearization of the Hamilton flow important for the more precise results (in terms of the precise expansion at the radial set) shown in this chapter. The non-trapping hypothesis is that $S = S_+ \cup S_-$ (disjoint union), with all bicharacteristics tending to the radial sets over $S_\pm$ as their affine parameter tends to $\pm\infty$, together with an additional topological condition that $\{v > 0\} = C_+ \cup C_-$ (disjoint union of open sets), with $S_\pm = \partial C_\pm$, with the latter assumption playing a role in the identification of the poles of the indicial operator below. The $C_\pm$ are called the future/past "hyperbolic caps," while $C_0 = \{v < 0\}$ is the "asymptotically de Sitter region," because in a natural manner g induces asymptotically hyperbolic metrics on $C_\pm$ and an asymptotically de Sitter metric on C_0.

Concretely then, in the case of non-trapping Lorentzian scattering metrics, the Fredholm framework applies with any fixed $r \in \mathbb{R}$, $r \neq -\operatorname{Im}\sigma_j$ for any resonance σ_j, with variable order s equal to $s = s_+ \in \mathbb{R}$ at the future radial sets (those over S_+) with $s_+ + r < 1/2$, and $s = s_-$ at the past radial sets (those over S_-) with $s_- + r > 1/2$. Indeed, $\hat{L}(\sigma)$ is meromorphic in

$$\{\sigma \in \mathbb{C} : s_+ - 1/2 > \operatorname{Im}\sigma > s_- - 1/2\},$$

with finitely many poles in strips when $\operatorname{Im}\sigma$ is restricted to compact subintervals, so r can be chosen as desired, and proves the Fredholm property of L, giving the forward solution if L is actually invertible (and otherwise the forward solution on a finite codimensional subspace). If instead one reverses these inequalities, one obtains the backward solution operator, which is the adjoint of the forward operator. There are also Feynman propagators, when one always propagates forward, resp. always propagates backwards, along the Hamilton flow, that is where s is greater than the threshold value on one component of the radial set over S_+ (source for forward, sink for backward propagation) and smaller on the other one, and similarly for the radial set over S_-. These were studied by Gell-Redman, Haber, and Vasy [12].

While we have not discussed actual invertibility, that is the triviality of $\operatorname{Ker} L$ and $\operatorname{Ker} L^*$, this in fact can be done quite easily for the backward and forward operators, using energy estimates on spaces with $|r|$ small, assuming that the poles of $\hat{L}(\sigma)^{-1}$ have an appropriate structure, and if one makes a time-like assumption for a boundary defining function near the closed interior of the future and past light cones at infinity. Indeed, in the case of the forward operator one first shows that the resonances of $\hat{L}(\sigma)$ are of three (potential) kinds: induced by resonances of the resolvent of the asymptotically hyperbolic Laplacian on the future cap, *negatives* of the resonances of the resolvent of the asymptotically hyperbolic Laplacian on the past cap, as well as possibly non-zero pure imaginary integers. Furthermore, the resonant state structure of $\hat{L}(\sigma)$ is such that the polar parts of the resonances induced by the resolvent of the asymptotically hyperbolic Laplacian on the future cap are all supported in the future cap, see [2], and the same holds for the potential resonances in $-i\mathbb{N}^+$ which do not correspond to asymptotically hyperbolic resonances (they are in fact differentiated delta distributions supported at the light cone). Thus, *assuming* that all the resonances of the past hyperbolic cap lie in the open lower half plane (so their negatives lie in the open upper half-plane), the argument of the proof of Proposition 5.37 shows that elements of $\operatorname{Ker} L$ on the 0-weighted space decay rapidly (faster than any power of ρ) near the closed past light cone, and then standard energy estimates show that it in fact vanishes near there, and thus globally, see [26]. Note that the resonances of the $(n-1)$-dimensional hyperbolic space actually satisfy this, indeed they lie in $-i(\frac{n-2}{2}+\mathbb{N})$ (they actually do not exist when n is even, in which case however $\hat{L}(\sigma)$ has the differentiated delta distribution resonances mentioned above; this corresponds to the strong Huygens principle), and thus the conclusion holds for Minkowski space (since for Minkowski space the induced asymptotically hyperbolic space *is* hyperbolic space) and its perturbations. Note also that a priori there are at most finitely many resonances of an asymptotically hyperbolic space in the closed upper half-plane corresponding to either L^2 eigenvalues or the bottom of the continuous spectrum (which would barely miss being in L^2). A similar result holds for $\operatorname{Ker} L^*$ if one assumes that the resonances of the future hyperbolic cap lie in the open lower half plane. Thus one has:

Theorem 5.12 ([2, 26]) *Suppose g is a non-trapping Lorentzian scattering metric, with M having a time-like boundary defining function near the closed* *interior of the future and past light cones at infinity and the asymptotically hyperbolic Laplacians on the future and past caps have all resonances in the open lower half-planes. Then all the hypotheses of Theorem 5.11 are satisfied for $|r|$ small and a suitable choice of s above threshold regularity at the past, below threshold regularity at the future, so the wave operator $\Box_g$ is invertible as a map $H_{\mathrm{b}}^{s,r+(n-2)/2}(M) \to H_{\mathrm{b}}^{s-2,r+(n+2)/2}(M)$, that is on forward-in-time (retarded) function spaces. Similar results hold for the backward-in-time (advanced) function spaces, with the role of future and past reversed.*

On the other hand, for the Feynman propagators on $\mathbb{R}^n$, one can find the poles of $\hat{L}(\sigma)$ directly, namely they are in $\pm i(\frac{n-2}{2} + \mathbb{N})$, so there are no poles near the real axis. Then the perturbation stability of the b-analysis means that this conclusion also holds for perturbations of Minkowski space, see [12], meaning that $L : \mathcal{X}^{s,r} \to \mathcal{Y}^{s-1,r}$ is Fredholm for $|r|$ small. In fact, the complex scaling (Wick rotation) argument of [12] also shows that $\operatorname{Ker} L, \operatorname{Ker} L^*$ are trivial for Minkowski space for $|r|$ small, so L is actually invertible, and thus the same holds for perturbations of Minkowski space.

Finally, for the Feynman propagator in general, a generalization of the positivity argument of [50] carried out in [51], as applied to $\hat{L}(\sigma)$ when σ is real, so $\hat{L}(\sigma)$ is a symmetric operator, shows that any element of the nullspace, which is a priori a conormal distribution associated to the light cones, is necessarily in $\mathcal{C}^\infty$, so in particular one has an element of the nullspace for the operator $\hat{L}(\sigma)$ on the forward and backward function spaces as well, thus a resonance for L. Thus, if we assume that the resonances of the asymptotically hyperbolic caps lie in the open lower half-plane as for the forward/backward problems, we deduce that any element of $\operatorname{Ker} \hat{L}(\sigma)$ is trivial, with the analogous statement for the kernel of the adjoint, so in fact there cannot be any Feynman resonances on the real axis. This gives the Fredholm statement for L, and thus $\Box_g$, for $|r|$ small. Furthermore, a variation of this positivity argument, again carried out in [51], shows that any element of $\operatorname{Ker} L$ is in $H_{\mathrm{b}}^{\infty,r}$, and thus is in the nullspace of L acting on the forward/backward spaces. Correspondingly, if one also assumes that there are time-like boundary defining functions near the asymptotically hyperbolic caps using Theorem 5.12 one concludes that all elements of $\operatorname{Ker} L$ are trivial, and the same holds for $\operatorname{Ker} L^*$. We thus have:

Theorem 5.13 ([12, 51]) *Suppose g is a non-trapping Lorentzian scattering metric, with M having a time-like boundary defining function near the closed interior of the future and past light cones at infinity and the asymptotically hyperbolic Laplacians on the future and past caps have all resonances in the open lower half planes. Then all the hypotheses of Theorem 5.11 are satisfied* *for $|r|$ small and a suitable choice of s with above threshold regularity at the sources, below threshold regularity at the sinks, so the wave operator $\Box_g$ is invertible as a map $H_b^{s,r+(n-2)/2}(M) \to H_b^{s-2,r+(n+2)/2}(M)$, i.e. on Feynman function spaces. A similar result holds on anti-Feynman function spaces, with the role of sources and sinks reversed.*

There are three additional complications for Kerr–de Sitter space, two of which also hold for de Sitter-like spaces from a similar perspective (as opposed to the perspective of extension across the boundary discussed in Section 5.5). The first requires a simple modification of Proposition 5.36 to saddle points of the Hamilton flow; in this case these are given by the conormal bundle of the event horizon at the boundary. The other is that the global structure of the spacetime is more complex in that the null-bicharacteristics do not tend to sources/sinks beyond the event horizon. One way to handle this is to impose complex absorption which gives rise to desired Fredholm statements, but without giving the precise support properties required for the identification of the forward/backward solutions unless one imposes some additional structure as in [48] where the Mellin transform properties far from the real axis were used very strongly. Another way is to add Cauchy hypersurfaces, which is less microlocal, but imposes the vanishing properties automatically; this was the approach of [26]. Finally, the third is trapping (this is the part that is not relevant for de Sitter-like spaces): there is an additional set in ${}^{\mathrm{b}}S^*_{\partial M}M$ which acts as a saddle manifold of the flow in a more complicated way than the radial set, for instance there are both stable and unstable directions within the boundary which means that it is a trapped set at high energies (semiclassical setting) even for the Mellin transformed normal operator. More precisely, this is normally hyperbolic trapping in the appropriate sense. Fortunately, the behavior of such trapping in the standard microlocal setting was extensively studied by Wunsch and Zworski [56], Nonnenmacher and Zworski [36], and Dyatlov [9, 10] recently, and one can apply their results here.

The propagation of singularities at saddles which are sources/sinks within the boundary, but have a stable/unstable manifold transversal to the boundary, is the statement that above the threshold regularity one can propagate estimates from outside the boundary into the boundary, while below the threshold regularity one can propagate estimates from the boundary into the interior. Concretely, the result is

Proposition 5.38 *([26, Proposition 2.1]) Suppose $P \in \Psi^m_{\mathrm{lb}}$, $L \subset \partial\overline{\mathbb{R}_{\exp}} \times \mathbb{R}^{n-1} \times \partial\overline{\mathbb{R}^n}$ is a compact smooth embedded submanifold, with*

$$\mp \mathsf{H}_{p,\mathrm{b},m}\rho_{\mathrm{fiber}} = \beta_0 \rho_{\mathrm{fiber}},\ \beta_0|_L > 0,$$

and

$$\pm \mathsf{H}_{p,\mathrm{b},m}\rho_{\mathrm{base}} = \beta_2\beta_0\rho_{\mathrm{base}},\ \beta_2|_L > 0.$$

Suppose further that

$$\tilde{p} = \sigma_{\mathrm{b},m-1}\left(\frac{1}{2i}(P - P^*)\right)$$

satisfies

$$\tilde{p} = \pm\beta_0\tilde{\beta}\rho_{\mathrm{fiber}}^{1-m},$$

and there is a quadratic defining function ρ_1 *of* L *in* $\mathrm{Char}(P) \subset \overline{\mathbb{R}_{\exp}} \times \mathbb{R}^{n-1} \times \partial\overline{\mathbb{R}^n}$ *such that*

$$\mp \mathsf{H}_{p,\mathrm{b},m}\rho_1 = \beta_1\rho_1 + F_2 + F_3,\ \beta_1 > 0,$$

where $F_2 \geq 0$ *and* F_3 *vanishes cubically at* L. *Let* $\mathcal{L}$ *be the stable* $(+)$ *or unstable* $(-)$ *submanifold of* L, *automatically transversal to* $\partial\overline{\mathbb{R}_{\exp}} \times \mathbb{R}^{n-1} \times \partial\overline{\mathbb{R}^n}$ *in* $\overline{\mathbb{R}_{\exp}} \times \mathbb{R}^{n-1} \times \partial\overline{\mathbb{R}^n}$; *the unstable, resp. stable, manifold of* L *on the other hand lies in* $\partial\overline{\mathbb{R}_{\exp}} \times \mathbb{R}^{n-1} \times \partial\overline{\mathbb{R}^n}$.

(i) *Suppose that* $s - \beta_2 r - (m-1)/2 + \tilde{\beta} > 0$ *on* L, *and* $s' + \beta_2 r - (m-1)/2 + \tilde{\beta} > 0$ *on* L. *Suppose* $u \in H_{\mathrm{b}}^{-N,r}$ *for some* N, r, $L \cap \mathrm{WF}_{\mathrm{b}}^{s-m+1,r}(Pu) = \emptyset$, $\mathrm{WF}_{\mathrm{b}}^{s',r}(u) \cap L = \emptyset$, *and suppose that* L *has a neighborhood* U *such that* $(U \setminus L) \cap \mathcal{L} \cap \mathrm{WF}_{\mathrm{b}}^{s,r}(u) = \emptyset$. *Then* $\mathrm{WF}_{\mathrm{b}}^{s,r}(u) \cap L = \emptyset$.

(ii) *Suppose that* $s - \beta_2 r - (m-1)/2 + \tilde{\beta} < 0$ *on* L. *If* $u \in H^{-N,r}$ *and* L *has a neighborhood* U *such that*

$$(U \setminus L) \cap (\partial\overline{\mathbb{R}_{\exp}} \times \mathbb{R}^{n-1} \times \partial\overline{\mathbb{R}^n}) \cap \mathrm{WF}_{\mathrm{b}}^{s,r}(u) = \emptyset$$

and $U \cap \mathrm{WF}_{\mathrm{b}}^{s-m+1,r}(Pu) = \emptyset$, *then* $L \cap \mathrm{WF}_{\mathrm{b}}^{s,r}(u) = \emptyset$.

This at once yields the Fredholm statement in the presence of saddle points of the flow and complex absorption (though not the trapping):

Theorem 5.14 *(cf.* [26, Proposition 2.3]*) Suppose* $P \in \mathrm{Diff}_{\mathrm{b}}^m(M)$, *with real principal symbol, and the bicharacteristic flow is non-trapping in the sense that all the (null) bicharacteristics, apart from the ones contained in the radial sets, which are saddle points of the kind described above (so exactly one of the stable and unstable manifolds is a subset of the boundary while the other is transversal to it), either reach complex absorption in finite time, or tend to radial points of the kind described above, in both the forward and in the backward direction along the Hamilton flow.*

Let $r \in \mathbb{R}$, *and suppose also that* s *is a variable order (possibly constant) monotone along the bicharacteristics with the property that if* s *is above the* *threshold regularity in a component of the characteristic set then estimates are propagated into ∂M (i.e. s is decreasing along bicharacteristics which tend to the radial set from outside ∂M), and if s is below threshold regularity in a component of the characteristic set then estimates are propagated out of ∂M. Suppose that $\hat{L}(\sigma)$ is invertible on the induced spaces on ∂M when* $\operatorname{Im}\sigma = -r$. *Then*

$$L : \mathcal{X}^{s,r} \to \mathcal{Y}^{s-m+1,r}$$

is Fredholm with

$$\mathcal{Y}^{s',r'} = H_{\mathrm{b}}^{s',r'},\ \mathcal{X}^{s',r'} = \{u \in H_{\mathrm{b}}^{s',r'} \ :\ Lu \in H_{\mathrm{b}}^{s'-m+1,r'}\}.$$

This now describes the microlocal analysis in a neighborhood of the static patch of de Sitter space, and indeed on the analogous region in asymptotically de Sitter spaces. The only missing ingredient for Kerr–de Sitter space is the analysis of trapping, which, as discussed above, is normally hyperbolic in this case, and was analyzed by Wunsch and Zworski [56], Nonnenmacher and Zworski [36], and Dyatlov [9, 10]. Note that on growing function spaces trapping is not an issue, and the standard non-trapping-type estimates hold, see [25], where borderline (no decay, but no growth) estimates are also discussed. These can be combined with Theorem 5.14 using gluing methods obtained by Datchev and Vasy [5, 6]. We refer to the discussion around Definition 2.16–Theorem 2.17 in [48] for details.

References

[1] D. Baskin, A. Vasy, and J. Wunsch. Asymptotics of scalar waves on long-range asymptotically Minkowski spaces. *Preprint, arXiv:1602.04795*, 2016.

[2] Dean Baskin, András Vasy, and Jared Wunsch. Asymptotics of radiation fields in asymptotically Minkowski space. *Amer. J. Math.*, 137(5):1293–1364, 2015.

[3] Michael Beals and Michael Reed. Microlocal regularity theorems for nonsmooth pseudodifferential operators and applications to nonlinear problems. *Trans. Amer. Math. Soc.*, 285(1):159–184, 1984.

[4] Xi Chen and Andrew Hassell. Resolvent and spectral measure on non-trapping asymptotically hyperbolic manifolds I: Resolvent construction at high energy. *Comm. Partial Differential Equations*, 41(3):515–578, 2016.

[5] Kiril Datchev and András Vasy. Gluing semiclassical resolvent estimates via propagation of singularities. *Int. Math. Res. Not. IMRN*, (23):5409–5443, 2012.

[6] Kiril Datchev and András Vasy. Propagation through trapped sets and semiclassical resolvent estimates. *Ann. Inst. Fourier (Grenoble)*, 62(6):2347–2377 (2013), 2012.

[7] J. J. Duistermaat. On Carleman estimates for pseudo-differential operators. *Invent. Math.*, 17:31–43, 1972.

[8] J. J. Duistermaat and L. Hörmander. Fourier integral operators. II. *Acta Math.*, 128(3-4):183–269, 1972.

[9] Semyon Dyatlov. Resonance projectors and asymptotics for r-normally hyperbolic trapped sets. *J. Amer. Math. Soc.*, 28(2):311–381, 2015.

[10] Semyon Dyatlov. Spectral gaps for normally hyperbolic trapping. *Ann. Inst. Fourier (Grenoble)*, 66(1):55–82, 2016.

[11] Semyon Dyatlov and Maciej Zworski. Dynamical zeta functions for Anosov flows via microlocal analysis. *Ann. Sci. Éc. Norm. Supér. (4)*, 49(3):543–577, 2016.

[12] Jesse Gell-Redman, Nick Haber, and András Vasy. The Feynman propagator on perturbations of Minkowski space. *Comm. Math. Phys.*, 342(1):333–384, 2016.

[13] Robert Geroch. Domain of dependence. *J. Mathematical Phys.*, 11:437–449, 1970.

[14] D. Grieser. Basics of the b-calculus. *Preprint, arXiv:math/0010314*, 2000.

[15] Colin Guillarmou. Meromorphic properties of the resolvent on asymptotically hyperbolic manifolds. *Duke Math. J.*, 129(1):1–37, 2005.

[16] Victor Guillemin and David Schaeffer. On a certain class of Fuchsian partial differential equations. *Duke Math. J.*, 44(1):157–199, 1977.

[17] Nick Haber and András Vasy. Propagation of singularities around a Lagrangian submanifold of radial points. *Bull. Soc. Math. France*, 143(4):679–726, 2015.

[18] A. Hassell, R. B. Melrose, and A. Vasy. Spectral and scattering theory for symbolic potentials of order zero. *Advances in Mathematics*, 181:1–87, 2004.

[19] A. Hassell, R. B. Melrose, and A. Vasy. Microlocal propagation near radial points and scattering for symbolic potentials of order zero. *Analysis and PDE*, 1:127–196, 2008.

[20] Ira Herbst and Erik Skibsted. Quantum scattering for potentials independent of $|x|$: asymptotic completeness for high and low energies. *Comm. Partial Differential Equations*, 29(3-4):547–610, 2004.

[21] Ira Herbst and Erik Skibsted. Absence of quantum states corresponding to unstable classical channels. *Ann. Henri Poincaré*, 9(3):509–552, 2008.

[22] P. Hintz and A. Vasy. Global analysis of quasilinear wave equations on asymptotically Kerr–de Sitter spaces. *Int. Math. Res. Notices*, 2016(17):5355–5426, 2016.

[23] P. Hintz and A. Vasy. The global non-linear stability of the Kerr–de Sitter family of black holes. *Preprint, arXiv:1606.04014*, 2016.

[24] Peter Hintz. Global analysis of quasilinear wave equations on asymptotically de Sitter spaces. *Ann. Inst. Fourier (Grenoble)*, 66(4):1285–1408, 2016.

[25] Peter Hintz and Andras Vasy. Non-trapping estimates near normally hyperbolic trapping. *Math. Res. Lett.*, 21(6):1277–1304, 2014.

[26] Peter Hintz and András Vasy. Semilinear wave equations on asymptotically de Sitter, Kerr–de Sitter and Minkowski spacetimes. *Anal. PDE*, 8(8):1807–1890, 2015.

[27] L. Hörmander. *The analysis of linear partial differential operators,* vol. 1-4. Springer-Verlag, Berlin and Heidelberg, 1983.

[28] Lars Hörmander. On the existence and the regularity of solutions of linear pseudo-differential equations. *Enseignement Math. (2)*, 17:99–163, 1971.

[29] Rafe R. Mazzeo and Richard B. Melrose. Meromorphic extension of the resolvent on complete spaces with asymptotically constant negative curvature. *J. Funct. Anal.*, 75(2):260–310, 1987.

[30] R. B. Melrose. *Spectral and scattering theory for the Laplacian on asymptotically Euclidian spaces*. Marcel Dekker, New York, 1994.

[31] R. B. Melrose. Lecture notes for '18.157: Introduction to microlocal analysis'. Available at http://math.mit.edu/~rbm/18.157-F09/18.157-F09.html, 2009.

[32] R. B. Melrose, A. Sá Barreto, and A. Vasy. Analytic continuation and semiclassical resolvent estimates on asymptotically hyperbolic spaces. *Comm. in PDEs*, 39(3):452–511, 2014.

[33] Richard B. Melrose. Transformation of boundary problems. *Acta Math.*, 147(3-4): 149–236, 1981.

[34] Richard B. Melrose. *The Atiyah-Patodi-Singer index theorem*, volume 4 of *Research Notes in Mathematics*. A K Peters Ltd., Wellesley, MA, 1993.

[35] Stéphane Nonnenmacher and Maciej Zworski. Quantum decay rates in chaotic scattering. *Acta Math.*, 203(2):149–233, 2009.

[36] Stéphane Nonnenmacher and Maciej Zworski. Decay of correlations for normally hyperbolic trapping. *Invent. Math.*, 200(2):345–438, 2015.

[37] Cesare Parenti. Operatori pseudo-differenziali in R^n e applicazioni. *Ann. Mat. Pura Appl. (4)*, 93:359–389, 1972.

[38] Xavier Saint Raymond. A simple Nash–Moser implicit function theorem. *Enseign. Math. (2)*, 35(3-4):217–226, 1989.

[39] I. M. Sigal and A. Soffer. N-particle scattering problem: asymptotic completeness for short range systems. *Ann. Math.*, 125:35–108, 1987.

[40] Plamen Stefanov, Gunther Uhlmann, and Andras Vasy. Boundary rigidity with partial data. *J. Amer. Math. Soc.*, 29(2):299–332, 2016.

[41] M. A. Šubin. Pseudodifferential operators in R^n. *Dokl. Akad. Nauk SSSR*, 196:316–319, 1971.

[42] Gunther Uhlmann and András Vasy. The inverse problem for the local geodesic ray transform. *Invent. Math.*, 205(1):83–120, 2016.

[43] André Unterberger. Résolution d'équations aux dérivées partielles dans des espaces de distributions d'ordre de régularité variable. *Ann. Inst. Fourier (Grenoble)*, 21(2):85–128, 1971.

[44] A. Vasy. Propagation of singularities in many-body scattering. *Ann. Sci. École Norm. Sup. (4)*, 34:313–402, 2001.

[45] A. Vasy. Propagation of singularities in many-body scattering in the presence of bound states. *J. Func. Anal.*, 184:177–272, 2001.

[46] András Vasy. Analytic continuation and high energy estimates for the resolvent of the Laplacian on forms on asymptotically hyperbolic spaces. *Adv. Math.*, 306:1019–1045, 2017.

[47] A. Vasy. *Microlocal analysis of asymptotically hyperbolic spaces and high energy resolvent estimates*, volume 60 of *MSRI Publications*. Cambridge University Press, Cambridge, 2012.

[48] A. Vasy. Microlocal analysis of asymptotically hyperbolic and Kerr–de Sitter spaces. *Inventiones Math.*, 194:381–513, 2013. With an appendix by S. Dyatlov.

[49] A. Vasy. Some recent advances in microlocal analysis. In *Proceedings of the ICM*, pp. 915–939, 2014.

[50] András Vasy. On the positivity of propagator differences. *Ann. Henri Poincaré*, 18(3):983–1007, 2017.

[51] A. Vasy and M. Wrochna. Quantum fields from global propagators on asymptotically Minkowski and extended de Sitter spacetime. *Preprint, arXiv:1512.08052*, 2015.

[52] András Vasy. Propagation of singularities for the wave equation on manifolds with corners. *Ann. of Math. (2)*, 168(3):749–812, 2008.

[53] András Vasy. The wave equation on asymptotically de Sitter-like spaces. *Adv. Math.*, 223(1):49–97, 2010.

[54] András Vasy. Resolvents, Poisson operators and scattering matrices on asymptotically hyperbolic and de Sitter spaces. *J. Spectr. Theory*, 4(4):643–673, 2014.

[55] Yiran Wang. *The resolvent of the Laplacian on non-trapping asymptotically hyperbolic manifolds*. ProQuest LLC, Ann Arbor, MI, 2015. Thesis (Ph.D.)–Purdue University, arXiv:1410.6936.

[56] Jared Wunsch and Maciej Zworski. Resolvent estimates for normally hyperbolic trapped sets. *Ann. Henri Poincaré*, 12(7):1349–1385, 2011.

[57] Maciej Zworski. Resonances for asymptotically hyperbolic manifolds: Vasy's method revisited. *J. Spectr. Theory*, 6(4):1087–1114, 2016.

[58] Maciej Zworski. *Semiclassical analysis*, volume 138 of *Graduate Studies in Mathematics*. American Mathematical Society, Providence, RI, 2012.

Department of Mathematics, Building 380, Stanford University, 450 Serra Mall, Stanford CA 94305-2125, USA.
E-mail address: andras@math.stanford.edu

inted in the United States
By Bookmasters

Printed in the United States
By Bookmasters